T0228538

Plasma and Fluid Turbulence: Theory and Modelling

Series in Plasma Physics

Series Editors:

Steve Cowley, Imperial College, UK
Peter Stott, CEA Cadarache, France
Hans Wilhelmsson, Chalmers University of Technology, Sweden

Other books in the series

The Interaction of High-Power Lasers with Plasmas
S Eliezer

Introduction to Dusty Plasma Physics
P K Shukla and A A Mamun

The Theory of Photon Acceleration
J T Mendonça

Laser Aided Diagnostics of Plasmas and Gases
K Muraoka and M Maeda

Reaction-Diffusion Problems in the Physics of Hot Plasmas
H Wilhelmsson and E Lazzaro

The Plasma Boundary of Magnetic Fusion Devices
P C Stangeby

Non-Linear Instabilities in Plasmas and Hydrodynamics
S S Moiseev, V N Oraevsky and V G Pungin

Collective Modes in Inhomogeneous Plasmas
J Weiland

Transport and Structural Formation in Plasmas
K Itoh, S-I Itoh and A Fukuyama

Tokamak Plasmas: A Complex Physical System
B B Kadomstev

Forthcoming titles in the series

Microscopic Dynamics of Plasmas and Chaos
Y Elskens and D Escande

Nonlinear Plasma Physics
P K Shukla

Series in Plasma Physics

Plasma and Fluid Turbulence: Theory and Modelling

Akira Yoshizawa
Institute of Industrial Science, University of Tokyo, Japan

Sanae-I Itoh
Research Institute for Applied Mechanics, Kyushu University, Japan

Kimitaka Itoh
National Institute for Fusion Science, Toki, Japan

CRC Press
Taylor & Francis Group
Boca Raton London New York

CRC Press is an imprint of the
Taylor & Francis Group, an **informa** business

First published 2003 by IOP Publishing Ltd

Published 2019 by CRC Press
Taylor & Francis Group
6000 Broken Sound Parkway NW, Suite 300
Boca Raton, FL 33487-2742

First issued in paperback 2019

ISBN 13: 978-0-367-45470-8 (pbk)
ISBN 13: 978-0-7503-0871-7 (hbk)

British Library Cataloguing-in-Publication Data

A catalogue record of this book is available from the British Library.

Library of Congress Cataloging-in-Publication Data are available

Cover Design: Victoria Le Billon

Typeset by Academic + Technical Typesetting, Bristol

This monograph is dedicated to the parents of the authors

Contents

Preface

Παντα ρει – All things flow
(Heraclitus)

A branch of science is qualified for a fundamental discipline if it is capable of finding new ways of perceiving structures in nature. Such attempts have been of historical interest to mankind in understanding, e.g., the elements that constitute matter. There is another fundamental and historical enigma: *All things flow*. What is the law that governs flows of all things? This question has been faced repeatedly when new observations of nature are added to human knowledge. Plasma is a state of matter and constitutes almost all matter in the universe. When a frontier of human knowledge reaches a new area of nature, flow and turbulence have been at the centre of the enigma. Understanding of turbulence in fluids and plasmas is basic to our knowledge of and insight into nature itself. For this, theory and modelling of turbulence are essential elements of the research of nature.

Basic science has been progressing simultaneously (or flying side by side) with technological success in enriching human life. Achievements in the twentieth century are breathtaking. Understanding of matter from elementary processes has progressed. For example, the knowledge of micro-scale Hamiltonian and equilibrium statistics has clarified properties of solid state and hence propelled the evolution of modern civilization through electronics. In the twenty-first century, systems away from thermal equilibrium are the main problems for a new engineering challenge and an evolution of human life. Plasmas are a central objective for such a new challenge, and the knowledge of them gives a key for material technology, space technology, biotechnology and so on. A large-scale structure formation in complicated media is a central concept of problems such as global climate, weather hazard, etc. In these fields, it is difficult to construct a method for fully predicting those phenomena from small-scale experiments. This is due to a highly nonequilibrium state of phenomena. Micro-scale and macro-scale dynamical processes are combined there, generating a state far from thermal equilibrium under normal circumstances. An understanding of turbulence and structural formation in inhomogeneous and nonequilibrium media needs the physics that combines the microscopic and macroscopic (global) viewpoints. It gives a key for resolving the above-mentioned new engineering problems as well.

In this monograph, turbulence phenomena and turbulent transport in fluids and plasmas are discussed, and methodologies and modelling for them are explained. Special attention is paid to structural formation and transitions, which are indispensable for forming a fundamental law describing a variety of flows. In addition, they provide necessary bases for engineering and technological application of knowledge in the area of nonequilibrium and turbulent matter. This area is characterized by two features: one is that the gradient is the order parameter of turbulence, and the other is that turbulent-driven transport and dissipation generate an observable global structure in a highly turbulent state. In addition, the long-range interaction of electromagnetic fields plays an essential role in plasmas.

This monograph is intended to show that a thread of ideas could consistently describe the turbulence, turbulent transport and the structural transitions. Some progress is shown in applications to neutral fluids, and others are found in plasmas. Common features in turbulence of fluids and plasmas are illustrated. Some particular differences of features are highlighted, e.g., through a role of long-range interactions through electromagnetic fields.

In Part I, a general introduction is made to turbulence and structural formation in fluids and plasmas. Methodologies for fluid turbulence are explained in Part II. Part III is devoted to subjects in magnetohydrodynamics, in particular dynamo problems. Plasma turbulence and transport are discussed in Part IV. Encyclopedic description is beyond the scope of this monograph, so that many essential processes are not covered and important references may be missed. This monograph is founded on the belief that the cooperation of theory and modelling with direct numerical simulation and experimental observations is indispensable for forming a firm understanding of the evolution of nature. The authors would be happy if this monograph could succeed in illustrating some interesting aspects of one of the most active research areas in modern physics.

Akira Yoshizawa, Sanae-I Itoh and Kimitaka Itoh
31 March 2002

Acknowledgments

This monograph contains a variety of topics. They range from electrically nonconducting turbulent flows to plasma turbulence encountered in thermonuclear fusion, through the intermediary of magnetohydrodynamic turbulence with special attention paid to planetary dynamo. In the course of understanding or studying those topics, the authors were inspired by valuable communication with many researchers.

The authors would like to extend cordial thanks to Prof S Murakami, Prof T Kobayashi, Prof Y Miyake, Prof H Kawamura, Prof Y Nagano, Prof N Kasagi, Dr T-H Shih, late Prof C G Speziale, Dr M-S Liou, Mr S Nisizima, late Prof N Takemitsu, Prof K Horiuti, and Dr H Fujiwara for valuable discussions on turbulence modelling of electrically nonconducting fluids, and to Dr R H Kraichnan, Dr J R Herring, Dr R Rubinstein, Prof T Nakano, Prof Y Shimomura, Prof F Hamba, and Dr M Okamoto for stimulating discussions on turbulence theories of electrically nonconducting fluids. The authors are grateful to Prof S Kato, Prof K Shibata, Prof T Yukutake, Prof K Miyamoto, Prof N Inoue, Dr N Yokoi, and Prof H Kato for valuable discussions on planetary and fusion dynamos.

The authors heartily acknowledge Prof A Fukuyama and Prof M Yagi for continuous collaboration on problems of plasma transport and structural formation, which has been an essential basis for this monograph. They are grateful to Dr M Azumi, Prof R Balescu, Dr J W Connor, Prof B Coppi, Prof P H Diamond, Prof R D Hazeltine, late Prof B B Kadomtsev, Dr J A Krommes, late Prof R Kubo, Prof A J Lichtenberg, Prof K Nishikawa, Prof T Ohkawa, Prof D Pfirsch, Prof M N Rosenbluth, Prof H Sanuki, Dr K C Shaing, Prof S Yoshikawa, Prof M Wakatani, Dr J Wesson for illuminating discussion on the plasma turbulence and transport. Valuable discussion of plasma confinement experiments with colleagues, in particular Dr J Campbell, Prof A Fujisawa, Prof G Fussman, Dr A Gibson, Prof J Hugill, Prof K Ida, Dr J Jacquinot, Prof M Keilhacker, late Prof H Maeda, Dr Y Miura, Dr P H Rebut, Prof H Soltwisch, Prof U Stroth, Prof F Wagner,

Prof R R Weynants, Prof G Wolf, Dr K-L Wong, Prof H Zohm, and members of ALCATOR C-Mod, ASDEX, ASDEX-U, CHS, D III-D, JET, JFT-2M, JT-60U, LHD, TEXTOR, TFTR and W7-AS are acknowledged.

The authors are grateful to Prof M Fujiwara and emeritus Prof A Iiyoshi of National Institute for Fusion Science, Prof A M Bradshaw, Prof K Pinkau, Prof F Wagner and Prof G Grieger of Max-Planck-Institut für Plasmaphysik, Prof K Lackner of European Fusion Development Agreement, Dr D Robinson of UKAEA Culham Laboratory, Dr V Chan of General Atomics, Prof R D Goldston and Prof W M Tang of Princeton Plasma Physics Laboratory, for the strong support of our continued international collaborations.

This work is partly supported by the Grant-in-Aid for Scientific Research of the Ministry of Education, Culture, Sports, Science and Technology Japan, by the collaboration programme of the National Institute for Fusion Science, by the collaboration programme of the Research Institute for Applied Mechanics of Kyushu University, and by the Research-Award Programme of Alexander von Humboldt-Stiftung.

Last but not least, the authors wish to express their thanks to Plasma Physics and Controlled Fusion and Institute of Physics Publishing for the help in preparing and completing this monograph.

Part I

General Introduction

Chapter 1

Introductory Remarks

The theoretical study of turbulence in electrically nonconducting fluids has developed very differently from its counterpart of plasma turbulence. In fluid flows, the growth of disturbances imposed on the laminar state of each flow is investigated on the basis of linear and nonlinear stability analyses. In these analyses, the effects of global boundary conditions such as wall boundaries often play an important role, compared with the stability analysis of plasma micro-instabilities. When the laminar state is unstable, it is not unusual for initially small disturbances to evolve rapidly into fully-developed turbulent states. In the vicinity of a solid wall, a steep gradient of velocity is generated and this inhomogeneity of velocity field plays an important role in supplying small-scale flow components with energy. A fully developed state of turbulence is sustained by the continuous supply of energy, for instance, through the imposition of pressure. In this situation, the fully developed state of turbulence that is distinct from the initial stage of growing disturbances may be studied.

In the long history of the theoretical study of turbulence, much attention has been paid to homogeneous, isotropic turbulence free from spatially-varying mean flow. Such turbulence is ideal and is difficult to realize in laboratory experiments. A flow state similar to isotropic turbulence may be observed in the small-scale components of motion in atmospheric and oceanic flows. Especially, the well-known $-5/3$ power law for the energy spectrum, which is derived with the aid of the concept of the inertial range, is a primary focus for the theoretical study of turbulence and greatly contributed to the formulation of statistical theoretical approaches. In other words, the structural simplicity intrinsic to isotropic fluid turbulence has greatly helped in the development of turbulence theory.

A big incentive for developing an inhomogeneous-turbulence theory on the basis of the accomplishments by isotropic-turbulence theories is turbulence modelling. In the study of real-world turbulent flows, nondimensional parameters such as Reynolds and Rayleigh numbers are very large, and the computer simulation based on the direct use of the fluid equations is not possible. As a result, the small-scale components of turbulent motion are eliminated, and their effects are taken into account through such concepts as turbulent or renormalized viscosity. This procedure is called turbulence modelling, and it is a very useful tool in the study of flows encountered in engineering and physical sciences although its applicability is not sufficient in the presence of a three-dimensional mean flow. A systematic approach to the study of inhomogeneous turbulence is indispensable for improving current turbulence models and constructing new ones. This situation has stimulated the study of inhomogeneous-turbulence theory.

In contrast to fluid turbulence, the fully-developed state of plasma turbulence is not usually a matter of theoretical concern, specifically, in fusion plasmas, for such a state is to be avoided for efficient plasma confinement. Here the primary focus of the theoretical study of plasma turbulence is on the growth of instabilities and the subsequent highly nonlinear regime which is subject to strong inhomogeneity. Turbulence in plasmas has several characteristic features. The first is that the fluctuation level becomes high through the instabilities driven by the inhomogeneity. The turbulent level and spectrum are greatly influenced by the spatial inhomogeneity and plasma configuration. Inhomogeneities exist for plasma parameters (e.g., density and temperature) as well as for the fields (e.g., magnetic field and radial electric field). These inhomogeneities couple together to drive and/or suppress instabilities and turbulent fluctuations. In particular, the anisotropy along and perpendicular to strong magnetic fields induces various shapes in fluctuations: fluctuations often have a very long correlation length along the magnetic field line and are quasi-two-dimensional. In addition, the mobilities of electrons and ions clearly differ. The inhomogeneities, the anisotropy due to a strong magnetic field, and the difference in ion and electron mobilities, all have a strong influence on the linear properties of the plasma waves as well as on the turbulent transport in plasmas. In many cases, instabilities develop into strong turbulence, so that the decorrelation rate caused by the nonlinear interactions is usually of the same order of or much larger than the damping rate (growth rate) of the linear eigenmode. Theoretical methods developed for fluid turbulence are helpful for the study of these phenomena. In some cases, however, only a few modes are excited, and an analysis based on weak turbulence suffices.

In the quest to understand anomalous transports in confined plasmas, investigation of turbulent fluctuations has been a central theme. In particular, after the discovery of high-confinement (H) modes in tokamaks, it was widely recognized that the plasma profiles vary and that the changes between them occur as sudden transitions. One of the keys to understanding the structural formation and transitions in plasmas is to study the mutual interactions among plasma inhomogeneities, electric-field structures and fluctuations. Advances in this type of theory are based on the development of turbulence theories in fluids and plasmas. It is important to survey the basis and recent development of turbulence theories in both fluids and plasmas in order to establish a perspective for future research on turbulence, turbulent transport and structural formation.

This book is composed of four parts. In Part I, some interesting phenomena related to fluid and plasma turbulence are introduced in the light of structure formation. Parts II and III are devoted to hydrodynamic and magnetohydrodynamic turbulence, respectively, and Part IV is concerned with plasma turbulence. In Part II, the basic concepts of turbulence are presented, and some analytical methods of analyzing turbulent flows are explained, with special emphasis on inhomogeneous turbulence. In Part III, the mean-field theory of dynamo is formulated on the basis of those analytical methods, and the magnetic-field generation and the feedback effect on fluid motion are discussed in the light of planetary and fusion phenomena. These discussions partly play a bridging role between Parts II and IV. In Part IV, various theoretical approaches to plasma turbulence are illustrated with examples of applications. Emphasis is put on how the theory of turbulence is applied to the system that is composed of components of different mobilities due to strong inhomogeneity and anisotropy. The connection with the current fluid turbulence theories is also sought in order to shed light on the similarity and difference between fluid and plasma turbulence. Readers of this book are supposed to be more familiar with plasma phenomena than fluid ones. In the exposition of fluid turbulence, therefore, more emphasis is put on the fundamental aspects.

Chapter 2

Structure Formation in Fluids and Plasmas

2.1 Flow in a Pipe

2.1.1 Enhancement of Mixing Effects Due to Turbulence

A representative inner flow of an electrically nonconducting fluid is a flow in a straight circular pipe that is driven by a pressure gradient. Its velocity profile is dependent on the Reynolds number R_e, which is defined by

$$R_e = \frac{U_B a}{\nu}. \tag{2.1}$$

Here U_B is the bulk velocity (the mean velocity across the pipe cross-section), a is the pipe radius, and ν is the kinematic viscosity (the viscosity divided by fluid density).

For $R_e \leq O(10^2)$, the flow is in the axial direction only and is called the Poiseuille flow whose velocity profile is parabolic [figure 2.1 (left)]. With increasing R_e, the Poiseuille flow becomes unstable and finally fully turbulent. The magnitude of R_e at which a laminar state becomes unstable is called a critical Reynolds number R_{eC} and depends on the intensity of a disturbance imposed on the flow. The linear stability analysis indicates that R_{eC} of the Poiseuille flow is infinite; that is, the flow is stable as long as the disturbance is infinitesimal [2.1].

A prominent feature of a fully developed turbulent pipe flow is the flattened mean velocity profile, as in figure 2.1 (right) [2.2]. In the case of a fixed flow rate, the increase in velocity fluctuations enhancing mixing effects gives rise to the increase in the momentum transfer from the center-line region towards the near-wall region, resulting in the loss of the parabolic profile. In this sense, turbulence is harmful to the sustainment of a distinct mean-flow structure.

In the context of the momentum transport due to turbulence, it is meaningful to consider the relationship of turbulence effects with the drag force experienced by a body immersed in a stream. In the case of a sphere

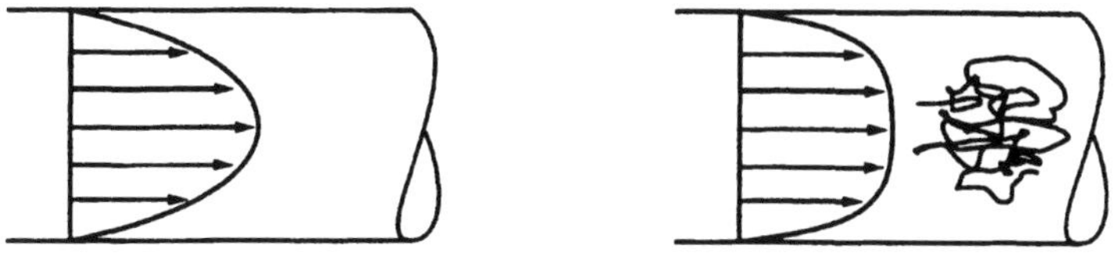

Figure 2.1. Pipe flow: parabolic profile of laminar flow (left); mean velocity profile of turbulent flow (right).

of radius a in a uniform flow U_∞, viscous effects are confined within a thin layer at the front surface for high R_e $(= U_\infty a/\nu)$ such as $O(10^4)$ (figure 2.2). This layer, called a boundary layer, usually remains laminar, but it eventually separates from the surface owing to the resistive-force effect. Behind the separation point, the flow becomes turbulent.

A rough-surface sphere immersed in a stream is observed to experience a smaller drag force than a smooth-surface sphere. In the case of a rough surface, the boundary layer becomes turbulent on the front side. A turbulent flow, in general, converts more kinetic energy into heat, compared with a laminar flow. The foregoing observation seems to contradict this general property of turbulence. Such apparent inconsistency may be resolved from the viewpoint of momentum transport. In the rough-surface case, a turbulent boundary layer causes a bigger drag force at the front surface as far as the boundary-layer region is concerned. In the region, however, more momentum is transported to the surface, as may be seen from the difference between the laminar and turbulent velocity profiles in a pipe; namely, the fluid near the surface is accelerated in the turbulent case, compared with the laminar case. As a result, the separation of the boundary layer is retarded and occurs behind the laminar separation point. Then the turbulent-flow region behind the sphere is narrowed, leading to a smaller drag force for the sphere.

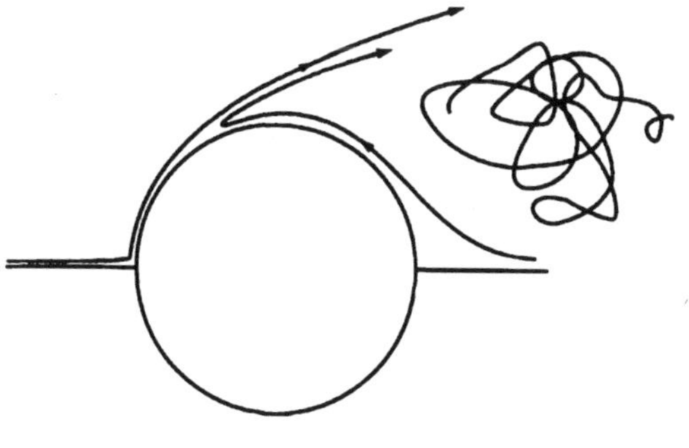

Figure 2.2. Sphere immersed in a stream.

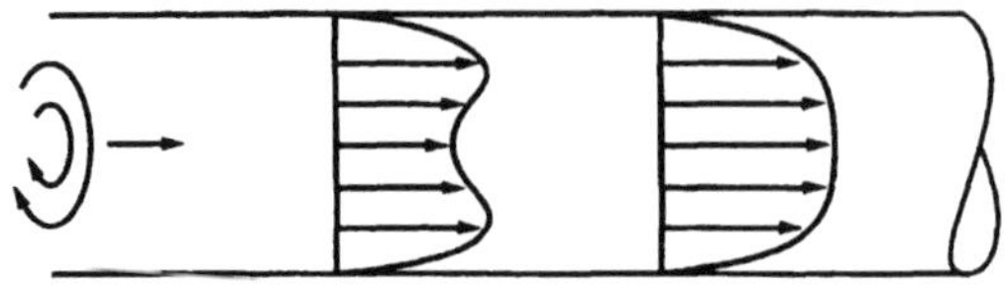

Figure 2.3. Retarded velocity profile in a turbulent swirling pipe flow.

2.1.2 Mean-Flow Structure Formation in Pipe Flows

In a circular-pipe flow, the occurrence of velocity fluctuations results in the loss of the parabolic velocity profile through enhanced mixing effects. We can mention, however, a number of flows in which distinct mean-flow structures persist against turbulent mixing effects or coexist with the latter. A typical instance is the turbulent swirling flow in a circular pipe that is driven by a pressure gradient [2.3]. There, swirling motion is imposed on the flow at the entrance of the pipe.

In a swirling pipe flow, the mean velocity is retarded near the center-line, and its axial-velocity profile becomes concave, as is depicted schematically in figure 2.3. With the increasing swirl intensity at the entrance, the flow direction near the center-line is reversed, resulting in the occurrence of recirculation zones (figure 2.4). The existence of such a retarded or reversed mean-flow structure is surprising in the light of enhanced mixing effects due to turbulence. This flow behavior is one of the properties characterizing a turbulent swirling pipe flow, and the exploration of the mechanism under which the distinct mean-flow profile coexists with turbulence effects is a challenging theme of turbulence study.

The importance of swirl effects is not limited to engineering flows such as a flow inside an engine. Tornadoes consist of strong rising and swirling motions. Flow structures in tornadoes are complicated, but they partially resemble a swirling pipe flow; namely, a reversed flow is observed inside a tornado. This fact shows the relevance of swirl effects to various kinds of flow phenomena.

The other instance of distinct mean-flow structure formation is a flow in a square-duct pipe [2.4]. In the laminar state, the flow is unidirectional along the central axis, but the turbulent state accompanies mean secondary flows

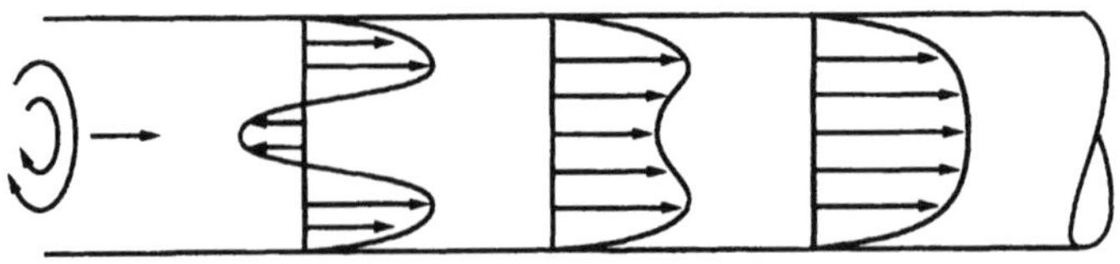

Figure 2.4. Reversed and retarded velocity profiles in a turbulent swirling pipe flow.

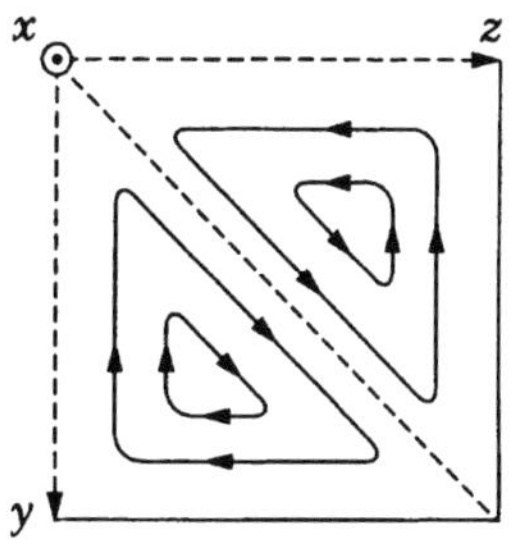

Figure 2.5. Secondary flows in a turbulent square-duct flow (one-fourth of the cross-section).

in the cross-section (figure 2.5). The persistence of such flows indicates the existence of a mechanism that keeps these circulating flows from diffusing and canceling with one another.

2.2 Magnetic-Field Generation by Turbulent Motion

Magnetic-field generation in stellar objects such as the earth and the sun is a structure-formation phenomenon in nature. The interiors of the earth and the sun are schematically shown in figures 2.6 and 2.7 (the radii of the earth and the sun are 6300 km and 700 000 km, respectively). The earth consists of the mantle, the outer core, and the inner core, whose primary ingredients are silicon, melted iron, and solid iron, respectively. Earth's magnetic field (the geomagnetic field) occurs from the motion of melted iron in the outer core. There the velocity is inferred to be $O(10^{-4})\,\mathrm{m\,s^{-1}}$

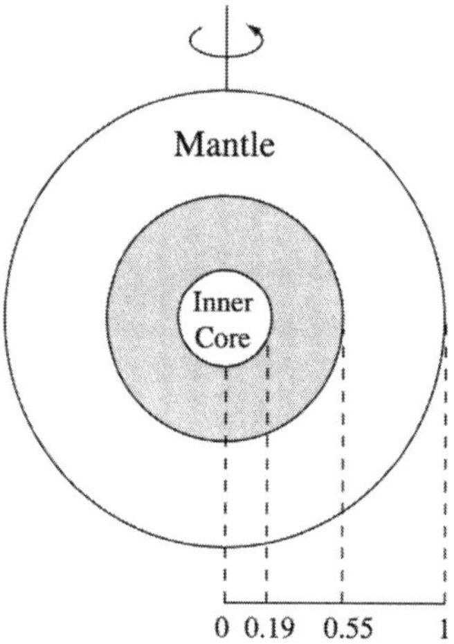

Figure 2.6. Interior of the earth.

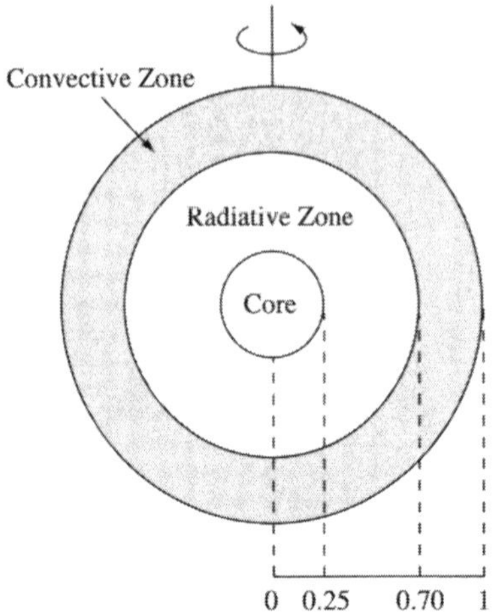

Figure 2.7. Interior of the sun.

from some geophysical observations, and the kinematic viscosity of melted iron is of the same order as for water [about $O(10^{-6})\,m^2\,s^{-1}$]. Then the Reynolds number R_e is estimated to be $O(10^8)$, suggesting that the fluid motion in the outer core is turbulent.

Some typical features of the geomagnetic field are summarized as follows [2.5, 2.6].

(E1) The main component of the geomagnetic field observed at the surface is the dipole field whose present axis is nearly along earth's rotation axis. The strength of the field is about a few Gauss (G).

(E2) The toroidal component is not observable at the surface since it is confined below the bottom of the mantle that is electrically non-conducting. The unobservable toroidal component is inferred to be of $O(10)$–$O(10^2)$ G, which is stronger than the dipole one.

(E3) The polarity of the geomagnetic field reverses irregularly in the interval of $O(10^5)$–$O(10^7)$ years, but the time necessary for the reversal is much shorter and is less than $O(10^4)$ years.

The formation of the distinct dipole field in a turbulent fluid motion is a central concern in the study of the generation mechanism of geomagnetic field or geodynamo. The situation that the toroidal field is not observable is a big stumbling block for the study of the geodynamo. The magnetic energy of 1 G per unit mass of iron is equivalent to the kinetic energy of velocity $O(10^{-3})\,m\,s^{-1}$. Then the energy of the dipole field of a few G is some hundred times the kinetic energy of the melted iron with an estimated velocity $O(10^{-4})\,m\,s^{-1}$. With the unobservable toroidal component included, the energy of the geomagnetic field is $O(10^4)$–$O(10^6)$ times the kinetic energy

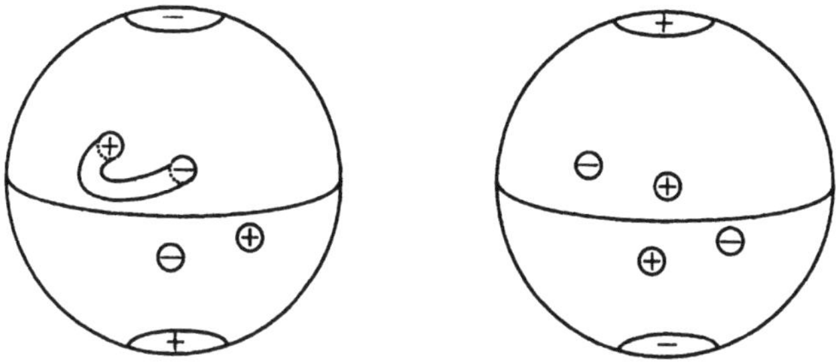

Figure 2.8. Sunspot's polarity rule.

of the fluid motion or the generator of the former. This leads us to a rather surprising conclusion that the geomagnetic field is generated quite efficiently by the turbulent fluid motion in the outer core. A proposed geodynamo model needs to address this point adequately. The persistence of one geomagnetic polarity is quite long, compared with the sun referred to below, which indicates that the dipole field along earth's rotation axis is very stable. The clarification of the mechanism for such stability is also an important theme of the geodynamo.

In contrast to the earth, the solar constitution is simple, and its constituents are hydrogen (90%) and helium (10%). All the solar energy arises from the thermonuclear fusion reaction in the core, and the resulting heat is supplied to the outermost or convective zone through the intermediate or radiative zone. The motion of highly ionized hydrogen gases in the convective zone is strongly turbulent. It is the origin of solar magnetic fields, whose most typical manifestation is sunspots. Inside the zone, the toroidal field aligned with the equator is generated, and loops of this field rise up owing to buoyancy effects and break through the photosphere above the convective zone. Their cross-sections are observed as pairs of sunspots (figure 2.8). The intensity of a large-sunspot magnetic field is a few kG, and the toroidal field with the intensity stronger by one order is inferred to be generated near the bottom of the convective zone.

Sunspots obey the well-known polarity rule, whose primary parts may be stated as follows [2.7, 2.8].

(S1) Sunspots are limited to the middle- to low-latitude region. The polarity reverses quite regularly, that is, in about 11 years. In figure 2.8, the left polarity turns into the right opposite polarity.

(S2) The polarity of the leading sunspot is coincident with the polarity of the polar field or the poloidal field near the pole of the hemisphere. The latter is a few G and is very weak, compared with sunspot's magnetic field.

In the relative intensity of poloidal to toroidal components, the solar poloidal field is very much weaker than its geomagnetic counterpart, as was noted above. It is important to clarify the relationship of this point with the regular polarity reversal.

At the early stage of the dynamo study, the kinematic approach plays a leading role. There the generation mechanism of magnetic fields is examined using the magnetic induction equation with a properly chosen fixed velocity field [2.9]. Afterwards, the computer simulation based on a combined system of fluid and magnetic-field equations became feasible with the remarkable advancement of computer capability. In reality, a large amount of information has already been obtained about the generation processes of magnetic fields in the presence of a highly three-dimensional motion of an electrically conducting fluid [2.10–2.12].

A computer simulation in the dynamo study does not take the place of an analytical method developing from the kinematic approach. The reason may be stated as follows.

(i) Nondimensional parameters characterizing the fluid motion in stellar objects are very large. They are the Taylor number (T_a), the Rayleigh number (R_a), etc., where T_a is the square of the ratio of Coriolis to molecular viscous forces, and R_a is the product of the Prandtl number (the ratio of kinematic viscosity to thermal diffusivity) and the square of the ratio of buoyancy to molecular viscous forces. In the current study of electrically nonconducting flows such as channel flow, the complete computer simulation capturing energy-dissipative components of motion is limited for $R_e \leq O(10^4)$. Then a computer simulation mimicking the situation close to the earth and the sun is not possible in the near future.

(ii) In a computer dynamo simulation, a large amount of numerical data is available in general. Those data are analyzed on the basis of computer graphics, and spatial and temporal properties of magnetic fields are investigated. The current geodynamo simulations may really show that the energy of generated magnetic fields is much larger than the kinetic energy of flow, as is consistent with the foregoing conjecture. These simulations, however, have not yet succeeded in revealing what is the key process in storing such a large amount of magnetic energy.

A representative analytical approach to dynamo is mean-field theory [2.13–2.16]. In the approach based on the application of ensemble averaging, attention is focused on a global behavior of magnetic fields at the cost of highly time-dependent properties. As a result, large nondimensional parameters are not a critical stumbling block for mean-field theory. Specifically, the theory is suitable for detecting the properties common to geodynamo, solar dynamo, etc., since the key dynamo processes are explored in mathematical but not numerical terms. From each merit of mean-field theory and a computer simulation, one is complementary to the other in the study of planetary dynamo.

2.3 Collimation of Jets

An astronomical instance of distinct global flow profiles in turbulence is the collimation of astronomical jets. High-mass objects such as active galactic nuclei, neutron stars, protostars, etc., are surrounded by gases in the form of a disk [2.17, 2.18]. The disk is called an accretion disk (figure 2.9). Gases accrete onto such objects while rotating, resulting in the release of the gravitational energy and the angular momentum. The gravitational energy is a primary source of energetic activities of those objects.

The process through which gases release the angular momentum is of great concern in understanding the physics of accretion-disk phenomena. Two mechanisms may be mentioned for the release of angular momentum. One is the angular-momentum transport by the turbulent motion of gases towards the outer part of the disk. The other is the release by ubiquitously observed bipolar jets that are composed of parts of accreting gases and are driven in the two directions normal to the disk. The jet speed is $O(10)$–$O(10^2)\ \mathrm{km\,s^{-1}}$ for protostars and several ten per cent of light speed for active galactic nuclei, respectively.

A noteworthy feature of the foregoing bipolar jets is the high collimation; namely, they keep a straight shape with an extremely small growth of jet width, compared with laboratory jets. For the mechanism of high collimation, there are two candidates. One is the suppression of jet growth due to

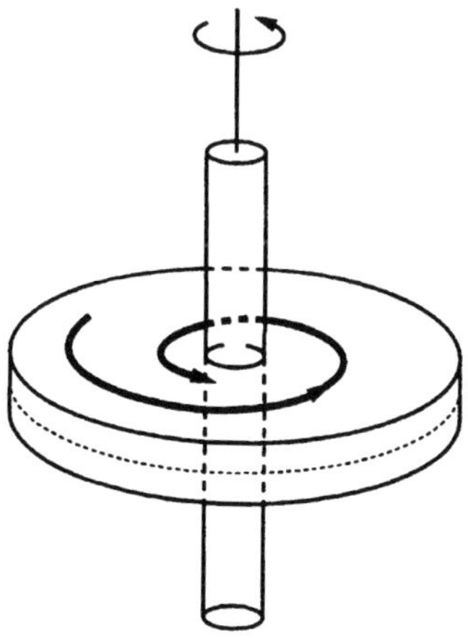

Figure 2.9. Accretion disk and bipolar jets.

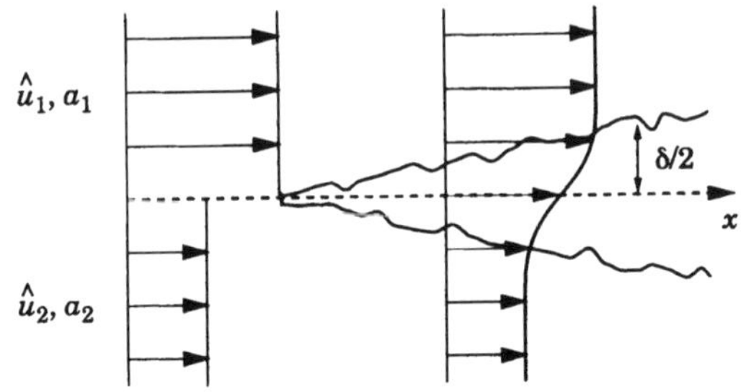

Figure 2.10. Turbulent free-shear layer flow.

high-Mach-number effects on turbulent flow, and the other is the confinement of gases by magnetic fields since accretion gases are often ionized owing to their high temperature.

The relationship of turbulence suppression with Mach-number effects is also a challenging theme in the study of electrically nonconducting flows related to aeronautical and mechanical engineering. This point has been examined most intensively for a free-shear layer flow (figure 2.10) arising from the merging of two free streams with different velocities. Its growth is highly dependent on their Mach number, and one of the important controlling parameters is the convective Mach number

$$M_C = \frac{\hat{u}_1 - \hat{u}_2}{a_1 + a_2}, \tag{2.2}$$

where $\hat{u}_n$ and a_n are the flow velocity and the sound velocity of each free stream, respectively [2.19, 2.20].

The growth rate of the layer width δ is normalized as

$$G_N = \frac{\delta'(x, M_C)}{\delta'(x, 0)} \tag{2.3}$$

where $\delta' = \mathrm{d}\delta/\mathrm{d}x$. Observations show that G_N decreases rapidly with increasing M_C (the observational data will be shown in §5.3.2). The strong suppression of the layer growth is tantamount to the suppression of turbulent momentum transport normal to the free streams. As a result, the distinct global velocity profile in the layer is persistent without spreading. The current observations, however, are limited to $M_C < 2$, and which of two limits

$$\lim_{M_C \to \infty} G_N = \begin{cases} \to 0 \\ \to G_\infty \quad (<1) \end{cases} \tag{2.4}$$

holds is not known.

When equation (2.4) is interpreted in the light of bipolar jets, vanishing G_N for large M_C signifies that the Mach-number effects play a critical role of collimating jets, specifically, in the case of protostar jets that are free from relativistic effects. On the contrary, nonvanishing G_N indicates that the confinement of ionized gases by magnetic effects is essential to the collimation. In the light of the latter, the study of the sustainment mechanism of magnetic fields is associated with the mean-field theory referred to in §2.2.

2.4 Magnetic Confinement of Plasmas

In plasmas consisting of ions and electrons, flow is also a subject of turbulence theory and modelling. High temperature plasmas have varieties in dynamics, compared with electrically nonconducting fluids. This is partly because of the collective response of plasma particles to fluctuating fields. The other reason is that the dynamics of plasma is strongly influenced by the interaction with an electromagnetic field which causes a long-range structure formation. The individuality of particles, which is beyond the fluid-like description of media, sometimes plays a crucial role in the dynamics and structure formation of plasmas; nevertheless, there are a lot of aspects in which plasmas share common physics with electrically nonconducting fluids. In what follows, some of plasma phenomena will be explained with the aid of examples from the magnetic confinement of high temperature plasmas.

2.4.1 Magnetic Confinement and Toroidal Plasmas

Plasmas have high electrical conductivity. In addition to mass flow, the electric current (relative velocity between electrons and ions) is generated in plasmas, and their structure is determined through the interaction with the self-generated electromagnetic field [2.21].

High temperature plasmas in laboratory experiments, in space and astrophysical circumstances as well, are often in such a state that they are confined by a strong magnetic field. A charged particle of species s is subject to the Lorentz force $e_s \boldsymbol{V}_s \times \boldsymbol{B}$, where e_s is the charge of species s, $\boldsymbol{V}_s$ is its velocity, and $\boldsymbol{B}$ is the magnetic field. As a result, the particle is subject to the circular motion around the magnetic field line with the Larmor radius (gyroradius) ρ_s as

$$\rho_s = \frac{m_s V_{s\perp}}{e_s |\boldsymbol{B}|}, \tag{2.5}$$

where m_s is the mass of species s, and $V_{s\perp}$ is the velocity component perpendicular to the magnetic field.

If $\rho_s \ll a$ (a is the system size of plasmas), the particle is confined in the perpendicular direction to the magnetic field. The cyclotron motion, however, does not impede the motion along the magnetic field line. If the magnetic field is constructed such that the magnetic field lines form a toroidal surface, charged particles do not escape from these field lines under the free motion in the direction of the magnetic field line. Such a surface is called

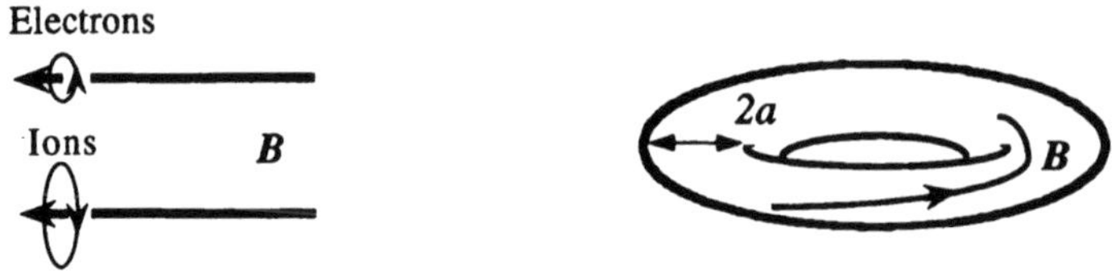

Figure 2.11. Larmour motion across the magnetic field line (left). Formation of a toroidal magnetic surface (right).

magnetic surface, and this scheme of confinement is called *toroidal magnetic confinement* (figure 2.11).

A toroidal magnetic surface is formed such that the magnetic field lines wind around a torus. The magnetic field has two components: toroidal and poloidal fields. In an axisymmetric plasma, the poloidal magnetic field is generated by the toroidal current. The most deeply explored confinement is based on the tokamak concept [2.22–2.24]. Alternatively, the poloidal magnetic field may be generated by external coil currents if the toroidal asymmetry is allowed. Such a configuration is called a stellarator or helical system [2.24].

The magnetic confinement of high temperature plasmas utilizes the formation of a set of nested magnetic surfaces. Figure 2.12 illustrates the nested toroidal magnetic surfaces. Each magnetic surface has the topology of a torus and is arranged not to intersect with other surfaces. The motion of charged particles along the magnetic field line is almost free, so that the parameters of plasma (number density, temperature, etc.) are nearly constant on the magnetic surface. On the contrary, the motion of charged particles across the magnetic field is constrained by the Larmour motion, so that the plasma parameters are not easily equilibrated between different magnetic surfaces. By this mechanism, the high temperature plasmas near the toroidal axis of the torus are isolated from the cold plasmas near the outer surfaces. This state sustained by external forces (e.g., those by external magnetic field coils) is in a dynamical equilibrium.

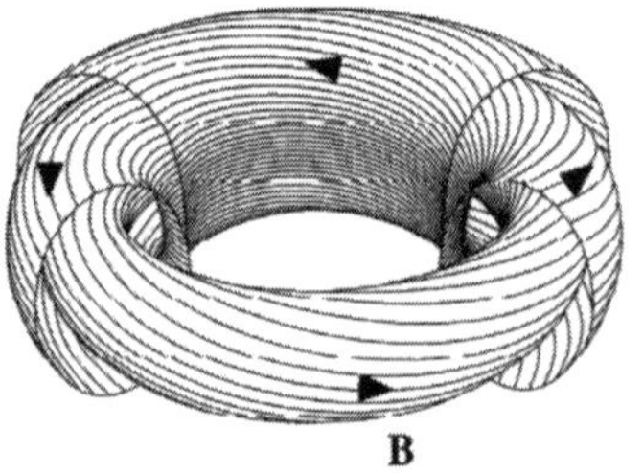

Figure 2.12. Nested toroidal magnetic surfaces. Arrows illustrate the magnetic field [2.27].

The method of maintaining the foregoing dynamical equilibrium has been a primary subject in the long history of plasma-confinement research. The high electrical conductivity of plasma generates the electric current on magnetic surfaces. This current affects the magnetic field, so that the self-consistent determination of the magnetic surfaces, together with the plasma current, needs to be investigated. A variety of the toroidal plasma equilibrium has been explored [2.22–2.26].

The dynamical-equilibrium state of toroidal plasmas is also observed in space and astrophysical plasmas. A topology of a torus is fundamental for high temperature equilibrium plasmas. If plasmas are confined in a limited region of space, the structure of the magnetic field must have a topology of a torus since the magnetic field is divergence free or $\nabla \cdot \boldsymbol{B} = 0$. The confined plasma subject to a strong influence of the magnetic field often and naturally forms a topology of torus.

2.4.2 Flows in Toroidal Plasmas

In the presence of nested magnetic surfaces, the flow of plasmas becomes strongly anisotropic. On magnetic surfaces, the flow is subject to little resistance, while the cross-field motion is strongly impeded. As a result, the flows of mass, charge, momentum, energy, etc., occur predominantly on magnetic surfaces. Various kinds of drift flows are induced by strong magnetic fields and radial gradients of number density, temperature, etc.

Flows of particle, charge, momentum, energy, etc., occur across the magnetic surfaces, although they are much smaller than those on the magnetic surfaces. The fluxes associated with gradients of these quantities have an essential role in determining the structure of plasmas, and the understanding of such mechanisms has been an important subject of plasma physics. In conventional arguments, fluxes are considered to be caused by such gradients. Which gradient and flux is considered to be a *cause* is, however, a matter of convention. It is described as a *gradient–flux relation*.

2.4.3 Topological Change of Magnetic Surfaces

The topology of magnetic surfaces is determined by the self-organized dynamics of plasmas since the plasma current induces the magnetic field, as was noted above. Therefore, if plasma is subject to a symmetry breaking perturbation, the induced perturbation current changes the magnetic field. In the presence of a perturbation current, the magnetic surfaces often lose the topology of nested toroidal surfaces, resulting in a large number of small tori. Those small tori are called *magnetic islands* (see figure 2.13). The change of topology is also a characteristic phenomenon in plasmas.

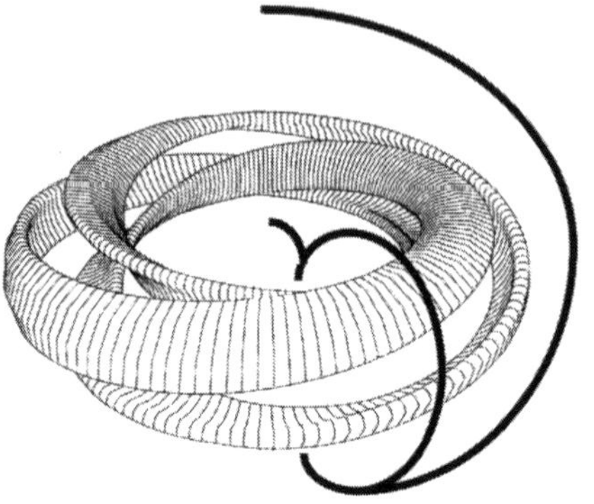

Figure 2.13. Formation of a magnetic island in a torus. A period-3, helically winding magnetic island is formed in a toroidal plasma. Magnetic surface with a simple toroidal topology is illustrated by a bold line [2.27].

2.5 Nonlinearity in Transport and Structural Transition

2.5.1 Nonlinear Gradient–Flux Relation

It is often observed that the relation between the gradient and flux is not linear. In daily experience, the flux of heat is proportional to the temperature gradient or the temperature difference between two locations. Such a linear relationship is known as Fick's law and is widely observed for a system in a thermodynamical equilibrium. In toroidal plasmas, however, the linearity is very often violated [2.28].

In general, the heat flux increases nonlinearly with the increase in the temperature gradient. Figure 2.14 illustrates an example of a nonlinear gradient–flux relationship. The thermal diffusivity is defined by the relation

$$\chi = \frac{-q_r}{n\nabla T}, \tag{2.6}$$

where q_r is the heat flux in the direction normal to the magnetic surface, n is the number density, and ∇T is the gradient of temperature T in the direction. If the relation in figure 2.14 holds, χ is an increasing function of q_r or ∇T. Here T is measured in units of Joule, i.e., $k_B T$ is abbreviated as T (k_B is the Boltzmann constant).

A nonlinear gradient–flux relationship is observed in a system with fluctuations. For plasmas, nonlinear gradient–flux relationships have been reported in, e.g. [2.29–2.31]. Electrically nonconducting fluids also show the increment of transport coefficients with increasing gradients. One example is the turbulent momentum transport in a pipe flow, and the deviation from the Hagen–Poiseuille law occurs, as was illustrated in

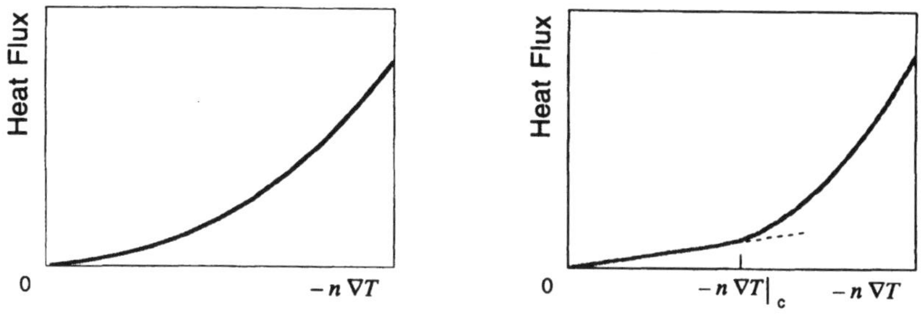

Figure 2.14. A nonlinear relationship between gradient and flux. On the left, the relation is always nonlinear. On the right, the nonlinearity sets in above a critical gradient.

§2.2.2; another is the thermal conductivity of fluid heated from the bottom. As the temperature gradient increases, the fluid motion turns into a turbulent state, and the thermal conductivity deviates from the value in a quiescent state. One motivation for turbulence research is to understand transport coefficients of various parameters in a turbulent state.

2.5.2 Bifurcation in Flow

Flow and structure in plasmas are subject to bifurcations. This fact is a primary motive force for modern plasma physics. When the electric field $\boldsymbol{E}$ is applied to plasmas in the direction perpendicular to the magnetic field, plasmas do not move in the direction of the electric field, but they perform the drift motion with the velocity

$$V_{E\times B} = \frac{\boldsymbol{E}\times\boldsymbol{B}}{B^2}. \tag{2.7}$$

The occurrence of this motion may be explained as follows. The electromagnetic force on plasma ions is written as $Z_i en(\boldsymbol{E} + \boldsymbol{V}\times\boldsymbol{B})$, where $Z_i e$ is the charge of ions. If other forces acting on plasma are much smaller than the electromagnetic force, the balance of force, which is written as

$$\boldsymbol{E} + \boldsymbol{V}\times\boldsymbol{B} = 0, \tag{2.8}$$

leads to the occurrence of the $\boldsymbol{E}\times\boldsymbol{B}$ flow in a stationary state.

When the radial electric field exists in toroidal plasmas, the plasma velocity is along the magnetic surface in the lowest-order approximation. Here, *radial electric field* means the component of electric field normal to the magnetic surface. Equation (2.8) allows various values of velocities, as is shown in figure 2.15. That is, the velocity component parallel to $\boldsymbol{B}$ is free under given $\boldsymbol{B}$ and $\boldsymbol{E}$. This freedom paves the way for the phase transition of plasma states.

In toroidal plasmas, the bifurcation of the $\boldsymbol{E}\times\boldsymbol{B}$ flow may take place for given $\boldsymbol{E}$. As is illustrated in figure 2.16, two types of flow exist on a magnetic

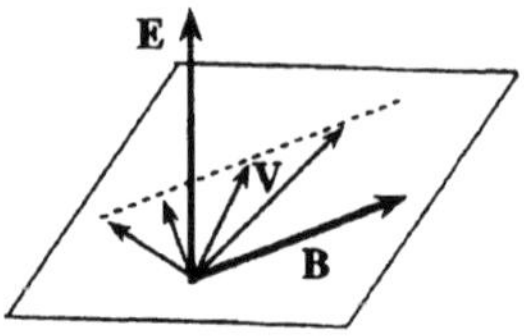

Figure 2.15. Velocities on the magnetic surface that satisfy the relation $\boldsymbol{E} + \boldsymbol{V} \times \boldsymbol{B} = 0$ for given $\boldsymbol{E}$ across the magnetic surface.

surface. One, type (a), is the flow winding helically around a torus. The other, type (b), is in the toroidal direction and does not wind around a torus. Both types of flow can satisfy equation (2.8) in the presence of the radial electric field (see [2.32, 2.33] for a review). The bifurcation between two types of flow in figures 2.16(a) and (b) has been really observed. In a short time interval, the flow can change its topology from (a) to (b) and vice versa. Such a bifurcation is also discussed in the light of turbulence in plasmas.

2.5.3 Bifurcation in Structural Formation

Bifurcation and rapid transition are also found in the structure associated with the changes of flux across magnetic surfaces. The gradient–flux relation could be either non-monotonic or even multi-valued.

One example is illustrated in figure 2.17, which shows the profile of the electrostatic potential ϕ on the minor cross-section of a torus. The electrostatic potential is peaked in one state (left), while it is flatter (or sometimes even concave) in the other state (right). In the case of the left profile, a strong radial electric field $E_r = -\partial\phi/\partial r$ (r is a minor radius of the torus) appears near the center of the plasma, and a strong flow occurs in relation to the field. The drastic change between two profiles occurs suddenly for a slight change of other plasma parameters. Therefore this change is regarded as a transition between two distinct profiles. The change can be repetitive and even periodic, although the external supply to confined plasma is kept

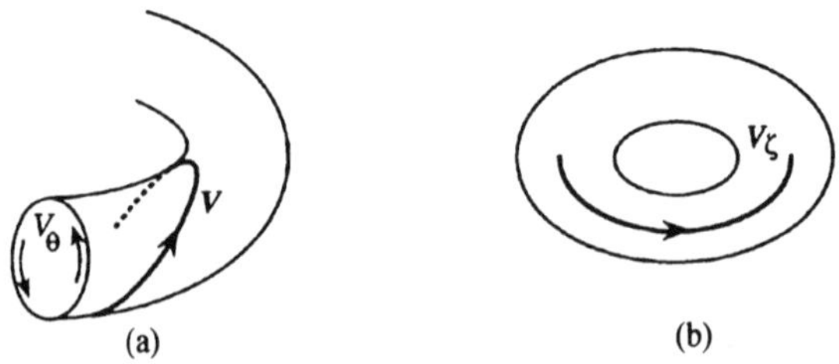

Figure 2.16. Two types of $\boldsymbol{E} \times \boldsymbol{B}$-flow velocity in the presence of a radial electric field in toroidal plasmas: one with poloidal velocity V_θ (a) and those without (b).

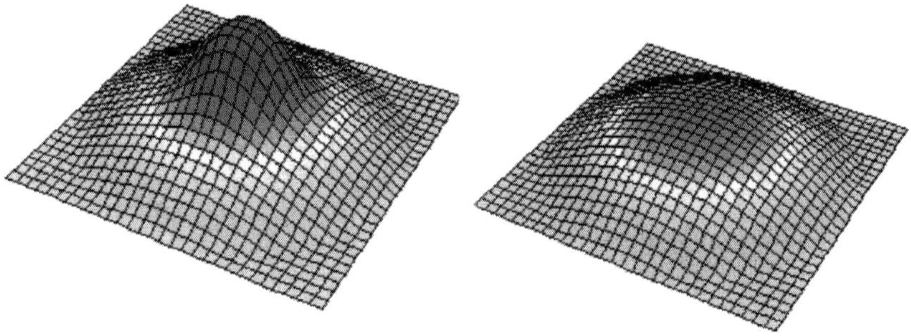

Figure 2.17. Profile of electrostatic potential on the cross-section of a toroidal plasma (based on the measurement on CHS device). Two types of profile are shown and a jump between them happens very abruptly: one is central-peaked (left) and the other is smooth-flat or concave (right) [2.34].

constant in time. The generation of self-organized oscillation in a large-scale structure is also common to high temperature plasmas.

The transition between distinct states of structures is considered as a bifurcation in the gradient–flux relation, as well. For plasma parameters subject to conservation relations, their temporal changes are related to those of the relevant fluxes. Such representative parameters are the number density n, the radial particle flux Γ_r, the charge density $e(Z_i n_i - n_e)$, the radial electric current J_r, the momentum density $m_i n_i V$ (angular momentum density), the momentum flux or the stress $\mathbf{\Pi}$, the internal energy nT, and the heat flux q_r. For instance, n obeys

$$\frac{\partial n}{\partial t} + \nabla \cdot \mathbf{\Gamma} = S, \tag{2.9}$$

where S is a source of particles. An abrupt change of the profile, which is of interest here, is induced by the rapid change of the flux $\mathbf{\Gamma}$ without a change of source. The relation $\Gamma_r(\nabla n, \ldots)$ might have a critical point at which $\Gamma_r(\nabla n, \ldots)$ has a very large change against a small variation of plasma parameters. A conceptual drawing is given in figure 2.18.

Among a variety of bifurcation phenomena in confined plasmas, the most famous is the H-mode transition [2.35]. Under the constant supply of energy, the pressure gradient of plasmas takes two distinct gradients (see figure 2.19). The gradient is weak in one state with high heat conductivity, and it is high in the other state with low heat conductivity. The former is called the *L-mode*, and the latter is the *H-mode*. The change between two states has been observed to be very rapid and is called the *L–H transition*. In the course of the transition between L- and H-modes, a self-organized oscillation has also been observed between the two states. As is typically represented by the L–H transition, confined plasmas show a large variety of structures and transitions. The region of steep gradient (like the edge

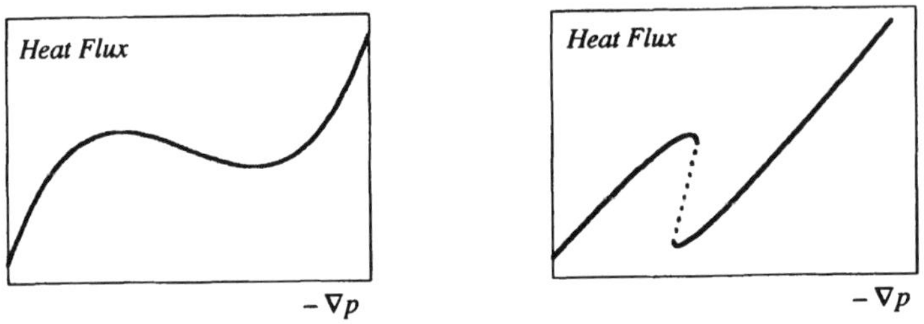

Figure 2.18. A conceptual drawing of the temperature gradient and heat flux, which can induce transition in plasma profiles.

region of H-mode in Figure 2.19(a)) is called the transport barrier. Transport barriers are realized not only near the plasma edge but also in the central region of plasma. In the latter case, it is called the internal transport barrier (ITB). Many observations of them are reported [in 2.31–2.38]. The turbulent transport, bifurcation and transitions have been central themes of plasma physics, and their studies have been reviewed [in 2.28, 2.39–2.53].

The nonlinear gradient–flux relation, including a transition, is considered to be induced by turbulence. The transport phenomena and related structures, which satisfy the linear gradient–flux relations such as Fick's law in heat transport and the Hagen–Poiseuille law in fluid dynamics, are governed by the molecular transport processes. In this case, a statistical theory of thermodynamical equilibrium may work well. In our perception of nature, such a linearity works in very limited circumstances even if it holds. A majority of events in nature are influenced by symmetry breaking fluctuations, turbulence, nonlinear transport relationships and abrupt transitions. One of the primary subjects of this monograph is to illustrate how

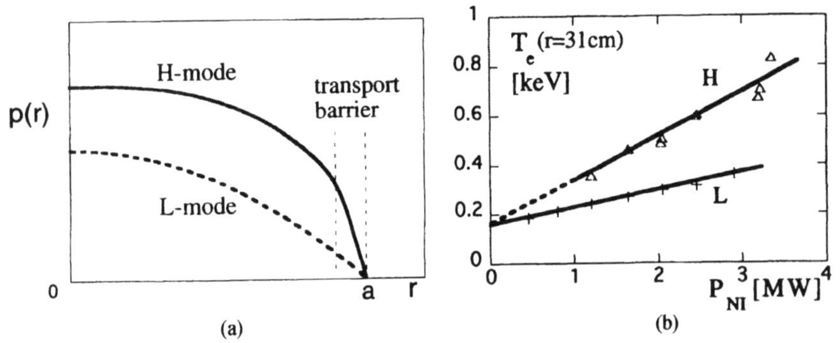

Figure 2.19. Two distinct pressure profiles before (dashed line) and after (solid line) for the L–H transition (a). Electron temperature near the plasma surface indicates the two states of plasma as a function of heating power (b). (b) is based on the result of the ASDEX experiments [2.35].

inhomogeneity-driven turbulence causes a dramatic dynamics and generates a distinctive structural formation. The examples referred to in the present chapter also indicate the characteristics of systems far from the thermodynamical equilibrium.

References

[2.1] Drazin P G and Reid W H 1981 *Hydrodynamic Stability* (Cambridge: Cambridge University Press)

[2.2] Hinze J O 1975 *Turbulence* (2nd edition) (New York: McGraw-Hill)

[2.3] Kitoh O 1991 *J. Fluid Mech.* **225** 445

[2.4] Melling A and Whitelaw J H 1976 *J. Fluid Mech.* **78** 289

[2.5] Melchior P 1986 *The Physics of the Earth's Core* (Oxford: Pergamon)

[2.6] Merrill R T, McElhinny M W and McFaden P L 1996 *The Magnetic Field of the Earth* (San Diego: Academic)

[2.7] Priest E 1982 *Solar Magnetohydrodynamics* (Dordrecht: Reidel)

[2.8] Wilson P W *Solar and Stellar Activity Cycles* (Cambridge: Cambridge University Press)

[2.9] Stix M 1981 *Solar Phys.* **74** 79

[2.10] Glatzmaier G A and Roberts P H 1995 *Phys. Earth Planet. Inter.* **91** 63

[2.11] Kageyama A and Sato T 1995 *Phys. Plasmas* **2** 1421

[2.12] Olson P, Christensen U and Glatzmaier G A 1999 *J. Geophys. Res.* **104** 10383

[2.13] Moffatt M K 1978 *Magnetic Field Generation in Electrically Conducting Fluids* (Cambridge: Cambridge University Press)

[2.14] Krause F and Rädler K-H 1980 *Mean-Field Magnetohydrodynamics and Dynamo Theory* (Oxford: Pergamon)

[2.15] Roberts P H 1990 *Astrophysical Fluid Dynamics* ed J-P Zahn and Zinn-Justin (Amsterdam: Elsevier) p 229

[2.16] Yoshizawa A 1998 *Hydrodynamic and Magnetohydrodynamic Turbulent Flows: Modelling and Statistical Theory* (Dordrecht: Kluwer)

[2.17] Begelman M C, Blandford R D and Rees M J 1984 *Rev. Mod. Phys.* **58** 55

[2.18] Ferrari A 1998 *Ann. Rev. Astron. Astrophys.* **36** 539

[2.19] Papamoshou D and Roshko A 1988 *J. Fluid Mech.* **197** 453

[2.20] Goebel S G and Dutton J C 1991 *AIAA J.* **29** 538

[2.21] Elementary processes in magnetized plasmas are explained in literature, e.g., Krall N A and Trivelpiece A W 1973 *Principles of Plasma Physics* (New York: McGraw-Hill); Stix T H 1962 *Theory of Plasma Waves* (New York: McGraw-Hill); Nishikawa K and Fukuyama A 2002 *Plasma Kinetic Theory* (Tokyo: Baihukan) (in Japanese)

[2.22] Yoshikawa S and Iiyoshi A 1972 *Introduction to Controlled Thermonuclear Reaction* (Tokyo: Kyoritu) (in Japanese)

[2.23] Wesson J A 1987 *Tokamaks* (Oxford: Oxford University Press)

[2.24] Miyamoto K 1976 *Plasma Physics for Nuclear Fusion* (Cambridge: MIT)

[2.25] White R 1989 *Theory of Tokamak Plasmas* (New York: North-Holland)

[2.26] Kadomtsev B B 1992 *Tokamak Plasma: A Complex Physical System* (Bristol: IOP Publishing Publishing)

[2.27] Courtesy of Dr D Düchs

[2.28] Itoh K, Itoh S-I and Fukuyama A 1999 *Transport and Structural Formation in Plasmas* (Bristol: IOP Publishing)

[2.29] Wagner F and Stroth U 1993 *Plasma Phys. Contr. Fusion* **35** 1321

[2.30] Lopes-Cardozo N 1995 *Plasma Phys. Contr. Fusion* **37** 799

[2.31] Stroth U 1998 *Plasma Phys. Contr. Fusion* **40** 9

[2.32] Burrel K H 1997 *Phys. Plasmas* **4** 1499

[2.33] Ida K 1998 *Plasma Phys. Contr. Fusion* **40** 1429

[2.34] Fujisawa A 2000 *J. Plasma Fusion Res.* **76** 335 (in Japanese)

[2.35] Wagner F *et al* 1982 *Phys. Rev. Lett.* **49** 1408; ASDEX Team 1989 *Nucl. Fusion* **29** 1959

[2.36] Yoshikawa S 1970 in *Methods of Experimental Physics* ed H R Griem and R H Loveberg (New York: Academic) vol 9, chapter 8; Liewer P C 1985 *Nucl. Fusion* **25** 543; Fonck R J *et al* 1992 *Plasma Phys. Contr. Fusion* **34** 1993; Wootton A *et al* 1992 *Plasma Phys. Contr. Fusion* **34** 2030

[2.37] Groebner R J 1993 *Phys. Fluids B* **5** 2343

[2.38] Zohm H 1996 *Plasma Phys. Contr. Fusion* **38** 105

[2.39] Kadomtsev B B 1965 *Plasma Turbulence* (New York: Academic Press)

[2.40] Ichimaru S 1973 *Basic Principles of Plasma Physics* (Reading: Benjamin)

[2.41] Itoh K 1994 *Plasma Phys. Contr. Fusion* **36** A307

[2.42] Itoh S-I *et al.* 1995 *J. Nucl. Materials* **220–222** 117

[2.43] Connor J W 1995 *Plasma Phys. Contr. Fusion* **37** A119

[2.44] Itoh K and Itoh S-I 1996 *Plasma Phys. Contr. Fusion* **38** 1

[2.45] Rozhansky Y and Tendler M 1996 *Plasma Rotation in Tokamaks* in *Reviews of Plasma Physics* ed B B Kadomtsev (New York: Consultants Bureau) vol 19, p 147

[2.46] Carreras B A 1997 *IEEE Trans.* **25** 1281

[2.47] Wakatani M 1998 *Plasma Phys. Contr. Fusion* **40** 597

[2.48] Connor J W 1998 *Plasma Phys. Contr. Fusion* **40** 531

[2.49] Krommes J A 1999 *Plasma Phys. Contr. Fusion* **41** A641

[2.50] Connor J W and Wilson H R 2000 *Plasma Phys. Contr. Fusion* **42** R1

[2.51] Terry P W 2000 *Rev. Mod. Phys.* **72** 109

[2.52] Krommes J A 2002 Fundamental statistical theories of plasma turbulence in magnetic fields *Phys. Reports*, in press

[2.53] Itoh S-I and Kawai Y (ed) 2002 *Bifurcation Phenomena in Plasmas* (Fukuoka: Kyushu University)

Part II

Fluid Turbulence

Nomenclature

a	Sound velocity
B_{ij}	Anisotropic tensor (anisotropic part of Reynolds stress)
C_S	Smagorinsky constant
C_V, C_P	Specific heats at constant volume and temperature
$D(k)$	Dissipation spectrum
$D_{ij}(\boldsymbol{k})$	Solenoidal tensor
$\mathrm{D}/\mathrm{D}t$	Lagrange derivative based on mean velocity
$\mathrm{D}/\mathrm{D}T$	Lagrange derivative based on mean velocity and slow variables
$E(k)$	Energy spectrum
F, $\langle f \rangle$	Ensemble mean of f
f'	Fluctuation of f in the ensemble or mass-weighted averaging
$\bar{f}^{\mathrm{F}}$	Grid-scale (GS) component of f
f''	Subgrid-scale (SGS) component of f
$\hat{f}$, $\{f\}_M$	Mass-weighted mean of f
$G_{ij}(\boldsymbol{k}; \tau, \tau')$	Green's or response function in wavenumber space
G_F	Filter function
H_K	Kinetic helicity
$\boldsymbol{H}_\theta$	Turbulent heat flux
$\boldsymbol{k}$	Wavenumber vector
K	Turbulent energy
K_O	Kolmogorov constant
K_θ	Temperature variance
l_D (k_D)	Energy dissipation length (wavenumber)
l_E (k_E)	Energy-containing length (wavenumber)
L_R	Reference length
M_C	Convective Mach number
M_T	Turbulent Mach number
p	Pressure or pressure per unit mass
P_K	Production rate of turbulent energy
P_{Rij}	Production rate of Reynolds stress
P_S	Production rate of SGS energy
$Q_{ij}(\boldsymbol{k}; \tau, \tau')$	Two-time velocity correlation function in wavenumber space
R	Gas constant per unit mass
R_e	Reynolds number
R_{ij}	Reynolds stress
$S(\tau)$	Unit step function
S_{ij}	Mean velocity-strain tensor
$\boldsymbol{T}_K$	Transport rate of turbulent energy
$\boldsymbol{u}$	Velocity
u_W	Friction velocity
U_R	Reference velocity
$\boldsymbol{X}$, T	Slow spatial and temporal variables

y_W	Wall coordinate
γ	Ratio of C_P to C_V
$\delta(\boldsymbol{k})$	Dirac delta function
δ_{ij}	Kronecker delta symbol
δ_S	Scale parameter
Δ_F	Filter width
$\boldsymbol{\nabla}, \boldsymbol{\nabla}_\xi, \boldsymbol{\nabla}_X$	Gradient vectors
ε	Dissipation rate of turbulent energy
ε_{ijk}	Alternating tensor
ε_S	Dissipation rate of SGS energy
ζ	Internal energy
θ	Temperature
κ	Karman constant
κ_θ	Heat conductivity
λ_θ	Temperature diffusivity
μ	Molecular viscosity
ν	Kinematic viscosity
ν_T	Turbulent viscosity
$\boldsymbol{\xi}, \tau$	Fast spatial and temporal variables
Π_{ij}	Pressure-strain correlation function
ρ	Density
τ_E	Characteristic time scale of energy-containing eddies
τ_{ij}	Molecular or SGS stress tensor
τ_{ij}^A	SGS stress tensor based on filter function G_A
$\boldsymbol{\omega}$	Vorticity
Ω_{ij}	Mean vorticity tensor

Chapter 3

Fundamentals of Fluid Turbulence

3.1 Fundamental Equations

We give a brief account of the derivation of fundamental equations [3.1, 3.2]. We take a fluid volume V with its surface denoted by S, as in figure 3.1 ($\boldsymbol{n}$ is the outward unit vector normal to S). For arbitrary V, we consider the temporal change of $\int_V \eta \, \mathrm{d}V$, where η represents a set of quantities per unit mass. The temporal change arises from the two effects as

$$\frac{\partial}{\partial t}\int_V \eta \, \mathrm{d}V = \int_S \eta(-u_n)\, \mathrm{d}S + S_\eta. \tag{3.1}$$

One is the inflow of η due to the fluid motion with velocity $\boldsymbol{u}$, and the other is the source (sink) S_η that generates (annihilates) η. As η, we adopt

$$\eta = [\rho, \rho\boldsymbol{u}, \rho(\tfrac{1}{2}\boldsymbol{u}^2 + \zeta)], \tag{3.2}$$

which are the density, the momentum, and the total energy, respectively (ζ is the internal energy).

For the density, we have neither source nor sink within the framework of Newtonian mechanics, and

$$\frac{\partial}{\partial t}\int_V \rho \, \mathrm{d}V = \int_S \rho(-u_n)\, \mathrm{d}S = -\int_V \nabla \cdot (\rho\boldsymbol{u})\, \mathrm{d}V, \tag{3.3}$$

where the second relation was derived from the Gauss integral theorem

$$\int_S A n_i \, \mathrm{d}S = \int_V \frac{\partial A}{\partial x_i}\, \mathrm{d}V \tag{3.4}$$

(A is assumed to be differentiable in V). As a result, we have

$$\frac{\partial \rho}{\partial t} + \nabla \cdot (\rho\boldsymbol{u}) = 0, \tag{3.5}$$

which is named the equation of continuity.

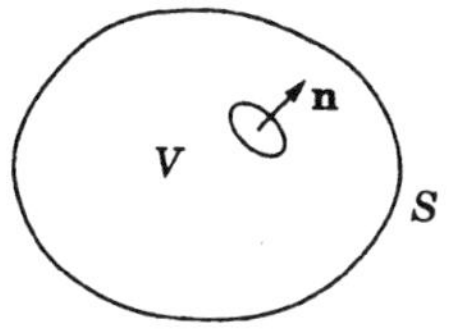

Figure 3.1. Fluid volume for conservation laws.

For the momentum, we have source and sink terms, that is, nonvanishing S_η. Their typical instance is a force exerted from the fluid outside V to the inside fluid across S. Such a force per unit surface is called the stress, which we denote by τ_{ij} in the Cartesian coordinates (x_i). At the surface normal to the i axis, τ_{ij} expresses the j component of the force exerted from the fluid at x_i+ to the fluid at x_i-. From the conservation of angular momentum, τ_{ij} is required to be symmetric, that is,

$$\tau_{ij} = \tau_{ji}. \tag{3.6}$$

The origin of τ_{ij} is the molecular motion constituting a fluid. We model τ_{ij} in terms of pressure and friction forces and write

$$\tau_{ij} = -p\delta_{ij} + \mu\left(\frac{\partial u_j}{\partial x_i} + \frac{\partial u_i}{\partial x_j}\right) + \mu_S \nabla \cdot \boldsymbol{u}\delta_{ij} \tag{3.7a}$$

or

$$\tau_{ij} = -p\delta_{ij} + \mu\left(\frac{\partial u_j}{\partial x_i} + \frac{\partial u_i}{\partial x_j} - \frac{2}{3}\nabla \cdot \boldsymbol{u}\delta_{ij}\right) + \mu_V \nabla \cdot \boldsymbol{u}\delta_{ij}, \tag{3.7b}$$

with $\mu_V = \mu_S + (\frac{2}{3})\mu$. In equation (3.7a), p is the pressure, and the second and third terms express the friction forces generated by the straining and dilatational motion of fluid, respectively. The coefficients μ and μ_S are the viscosity and the second viscosity, respectively, whereas μ_V is called the bulk viscosity. In fluid motion, it is difficult to distinguish among the diagonal components of equation (3.7) by observations. Then p is often defined by $-\tau_{ii}/3$, which corresponds to $\mu_V = 0$. In what follows, we choose vanishing μ_V. This choice is familiar, but its physical basis is not so clear.

Using τ_{ij}, we may write the equation for the momentum as

$$\frac{\partial}{\partial t}\int_V \rho u_i \,\mathrm{d}V = -\int_S \rho u_i u_n \,\mathrm{d}S + \int_S n_j \tau_{ji} \,\mathrm{d}S. \tag{3.8}$$

Here the second term expresses the i component of the force exerted from the outer to inner fluids. From equation (3.4), we have

$$\frac{\partial}{\partial t}\rho u_i + \frac{\partial}{\partial x_j}\rho u_i u_j = \frac{\partial \tau_{ji}}{\partial x_j}, \tag{3.9a}$$

that is,

$$\frac{\partial}{\partial t}\rho u_i + \frac{\partial}{\partial x_j}\rho u_i u_j = \rho\left(\frac{\partial}{\partial t} + \boldsymbol{u}\cdot\boldsymbol{\nabla}\right)u_i$$

$$= -\frac{\partial p}{\partial x_i} + \frac{\partial}{\partial x_j}\left[\mu\left(\frac{\partial u_j}{\partial x_i} + \frac{\partial u_i}{\partial x_j} - \frac{2}{3}\boldsymbol{\nabla}\cdot\boldsymbol{u}\delta_{ij}\right)\right], \tag{3.9b}$$

from equations (3.5) and (3.7) with vanishing μ_V. Equation (3.9b) is called the Navier–Stokes equation.

In the equation for the total energy, we may mention two sources; one is the inflow of heat, and the other is the work done through τ_{ij}. We include them and write

$$\frac{\partial}{\partial t}\int_V \rho(\tfrac{1}{2}\boldsymbol{u}^2 + \zeta)\,\mathrm{d}V = \int_S \rho(\tfrac{1}{2}\boldsymbol{u}^2 + \zeta)(-u_n)\,\mathrm{d}S$$

$$+ \int_S (-q_n)\,\mathrm{d}S + \int_S u_i n_j \tau_{ji}\,\mathrm{d}S, \tag{3.10}$$

where $\boldsymbol{q}$ is the heat flux due to molecular motion

$$\boldsymbol{q} = -\kappa_\theta \boldsymbol{\nabla}\theta \tag{3.11}$$

(θ is the temperature, and κ_θ is the heat conductivity). We substitute equations (3.7) and (3.11) into equation (3.10), and use equation (3.9b). As a result, we have

$$\frac{\partial}{\partial t}\rho\zeta + \boldsymbol{\nabla}\cdot(\rho\zeta\boldsymbol{u}) = \rho\left(\frac{\partial}{\partial t} + \boldsymbol{u}\cdot\boldsymbol{\nabla}\right)\rho\zeta = \boldsymbol{\nabla}\cdot(\kappa_\theta\boldsymbol{\nabla}\theta) - p\boldsymbol{\nabla}\cdot\boldsymbol{u} + \phi_D. \tag{3.12}$$

Here ϕ_D is the dissipation function

$$\phi_D = \mu\frac{\partial u_j}{\partial x_i}\left(\frac{\partial u_j}{\partial x_i} + \frac{\partial u_i}{\partial x_j} - \frac{2}{3}\boldsymbol{\nabla}\cdot\boldsymbol{u}\delta_{ij}\right) = \frac{1}{2}\mu\left(\frac{\partial u_j}{\partial x_i} + \frac{\partial u_i}{\partial x_j} - \frac{2}{3}\boldsymbol{\nabla}\cdot\boldsymbol{u}\delta_{ij}\right)^2$$

$$= \frac{1}{2}\mu\left[\left(\frac{\partial u_j}{\partial x_i} + \frac{\partial u_i}{\partial x_j}\right)^2 - \frac{4}{3}(\boldsymbol{\nabla}\cdot\boldsymbol{u})^2\right]. \tag{3.13}$$

From the second relation, we may confirm the positiveness of ϕ_D.

Finally, we add the thermodynamic relation

$$\zeta = C_V\theta, \qquad p = R\rho\theta = (\gamma - 1)\rho\zeta, \tag{3.14}$$

where C_V is the specific heat at constant volume, R is the gas constant per unit mass, and γ is the ratio of the specific heat at constant pressure, C_P, to C_V.

For constant ρ, we have

$$\boldsymbol{\nabla}\cdot\boldsymbol{u} = 0, \tag{3.15}$$

$$\frac{\partial u_i}{\partial t}+\frac{\partial}{\partial x_j}u_i u_j=\left(\frac{\partial}{\partial t}+\boldsymbol{u}\cdot\nabla\right)u_i=-\frac{1}{\rho}\frac{\partial p}{\partial x_i}+\nu\nabla^2 u_i, \tag{3.16}$$

$$\frac{\partial \theta}{\partial t}+\nabla\cdot(\theta\boldsymbol{u})=\lambda_\theta\nabla^2\theta, \tag{3.17}$$

where ν $(=\mu/\rho)$ is the kinematic viscosity, and λ_θ $[=\kappa_\theta/(\rho C_{\mathrm{P}})]$ is the heat diffusivity. We apply equation (3.15), and have

$$\nabla^2 p=-2\rho\frac{\partial u_j}{\partial x_i}\frac{\partial u_i}{\partial x_j}. \tag{3.18}$$

Equation (3.18) indicates that the pressure is a nonlocal quantity since pressure effects in an incompressible fluid spread with an infinite sound velocity.

3.2 Averaging Procedures

Fluid motion is roughly divided into three states: a laminar state, a turbulent state, and a transitional state connecting the former two. In chapters 3–5 attention is focused on turbulent states of electrically nonconducting and conducting fluids. This treatment, however, does not lessen the importance of a transitional phase, but is due to the fact that flow phenomena in engineering and nature are explored in the light of strong velocity fluctuations.

In the theoretical study of turbulence, an averaging procedure plays a key role. Its use is not merely due to the difficulty of simultaneously treating all components of motion included in a flow. By this procedure, we can often abstract some global characteristics of turbulent flow in a clear mathematical form.

As typical averaging procedures, we may mention the time, volume, and ensemble averaging, which are defined as

$$\langle f\rangle_T(\boldsymbol{x},t)=\frac{1}{\Delta T}\int_{-\Delta T/2}^{\Delta T/2}f(\boldsymbol{x},t+\tau)\,\mathrm{d}\tau, \tag{3.19}$$

$$\langle f\rangle_V(\boldsymbol{x},t)=\frac{1}{\Delta V}\int_{\Delta V}f(\boldsymbol{x}+\boldsymbol{\xi},t)\,\mathrm{d}\boldsymbol{\xi}, \tag{3.20}$$

$$\langle f\rangle(\boldsymbol{x},t)=\lim_{N\to\infty}\frac{1}{N}\sum_{\alpha=1}^{N}f_\alpha(\boldsymbol{x},t), \tag{3.21}$$

respectively. Here f is a function of time t and location $\boldsymbol{x}$, ΔT and ΔV are the time interval and the volume for averaging, respectively, and f_α is the value of f that is obtained from the α-th observation in the repeated experiments with a fixed external condition.

Of the three averaging procedures, the ensemble averaging suffers from the least ambiguity from a mathematical viewpoint. In decaying turbulence

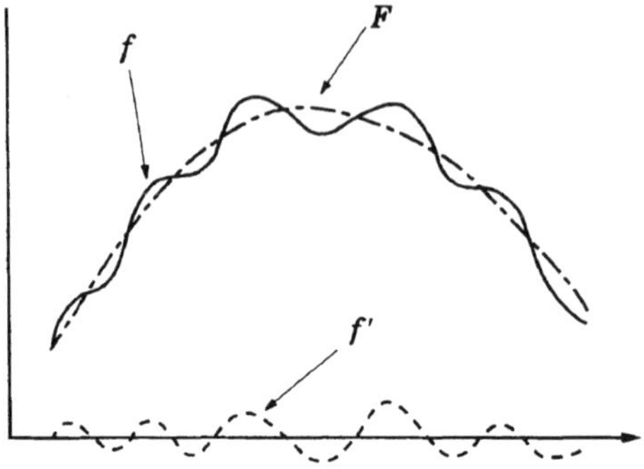

Figure 3.2. Variations of mean and fluctuation.

generated by the imposition of disturbances at an initial time, $\langle f \rangle_T$ is sensitive to the choice of ΔT. A similar situation holds for $\langle f \rangle_V$ in flows whose mean value is dependent on location. From these reasons, we adopt the ensemble averaging procedure in chapters 3–5. In observations it is difficult to repeat experiments so many times that the proper ensemble mean value of f may be obtained. We often combine the ensemble averaging with the time averaging. For instance, the mean velocity in a turbulent pipe flow is obtained from the time averaging as well as the ensemble averaging along the pipe axis.

Under the ensemble averaging, we divide a quantity f into the mean F and the fluctuation around it, f', as

$$f = F + f', \qquad F = \langle f \rangle. \tag{3.22}$$

In general, the spatial and temporal variations of F are much slower than those of f', as is illustrated in figure 3.2.

3.3 Ensemble-Mean Equations

3.3.1 Mean-Field Equations

In chapters 3–5 on fully-developed turbulence of electrically nonconducting fluids and chapters 6–10 on magnetohydrodynamic turbulence, we focus attention on the case of constant density, except §5.3 dealing with modelling of high-Mach-number effects on turbulence. Then we introduce the pressure per unit mass

$$\frac{p}{\rho} \to p. \tag{3.23}$$

As f, we adopt the velocity $\boldsymbol{u}$, the pressure p, the vorticity $\boldsymbol{\omega}$ $(= \nabla \times \boldsymbol{u})$, and the temperature θ, and write

$$f = (\boldsymbol{u}, p, \boldsymbol{\omega}, \theta), \qquad F = (\boldsymbol{U}, P, \boldsymbol{\Omega}, \Theta), \qquad f' = (\boldsymbol{u}', p', \boldsymbol{\omega}', \theta'). \tag{3.24}$$

From the averaging of equations (3.15) and (3.16), we have

$$\nabla \cdot \boldsymbol{U} = 0, \tag{3.25}$$

$$\frac{\mathrm{D}U_i}{\mathrm{D}t} \equiv \left(\frac{\partial}{\partial t} + \boldsymbol{U} \cdot \nabla \right) U_i = -\frac{\partial P}{\partial x_i} + \frac{\partial}{\partial x_j}(-R_{ij}) + \nu \nabla^2 U_i. \tag{3.26}$$

In equation (3.26), R_{ij} is called the Reynolds stress, which is defined by

$$R_{ij} = \langle u_i' u_j' \rangle, \tag{3.27}$$

and expresses the momentum transport due to fluctuations. Similarly, we have

$$\frac{\mathrm{D}\Theta}{\mathrm{D}t} = \nabla \cdot (-\boldsymbol{H}_\theta + \lambda_\theta \nabla \Theta). \tag{3.28}$$

Here $\boldsymbol{H}_\theta$ is the turbulent heat flux

$$\boldsymbol{H}_\theta = \langle \boldsymbol{u}' \theta' \rangle \tag{3.29}$$

and signifies the enhancement of heat transfer due to fluctuations.

3.3.2 Turbulence Equations

3.3.2.1 Reynolds-stress equation

In order to close equations (3.25) and (3.26), we need to relate the Reynolds stress R_{ij} to the mean velocity $\boldsymbol{U}$. For this purpose, we have two approaches. One is to relate these two quantities in an algebraic manner. The other is to close equation (3.26) in combination with the equation for R_{ij}. To understand the characteristics of the latter equation is also useful for the former approach.

The velocity fluctuation $\boldsymbol{u}'$ obeys

$$\nabla \cdot \boldsymbol{u}' = 0, \tag{3.30}$$

$$\frac{\mathrm{D}u_i'}{\mathrm{D}t} = -u_j' \frac{\partial U_i}{\partial x_j} - \frac{\partial}{\partial x_j}(u_i' u_j' - R_{ij}) - \frac{\partial p'}{\partial x_i} + \nu \nabla^2 u_i', \tag{3.31}$$

from equations (3.15), (3.16), (3.25), and (3.26). Then the equation for p' is

$$\nabla^2 p' = -2 \frac{\partial u_j'}{\partial x_i} \frac{\partial U_i}{\partial x_j} - \frac{\partial^2}{\partial x_i \partial x_j}(u_i' u_j' - R_{ij}). \tag{3.32}$$

From equation (3.31), we have

$$\frac{\mathrm{D}R_{ij}}{\mathrm{D}t} = P_{Rij} + \Pi_{ij} - \varepsilon_{ij} + \frac{\partial}{\partial x_l}\left(T_{Rijl} + \nu \frac{\partial R_{ij}}{\partial x_l} \right). \tag{3.33}$$

Here P_{Rij}, Π_{ij}, ε_{ij}, and T_{Rijl} are called the production, pressure–strain, dissipation, and transport terms, respectively, and are defined by

$$P_{Rij} = -R_{jk}\frac{\partial U_i}{\partial x_k} - R_{ik}\frac{\partial U_j}{\partial x_k}, \tag{3.34}$$

$$\Pi_{ij} = \left\langle p'\left(\frac{\partial u'_j}{\partial x_i} + \frac{\partial u'_i}{\partial x_j}\right)\right\rangle, \tag{3.35}$$

$$\varepsilon_{ij} = 2\nu\left\langle \frac{\partial u'_i}{\partial x_k}\frac{\partial u'_j}{\partial x_k}\right\rangle, \tag{3.36}$$

$$T_{Rijl} = -(\langle u'_i u'_j u'_l\rangle + \langle p' u'_j\rangle \delta_{il} + \langle p' u'_i\rangle \delta_{jl}) \tag{3.37}$$

(subscript R denotes Reynolds).

In the method using equations (3.26) and (3.33) in their combination, P_{Rij} is written in terms of $\boldsymbol{U}$ and R_{ij}, and no further mathematical procedure is necessary. The relationship of Π_{ij}, ε_{ij}, and T_{Rijl} with $\boldsymbol{U}$ and R_{ij}, however, is not known in advance. A mathematical procedure for constructing this relationship is usually called turbulence modelling, irrespective of heuristic and theoretical methods. The feature of equation (3.33) will be referred to after the discussion on the equation for the turbulent energy that is its specific form.

3.3.2.2 Turbulent-energy equation

A typical index giving the intensity of velocity fluctuations is the turbulent energy

$$K = \langle \tfrac{1}{2}\boldsymbol{u}'^2\rangle = \tfrac{1}{2}R_{ii}, \tag{3.38}$$

using which the variance and standard deviation of $\boldsymbol{u}'$ are written as $2K$ and $\sqrt{2K}$, respectively. From equation (3.33), K obeys

$$\frac{\mathrm{D}K}{\mathrm{D}t} = P_K - \varepsilon + \nabla\cdot(\boldsymbol{T}_K + \nu\nabla K), \tag{3.39}$$

where

$$P_K = -R_{ij}\frac{\partial U_j}{\partial x_i}, \tag{3.40}$$

$$\varepsilon = \nu\left\langle\left(\frac{\partial u'_j}{\partial x_i}\right)^2\right\rangle, \tag{3.41}$$

$$\boldsymbol{T}_K = -\langle(\tfrac{1}{2}\boldsymbol{u}'^2 + p')\boldsymbol{u}'\rangle. \tag{3.42}$$

Equations (3.40)–(3.42) correspond to equations (3.34), (3.36), and (3.37), respectively, and are called the production, dissipation, and transport terms of turbulent energy.

In order to see the physical meaning of equation (3.39), we integrate it over the whole fluid region and have

$$\frac{\partial}{\partial t}\int_V K\,\mathrm{d}V = \int_V P_K\,\mathrm{d}V - \int_V \varepsilon\,\mathrm{d}V + \int_S (-K\boldsymbol{U} + \boldsymbol{T}_K + \nu\nabla K)\cdot\boldsymbol{n}\,\mathrm{d}S, \tag{3.43}$$

from the Gauss integral theorem, equation (3.4). In the absence of mean flow, P_K disappears, and equation (3.43) is reduced to

$$\frac{\partial}{\partial t}\int_V K\,\mathrm{d}V = -\int_V \varepsilon\,\mathrm{d}V + \int_S (\boldsymbol{T}_K + \nu\nabla K)\cdot\boldsymbol{n}\,\mathrm{d}S. \tag{3.44}$$

On the right-hand side of equation (3.44), the first term with the minus sign included is negative and expresses the conversion of kinetic energy to heat. The second term represents the flow of turbulent energy across the surface. In the absence of the energy inflow, K vanishes eventually: that is, turbulence decays. This fact suggests that P_K is closely related to the sustainment of a turbulent state or the supply of K.

We see how energy is supplied to velocity fluctuations. For this purpose, we consider the mean-flow counterpart of equation (3.43), which is written as

$$\frac{\partial}{\partial t}\int_V \frac{1}{2}\boldsymbol{U}^2\,\mathrm{d}V = \int_V (-P_K)\,\mathrm{d}V - \int_V \nu\left(\frac{\partial U_j}{\partial x_i}\right)^2\mathrm{d}V + \int_S n_i\left[-PU_i - R_{ij}U_j + \nu\frac{\partial}{\partial x_i}\left(\frac{1}{2}\boldsymbol{U}^2\right)\right]\mathrm{d}S, \tag{3.45}$$

from equation (3.26). We should notice that P_K also appears in equation (3.45), but with the minus sign attached. From the discussion on equation (3.44), a turbulent state is sustained as long as

$$\int_V P_K\,\mathrm{d}V > 0 \tag{3.46}$$

(this condition does not exclude the local negativeness of P_K). Equation (3.46) signifies that $\int_V(-P_K)\,\mathrm{d}V$ plays a role of draining energy from the mean flow and supplying it to the fluctuating field. The energy transfer from large- to small-scale components of motion is generally called the energy cascade.

How is energy supplied to the mean flow? In equation (3.45), the second term is the dissipation due to mean flow, which is much smaller than ε, except when close to a solid wall. In the third term, the P-related part

$$\int_S \boldsymbol{U}\cdot(-P\boldsymbol{n})\,\mathrm{d}S \tag{3.47}$$

expresses the work done on the fluid by the pressure and is a primary energy source of mean flow. In a turbulent pipe flow, for instance, equation (3.47) corresponds to the work by the pressure difference between two cross-sections.

From the above discussions, we have the following picture of the energy cascade. Energy is first supplied to the mean flow through the imposition of external forces such as pressure. The energy is supplied to the fluctuating field through P_K and is eventually converted to heat by molecular viscous effects. Such a clear picture arises from the conservation property of kinetic energy in the absence of viscous effects. In reality, we have

$$\frac{\partial}{\partial t}\int_V \left(\frac{1}{2}\boldsymbol{u}^2\right) \mathrm{d}V = -\int_V \nu\left(\frac{\partial u_j}{\partial x_i}\right)^2 \mathrm{d}V + \int_S \left[-\left(\frac{1}{2}\boldsymbol{u}^2 + p\right)\boldsymbol{u} + \nu\nabla\left(\frac{1}{2}\boldsymbol{u}^2\right)\right]\cdot\boldsymbol{n}\,\mathrm{d}S, \quad (3.48)$$

from equation (3.16), and a total amount of kinetic energy is conserved in the absence of viscous effects so long as no energy is supplied or lost across a boundary. This picture does not hold for a quantity whose total amount is not conserved in the absence of viscous effects.

The production rate P_K is linearly dependent on the mean velocity gradient. This fact corresponds to the observational finding that K is often large in the vicinity of a solid wall where the mean velocity itself is small, but its gradient becomes steep.

3.3.2.3 *Mechanism of turbulent-energy distribution*

Let us come back to the Reynolds stress equation (3.33). From equation (3.16), we have

$$\frac{\partial}{\partial t}u_i u_j = -p\left(\frac{\partial u_j}{\partial x_i} + \frac{\partial u_i}{\partial x_j}\right) - 2\nu\frac{\partial u_i}{\partial x_l}\frac{\partial u_j}{\partial x_l} + \frac{\partial}{\partial x_k}\left(-u_i u_j u_k - u_i p\delta_{jk} - u_j p\delta_{ik} + \nu\frac{\partial}{\partial x_k}u_i u_j\right). \quad (3.49)$$

On the right-hand side, the occurrence of the first term signifies that $\int_V u_i u_j \,\mathrm{d}V$ is not conserved even in the absence of viscous effects and the supply or loss of $u_i u_j$ across a boundary. The first term enters equation (3.33) as Π_{ij} [equation (3.35)]. The solenoidal (incompressible) condition (3.30) gives

$$\Pi_{ii} = 0, \quad (3.50)$$

which indicates that Π_{ij} does not contribute directly to the variation of the sum of

$$\langle u_x'^2\rangle, \qquad \langle u_y'^2\rangle, \qquad \langle u_z'^2\rangle, \quad (3.51)$$

but to each variation.

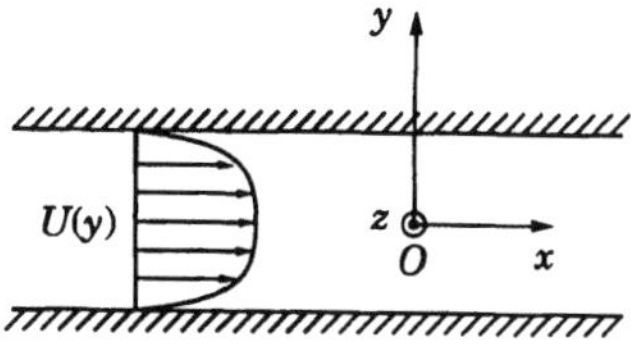

Figure 3.3. Channel flow.

In order to understand the foregoing important role of Π_{ij}, we consider a turbulent channel flow (figure 3.3). There the mean velocity $\boldsymbol{U}$ and the turbulence quantities such as expression (3.51) are dependent on y only, and are written as

$$\boldsymbol{U} = [U(y), 0, 0], \qquad \boldsymbol{u}' = (u', v', w'). \tag{3.52}$$

In the flow, R_{ij} is reduced to

$$\{R_{ij}\} = \begin{pmatrix} \langle u'^2 \rangle & \langle u'v' \rangle & 0 \\ \langle u'v' \rangle & \langle v'^2 \rangle & 0 \\ 0 & 0 & \langle w'^2 \rangle \end{pmatrix}. \tag{3.53}$$

For instance, $\langle u'w' \rangle$ vanishes since w' for fixed u' may take both signs on the equal probability owing to the geometrical symmetry of flow.

From equation (3.26), U obeys

$$0 = G - \frac{\mathrm{d}\langle u'v' \rangle}{\mathrm{d}y} + \nu \frac{\mathrm{d}^2 U}{\mathrm{d}y^2}. \tag{3.54}$$

In what follows, we put

$$G = -\frac{\partial P}{\partial x} = \text{const} \quad (>0). \tag{3.55}$$

For expression (3.51), equation (3.33) gives

$$0 = -2\langle u'v' \rangle \frac{\mathrm{d}U}{\mathrm{d}y} + \Pi_{xx} - \varepsilon_{xx} + \frac{\mathrm{d}T_{Rxxy}}{\mathrm{d}y} + \nu \frac{\mathrm{d}^2 \langle u'^2 \rangle}{\mathrm{d}y^2}, \tag{3.56}$$

$$0 = \Pi_{yy} - \varepsilon_{yy} + \frac{\mathrm{d}T_{Ryyy}}{\mathrm{d}y} + \nu \frac{\mathrm{d}^2 \langle v'^2 \rangle}{\mathrm{d}y^2}, \tag{3.57}$$

$$0 = \Pi_{zz} - \varepsilon_{zz} + \frac{\mathrm{d}T_{Rzzy}}{\mathrm{d}y} + \nu \frac{\mathrm{d}^2 \langle w'^2 \rangle}{\mathrm{d}y^2}. \tag{3.58}$$

For $\langle u'v' \rangle$, we have

$$0 = -\langle v'^2 \rangle \frac{\mathrm{d}U}{\mathrm{d}y} + \Pi_{xy} - \varepsilon_{xy} + \frac{\mathrm{d}T_{Rxyy}}{\mathrm{d}y} + \nu \frac{\mathrm{d}^2 \langle u'v' \rangle}{\mathrm{d}y^2}. \tag{3.59}$$

The integration of equation (3.54) leads to

$$\langle u'v'\rangle = Gy + \nu \frac{dU}{dy} + C, \tag{3.60}$$

with C as a constant. We apply the noslip condition at the lower wall ($y = -H$), and have

$$\langle u'v'\rangle = G(y+H) + \nu \frac{dU}{dy} - \left(\nu \frac{dU}{dy}\right)_{y=-H}. \tag{3.61}$$

The noslip condition at the upper wall ($y = H$) to equation (3.61) and the symmetry of U concerning y result in

$$\left(\nu \frac{dU}{dy}\right)_{y=-H} \equiv u_W^2 = GH. \tag{3.62}$$

Here u_W has the dimension of velocity and is called the wall friction velocity, which is one of the quantities characterizing a turbulent flow near a solid wall. The combination of equation (3.61) with equation (3.62) gives

$$\langle u'v'\rangle = Gy + \nu \frac{dU}{dy}. \tag{3.63}$$

Let us consider the physical meaning of equations (3.56)–(3.59) in the region except the close vicinity of two walls where molecular viscous effects are dominant. Of equations (3.56)–(3.58), the energy production term related to dU/dy appears only in equation (3.56) for $\langle u'^2\rangle$. We substitute equation (3.63) with the ν-related term dropped into equation (3.56), and have

$$0 = -2Gy \frac{dU}{dy} + \Pi_{xx} - \varepsilon_{xx} + \frac{dT_{Rxxy}}{dy}. \tag{3.64}$$

From

$$\frac{dU}{dy} > 0 \quad (-H < y < 0), \qquad \frac{dU}{dy} < 0 \quad (0 < y < H), \tag{3.65}$$

the first or production term in equation (3.64) is positive except the center-line, indicating that the energy is supplied from U to $\langle u'^2\rangle$.

We examine how the energy supplied from U is distributed to $\langle v'^2\rangle$ and $\langle w'^2\rangle$. We drop the molecular-viscosity term important in the close vicinity of the walls, and integrate equations (3.57) and (3.58) with respect to y. Each third term in these equations gives no contribution from the noslip boundary condition at the walls. This fact indicates that the first term in each equation is nearly balanced with the second term. The latter with the minus sign included is negative. We have

$$\Pi_{xx} < 0, \qquad \Pi_{yy} > 0, \qquad \Pi_{zz} > 0. \tag{3.66}$$

If $\Pi_{xx} > 0$, at least one of Π_{yy} and Π_{zz} needs to be negative from

$$\Pi_{xx} + \Pi_{yy} + \Pi_{zz} = 0. \tag{3.67}$$

For $\Pi_{yy} < 0$, there is no mechanism for sustaining $\langle v'^2 \rangle$ in equation (3.57) from the above discussion. Entirely the same situation holds for Π_{zz}. From equation (3.66), the energy supplied from $\boldsymbol{U}$ to u' is distributed to v' and w' through Π_{yy} and Π_{zz}. In other words, Π_{yy} and Π_{zz} play a role in producing $\langle v'^2 \rangle$ and $\langle w'^2 \rangle$ in channel turbulence.

In equation (3.59) the first term is

$$-\langle v'^2 \rangle \frac{\mathrm{d}U}{\mathrm{d}y} > 0 \quad (y > 0), \qquad -\langle v'^2 \rangle \frac{\mathrm{d}U}{\mathrm{d}y} < 0 \quad (y < 0). \tag{3.68}$$

The sign of each of equations (3.68) coincides with the sign of the first term in equation (3.63) in the upper and lower halves of the channel, respectively. This fact expresses that the shear–stress component of the Reynolds stress, $\langle u'v' \rangle$, continues to be generated in the presence of nonvanishing $\mathrm{d}U/\mathrm{d}y$.

Observations of a channel flow show that

$$\langle u'^2 \rangle > \langle w'^2 \rangle > \langle v'^2 \rangle. \tag{3.69}$$

That is, the fluctuation in the direction of mean flow is strongest, and the fluctuation normal to a wall is weakest. The former is understandable from the reason that u' is parts of the flow component possessing the mean motion. The reason for $\langle v'^2 \rangle < \langle w'^2 \rangle$ is stated as follows. The shear–stress component $\langle u'v' \rangle$ is necessary for sustaining $\langle u'^2 \rangle$ that is the origin of $\langle v'^2 \rangle$ and $\langle w'^2 \rangle$. For the production of $\langle u'v' \rangle$, $\langle v'^2 \rangle$ is necessary, as is seen from the first term in equation (3.59), and eventually contributes to the production of $\langle u'^2 \rangle$ [equation (3.56)]. This process is the reason for $\langle v'^2 \rangle < \langle w'^2 \rangle$.

3.3.2.4 Turbulence equations related to temperature and vorticity

A quantity characterizing the intensity of temperature fluctuation is the temperature variance:

$$K_\theta = \langle \theta'^2 \rangle. \tag{3.70}$$

From equations (3.17) and (3.28), θ' obeys

$$\frac{\mathrm{D}\theta'}{\mathrm{D}t} = -(\boldsymbol{u}' \cdot \nabla)\Theta - \nabla \cdot (\boldsymbol{u}'\theta' - \boldsymbol{H}_\theta) + \lambda_\theta \nabla^2 \theta'. \tag{3.71}$$

Then we have

$$\frac{\mathrm{D}K_\theta}{\mathrm{D}t} = P_\theta - \varepsilon_\theta + \nabla \cdot (\boldsymbol{T}_\theta + \lambda_\theta \nabla K_\theta). \tag{3.72}$$

Here P_θ, ε_θ, and $\boldsymbol{T}_\theta$ are the counterparts of each term on the right-hand side of equation (3.39), which are defined by

$$P_\theta = -2\boldsymbol{H}_\theta \cdot \nabla\Theta, \tag{3.73}$$

$$\varepsilon_\theta = 2\lambda_\theta\left\langle\left(\frac{\partial\theta'}{\partial x_i}\right)^2\right\rangle, \tag{3.74}$$

$$\boldsymbol{T}_\theta = -\langle \boldsymbol{u}'\theta'^2\rangle. \tag{3.75}$$

The clear resemblance between equations (3.39) and (3.72) is due to the conservation property concerning the equation for θ^2,

$$\frac{\partial\theta^2}{\partial t} = -2\lambda_\theta\left(\frac{\partial\theta}{\partial x_i}\right)^2 + \nabla\cdot(-\theta^2\boldsymbol{u} + \lambda_\theta\nabla\theta^2). \tag{3.76}$$

Temperature fluctuation is large in the region with a large temperature gradient, as is seen from equation (3.73).

From equation (3.71) we have

$$\frac{\mathrm{D}H_{\theta i}}{\mathrm{D}t} = P_{\theta i} + \Pi_{\theta i} - \varepsilon_{\theta i} + \frac{\partial T_{\theta ij}}{\partial x_j}, \tag{3.77}$$

where the right-hand side corresponds to that of equation (3.33) and is given by

$$P_{\theta i} = -H_{\theta j}\frac{\partial U_i}{\partial x_j} - R_{ij}\frac{\partial\Theta}{\partial x_j}, \tag{3.78}$$

$$\Pi_{\theta i} = \left\langle p'\frac{\partial\theta'}{\partial x_i}\right\rangle, \tag{3.79}$$

$$\varepsilon_{\theta i} = (\nu + \lambda_\theta)\left\langle\frac{\partial u_i'}{\partial x_j}\frac{\partial\theta'}{\partial x_j}\right\rangle, \tag{3.80}$$

$$T_{\theta ij} = -\langle u_i'u_j'\theta'\rangle - \langle p'\theta'\rangle\delta_{ij} + \nu\left\langle\theta'\frac{\partial u_i'}{\partial x_j}\right\rangle + \lambda_\theta\left\langle u_i'\frac{\partial\theta'}{\partial x_j}\right\rangle, \tag{3.81}$$

respectively.

Finally, we refer to the helicity $\boldsymbol{u}\cdot\boldsymbol{\omega}$. It is a scalar, but its sign is changed under the reflection of coordinates, $\boldsymbol{x} \to -\boldsymbol{x}$. Such a scalar is called a pseudoscalar. The helicity $\boldsymbol{u}\cdot\boldsymbol{\omega}$ obeys

$$\frac{\partial\boldsymbol{u}\cdot\boldsymbol{\omega}}{\partial t} = -2\nu\frac{\partial u_j}{\partial x_i}\frac{\partial\omega_j}{\partial x_i} + \nabla\cdot[-(\boldsymbol{u}\cdot\boldsymbol{\omega})\boldsymbol{u} + (\tfrac{1}{2}\boldsymbol{u}^2 - p)\boldsymbol{\omega} + \nu\nabla(\boldsymbol{u}\cdot\boldsymbol{\omega})]. \tag{3.82}$$

Equation (3.82) shows that the total amount of helicity is conserved so long as there are no viscous effects and neither helicity supply nor loss across a boundary. As a result, the turbulent kinetic helicity

$$H_{\mathrm{K}} = \langle \boldsymbol{u}' \cdot \boldsymbol{\omega}' \rangle \tag{3.83}$$

obeys the same type of equation as equations (3.39) and (3.72):

$$\frac{\mathrm{D}H_{\mathrm{K}}}{\mathrm{D}t} = P_H - \varepsilon_H + \nabla \cdot (\boldsymbol{T}_H + \nu \nabla H_{\mathrm{K}}), \tag{3.84}$$

where

$$P_H = -(\langle \boldsymbol{u}' \times \boldsymbol{\omega}' \rangle - \nabla K) \cdot \boldsymbol{\Omega} - R_{ij} \frac{\partial \Omega_i}{\partial x_j} = \frac{\partial R_{ij}}{\partial x_j} \Omega_i - R_{ij} \frac{\partial \Omega_i}{\partial x_j}, \tag{3.85}$$

$$\varepsilon_H = 2\nu \left\langle \frac{\partial u'_j}{\partial x_i} \frac{\partial \omega'_j}{\partial x_i} \right\rangle, \tag{3.86}$$

$$\boldsymbol{T}_H = -\langle (\boldsymbol{u}' \cdot \boldsymbol{\omega}') \boldsymbol{u}' \rangle + \left\langle \left(\frac{\boldsymbol{u}'^2}{2} - p' \right) \boldsymbol{\omega}' \right\rangle. \tag{3.87}$$

In deriving the second relation of equation (3.85), use has been made of

$$(\boldsymbol{u}' \times \boldsymbol{\omega}')_i = -\frac{\partial u'_i u'_j}{\partial x_j} + \frac{\partial}{\partial x_i} \left(\frac{1}{2} \boldsymbol{u}'^2 \right). \tag{3.88}$$

The turbulent energy K and the temperature variance K_θ are generated in the region with large velocity and temperature gradients, respectively. On the other hand, the generator of the turbulent helicity H_{K} is the mean vorticity and its gradient, as is seen from equation (3.85). This fact suggests that H_{K} is an important quantity characterizing turbulent flows possessing global vortical motion.

3.4 Homogeneous Turbulence

In this section we give a brief account of fundamental concepts of homogeneous turbulence, specifically, isotropic turbulence with no mean flow [3.3–3.6]. There some of the important properties of turbulence are lost. Their typical instance is the production mechanism of turbulent energy that is discussed in §3.3. At its cost, the energy-cascade process may be investigated in detail, irrespective of the difference among energy production mechanisms. The purpose of the following discussions is to give the least knowledge about isotropic turbulence that is useful for understanding inhomogeneous turbulence or turbulence with mean flow.

In this section we simply write the fluctuating part of velocity as $\boldsymbol{u}$ since our attention is focused on the case of no mean flow.

3.4.1 Fundamental Concepts

3.4.1.1 Homogeneity

As important statistics related to velocity fluctuations, we may mention velocity correlation functions such as

$$Q_{ij}(\boldsymbol{x}, \boldsymbol{x}'; t, t') = \langle u_i(\boldsymbol{x}, t) u_j(\boldsymbol{x}', t') \rangle, \tag{3.89}$$

$$Q_{ijk}(\boldsymbol{x}, \boldsymbol{x}', \boldsymbol{x}''; t, t', t'') = \langle u_i(\boldsymbol{x}, t) u_j(\boldsymbol{x}', t') u_k(\boldsymbol{x}'', t'') \rangle, \tag{3.90}$$

which are called the second- and third-order correlation functions, respectively. Using the difference vectors of two locations

$$\boldsymbol{r} = \boldsymbol{x}' - \boldsymbol{x}, \qquad \boldsymbol{r}' = \boldsymbol{x}'' - \boldsymbol{x}, \tag{3.91}$$

we rewrite equations (3.89) and (3.90) as

$$Q_{ij}(\boldsymbol{x}, \boldsymbol{x} + \boldsymbol{r}; t, t') = \langle u_i(\boldsymbol{x}, t) u_j(\boldsymbol{x} + \boldsymbol{r}, t') \rangle, \tag{3.92}$$

$$Q_{ijk}(\boldsymbol{x}, \boldsymbol{x} + \boldsymbol{r}, \boldsymbol{x} + \boldsymbol{r}'; t, t', t'') = \langle u_i(\boldsymbol{x}, t) u_j(\boldsymbol{x} + \boldsymbol{r}, t') u_k(\boldsymbol{x} + \boldsymbol{r}', t'') \rangle. \tag{3.93}$$

In the case that correlation functions such as equations (3.92) and (3.93) are not dependent on $\boldsymbol{x}$, that is

$$Q_{ij}(\boldsymbol{x}, \boldsymbol{x} + \boldsymbol{r}; t, t') = Q_{ij}(\boldsymbol{0}, \boldsymbol{r}; t, t'), \tag{3.94}$$

$$Q_{ijk}(\boldsymbol{x}, \boldsymbol{x} + \boldsymbol{r}, \boldsymbol{x} + \boldsymbol{r}'; t, t', t'') = Q_{ijk}(\boldsymbol{0}, \boldsymbol{r}, \boldsymbol{r}'; t, t', t''), \tag{3.95}$$

the turbulence field is called homogeneous (**0** denotes the position of the origin). Then the one-point correlation function is spatially constant or

$$\frac{\partial}{\partial x_l} Q_{ij}(\boldsymbol{x}, \boldsymbol{x}; t, t') = \frac{\partial}{\partial x_l} Q_{ijk}(\boldsymbol{0}, \boldsymbol{0}; t, t') = 0. \tag{3.96}$$

A useful mathematical tool for analyzing homogeneous turbulence is the Fourier representation of field, which is defined as

$$\boldsymbol{u}(\boldsymbol{x}, t) = \int \boldsymbol{u}(\boldsymbol{k}, t)\, \mathrm{e}^{-\mathrm{i}\boldsymbol{k}\cdot\boldsymbol{x}}\, \mathrm{d}\boldsymbol{k} \qquad \left(\int \mathrm{d}\boldsymbol{k} = \int_{-\infty}^{\infty} \mathrm{d}k_x \int_{-\infty}^{\infty} \mathrm{d}k_y \int_{-\infty}^{\infty} \mathrm{d}k_z \right), \tag{3.97a}$$

$$\boldsymbol{u}(\boldsymbol{k}, t) = \frac{1}{(2\pi)^3} \int \boldsymbol{u}(\boldsymbol{x}, t)\, \mathrm{e}^{\mathrm{i}\boldsymbol{k}\cdot\boldsymbol{x}}\, \mathrm{d}\boldsymbol{x} \qquad \left(\int \mathrm{d}\boldsymbol{x} = \int_{-\infty}^{\infty} \mathrm{d}x \int_{-\infty}^{\infty} \mathrm{d}y \int_{-\infty}^{\infty} \mathrm{d}z \right). \tag{3.97b}$$

We consider the Fourier representation of the second-order correlation function $Q_{ij}(\boldsymbol{x}, \boldsymbol{x}'; t, t')$. From equation (3.97b), we have

$$\begin{aligned} \langle u_i(\boldsymbol{k}, t) u_j(\boldsymbol{k}', t') \rangle &= \frac{1}{(2\pi)^6} \iint \langle u_i(\boldsymbol{x}, t) u_j(\boldsymbol{x}', t') \rangle\, \mathrm{e}^{\mathrm{i}(\boldsymbol{k}\cdot\boldsymbol{x} + \boldsymbol{k}'\cdot\boldsymbol{x}')}\, \mathrm{d}\boldsymbol{x}\, \mathrm{d}\boldsymbol{x}' \\ &= \frac{1}{(2\pi)^3} \delta(\boldsymbol{k} + \boldsymbol{k}') \int Q_{ij}(\boldsymbol{0}, \boldsymbol{r}; t, t')\, \mathrm{e}^{\mathrm{i}\boldsymbol{k}\cdot\boldsymbol{r}}\, \mathrm{d}\boldsymbol{r}, \end{aligned} \tag{3.98}$$

where use has been made of the Fourier representation for the Dirac delta function

$$\delta(\boldsymbol{k}) = \frac{1}{(2\pi)^3} \int \exp(\pm \mathrm{i}\boldsymbol{k} \cdot \boldsymbol{x})\, \mathrm{d}\boldsymbol{x}. \tag{3.99}$$

Equation (3.98) signifies that $\langle u_i(\boldsymbol{k}, t) u_j(\boldsymbol{k}', t') \rangle$ is proportional to $\delta(\boldsymbol{k} + \boldsymbol{k}')$. We define its proportional coefficient by

$$\frac{\langle u_i(\boldsymbol{k}, t) u_j(\boldsymbol{k}', t') \rangle}{\delta(\boldsymbol{k} + \boldsymbol{k}')}. \tag{3.100}$$

This expression includes $\delta(\boldsymbol{k} + \boldsymbol{k}')$ in its denominator and is not correct in the strict mathematical sense of generalized functions [3.7], but the use of such definition gives rise to no mathematical errors in final results.

With the foregoing point in mind, we have

$$Q_{ij}(\boldsymbol{0}, \boldsymbol{r}; t, t') = \int \frac{\langle u_i(\boldsymbol{k}, t) u_j(\boldsymbol{k}', t') \rangle}{\delta(\boldsymbol{k} + \boldsymbol{k}')} \mathrm{e}^{-\mathrm{i}\boldsymbol{k} \cdot \boldsymbol{r}}\, \mathrm{d}\boldsymbol{k} \tag{3.101}$$

as the inverse of equation (3.98). The turbulent energy K defined by equation (3.38) is written as

$$K = \frac{1}{2} \int \frac{\langle u_i(\boldsymbol{k}, t) u_i(\boldsymbol{k}', t) \rangle}{\delta(\boldsymbol{k} + \boldsymbol{k}')}\, \mathrm{d}\boldsymbol{k}. \tag{3.102}$$

3.4.1.2 *Isotropy*

In the case that statistical properties of turbulent field is invariant concerning the rotation of a coordinate system, the field is called isotropic. In isotropic turbulence, three turbulence intensities in expression (3.51) become

$$\langle u_x'^2 \rangle = \langle u_y'^2 \rangle = \langle u_z'^2 \rangle, \tag{3.103}$$

in contrast to expression (3.69) in a channel flow. In the flow, the x and z directions are geometrically equivalent, unlike the y direction. The equivalence, however, is broken once the mean flow occurs in the x direction, resulting in the anisotropy given by expression (3.69).

We consider equation (3.103) in wavenumber space. As is typically seen in a channel flow, isotropy is broken by the preferential directivity due to a boundary and a mean flow. On the contrary, the field free from these factors inevitably remains isotropic. In the wavenumber space (k_1, k_2, k_3), the second-order tensors free from such factors are

$$\delta_{ij}, \qquad \varepsilon_{ijl} k_l, \qquad k_i k_j, \tag{3.104}$$

where δ_{ij} and ε_{ijl} are the Kronecker delta symbol and alternating tensor, respectively. From equation (3.104) we may write

$$\frac{\langle u_i(\boldsymbol{k},t)u_j(\boldsymbol{k}',t')\rangle}{\delta(\boldsymbol{k}+\boldsymbol{k}')} = \delta_{ij}Q(k;t,t') + Q'(k;t,t')k_ik_j + \frac{\mathrm{i}}{2}\varepsilon_{ijl}\frac{k_l}{k^2}\Gamma_{\mathrm{K}}(k;t,t'), \tag{3.105}$$

using Q, Q', and Γ_{K} that are functions of $k = |\boldsymbol{k}|$.

From the wavenumber representation of the solenoidal condition

$$\boldsymbol{k}\cdot\boldsymbol{u}(\boldsymbol{k},t) = 0, \tag{3.106}$$

equation (3.105) gives

$$Q'(k;t,t') = -\frac{1}{k^2}Q(k;t,t'), \tag{3.107}$$

resulting in

$$\frac{\langle u_i(\boldsymbol{k},t)u_j(\boldsymbol{k}',t')\rangle}{\delta(\boldsymbol{k}+\boldsymbol{k}')} = D_{ij}(\boldsymbol{k})Q(k;t,t') + \frac{\mathrm{i}}{2}\varepsilon_{ijl}\frac{k_l}{k^2}\Gamma_{\mathrm{K}}(k;t,t'). \tag{3.108}$$

Here $D_{ij}(\boldsymbol{k})$ is the solenoidal operator defined by

$$D_{ij}(\boldsymbol{k}) = \delta_{ij} - \frac{k_ik_j}{k^2}, \tag{3.109}$$

and obeys

$$k_iD_{ij}(\boldsymbol{k}) = k_jD_{ij}(\boldsymbol{k}) = 0. \tag{3.110}$$

From equation (3.108) we may write

$$R_{ij} = \int \frac{\langle u_i(\boldsymbol{k},t)u_j(\boldsymbol{k}',t)\rangle}{\delta(\boldsymbol{k}+\boldsymbol{k}')}\,\mathrm{d}\boldsymbol{k} = \int D_{ij}(\boldsymbol{k})Q(k;t,t)\,\mathrm{d}\boldsymbol{k} = \left(\frac{2}{3}\int Q(k;t,t)\,\mathrm{d}\boldsymbol{k}\right)\delta_{ij}, \tag{3.111}$$

where use has been made of

$$\int_{S(k)} \frac{k_ik_j}{k^2}\,\mathrm{d}S = \frac{1}{3}\delta_{ij}\int_{S(k)}\mathrm{d}S \tag{3.112}$$

[$S(k)$ is the spherical surface with radius k]. Then the turbulent energy K [equation (3.102)] is

$$K = \int Q(k;t,t)\,\mathrm{d}\boldsymbol{k} = \int E(k;t,t)\,\mathrm{d}k. \tag{3.113}$$

Here E is called the energy spectrum which expresses the energy density between the wavenumber magnitude k and $k+\mathrm{d}k$, and is given by

$$E(k;t,t) = 4\pi k^2Q(k;t,t) = 2\pi k^2\frac{\langle u_i(\boldsymbol{k},t)u_i(\boldsymbol{k}',t)\rangle}{\delta(\boldsymbol{k}+\boldsymbol{k}')}. \tag{3.114}$$

The energy spectrum E is the most fundamental statistical quantity in isotropic turbulence and will play a central role in the following discussions.

In equation (3.108) the second term does not contribute to equation (3.113) since it is an odd function of $\boldsymbol{k}$. Specifically, Γ_{K} is a pseudo-scalar or a scalar whose sign changes under the reflection of a coordinate system; namely,

$$\boldsymbol{k} \to -\boldsymbol{k} \Leftrightarrow H_{\mathrm{K}} \to -H_{\mathrm{K}}. \tag{3.115}$$

In §3.3.2.4 we referred to the turbulent helicity H_{K} [(3.83)]. We examine its relationship with Γ_{K}. We use equation (3.108), and evaluate H_{K} as

$$\begin{aligned} H_{\mathrm{K}} = \varepsilon_{ijl}\left\langle u_i \frac{\partial u_l}{\partial x_j}\right\rangle &= -\mathrm{i}\varepsilon_{ijl} \iint k'_j \langle u_i(\boldsymbol{k},t) u_l(\boldsymbol{k}',t)\rangle \, \mathrm{e}^{-\mathrm{i}(\boldsymbol{k}+\boldsymbol{k}')\cdot\boldsymbol{x}} \, \mathrm{d}\boldsymbol{k}\, \mathrm{d}\boldsymbol{k}' \\ &= \int \Gamma_{\mathrm{K}}(k;t,t)\, \mathrm{d}\boldsymbol{k} = \int 4\pi k^2 \Gamma_{\mathrm{K}}(k;t,t)\, \mathrm{d}k. \end{aligned} \tag{3.116}$$

As a result, $4\pi k^2 \Gamma_{\mathrm{K}}$ is the helicity spectrum.

3.4.1.3 Fourier representation of Navier–Stokes equation

We apply the Fourier representation (3.97) to equation (3.16) for an incompressible fluid [note expression (3.23)]. From

$$\begin{aligned} \frac{\partial}{\partial x_j} u_j u_i &= \frac{\partial}{\partial x_j}\left(\int u_j(\boldsymbol{p},t)\, \mathrm{e}^{-\mathrm{i}\boldsymbol{p}\cdot\boldsymbol{x}} \, \mathrm{d}\boldsymbol{p} \int u_i(\boldsymbol{q},t)\, \mathrm{e}^{-\mathrm{i}\boldsymbol{q}\cdot\boldsymbol{x}}\, \mathrm{d}\boldsymbol{q}\right) \\ &= -\mathrm{i} \iint (p_j + q_j) u_j(\boldsymbol{p},t) u_i(\boldsymbol{q},t)\, \mathrm{e}^{-\mathrm{i}(\boldsymbol{p}+\boldsymbol{q})\cdot\boldsymbol{x}}\, \mathrm{d}\boldsymbol{p}\, \mathrm{d}\boldsymbol{q}, \end{aligned} \tag{3.117}$$

we have

$$\begin{aligned} &\frac{\partial u_i(\boldsymbol{k},t)}{\partial t} - \mathrm{i}k_j \iint u_j(\boldsymbol{p},t) u_i(\boldsymbol{q},t) \delta(\boldsymbol{k}-\boldsymbol{p}-\boldsymbol{q})\, \mathrm{d}\boldsymbol{p}\, \mathrm{d}\boldsymbol{q} \\ &\quad = \mathrm{i}k_i p(\boldsymbol{k},t) - \nu k^2 u_i(\boldsymbol{k},t). \end{aligned} \tag{3.118}$$

We apply equations (3.106) to equation (3.118), and have

$$p(\boldsymbol{k},t) = -\frac{k_i k_j}{k^2} \iint u_j(\boldsymbol{p},t) u_i(\boldsymbol{q},t) \delta(\boldsymbol{k}-\boldsymbol{p}-\boldsymbol{q})\, \mathrm{d}\boldsymbol{p}\, \mathrm{d}\boldsymbol{q}, \tag{3.119}$$

which reduces equation (3.118) to

$$\frac{\partial u_i(\boldsymbol{k},t)}{\partial t} + \nu k^2 u_i(\boldsymbol{k},t) = \mathrm{i}k_j D_{il}(\boldsymbol{k}) \iint u_j(\boldsymbol{p},t) u_l(\boldsymbol{q},t) \delta(\boldsymbol{k}-\boldsymbol{p}-\boldsymbol{q})\, \mathrm{d}\boldsymbol{p}\, \mathrm{d}\boldsymbol{q}, \tag{3.120a}$$

or

$$\frac{\partial u_i(\boldsymbol{k},t)}{\partial t} + \nu k^2 u_i(\boldsymbol{k},t) = \mathrm{i}M_{ijl}(\boldsymbol{k}) \iint u_j(\boldsymbol{p},t)u_l(\boldsymbol{q},t)\delta(\boldsymbol{k}-\boldsymbol{p}-\boldsymbol{q})\,\mathrm{d}\boldsymbol{p}\,\mathrm{d}\boldsymbol{q}, \tag{3.120b}$$

where

$$M_{ijl}(\boldsymbol{k}) = \tfrac{1}{2}[k_j D_{il}(\boldsymbol{k}) + k_l D_{ij}(\boldsymbol{k})]. \tag{3.121}$$

Concerning $D_{ij}(\boldsymbol{k})$ and $M_{ijl}(\boldsymbol{k})$, we have the relations

$$\begin{aligned} &k_i M_{ijl}(\boldsymbol{k}) = 0, \qquad D_{il}(\boldsymbol{k})D_{lj}(\boldsymbol{k}) = D_{ij}(\boldsymbol{k}), \\ &D_{im}(\boldsymbol{k})M_{mjl}(\boldsymbol{k}) = M_{ijl}(\boldsymbol{k}). \end{aligned} \tag{3.122}$$

Entirely similarly, equation (3.17) for the temperature θ is rewritten as

$$\frac{\partial \theta(\boldsymbol{k},t)}{\partial t} + \nu k^2 \theta(\boldsymbol{k},t) = \mathrm{i}k_i \iint u_i(\boldsymbol{p},t)\theta(\boldsymbol{q},t)\delta(\boldsymbol{k}-\boldsymbol{p}-\boldsymbol{q})\,\mathrm{d}\boldsymbol{p}\,\mathrm{d}\boldsymbol{q}. \tag{3.123}$$

In equation (3.97), $\boldsymbol{u}(\boldsymbol{k})$ may be interpreted as the amplitude of a wave with wavenumber $\boldsymbol{k}$. The right-hand side of equation (3.120) expresses the generation or annihilation rate of a wave with wavenumber $\boldsymbol{k}$ through the interaction between two waves with wavenumbers $\boldsymbol{p}$ and $\boldsymbol{q}$. The first-order dependence of $M_{ijl}(\boldsymbol{k})$ on $\boldsymbol{k}$ is due to the dependence of $(\partial/\partial x_j)u_i u_j$ on the first-order derivative that originates from the flow of medium.

3.4.1.4 Energy-spectrum equation

In homogeneous turbulence with no mean flow, equation (3.39) is reduced to the simplest form

$$\frac{\mathrm{d}K}{\mathrm{d}t} = -\varepsilon, \tag{3.124}$$

where the last term vanishes due to homogeneity. In order to see the relationship of equation (3.124) with the energy transfer in the $\boldsymbol{k}$ space, we construct the equation for the energy spectrum $E(k)$ from equation (3.120b). It is given by

$$\frac{\partial}{\partial t}E(k) = -2\nu k^2 E(k) + T(k) \tag{3.125}$$

(the dependence on time t is omitted, except when necessary). Here $T(k)$ is written as

$$\begin{aligned} T(k) = 2\pi k^2 &\iint T_1(\boldsymbol{k}\,|\,\boldsymbol{p},\boldsymbol{q})\delta(\boldsymbol{k}-\boldsymbol{p}-\boldsymbol{q})\,\mathrm{d}\boldsymbol{p}\,\mathrm{d}\boldsymbol{q} \\ + 2\pi k^2 &\iint T_2(\boldsymbol{k}\,|\,\boldsymbol{p},\boldsymbol{q})\delta(\boldsymbol{k}'-\boldsymbol{p}-\boldsymbol{q})\,\mathrm{d}\boldsymbol{p}\,\mathrm{d}\boldsymbol{q}, \end{aligned} \tag{3.126}$$

with

$$T_1(\boldsymbol{k}\,|\,\boldsymbol{p},\boldsymbol{q}) = \mathrm{i}M_{ijl}(\boldsymbol{k})\,\frac{\langle u_i(\boldsymbol{k}';t)u_j(\boldsymbol{p};t)u_l(\boldsymbol{q};t)\rangle}{\delta(\boldsymbol{k}+\boldsymbol{k}')}, \tag{3.127a}$$

$$T_2(\boldsymbol{k}\,|\,\boldsymbol{p},\boldsymbol{q}) = \mathrm{i}M_{ijl}(\boldsymbol{k}')\,\frac{\langle u_i(\boldsymbol{k};t)u_j(\boldsymbol{p};t)u_l(\boldsymbol{q};t)\rangle}{\delta(\boldsymbol{k}+\boldsymbol{k}')}. \tag{3.127b}$$

To represent ε [equation (3.41)] in the $\boldsymbol{k}$ space, we substitute equation (3.97a) into equation (3.41), and use equation (3.108). Then we have

$$\varepsilon = -\nu\iint k_i k'_i\langle u_j(\boldsymbol{k})u_j(\boldsymbol{k}')\rangle\,\mathrm{e}^{-\mathrm{i}(\boldsymbol{k}+\boldsymbol{k}')\cdot\boldsymbol{x}}\,\mathrm{d}\boldsymbol{k}\,\mathrm{d}\boldsymbol{k}' = 2\nu\int k^2Q(k)\,\mathrm{d}\boldsymbol{k}. \tag{3.128}$$

Equation (3.128) is rewritten as

$$\varepsilon = \int D(k)\,\mathrm{d}k, \tag{3.129}$$

where the dissipation spectrum $D(k)$ is defined by

$$D(k) = 2\nu k^2E(k), \tag{3.130}$$

using the energy spectrum $E(k)$. We integrate equation (3.125) with respect to k and use equation (3.129). From the comparison with equation (3.124), we have

$$\int_0^\infty T(k)\,\mathrm{d}k = 0. \tag{3.131}$$

We consider the physical meaning of equation (3.125). On its right-hand side, $2\nu k^2E(k)$ is nonnegative and expresses the conversion rate of kinetic energy to heat. The second term expresses the energy change caused by the interaction between waves $\boldsymbol{p}$ and $\boldsymbol{q}$, and is not linked directly with an external energy source. In general, $E(k)$ has a peak in the low-k region corresponding to large spatial scales, as in figure 3.4, for energy is usually supplied from a mean flow whose characteristic wavenumber is low. Moreover energy dissipation is vigorous in the large-k region with small spatial scales, owing to the factor k^2 in equation (3.130). The relative location of $E(k)$ and $D(k)$

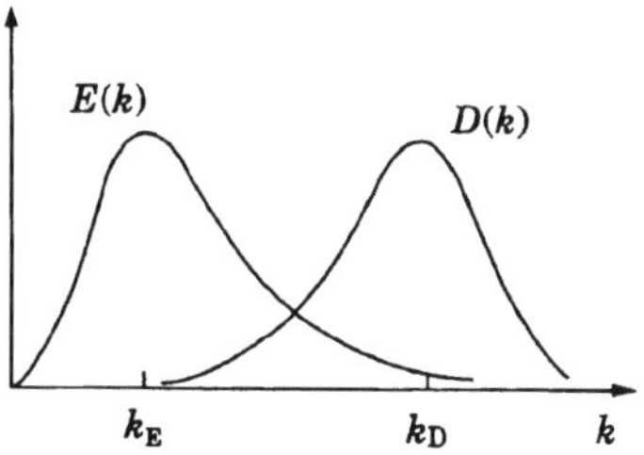

Figure 3.4. Energy and dissipation spectra.

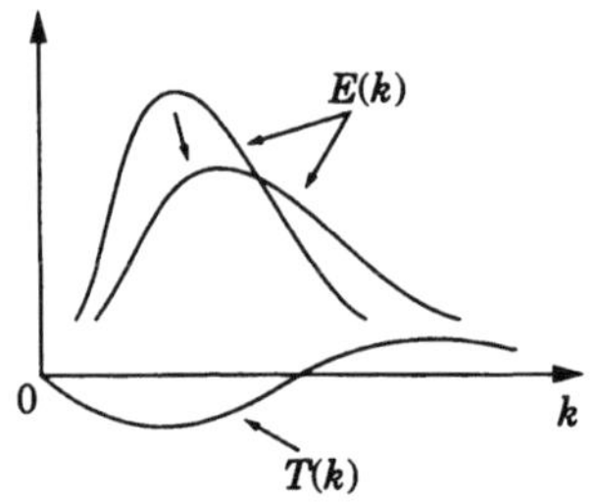

Figure 3.5. Energy transfer function.

is illustrated in figure 3.4 (k_E and k_D are the wavenumbers characterizing $E(k)$ and $D(k)$, respectively, which will be specified later).

In equation (3.125) we have no energy source such as mean flow. Then the energy dissipated as heat needs to be supplemented through the energy decrease in the low-k part of $E(k)$, as in figure 3.5. The decrease in $E(k)$ is due to negative $T(k)$ since the contribution of the low-k part to $D(k)$ is small at high Reynolds numbers. In the high-k region, $E(k)$ generally increases in addition to the occurrence of energy dissipation, indicating that $T(k)$ is positive in the high-k region. These properties of $T(k)$ are illustrated in figure 3.5. The function $T(k)$, which plays a role of transferring energy from the low- to the high-k region, is called the energy transfer function.

3.4.2 Kolmogorov's Scaling Law

3.4.2.1 Inertial range and Kolmogorov spectrum

In general, energy is stored in the low-wavenumber (k) region, and energy dissipation occurs in the high-k region. The regions characterized by wavenumbers k_E and k_D are called the energy-containing region (R_E) and the energy dissipation region (R_D), respectively. In the case that

$$\frac{k_E}{k_D} \ll 1, \tag{3.132}$$

there exists the region between R_E and R_D in which little energy resides and little energy dissipation occurs. Then the energy dissipated in R_D cannot be supplemented by the decrease in the energy of the intermediate region. The energy supplied by this region to R_D is drained from R_E through the energy transfer function $T(k)$. In this picture of energy transfer, such an intermediate region is called the inertial range or region (R_I). As a result, the whole wavenumber range is roughly divided into three parts

$$R_E \quad (k \cong k_E), \qquad R_I \quad (k_E \ll k \ll k_D), \qquad R_D \quad (k \cong k_D). \tag{3.133}$$

The degree of separation of these three regions is highly dependent on the magnitude of k_E/k_D.

We make discussions on the premise that equation (3.132) holds. The energy decrease in R_E, which is given by

$$\varepsilon_T = -\frac{d}{dt} \int_{k \le O(k_E)} E(k)\, dk, \tag{3.134}$$

is transferred to R_I and is eventually dissipated in R_D. In the case that the energy spectra in R_I and R_D are nearly constant in time, we mention that turbulence is in equilibrium state, in which

$$\varepsilon = \varepsilon_T \tag{3.135}$$

holds.

The foregoing discussions indicate that $E(k)$ in R_I, which is originally linked with ε_T, can be represented using k and ε, namely

$$E(k) = F(k, \varepsilon). \tag{3.136}$$

We denote the dimension of length and time by $[L]$ and $[T]$, respectively. From

$$k = [L]^{-1}, \qquad E = [L]^3[T]^{-2}, \qquad \varepsilon = [L]^2[T]^{-3}, \tag{3.137}$$

we write

$$E(k) = K_O \varepsilon^{2/3} k^{-5/3}. \tag{3.138}$$

Equation (3.138) is the Kolmogorov spectrum, and K_O is named the Kolmogorov constant. The scaling of various statistics in R_I in terms of k and ε is called Kolmogorov's scaling [3.8–3.10].

The Kolmogorov spectrum (3.138) was derived from the assumption about the distinct separation between R_E and R_D. As will be referred to below, such a situation is realized in flows at high Reynolds numbers. The energy spectrum quite close to equation (3.138) was first confirmed from observations of atmospheric and oceanic turbulent flows whose Reynolds numbers are very high owing to their large spatial scales. The magnitude of K_O differs considerably from one observation to the other and scatters around 1.5 [3.11, 3.12]. This situation is partly because of the difficulty in accurately measuring ε.

The energy cascade from large- to small-scale components of flow may be interpreted as the process in which large eddies are broken and small eddies are generated. Wavenumber k corresponds to eddies with a spatial scale $2\pi/k$. The lifetime of such an eddy, $\tau(k)$, is written as

$$\tau(k) = C_\tau \varepsilon^{-1/3} k^{-2/3}, \tag{3.139}$$

in entirely the same manner as for equation (3.138), where C_τ is a numerical factor. Equation (3.139) signifies that a higher-k component of flow has a shorter lifetime; namely, small eddies decay fast.

3.4.2.2 *Physical implications of the inertial range*

The energy spectrum in the inertial range has a very simple expression, that is, equation (3.138). This is due to its role that the range drains energy from R_E and transfers it to R_D without any energy generation and dissipation. For this role, the physical importance of the Kolmogorov spectrum seems to be minor at first glance. The intermediate property intrinsic to the spectrum, however, signifies that equation (3.138) possesses both the properties of the spectra in R_E and R_D, although they are partial. In fluid mechanics, it is rare that such an intermediate property can be found in an analytical manner. For instance, laminar flows at low and high Reynolds numbers can be examined analytically using the Stokes and boundary-layer approximations, respectively, but flows at intermediate Reynolds numbers are beyond the scope of an analytical approach.

We first examine the R_E side of equation (3.138). It is not integrable in the limit $k \to 0$. Noting that most of the energy resides near k_E, we write

$$K = \int_0^\infty E(k)\,dk \cong K_O \varepsilon^{2/3} \int_{k_E}^\infty k^{-5/3}\,dk. \tag{3.140}$$

Here the lower limit of the integral signifies $O(k_E)$, so the numerical factor found from equation (3.140) is mathematically unimportant and will be dropped below. From equation (3.140), we have

$$K = \varepsilon^{2/3} k_E^{-2/3}, \tag{3.141}$$

which gives

$$k_E = \frac{\varepsilon}{K^{3/2}}, \tag{3.142a}$$

$$l_E \equiv \frac{2\pi}{k_E} = \frac{K^{3/2}}{\varepsilon}. \tag{3.142b}$$

The magnitude of velocity fluctuation characterizing R_E is

$$u_E = \sqrt{K}. \tag{3.143}$$

We use equations (3.141) and (3.143) and have the characteristic time scale in R_E, τ_E, as

$$\tau_E \equiv \frac{l_E}{u_E} = \frac{K}{\varepsilon}, \tag{3.144}$$

which is coincident with the expression obtained by substituting equation (3.142a) into equation (3.139). From this coincidence, τ_E is the lifetime of the eddy possessing most turbulent energy.

Next we investigate the relationship of equation (3.138) with R_D. We replace the upper limit of the integral with the wavenumber characterizing

R_D, k_D, and write

$$\varepsilon = \nu \int_0^{k_D} D(k)\,dk = \nu\varepsilon^{2/3} k_D^{4/3}. \tag{3.145}$$

Then we have

$$k_D = \left(\frac{\varepsilon}{\nu^3}\right)^{1/4}, \tag{3.146a}$$

$$l_D \equiv \frac{2\pi}{k_D} = \left(\frac{\nu^3}{\varepsilon}\right)^{1/4}. \tag{3.146b}$$

On the basis of the foregoing findings, we discuss equation (3.132) concerning the existence of R_I. From equations (3.142) and (3.146), equation (3.132) is reduced to

$$\frac{k_E}{k_D} = \frac{l_D}{l_E} = R_{eT}^{-3/4} \ll 1. \tag{3.147}$$

Here the nondimensional parameter R_{eT} is defined by

$$R_{eT} = \frac{K^2}{\nu\varepsilon}, \tag{3.148}$$

which indicates that K^2/ε is of the same dimension as ν. We use equations (3.142b) and (3.143), and write

$$\nu_T = l_E u_E = \frac{K^2}{\varepsilon}. \tag{3.149}$$

Equation (3.149) may be regarded as the viscosity arising from velocity fluctuations and is called the turbulent or eddy viscosity (we should note that the dimension of viscosity is equal to the dimension of the product of length and velocity). Using equation (3.149), we rewrite R_{eT} as

$$R_{eT} = \frac{l_E u_E}{\nu} = \frac{\nu_T}{\nu}, \tag{3.150}$$

which represents the ratio of the turbulent to molecular viscosity and may be called the turbulent Reynolds number. Equation (3.147) indicates that the turbulent viscosity ν_T is much larger than the molecular viscosity ν. This is a necessary condition for the existence of R_I.

Finally, we discuss the relationship with a turbulent flow having mean flow. From equation (3.135), ε is equivalent to the energy-supply or injection rate from R_E, ε_T, in an equilibrium state of turbulence. In the presence of mean flow, energy is supplied by the production term P_K [equation (3.40)] that is directly linked with mean flow. Then ε is estimated using the reference length L_R and reference velocity U_R characterizing mean flow, resulting in

$$\varepsilon = \frac{U_R^3}{L_R}. \tag{3.151}$$

From equations (3.146b) and (3.151), we have

$$\frac{l_D}{L_R} = R_e^{-3/4}, \tag{3.152}$$

where R_e is the Reynolds number based on L_R and U_R, such as equation (2.1). Equation (3.152) shows that energy dissipation occurs at very small scales in the case of high Reynolds numbers.

3.4.3 Failure of Kolmogorov's Scaling

The essence of Kolmogorov's scaling on which equation (3.138) is founded lies in the use of k and ε. As has already been stated, spectra quite similar to equation (3.138) may be observed in atmospheric and oceanic flows at very high Reynolds numbers. The improvements to computers also makes possible the numerical test of the spectrum at moderate Reynolds numbers. Such texts are affirmative, but the small deviation from the $k^{-5/3}$ form cannot be ruled out. Kolmogorov's scaling is not limited to the second-order statistics resulting in equation (3.138), and its validity needs to be examined for higher-order statistics.

3.4.3.1 Intermittency effects

For investigating the validity of Kolmogorov's scaling, we consider the nth-order statistics of the velocity difference at a distance r in physical space, which is given by

$$C_n\{\boldsymbol{u}\} \equiv \langle |\Delta \boldsymbol{u}(\boldsymbol{r})|^n \rangle, \tag{3.153}$$

where

$$\Delta \boldsymbol{u}(\boldsymbol{r}) = \boldsymbol{u}(\boldsymbol{x} + \boldsymbol{r}) - \boldsymbol{u}(\boldsymbol{x}). \tag{3.154}$$

Kolmogorov's scaling leads to

$$C_n\{\boldsymbol{u}\} = c_n (\varepsilon r)^{n/3}, \tag{3.155}$$

with a numerical coefficient c_n, and $C_2\{\boldsymbol{u}\}$ corresponds to the Kolmogorov spectrum (3.138).

The deviation of $C_n\{\boldsymbol{u}\}$ from equation (3.155) becomes prominent with increasing n [3.10, 3.13]. In order to see its cause, we introduce the instantaneous counterpart of ε or

$$\varepsilon' = \nu \left(\frac{\partial u_j}{\partial x_i} \right)^2, \tag{3.156}$$

and write

$$|\Delta \boldsymbol{u}(\boldsymbol{r})|^n \propto (\varepsilon' r)^{n/3} \tag{3.157}$$

in the light of Kolmogorov's scaling.

We divide ε' into the ensemble mean ε and the deviation around it, $\delta\varepsilon'$, as

$$\varepsilon' = \varepsilon + \delta\varepsilon'. \tag{3.158}$$

From equations (3.157) and (3.158), we have

$$C_n\{\boldsymbol{u}\} \propto (\varepsilon r)^{n/3} \left\langle \left(1 + \frac{\delta\varepsilon'}{\varepsilon}\right)^{n/3} \right\rangle. \tag{3.159}$$

Equation (3.155) is valid for

$$\left|\frac{\delta\varepsilon'}{\varepsilon}\right| \ll 1. \tag{3.160}$$

Neither observation nor numerical simulation of isotropic turbulence supports equation (3.160), which indicates that ε' is highly intermittent. Namely, the region with large ε' is spatially spotty, and $|\delta\varepsilon'/\varepsilon|$ may become large there. Such intermittency effects become stronger with increasing n. This is the reason why Kolmogorov's scaling breaks down for large n.

The foregoing discussions indicate that the form of $C_n\{\boldsymbol{u}\}$ is highly dependent on the probability distribution function of ε', $P\{\varepsilon'\}$ [3.14, 3.15]. In reality, various models for $P\{\varepsilon'\}$ have been proposed and are still under an intensive study. The prototype of those models is the log-normal distribution function, in which $\log\varepsilon'$ obeys the Gaussian distribution, that is,

$$P\{\log\varepsilon'\} = \frac{1}{\sqrt{2\pi}\sigma_{l\varepsilon}} \exp\left(-\frac{(\log\varepsilon' - m_{l\varepsilon})^2}{2\sigma_{l\varepsilon}^2}\right), \tag{3.161}$$

where

$$m_{l\varepsilon} = \langle \log\varepsilon' \rangle, \tag{3.162a}$$

$$\sigma_{l\varepsilon} = \sqrt{\langle (\log\varepsilon' - \langle \log\varepsilon' \rangle)^2 \rangle}. \tag{3.162b}$$

Equation (3.161) gives

$$\langle \varepsilon'^{n/3} \rangle = \int_{-\infty}^{\infty} \varepsilon'^{n/3} P\{\log\varepsilon'\}\, \mathrm{d}(\log\varepsilon') = \exp\left(\frac{n}{3} m_{l\varepsilon} + \frac{n^2}{18}\sigma_{l\varepsilon}^2\right). \tag{3.163}$$

We apply equation (3.163) to equations (3.153) and (3.157), and have

$$C_n\{\boldsymbol{u}\} \propto (\varepsilon r)^{n/3} \exp\left(\frac{n(n-3)}{18}\sigma_{l\varepsilon}^2\right). \tag{3.164}$$

We should note that fluctuation effects of ε' identically vanish for $n = 3$. This finding is coincident with the conclusion that may be derived from the analysis of the Navier–Stokes equation. For more details of intermittency effects, readers may consult [3.10, 3.13, 3.16] and the works cited therein, and [3.17] for a novel attempt based on the Tsallis entropy.

In the context of the Kolmogorov spectrum (3.138), the intermittency effect is observed as

$$E(k) \propto \varepsilon^{2/3} k^{-5/3} (k l_{\mathrm{E}})^{-\mu_{\mathrm{I}}}, \tag{3.165}$$

where l_{E} is a length scale characterizing the energy-containing region [see equation (3.142b)], and the numerical factor μ_{I} is

$$\mu_{\mathrm{I}} \cong 0.06. \tag{3.166}$$

This intermittency effect on $E(k)$ is not very important, at least in the analysis of turbulent transports such as the Reynolds stress and the turbulent heat flux. The situation does not merely arise from the smallness of μ_{I}, but it is related to the intrinsic property of the intermittency effect. The effect originates from the small-scale components of motion contributing to the energy dissipation, whereas the transports of momentum and heat are caused by their larger-scale counterparts (this point will become clear in the discussions of chapter 5). In fact, the μ_{I} effect disappears in equation (3.165) in the limit of $k^{-1} \to l_{\mathrm{E}}$.

Equation (3.157) is similar to Kolmogorov's scaling such as equation (3.136), except the fluctuation of ε'. In turbulent flows observed in engineering and nature, ε_{T} [equation (3.134)] is closely related to the production rate P_K [equation (3.40)]. The mean velocity $\boldsymbol{U}$ on which P_K explicitly depends is often subject to rapid temporal and spatial changes. Such a situation is far from the equilibrium state assumed in equation (3.135). The effect of deviation from an equilibrium state on ε_{T}, that is, the nonequilibrium effect, has a greater influence on the turbulent transports of momentum and heat since the effect is linked with large-scale components of motion [3.18].

3.4.3.2 Shell model

In the foregoing discussions on the Kolmogorov scaling and its failure, no direct use is made of the Navier–Stokes equation (3.16), although the concept of the energy cascade is its important property. A dynamic approach to intermittency effects is the use of shell models [3.19–3.22]. In this approach, we divide the wavenumber space into the successive spherical shells, whose radii are denoted by

$$k_n = k_0 q^n, \tag{3.167}$$

where k_0 is the reference wavenumber, $q > 1$, and n is a positive integer. By V_n in the range $k_n < k < k_{n+1}$, we denote a one-dimensional model for $\boldsymbol{u}(\boldsymbol{k})$ obeying equation (3.120). Under the constraint $\boldsymbol{k} = \boldsymbol{p} + \boldsymbol{q}$, $\boldsymbol{u}(\boldsymbol{k})$ interacts with all other modes. In shell models, however, the interaction is usually restricted to the nearest and second-nearest interactions.

The representative instance of shell models is

$$\frac{dV_n}{dt} + \nu k^2 V_n = \mathrm{i}(c_{n1} V_{n+1} V_{n+2} + c_{n2} V_{n-1} V_{n+1} + c_{n3} V_{n-1} V_{n-2})^* + f_n, \tag{3.168}$$

where the c_{nm} are real constants, f_n is the external force maintaining a turbulent state, and A^* denotes the complex conjugate of A [3.22]. Equation (3.168) does not correspond to equation (3.120) straightforwardly, but the former possesses some properties similar to the latter. For instance, the energy, which is defined by $\sum_n V_n^2/2$, is conserved in the absence of ν and f_n, under a proper choice of c_{nm}. One of the interesting consequences of equation (3.168) is the intermittency effect on the structure function

$$S_p(k_n) = \langle |V_n|^p \rangle, \tag{3.169}$$

which is the shell-model counterpart of $C_n\{\boldsymbol{u}\}$ [equation (3.153)]. The numerical computation of equation (3.168) shows the following. The scaling property close to the Kolmogorov scaling holds for small p, whereas the intermittency effect becomes stronger with increasing p.

3.4.4 Two-Dimensional Turbulence

As the other example of homogeneous turbulence, we may mention two-dimensional turbulence exhibiting the inverse cascade of energy and the formation of coherent vortices. Two-dimensional turbulence is considered to provide an idealized model for geophysical and atmospheric flows. These flows subject to frame rotation are often confined in a thin layer, and the fluid motion is characterized by a large horizontal scale length and a short vertical counterpart. Here we should note that fluctuations are still three-dimensional even when the mean or large-scale flow may be regarded as two-dimensional. Then a purely two-dimensional turbulence is not realized in nature, but the presence of a strong rotation may suppress the fluid motion along the axis, resulting in a pseudo-two-dimensional turbulence. The two-dimensional treatment of turbulence is useful as a first step towards modelling geophysical and atmospheric turbulence subject to strong frame rotation [3.23].

From equation (3.16), we have the equation for the vorticity $\boldsymbol{\omega}$

$$\left(\frac{\partial}{\partial t} + \boldsymbol{u}\cdot\nabla\right)\boldsymbol{\omega} = (\boldsymbol{\omega}\cdot\nabla)\boldsymbol{u} + \nu\nabla^2\boldsymbol{\omega}, \tag{3.170}$$

with the solenoidal condition $\nabla\cdot\boldsymbol{\omega} = 0$. The first term on the right-hand side of equation (3.170) represents the vorticity amplification due to stretching of vortex lines and is called the vortex-stretching term. Vortex stretching is related to the mechanism of vorticity maintenance and energy transfer in turbulent motion.

In a two-dimensional flow, effects of vortex stretching disappear since

$$(\boldsymbol{\omega}\cdot\nabla)\boldsymbol{u} = 0. \tag{3.171}$$

From this property, vorticity is conserved along fluid motion in the inviscid limit ($\nu \to 0$). Vanishing vortex stretching and the resultant vorticity conservation are the properties intrinsic to two-dimensional fluid motion. An important quantity characterizing two-dimensional turbulence is the enstrophy defined by $\omega^2/2$. From equation (3.170), a total amount of enstrophy obeys

$$\frac{\partial}{\partial t}\int_V \frac{1}{2}\omega^2 \,\mathrm{d}V = -\int_V \nu\left(\frac{\partial \omega_j}{\partial x_i}\right)^2 \mathrm{d}V + \int_S \left\{\left[-\frac{1}{2}\omega^2\boldsymbol{u} + \nu\nabla\left(\frac{1}{2}\omega^2\right)\right]\cdot\boldsymbol{n}\right\}\mathrm{d}S, \tag{3.172}$$

whereas its energy counterpart is given by equation (3.48). From those two equations, the total amounts of energy and enstrophy in the absence of viscosity are conserved in two-dimensional turbulence so long as there are no net energy and enstrophy flows across a boundary.

In homogeneous isotropic turbulence, the energy and enstrophy spectra are defined as

$$K = \int E_{2\mathrm{D}}(k)\,\mathrm{d}k, \tag{3.173}$$

$$K_\omega \equiv \langle \tfrac{1}{2}\omega^2\rangle = \int k^2 E_{2\mathrm{D}}(k)\,\mathrm{d}k. \tag{3.174}$$

The foregoing dual conservation property leads to the inverse energy cascade, that is, the energy transfer from smaller to larger scales, resulting in the formation of flow structure in two-dimensional turbulence. Readers may consult [3.25–3.27] in the context of atmospheric and oceanic turbulence researches.

3.5 Production and Diffusion Characteristics of Turbulent Energy

Turbulent flows with mean velocity are generally called inhomogeneous or shear turbulence, in contrast to homogeneous turbulence mentioned in §3.4. A prominent feature of inhomogeneous turbulence is the production mechanism of turbulent energy given by equation (3.40). It is similar to its Reynolds-stress counterpart, equation (3.34). Therefore knowledge of the characteristics of the energy production is also useful for the investigation into the Reynolds stress, for instance, turbulence modelling explained in detail in chapter 4.

There are a variety of turbulent flows in real-world phenomena, as was referred to in chapter 2. Those flows are roughly classified into two types [2.2, 3.28, 3.29]. One is wall-bounded flows on which solid boundaries have a great influence through molecular viscous effects. Their typical instances are a pipe flow, a channel flow, a boundary-layer flow along a wall, etc. The other is jets ejected from circular and plane nozzles to surrounding fluids, a wake (a flow behind a body immersed in a stream), etc. They are generally called free-shear flows.

In fully-developed regions far downstream from entrances of a pipe and a channel, there are no spatial variations of mean flow and turbulence properties in the mean-flow direction. In a wake and a jet, however, these quantities change in the downstream direction. In what follows, we shall give an overview of the difference between these two types of flows from a viewpoint of the turbulent-energy equation.

We cast equation (3.39) into the form

$$\frac{\partial K}{\partial t} = A_K + P_K - \varepsilon + \nabla \cdot (\boldsymbol{T}_K + \nu \nabla K). \tag{3.175}$$

Here we call each term on the right-hand side

$$\begin{array}{ll} \text{Advection } (A), & A_K \ (= -(\boldsymbol{U}\cdot\nabla)K) \\ \text{Production } (P), & P_K \\ \text{Dissipation } (D), & -\varepsilon \\ \text{Diffusion } (T), & \nabla \cdot (\boldsymbol{T}_K + \nu\nabla K), \end{array} \tag{3.176}$$

respectively.

In the fully-developed state of pipe and channel flows, we have

$$\frac{\partial K}{\partial t} = A_K = 0, \tag{3.177}$$

leading to the balance among three effects P, D, and T. In a pipe flow, the profiles of the mean axial velocity (U) and K are depicted schematically in figure 3.6. The production P, which vanishes at the wall, becomes large near the wall owing to the large gradient of U. Near the center-line, P becomes small from the symmetry of U. The resulting profile of P is illustrated in figure 3.7. The dissipation D is largest at the wall and decreases towards the center-line, but it does not vanish there. The difference between P and D near the center-line is balanced by T. The molecular effect $\nu\nabla^2 K$ in T becomes important only near the wall. In short, K is generated by P near

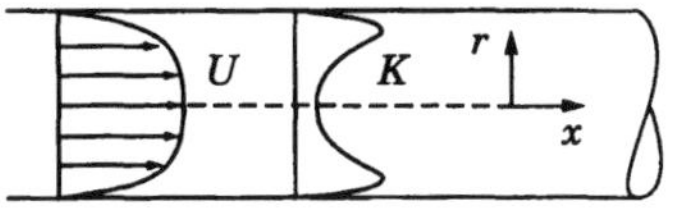

Figure 3.6. Turbulent energy in a pipe flow.

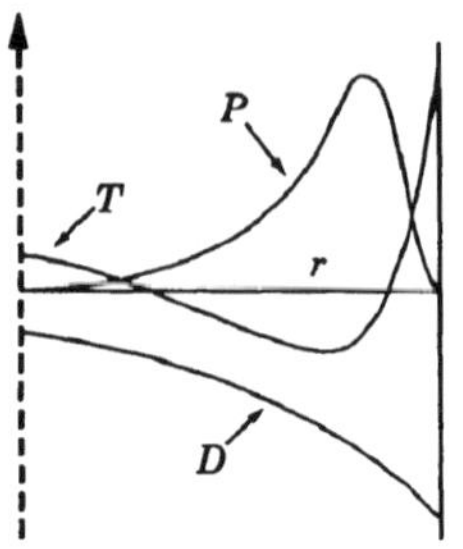

Figure 3.7. Turbulent-energy budget in a pipe flow.

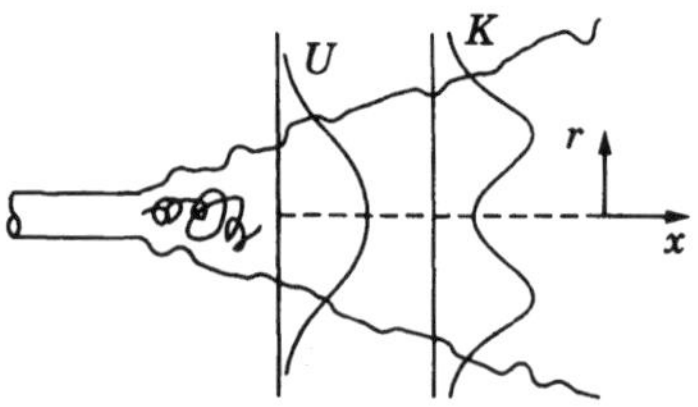

Figure 3.8. Turbulent energy in a circular jet.

the wall and is diffused towards the center-line. As a result, the general profile of K may be understood from the profile of P, specifically, that of U.

Next, we consider a circular jet from a nozzle as an instance of free-shear turbulent flows (figure 3.8). In a pipe flow, the state with $P_K \cong \varepsilon$ is observed in most of the flow region. A similar tendency is also seen in a circular jet, as is illustrated in figure 3.9. The critical difference between pipe and jet flows is that A is comparable to P and D in a jet, resulting in the greater importance of T. The behavior of T is nearly opposite to that of A. The gradient of U is largest a little distance from the center-line. Then P and resulting K are also largest there, and the latter is diffused towards the edge of the jet. This is the reason why T becomes large near the edge of the jet.

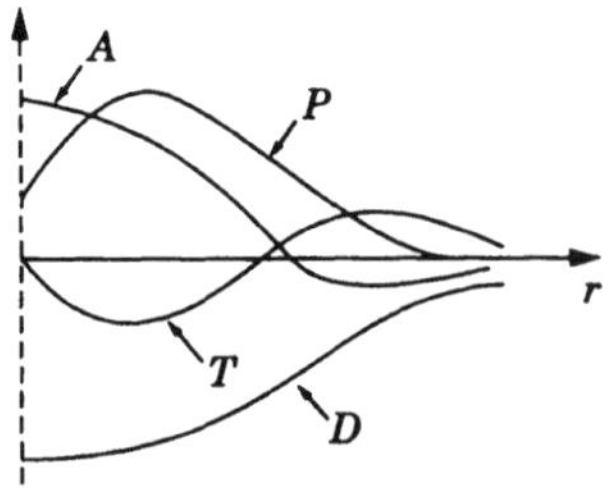

Figure 3.9. Turbulent-energy budget in a circular jet.

References

[3.1] Batchelor G K 1967 *An Introduction to Fluid Dynamics* (Cambridge: Cambridge University Press)

[3.2] Landau L D and Lifshitz E M 1959 *Fluid Mechanics* (Oxford: Pergamon)

[3.3] Batchelor G K 1971 *Theory of Homogeneous Turbulence* (2nd edition) (Cambridge: Cambridge University Press)

[3.4] Leslie D C 1974 *Developments in the Theory of Turbulence* (Oxford: Clarendon)

[3.5] McComb W D 1990 *The Physics of Fluids* (Oxford: Clarendon)

[3.6] Leseiur M 1997 *Turbulence in Fluids* (3rd edition) (Dordrecht: Kluwer)

[3.7] Lighthill M J 1970 *Fourier Representation and Generalized Functions* (Cambridge: Cambridge University Press)

[3.8] Kolmogorov A N 1941 *C. R. Acad. Sci. USSR* **30** 301

[3.9] Kolmogorov A N 1941 *C. R. Acad. Sci. USSR* **32** 16

[3.10] Frisch U 1995 *Turbulence* (Cambridge: Cambridge University Press)

[3.11] Paquin J E and Pond S 1971 *J. Fluid Mech.* **50** 257

[3.12] Williams R W and Paulson C A 1971 *J. Fluid Mech.* **83** 547

[3.13] Meneveau C and Screenivasan K R 1991 *J. Fluid Mech.* **224** 429

[3.14] Oboukhov A M 1962 *J. Fluid Mech.* **13** 77

[3.15] Kolmogorov A N 1962 *J. Fluid Mech.* **13** 82

[3.16] Fukayama D, Oyamada T, Nakano T, Gotoh T and Yamamoto K 2000 *J. Phys. Soc. Jpn.* **69** 701

[3.17] Arimitsu T and Arimitsu N 2001 *Prog. Theor. Phys.* **105** 355

[3.18] Yoshizawa A 1994 *Phys. Rev. E.* **49** 4065

[3.19] Desnyansky V N and Novikov E A 1974 *Akad. Nauk SSSR Fiz. Atoms. Okeana* **10** 127

[3.20] Gledzer E B 1973 *Phys. Dokl.* **18** 26

[3.21] Yamada M and Ohkitani K 1987 *J. Phys. Soc. Jpn.* **56** 4210

[3.22] Ohkitani K and Yamada M 1989 *Prog. Theor. Phys.* **81** 239

[3.23] Kraichnan R H and Montgomery D 1980 *Rep. Prog. Phys.* **43** 547

[3.23] Kraichnan R H 1967 *Phys. Fluids* **10** 417

[3.24] Batchelor G K 1969 *Phys. Fluids Suppl.* **12** II-233

[3.25] Pedlosky J 1979 *Geophysical Fluid Dynamics* (New York: Springer)

[3.26] Williams G P 1978 *J. Atom. Sci.* **35** 1399

[3.27] Yoden S and Yamada M 1993 *J. Atom. Sci.* **50** 631

[3.28] Tennekes H and Lumley J L 1972 *A First Course in Turbulence* (Cambridge: MIT Press)

[3.29] Townsend A A 1976 *The Structure of Turbulent Shear Flow* (2nd edition) (Cambridge: Cambridge University Press)

Chapter 4

Heuristic Turbulence Modelling

4.1 Approaches to Turbulence

The mathematical basis of the equations for electrically nonconducting flows is firm as long as they are applied to phenomena with fluid velocity much lower than light speed. In this sense, the numerical simulation (experiment) of these equations seems most suitable for the exploration of flow properties. In the study of turbulent flows encountered in engineering and scientific phenomena, however, the feasibility of the approach is not so large. Such a situation arises from a great number of spatial scales included in turbulent flows at high Reynolds numbers, as will be discussed below in a quantitative manner.

The numerical method resolving all flow components ranging from mean to energy-dissipating motion is called direct numerical simulation (DNS) [4.1, 4.2]. The important point to be noted in this method is that the interval of computational grids, h, in discretizing the fluid equations is comparable to the energy dissipation scale l_D [equation (3.146b)]; namely, we need to take

$$h \cong l_D. \tag{4.1}$$

We consider a channel flow referred to in §3.3.2.3 (figure 3.3). In the DNS, the computational domain is a rectangular prism whose three sides are the channel width ($2H_0 \to H$) and a few times that in the other two directions, resulting in the total number of grid points

$$N = O\left[\left(\frac{H_0 \to H}{h}\right)^3\right] = O\left[\left(\frac{H_0 \to H}{l_D}\right)^3\right]. \tag{4.2}$$

As the reference velocity characterizing a channel flow, we choose the bulk velocity U_B (the mean velocity across the channel cross-section). From equation (3.152) with $L_R = H_0 \to H$, we have

$$N = O(R^{9/4}), \tag{4.3}$$

where R_e is the Reynolds number defined by

$$R_e = \frac{U_B H_0 \to H}{\nu}. \tag{4.4}$$

The constraint (4.3) is a big stumbling block for DNS of turbulent flows. The largest R_e in the current DNS of channel flows is $O(10^4)$. For water ($\nu \cong 0.01\,\mathrm{cm^2\,s^{-1}}$), this R_e corresponds to $H_0 \to H \cong O(10)\,\mathrm{cm}$ and $U_B = O(10^2)\,\mathrm{cm\,s^{-1}}$. We may see that the strict DNS approach to turbulent flows in engineering and scientific phenomena is quite limited. Mathematical devices leading to the drastic decrease in components of motion to be treated is indispensable in addition to the effort towards the improvement of computational efficiency.

As a representative method of reducing flow components to be dealt with, we have two ways. One is to focus attention on the components of motion linked with mean flow, and the other is to scrutinize small-scale properties such as intermittency effects on velocity correlation functions by putting aside the production mechanism of turbulent energy (recall the discussion in §3.4.3).

The typical approach in the former category is turbulence modelling [2.16, 4.3–4.7]. Its fundamental concept has already been introduced in §3.3. The method is roughly divided into two classes, according to whether equation (3.33) for the Reynolds stress R_{ij} is directly utilized or not. In the case where the equation is not used, we express R_{ij} itself in terms of the mean velocity and turbulence quantities such as the turbulent energy in an algebraic manner. This method is called algebraic turbulence modelling. The method using equation (3.33) is named second-order modelling, and modelling the pressure–strain correlation function Π_{ij} [equation (3.35)] is its key mathematical point, as was explained in §3.3.2.3.

Turbulence modelling has been developed on the basis of dimensional and tensor analyses in the combination of invariance and realizability principles (Galilean invariance of models, positive definiteness of turbulence intensities, etc.). Such modelling may be called heuristic turbulence modelling and has been applied to a variety of turbulent flows in mechanical and aeronautical engineering, meteorology, etc., that are beyond the scope of DNS.

Turbulence modelling is not merely useful to understanding large-scale properties of turbulent flows for practical purposes. In the study of turbulent flows by DNS, the important task is how we can abstract basic properties from huge numerical data. For examining the mechanism of turbulent transports of momentum, heat, etc., by use of DNS data, equation (3.33) provides an important mathematical basis. In this context, the knowledge accumulating in the study of turbulence modelling provides a useful guide line in arranging numerical data.

An intermediate approach between DNS and turbulence modelling is large eddy simulation (LES). In the approach, we make full use of computer capability and model the components of motion whose spatial scales are smaller than the interval of computational grids. Such modelling is distinguished from the modelling based on ensemble averaging and is named subgrid-scale (SGS) modelling. In this approach, the mechanism of energy dissipation is replaced with an SGS model, and the range of Reynolds numbers that can be treated is enlarged. In the context of the arrangement of computed results, LES is essentially the same as DNS, and the knowledge about the ensemble-averaged equations is a useful tool.

In the remainder of chapter 4 we shall discuss turbulence modelling from a heuristic viewpoint.

4.2 Algebraic Turbulence Modelling

4.2.1 Modelling of Reynolds Stress

The anisotropy of R_{ij} (Reynolds stress), which is characterized by the anisotropic tensor

$$B_{ij} = [R_{ij}]_{\mathrm{D}} \equiv R_{ij} - \tfrac{2}{3}K\delta_{ij}, \tag{4.5}$$

is one of the essential properties of inhomogeneous turbulence (D denotes the deviatoric part). As may be seen from the discussion on channel turbulence of §3.3.2.3, the spatial variation of the mean velocity $\boldsymbol{U}$ is characterized by nonvanishing $\partial U_j/\partial x_i$ and is a cause of the anisotropy of turbulence [see equation (3.69)]. We introduce the mean velocity-strain and vorticity tensors

$$S_{ij} = \frac{\partial U_j}{\partial x_i} + \frac{\partial U_i}{\partial x_j}, \tag{4.6}$$

$$\Omega_{ij} = \frac{\partial U_j}{\partial x_i} - \frac{\partial U_i}{\partial x_j}, \tag{4.7}$$

and write $\partial U_j/\partial x_i$ as

$$\frac{\partial U_j}{\partial x_i} = \frac{1}{2}(S_{ij} + \Omega_{ij}). \tag{4.8}$$

Here Ω_{ij} is related to the mean vorticity $\boldsymbol{\Omega}$ $(= \nabla \times \boldsymbol{U})$ as

$$\Omega_{ij} = \varepsilon_{ijk}\Omega_k. \tag{4.9}$$

A hint about how R_{ij} may be modelled is obtained from the relationship between the spatial profiles of K (turbulent energy) and its generator P_K [equation (3.40)]. In §3.5, we saw that the former profile resembles the latter, as in figures 3.6 and 3.7. The generator of R_{ij}, P_{Rij}, is defined by equation (3.34) and is related to P_K as

$$P_K = \tfrac{1}{2}P_{Rii}. \tag{4.10}$$

This fact suggests that the spatial profile of R_{ij} may be also inferred from that of P_{Rij}.

From equation (3.33), B_{ij} obeys

$$\frac{\mathrm{D}B_{ij}}{\mathrm{D}t} = [P_{Rij}]_{\mathrm{D}} + \Pi_{ij} - [\varepsilon_{ij}]_{\mathrm{D}} + \left[\frac{\partial}{\partial x_l}\left(T_{Rijl} + \nu\frac{\partial R_{ij}}{\partial x_l}\right)\right]_{\mathrm{D}}. \tag{4.11}$$

We substitute equations (4.5) and (4.8) into the first or production term of equation (4.11), and have

$$[P_{Rij}]_{\mathrm{D}} = -\tfrac{2}{3}KS_{ij} - \tfrac{1}{2}[B_{jk}S_{ki} + B_{ik}S_{kj}]_{\mathrm{D}} - \tfrac{1}{2}(B_{jk}\Omega_{ki} + B_{ik}\Omega_{kj}), \tag{4.12}$$

where we should note that

$$[B_{jk}\Omega_{ki} + B_{ik}\Omega_{kj}]_{\mathrm{D}} = B_{jk}\Omega_{ki} + B_{ik}\Omega_{kj}. \tag{4.13}$$

Assuming that the spatial profile of B_{ij} is similar to that of its generator $[P_{Rij}]_{\mathrm{D}}$, we write

$$B_{ij} = \tau_B[P_{Rij}]_{\mathrm{D}}. \tag{4.14}$$

Here τ_B is the proportional coefficient with dimension of time.

How should τ_B be expressed? As was mentioned in §3.3.2.2, energy is supplied from mean flow to velocity fluctuation through P_K, eventually being dissipated as heat. The energy transfer process from large- to small-scale components of motion is called the energy cascade. It may be interpreted as the process in which large eddies break up and small eddies occur. The lifetime of an eddy with the scale $2\pi/k$ is given by equation (3.139). The lifetime of the energy-containing eddies that are the main contributor to K is given by τ_{E} [equation (3.144)]. The anisotropic part of R_{ij}, B_{ij}, also arises from those energy-containing eddies interacting with mean flow. Then we take

$$\tau_B = C_B\tau_{\mathrm{E}} = C_B\frac{K}{\varepsilon}, \tag{4.15}$$

where C_B is a numerical factor. Summarizing equations (4.12), (4.14), and (4.15), we have [4.8–4.10]

$$B_{ij} = -\frac{2C_B}{3}\frac{K^2}{\varepsilon}S_{ij} - \frac{C_B}{2}\frac{K}{\varepsilon}[B_{jk}S_{ki} + B_{ik}S_{kj}]_{\mathrm{D}} - \frac{C_B}{2}\frac{K}{\varepsilon}(B_{jk}\Omega_{ki} + B_{ik}\Omega_{kj}). \tag{4.16}$$

The simplest method for expressing B_{ij} explicitly in terms of $\boldsymbol{U}$, K, and ε is to solve equation (4.16) in an iterative manner based on the first term on the right-hand side. The first iteration gives

$$B_{ij} = -\nu_{\mathrm{T}}S_{ij} + \eta_1^{(2)}[S_{ik}S_{kj}]_{\mathrm{D}} + \eta_2^{(2)}(S_{jk}\Omega_{ki} + S_{ik}\Omega_{kj}). \tag{4.17}$$

Dimensional coefficients ν_{T} and $\eta_n^{(2)}$ $(n = 1, 2)$ are written as

$$\nu_{\mathrm{T}} = C_\nu \frac{K^2}{\varepsilon}, \tag{4.18}$$

$$\eta_n^{(2)} = C_n^{(2)} \frac{K^3}{\varepsilon^2}, \tag{4.19}$$

where C_ν and $C_n^{(2)}$ $(n = 1, 2)$ are newly introduced numerical constants (parenthesized superscript 2 in $\eta_n^{(2)}$ means that it is related to the terms of the second order in S_{ij} and Ω_{ij}). In equation (4.17) the first term is specifically called the turbulent-viscosity representation. Equation (4.18) is nothing but equation (3.149), and the physical meaning of ν_{T} will become clear in §4.2.4.

We use equation (4.17), and perform the second iteration, which results in

$$\begin{aligned} B_{ij} = &-\nu_{\mathrm{T}} S_{ij} + \eta_1^{(2)} [S_{ik} S_{kj}]_{\mathrm{D}} + \eta_2^{(2)} (S_{jk} \Omega_{ki} + S_{ik} \Omega_{kj}) \\ &+ \eta_1^{(3)} (S_{ik} S_{kl} \Omega_{lj} + S_{jk} S_{kl} \Omega_{li}) + \eta_2^{(3)} [S_{ik} \Omega_{kl} \Omega_{lj} + S_{jk} \Omega_{kl} \Omega_{li}]_{\mathrm{D}} \\ &+ \eta_3^{(3)} [S_{ik} S_{kl} S_{lj}]_{\mathrm{D}}, \end{aligned} \tag{4.20}$$

where

$$\eta_n^{(3)} = C_n^{(3)} \frac{K^4}{\varepsilon^3} \qquad (n = 1, 2), \tag{4.21}$$

and we should note that $S_{ik} S_{kl} \Omega_{li} = 0$ [$C_n^{(3)}$ $(n = 1, 2)$ are model constants].

For an arbitrary tensor of rank two, A_{ij}, we have the Cayley–Hamilton formula [4.3]

$$A_{il} A_{lm} A_{mj} = I_A A_{il} A_{lj} - II_A A_{ij} + III_A \delta_{ij}, \tag{4.22}$$

where

$$I_A = A_{mm}, \tag{4.23a}$$

$$II_A = \tfrac{1}{2} (A_{mm} A_{nn} - A_{mn} A_{nm}), \tag{4.23b}$$

$$III_A = \tfrac{1}{6} (A_{ll} A_{mm} A_{nn} - 3 A_{ll} A_{mn} A_{nm} + 2 A_{lm} A_{mn} A_{nl}). \tag{4.23c}$$

Then we rewrite

$$[S_{il} S_{lk} S_{kj}]_{\mathrm{D}} = \tfrac{1}{2} S_{mn}^2 S_{ij}, \tag{4.24}$$

under the solenoidal condition (3.25). From equation (4.24), the last term on the right-hand side of equation (4.20) may be absorbed into the first term as

$$-\nu_{\mathrm{T}} S_{ij} + \eta_3^{(3)} [S_{ik} S_{kl} S_{lj}]_{\mathrm{D}} = -C_\nu \left[1 - \frac{C_3^{(3)}}{C_\nu} \left(\frac{K}{\varepsilon} S_{mn} \right)^2 \right] \frac{K^2}{\varepsilon} S_{ij}. \tag{4.25}$$

As a result, the numerical constant in the first term becomes a functional of S_{ij}, K, and ε.

On repeating the above iteration procedure in equation (4.16), resulting expressions take the same as equation (4.20), but their numerical coefficients C_ν, $\eta_n^{(2)}$, and $\eta_n^{(3)}$ $(n = 1, 2)$ are changed to functionals of

$$\left(\frac{K}{\varepsilon} S_{lm}\right)^2, \quad \left(\frac{K}{\varepsilon} \Omega_{lm}\right)^2, \quad \left(\frac{K}{\varepsilon} S_{lm}\right)\left(\frac{K}{\varepsilon} S_{mn}\right)\left(\frac{K}{\varepsilon} S_{nl}\right),$$

$$\left(\frac{K}{\varepsilon} S_{lm}\right)\left(\frac{K}{\varepsilon} \Omega_{mn}\right)\left(\frac{K}{\varepsilon} \Omega_{nl}\right). \tag{4.26}$$

A method for resolving equations such as equation (4.16) in a noniterative manner has also been studied, and various explicit algebraic models have been proposed [4.11–4.14].

4.2.2 Modelling of Heat Flux

The method of modelling R_{ij} is straightforwardly applicable to modelling the heat flux $\boldsymbol{H}_\theta$ [equation (3.29)]. We rewrite equation (3.78) as

$$P_{\theta i} = -\tfrac{2}{3} K \frac{\partial \Theta}{\partial x_i} - \tfrac{1}{2}(S_{ij} + \Omega_{ij}) H_{\theta j} - B_{ij} \frac{\partial \Theta}{\partial x_j}, \tag{4.27}$$

Following equation (4.14), we write

$$H_{\theta i} = \tau_\theta P_{\theta i}, \tag{4.28}$$

where the characteristic time τ_θ is

$$\tau_\theta = C_\theta \tau_{\mathrm{E}} = C_\theta \frac{K}{\varepsilon}, \tag{4.29}$$

with C_θ as a numerical factor, entirely similar to equation (4.15). From equations (4.27) and (4.29), we have

$$H_{\theta i} = -\frac{2 C_\theta}{3} \frac{K^2}{\varepsilon} \frac{\partial \Theta}{\partial x_i} - \frac{C_\theta}{2} \frac{K}{\varepsilon} (S_{ij} + \Omega_{ij}) H_{\theta j} - C_\theta \frac{K}{\varepsilon} B_{ij} \frac{\partial \Theta}{\partial x_j}. \tag{4.30}$$

The leading term on the right-hand side of equation (4.30) corresponds to that in equation (4.16) and is called the turbulent-diffusivity representation. In the first iteration, we use these leading terms and have

$$H_{\theta i} = -\nu_{\mathrm{T}\theta} \frac{\partial \Theta}{\partial x_i} + \left(\zeta_1^{(2)} S_{ij} + \zeta_2^{(2)} \Omega_{ij}\right) \frac{\partial \Theta}{\partial x_j}. \tag{4.31}$$

Dimensional coefficients $\nu_{\mathrm{T}\theta}$ and $\zeta_n^{(2)}$ $(n = 1, 2)$ are defined by

$$\nu_{\mathrm{T}\theta} = \frac{\nu_{\mathrm{T}}}{\sigma_\theta}, \tag{4.32}$$

$$\zeta_n^{(2)} = C_{\theta n}^{(2)} \frac{K^3}{\varepsilon^2}, \tag{4.33}$$

where σ_θ and $C_{\theta n}^{(2)}$ $(n = 1, 2)$ are numerical constants.

The foregoing modelling procedure is helpful for understanding buoyancy effects on R_{ij}. Under the Boussinesq approximation, equation (3.31) is replaced with

$$\frac{Du_i'}{Dt} = -u_j'\frac{\partial U_i}{\partial x_j} - \frac{\partial}{\partial x_j}(u_i'u_j' - R_{ij}) - \frac{1}{\rho}\frac{\partial p'}{\partial x_i} - \alpha_T \theta' g_i + \nu \nabla^2 u_i, \qquad (4.34)$$

where $\boldsymbol{g}$ is the vector of gravitational acceleration, and α_T is the coefficient of thermal expansion. We incorporate this buoyancy effect into equation (4.12). Then equation (4.16) is replaced with

$$B_{ij} = -\frac{2C_B}{3}\frac{K^2}{\varepsilon}S_{ij} - \alpha_T[H_{\theta i}g_j + H_{\theta j}g_i]_D$$
$$-\frac{C_B}{2}\frac{K}{\varepsilon}[B_{jk}S_{ki} + B_{ik}S_{kj}]_D - \frac{C_B}{2}\frac{K}{\varepsilon}(B_{jk}\Omega_{ki} + B_{ik}\Omega_{kj}). \qquad (4.35)$$

We substitute the first term in equation (4.31) into the second term in equation (4.35), and have

$$B_{ij} = -\nu_T S_{ij} + \eta_B^{(2)}\left[g_j\frac{\partial \Theta}{\partial x_i} + g_i\frac{\partial \Theta}{\partial x_j}\right]_D, \qquad (4.36)$$

where the third and fourth terms in equation (4.35) were neglected for simplicity of discussion, and

$$\eta_B^{(2)} = C_B^{(2)}\alpha_T\frac{K^3}{\varepsilon^2}, \qquad (4.37)$$

with $C_B^{(2)}$ as a numerical constant. Equation (4.36) indicates that the buoyancy effect on R_{ij} occurs in the combination of $\nabla\Theta$ and $\boldsymbol{g}$.

4.2.3 Modelling of Turbulence Equations

In §4.2.1 and §4.2.2, R_{ij} and $\boldsymbol{H}_\theta$ were related to $\boldsymbol{U}$, Θ, K, and ε. The occurrence of the last two turbulence quantities is linked with the choice of the time scales τ_T [equation (4.15)] and τ_θ [equation (4.29)]. Use of other quantities such as the characteristic length of energy-containing eddies, l_E [equation (3.142b)], is also allowable. In this context, the critical point is whether the equations for those turbulence quantities with firm mathematical and physical bases can be constructed or not.

Discussions on conservation properties were made in §3.3.2.2. In the case of incompressible flow, a total amount of kinetic energy is conserved in the absence of molecular viscous effects, resulting in equation (3.39) for K with a clear mathematical structure. Therefore it is appropriate to adopt K as one of the quantities characterizing a turbulent state. The dissipation rate ε enters equation (3.39). Here we have two ways. One is to adopt ε as the other turbulence quantity, and the other is to choose a turbulence quantity other than ε, for instance, l_E. For the choice of l_E, ε is modelled in the form proportional to $K^{3/2}/\varepsilon$ [see equation (3.142b)], and a model equation for l_E is sought [4.9]. For these two quantities, however, we do not have the conservation properties comparable to that for K. This point is a stumbling block for turbulence modelling. In the current turbulent modelling, ε is preferentially selected as the second turbulent quantity since its information may be obtained from the DNS of several fundamental turbulent flows.

4.2.3.1 Turbulent-energy equation

Once K and ε have been chosen as the characteristic turbulence quantities, what remains to be modelled in the turbulent-energy equation (3.39) is the transport term $\boldsymbol{T}_K$ [equation (3.42)]. We divide it into the velocity- and pressure-transport parts, $\boldsymbol{T}_{KV}$ and $\boldsymbol{T}_{KP}$, as

$$\boldsymbol{T}_K = \boldsymbol{T}_{KV} + \boldsymbol{T}_{KP}, \tag{4.38}$$

where

$$\boldsymbol{T}_{KV} = -\left\langle \frac{\boldsymbol{u}'^2}{2} \boldsymbol{u}' \right\rangle, \tag{4.39}$$

$$\boldsymbol{T}_{KP} = -\langle p' \boldsymbol{u}' \rangle. \tag{4.40}$$

The pressure fluctuation p' obeys equation (3.32), which may be formally integrated as

$$p' = p'_{V1} + p'_{V2} + p'_S, \tag{4.41}$$

each of which is written as

$$p'_{V1} = -2 \int_V G(\boldsymbol{x} - \boldsymbol{x}') \frac{\partial U_i(\boldsymbol{x}')}{\partial x_j} \frac{\partial u'_j(\boldsymbol{x}')}{\partial x_i} \, \mathrm{d}V, \tag{4.42a}$$

$$p'_{V2} = -\int_V G(\boldsymbol{x} - \boldsymbol{x}') \frac{\partial^2}{\partial x_i \, \partial x_j} [u'_i(\boldsymbol{x}') u'_j(\boldsymbol{x}') - R_{ij}(\boldsymbol{x}')] \, \mathrm{d}V, \tag{4.42b}$$

$$p'_S = \int_S \left(G(\boldsymbol{x} - \boldsymbol{x}') \frac{\partial p'(\boldsymbol{x}')}{\partial n} - p'(\boldsymbol{x}') \frac{\partial G(\boldsymbol{x} - \boldsymbol{x}')}{\partial n} \right) \mathrm{d}S. \tag{4.42c}$$

Here V and S denote a whole fluid region and its surrounding surface, respectively, $\boldsymbol{n}$ is the outward unit vector normal to S, and $G(\boldsymbol{x} - \boldsymbol{x}')$ is the Green's function subject to

$$\nabla^2 G(\boldsymbol{x} - \boldsymbol{x}') = \delta(\boldsymbol{x} - \boldsymbol{x}'), \tag{4.43}$$

and is given by

$$G(\boldsymbol{x} - \boldsymbol{x}') = -\frac{1}{4\pi} \frac{1}{|\boldsymbol{x} - \boldsymbol{x}'|}. \tag{4.44}$$

In heuristic turbulence modelling, $\boldsymbol{T}_{KP}$ is usually neglected from the following conjecture. In the motion of an incompressible fluid, p' is a nonlocal quantity, as can be seen from equations (4.41) and (4.42). The contributions of p' to $\boldsymbol{T}_{KP}$ come from a whole fluid region and tend to cancel each other. As a result, $\boldsymbol{T}_{KP}$ becomes smaller than $\boldsymbol{T}_{KV}$ expressing a local transport effect. Such a conjecture is correct in one situation, but not in another. This point will be remarked on below equation (4.51).

The velocity transport $\boldsymbol{T}_{KV}$ represents the energy transport rate by velocity fluctuation. A convenient method for modelling such a quantity is the gradient-diffusion approximation. We now divide a quantity ϕ into the mean Φ and the fluctuation around it, ϕ':

$$\phi = \Phi + \phi', \tag{4.45}$$

and consider

$$\boldsymbol{H}_\phi = \langle \boldsymbol{u}' \phi' \rangle. \tag{4.46}$$

In general, ϕ' in the region with large Φ is larger than its small-Φ counterpart. We compare the process in which ϕ' is advected from large- to small-Φ regions with its inverse process. It is probable that the former case gives rise to a larger transport of ϕ', that is, ϕ' is transported from large- to small-Φ regions in net. This inference suggests

$$\boldsymbol{H}_\phi = -\nu_\phi \nabla \Phi, \tag{4.47}$$

with ν_ϕ as a diffusion coefficient. This modelling is called the gradient-diffusion approximation.

In equation (4.31) for $\boldsymbol{H}_\theta$, the first term corresponds to equation (4.47). For R_{ij}, $u_i' u_j'$ is the transport rate of u_j' by u_i' and vice versa. The application of equation (4.47) to $u_i' u_j'$ results in the first term of equation (4.17) or (4.20). As is seen from these instances, we may see that equation (4.47) can capture parts of the transport processes by velocity fluctuation.

We apply equation (4.47) straightforwardly to $\boldsymbol{T}_{KV}$ [equation (4.39)], and have

$$\boldsymbol{T}_{KV} = \nu_K \nabla (\langle \tfrac{1}{2} \boldsymbol{u}^2 \rangle) = \nu_K \nabla (\tfrac{1}{2} \boldsymbol{U}^2 + K). \tag{4.48}$$

This modelling, however, does not obey the Galilean-invariance principle owing to $\boldsymbol{U}^2/2$. We drop the $\boldsymbol{U}^2$-related part, and write

$$\boldsymbol{T}_{KV} = \nu_K \nabla K. \tag{4.49}$$

Under the approximation of dropping $\boldsymbol{T}_{KP}$, $\boldsymbol{T}_K$ is written as

$$\boldsymbol{T}_K = \frac{\nu_\mathrm{T}}{\sigma_K} \nabla K, \tag{4.50}$$

where ν_T is defined by equation (4.18), and σ_K is a nondimensional coefficient.

Equation (4.50) signifies that $\boldsymbol{T}_K$ arises from ∇K. On the other hand, ε was chosen as a quantity characterizing a turbulent state, besides K. From their close connection through equation (3.39), the importance of ε is supposed to be equal to that of K in describing a turbulent state. It is highly probable that $\nabla\varepsilon$ also has an influence on $\boldsymbol{T}_K$. From this viewpoint, $\boldsymbol{T}_K$ is modelled as

$$\boldsymbol{T}_{\mathrm{K}} = \frac{\nu_{\mathrm{T}}}{\sigma_{\mathrm{K}}} \nabla K - \frac{\nu_{\mathrm{T}}}{\sigma_{\mathrm{K}\varepsilon}} \frac{K}{\varepsilon} \nabla\varepsilon, \tag{4.51}$$

with $\sigma_{\mathrm{K}\varepsilon}$ as the other nondimensional coefficient. Here the negative sign is attached to the second part since the increase in ε tends to reduce K from equation (3.39).

In the heuristic modelling leading to equation (4.50), we assumed that $\boldsymbol{T}_{KP}$ is negligible, compared with $\boldsymbol{T}_{KV}$. This is not self-evident. In fact, we encounter two different flow situations; one situation is consistent with this assumption (a channel flow), and the other is inconsistent (a flow around a bluff body). A prominent feature of the latter is that the mean flow is subject to the rapid change in the flow direction. In order to seek a cause of such discrepancy, it is indispensable to analyze $\boldsymbol{T}_{KV}$ and $\boldsymbol{T}_{KP}$ by a theoretical method based on the Navier–Stokes equation. This point, as well as the occurrence of the $\nabla\varepsilon$ transport effect, will be discussed in §5.3.1.

4.2.3.2 Dissipation-rate equation

From equation (3.31) for $\boldsymbol{u}'$, the equation for ε obeys

$$\begin{aligned}\frac{\mathrm{D}\varepsilon}{\mathrm{D}t} = & -2\nu\left\langle \frac{\partial u_i'}{\partial x_j}\frac{\partial u_i'}{\partial x_l}\frac{\partial u_j'}{\partial x_l}\right\rangle - 2\left\langle \left(\nu\frac{\partial^2 u_i'}{\partial x_j \partial x_l}\right)^2\right\rangle \\ & -2\nu\left\langle \frac{\partial u_i'}{\partial x_l}\frac{\partial u_j'}{\partial x_l} + \frac{\partial u_l'}{\partial x_i}\frac{\partial u_l'}{\partial x_j}\right\rangle\frac{\partial U_i}{\partial x_j} - 2\nu\left\langle u_j'\frac{\partial u_i'}{\partial x_l}\right\rangle\frac{\partial^2 U_i}{\partial x_j \partial x_l} \\ & + \frac{\partial}{\partial x_i}\left(-\nu\left\langle u_i'\left(\frac{\partial u_j'}{\partial x_l}\right)^2\right\rangle - 2\nu\left\langle\frac{\partial p'}{\partial x_j}\frac{\partial u_i'}{\partial x_j}\right\rangle + \nu\frac{\partial\varepsilon}{\partial x_i}\right). \end{aligned} \tag{4.52}$$

Equation (4.52) is very complicated, compared with equation (3.39) for K. This situation arises from the fact that there is no conservation property concerning $\nu(\partial u_j/\partial x_i)^2$, as has already been remarked in §3.3.2.2.

We examine the behavior of each term in equation (4.52). In correspondence to the spatial length of energy-dissipating eddies,

$$l_{\mathrm{D}} = \nu^{3/4}\varepsilon^{-1/4}, \tag{4.53}$$

[see equation (3.146b)], we introduce its related velocity

$$u_{\mathrm{D}} = \nu^{1/4}\varepsilon^{1/4}. \tag{4.54}$$

Using these two quantities, we may estimate

$$\{\varepsilon\}_{l_\mathrm{D},u_\mathrm{D}} = \left\{\nu\left\langle\left(\frac{\partial u_j'}{\partial x_l}\right)^2\right\rangle\right\}_{l_\mathrm{D},u_\mathrm{D}} = \nu\left(\frac{u_\mathrm{D}}{l_\mathrm{D}}\right)^2 = \varepsilon, \tag{4.55}$$

which guarantees the use of equation (4.54) as the dissipation-related velocity scale.

We make an estimate of the first four terms on the right-hand side of equation (4.52), and have

$$\left\{\nu\left\langle\frac{\partial u_i'}{\partial x_j}\frac{\partial u_j'}{\partial x_l}\frac{\partial u_j'}{\partial x_l}\right\rangle\right\}_{l_\mathrm{D},u_\mathrm{D}} = \varepsilon^{3/2}\nu^{-1/2}, \tag{4.56}$$

$$\left\{2\left\langle\left(\nu\frac{\partial^2 u_i'}{\partial x_j \partial x_l}\right)^2\right\rangle\right\}_{l_\mathrm{D},u_\mathrm{D}} = \varepsilon^{3/2}\nu^{-1/2}, \tag{4.57}$$

$$\left\{\nu\left\langle\frac{\partial u_i'}{\partial x_l}\frac{\partial u_j'}{\partial x_l} + \frac{\partial u_l'}{\partial x_i}\frac{\partial u_l'}{\partial x_j}\right\rangle\right\}_{l_\mathrm{D},u_\mathrm{D}} = \varepsilon, \tag{4.58}$$

$$\left\{\nu\left(\left\langle u_j'\frac{\partial u_j'}{\partial x_l}\right\rangle\right)^2\right\}_{l_\mathrm{D},u_\mathrm{D}} = \nu^{3/4}\varepsilon^{3/4}. \tag{4.59}$$

The last term in equation (4.52) is written in a divergence form. After the integration over a whole fluid region, it expresses the contribution across a boundary. There the $\nu(\partial\varepsilon/\partial x_i)$-related part may become important near a solid wall.

In real-world turbulent flows, their Reynolds number is usually very large. Then we consider the case of $\nu \to 0$. In the limit, equations (4.56) and (4.57) diverge. In reality, these two terms nearly balance with each other, and their residual constitutes an important part of the equation for ε. Here we should note that equation (4.56) is negative in this limit since equation (4.57) is always positive. Equation (4.58) is also nonvanishing in this limit and is considered to be important near a solid boundary since it is linked with mean-velocity gradients [note the third term in equation (4.52)].

From the foregoing discussions we may understand that the mathematical structure of the ε equation is entirely different from that of equation (3.39) for K. Its modelling is beyond the scope of the simple method applied to $\boldsymbol{R}_{ij}$, $\boldsymbol{T}_K$, etc. In the current heuristic modelling, we adopt

$$\frac{\mathrm{D}\varepsilon}{\mathrm{D}t} = C_{\varepsilon 1}\frac{\varepsilon}{K}P_K - C_{\varepsilon 2}\frac{\varepsilon^2}{K} + \nabla\cdot\left(\frac{\nu_\mathrm{T}}{\sigma_\varepsilon}\nabla\varepsilon\right) + \nu\nabla^2\varepsilon, \tag{4.60}$$

with $C_{\varepsilon 1}$ and $C_{\varepsilon 2}$ as model constants [4.15]. As is seen from

$$C_{\varepsilon 1}\frac{\varepsilon}{K}P_K - C_{\varepsilon 2}\frac{\varepsilon^2}{K} = C_{\varepsilon 1}\frac{\varepsilon}{K}\left(P_\mathrm{K} - \frac{C_{\varepsilon 2}}{C_{\varepsilon 1}}\varepsilon\right), \tag{4.61}$$

the structure of equation (4.60) is very similar to equation (3.39), which indicates that ε is large in a region with large K. This situation is usually consistent with observations. Modelling the equation for ε is still an important theme for the study of turbulence modelling [4.16].

4.2.4 The Simplest Algebraic Model

On the basis of the results obtained in §4.2.1–§4.2.3, we give the simplest model for grasping an overview of turbulence modelling. This model is called the standard $K - \varepsilon$ model in the terminology of turbulence modelling. The mean-flow equation is

$$\frac{\mathrm{D}U_i}{\mathrm{D}t} = -\frac{\partial P}{\partial x_i} + \frac{\partial}{\partial x_j}(-R_{ij}) + \nu \nabla^2 U_i. \tag{4.62}$$

The Reynolds stress R_{ij} is modelled as

$$R_{ij} = \tfrac{2}{3} K \delta_{ij} - \nu_{\mathrm{T}} S_{ij}. \tag{4.63}$$

Here ν_{T} is called the turbulent or eddy viscosity, which is given by

$$\nu_{\mathrm{T}} = C_\nu \frac{K^2}{\varepsilon}. \tag{4.64}$$

Its physical meaning is clear from

$$\frac{\partial}{\partial x_j}(-R_{ij}) + \nu \nabla^2 U_i = \frac{\partial}{\partial x_i}\left(-\frac{2}{3}K\right) + \frac{\partial}{\partial x_j}\left[(\nu + \nu_{\mathrm{T}})\left(\frac{\partial U_j}{\partial x_i} + \frac{\partial U_i}{\partial x_j}\right)\right], \tag{4.65}$$

and ν_{T} expresses the enhancement of diffusion effects due to turbulence. As a result, equation (4.63) is named the turbulent-viscosity representation for the Reynolds stress.

The equations for K and ε constituting ν_{T} are given by

$$\frac{\mathrm{D}K}{\mathrm{D}t} = P_K - \varepsilon + \nabla \cdot \left(\frac{\nu_{\mathrm{T}}}{\sigma_K} \nabla K\right) + \nu \nabla^2 K, \tag{4.66}$$

$$\frac{\mathrm{D}\varepsilon}{\mathrm{D}t} = C_{\varepsilon 1} \frac{\varepsilon}{K} P_K - C_{\varepsilon 2} \frac{\varepsilon^2}{K} + \nabla \cdot \left(\frac{\nu_{\mathrm{T}}}{\sigma_\varepsilon} \nabla \varepsilon\right) + \nu \nabla^2 \varepsilon, \tag{4.67}$$

with

$$P_K = -R_{ij} \frac{\partial U_j}{\partial x_i}. \tag{4.68}$$

The mean temperature equation is

$$\frac{\mathrm{D}\Theta}{\mathrm{D}t} = \nabla \cdot (-\boldsymbol{H}_\theta) + \lambda_\theta \nabla^2 \Theta. \tag{4.69}$$

The turbulent heat flux $\boldsymbol{H}_\theta$ is modelled as

$$\boldsymbol{H}_\theta = -\frac{\nu_\mathrm{T}}{\sigma_\theta}\nabla\Theta, \tag{4.70}$$

where $\nu_\mathrm{T}/\sigma_\theta$, which corresponds to ν_T in R_{ij}, is named the turbulent heat diffusivity.

The standard choice of model constants in the $K-\varepsilon$ model is

$$\begin{aligned} &C_\nu = 0.09, \quad &\sigma_K = 1, \quad &C_{\varepsilon 1} = 1.4, \\ &C_{\varepsilon 2} = 1.9, \quad &\sigma_\varepsilon = 1.3, \quad &\sigma_\theta = 0.7\text{–}1. \end{aligned} \tag{4.71}$$

This choice will be referred to later in relation to the application to some typical turbulent flows.

4.2.5 Investigation into Some Representative Turbulent Flows

4.2.5.1 Grid turbulence

A turbulent state mimicking homogeneous turbulence is observed in grid turbulence, which is generated behind a grid composed of bars. Its turbulent intensities decay with the distance from the grid, but the mean velocity changes much more slowly [we should recall equation (3.139) signifying that large eddies decay more slowly than small eddies]. We write the mean velocity

$$\boldsymbol{U} = (U, 0, 0), \tag{4.72}$$

with U kept as constant (the x direction is normal to the grid).

Under equation (4.72), equations (4.66) and (4.67) are reduced to

$$\frac{\mathrm{d}K}{\mathrm{d}\tau} = -\varepsilon, \tag{4.73}$$

$$\frac{\mathrm{d}\varepsilon}{\mathrm{d}\tau} = -C_{\varepsilon 2}\frac{\varepsilon^2}{K}, \tag{4.74}$$

where τ is defined by

$$\tau = \frac{x}{U} \tag{4.75}$$

and has the dimension of time.

Considering that K and ε decay with τ, we assume the similarity form

$$\frac{K}{K_0} = \left(\frac{\tau}{\tau_\mathrm{R}} + 1\right)^{-m}, \tag{4.76}$$

$$\frac{\varepsilon}{\varepsilon_0} = \left(\frac{\tau}{\tau_\mathrm{R}} + 1\right)^{-n}, \tag{4.77}$$

with

$$\tau_\mathrm{R} = \frac{x_R}{U}. \tag{4.78}$$

Here x_R is the reference length related to the geometrical characteristics of a grid such as the radius of bars and the distance between them. Moreover K_0 and ε_0 are the values of K and ε at $\tau = 0$, but they do not correspond to the values at the actual position of the grid since the similarity expressions (4.76) and (4.77) do not hold near the grid. Then K_0 and ε_0 should be regarded as the values at the virtual origin.

We substitute equations (4.76) and (4.77) into equations (4.73) and (4.74), and have

$$n = m + 1, \tag{4.79}$$

$$C_{\varepsilon 2} = \frac{n}{m} = 1 + \frac{1}{m}. \tag{4.80}$$

The choice of

$$m = 1.1\text{–}1.3 \tag{4.81}$$

is consistent with the observed decay law. Equation (4.81) leads to

$$C_{\varepsilon 2} = 1.8\text{–}1.9, \tag{4.82}$$

which corresponds to $C_{\varepsilon 2}$ in equation (4.71).

4.2.5.2 Channel turbulence

In wall-bounded turbulence represented by channel and pipe flows, one of the prominent properties is the logarithmic velocity profile, which is related to a flattened velocity profile referred to in §2.1.1. In channel turbulence (figure 3.3), the shear-stress part R_{xy} is given by

$$R_{xy} = Gy + \nu \frac{\mathrm{d}U}{\mathrm{d}y}, \tag{4.83}$$

from equation (3.63). In order to see the flow behavior near a lower wall, we adopt

$$Y = y + H. \tag{4.84}$$

Then equation (4.83) is reduced to

$$R_{xy} - GY = \nu \frac{\mathrm{d}U}{\mathrm{d}Y} - GH. \tag{4.85}$$

As was noted in relation to equation (3.62), an important quantity characterizing wall-bounded flows is the wall friction velocity

$$u_W = \sqrt{\nu \left(\frac{\mathrm{d}U}{\mathrm{d}Y}\right)_{Y=0}} = \sqrt{GH}. \tag{4.86}$$

This quantity plays the role of reference velocity in discussing near-wall turbulence properties.

From the continuity equation

$$\frac{\partial u'}{\partial x}+\frac{\partial v'}{\partial Y}+\frac{\partial w'}{\partial z}=0, \tag{4.87}$$

the behavior of $\boldsymbol{u}'$ near the lower wall is given by

$$u'=O(Y), \qquad v'=O(Y^2), \qquad w'=O(Y), \tag{4.88}$$

which shows

$$R_{xy}=O(Y^3). \tag{4.89}$$

In equation (4.85), we neglect the contributions of $O(Y^n)$ $(n \geq 1)$. Then only the right-hand side survives, and its integration gives the velocity profile valid in the close vicinity of the wall,

$$\frac{U}{u_{\mathrm{W}}}=y_{\mathrm{W}}, \tag{4.90}$$

where y_{W} is called the wall-unit coordinate and is defined as

$$y_{\mathrm{W}}=\frac{u_{\mathrm{W}}Y}{\nu}. \tag{4.91}$$

The wall coordinate y_{W} is the Reynolds number based on u_{W} and the distance from the wall, Y.

Next, we examine the region described by

$$y_{\mathrm{W}} \gg 1, \qquad \frac{Y}{H} \ll 1, \tag{4.92a}$$

which is equivalent to

$$\frac{1}{R_{eW}} \ll \frac{Y}{H} \ll 1, \tag{4.92b}$$

where $R_{e\mathrm{W}}$ is the Reynolds number based on u_{W} and H, that is,

$$R_{e\mathrm{W}}=\frac{u_{\mathrm{W}}H}{\nu}. \tag{4.93}$$

In channel turbulence whose Reynolds number based on the center-line velocity is $O(10^3)$ or more, u_{W} is $O(10^{-1})$ times the center-line velocity, and the region designated by equation (4.92) exists.

For the region obeying equation (4.92), it is plausible to rewrite equation (4.85) as

$$R_{xy}=-GH=-u_{\mathrm{W}}^2. \tag{4.94}$$

We substitute equations (4.63) and (4.64) into equation (4.94), and have

$$\frac{\mathrm{d}U}{\mathrm{d}Y}=\frac{1}{C_\nu}\frac{\varepsilon u_{\mathrm{W}}^2}{K^2}. \tag{4.95}$$

The turbulent-energy equation (4.66) is reduced to

$$P_K - \varepsilon + \frac{d}{dY}\left(\frac{\nu_T}{\sigma_K}\frac{dK}{dY}\right) = 0, \tag{4.96}$$

under $D/Dt = 0$ and the condition (4.92). From the discussion in §3.5, the diffusion term is small in the region except the center-line (see figure 3.7), resulting in

$$P_K \cong \varepsilon. \tag{4.97}$$

We substitute equation (4.68) into equation (4.97), and make use of equation (4.94). Then we have

$$\frac{dU}{dY} = \frac{\varepsilon}{u_W^2}. \tag{4.98}$$

The comparison between equations (4.95) and (4.98) leads to

$$\frac{K}{u_W^2} = \frac{1}{\sqrt{C_\nu}}. \tag{4.99}$$

Observations of channel flow show

$$\frac{|R_{xy}|}{K} = \frac{u_W^2}{K} \cong 0.3, \tag{4.100}$$

where use has been made of equation (4.94). The resulting constancy of K guarantees the neglect of the diffusion term in equation (4.96). From equations (4.99) and (4.100), we have

$$C_\nu = 0.09, \tag{4.101}$$

which is included in equation (4.71).

As may be seen from equation (4.95) or (4.98), we need to estimate ε to determine the profile of U. For this purpose, we use equation (3.142b) for the length scale of energy-containing eddies. Under the condition (4.92b), the reference length characterizing energy-containing eddies, l_E, is not H but the distance from the wall, Y. On the basis of this consideration, we write equation (3.142b) as

$$\frac{K^{3/2}}{\varepsilon} = \frac{\kappa}{C_\nu^{3/4}} Y, \tag{4.102}$$

with κ as a newly introduced numerical constant. From equations (4.99) and (4.102), we have

$$\varepsilon = \frac{1}{\kappa}\frac{u_W^3}{Y}. \tag{4.103}$$

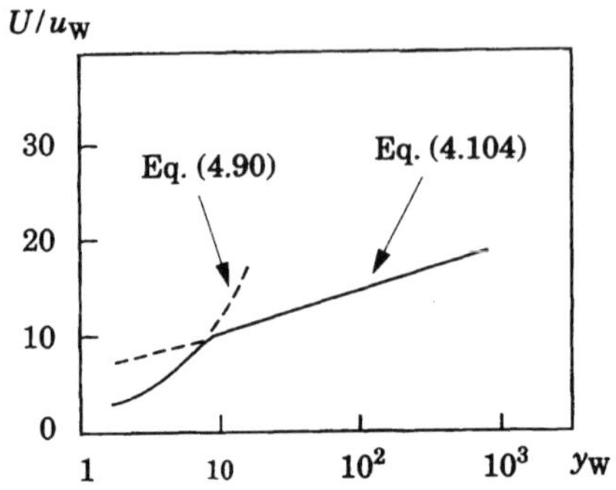

Figure 4.1. Mean velocity in channel flow.

We substitute equation (4.103) into equation (4.98), and integrate the resulting expression, having

$$\frac{U}{u_W} = \frac{1}{\kappa}\log Y + A, \tag{4.104}$$

where A is an integral constant.

Equation (4.104) is the so-called logarithmic velocity profile, which is ubiquitously observed near a solid boundary. The constant κ is the Karman constant, which is

$$\kappa = 0.4 \tag{4.105}$$

from observations. The other constant A changes according to the condition of a wall and is given by

$$A \cong 5.0 \tag{4.106}$$

in the case of a smooth wall. The logarithmic profile (4.104) is shown in figure 4.1, with the near-wall profile (4.90).

The dissipation-rate equation (4.67) is reduced to

$$C_{\varepsilon 1}\frac{\varepsilon}{K}P_K - C_{\varepsilon 2}\frac{\varepsilon^2}{K} + \frac{\mathrm{d}}{\mathrm{d}Y}\left(\frac{\nu_T}{\sigma_\varepsilon}\frac{\mathrm{d}\varepsilon}{\mathrm{d}Y}\right) = 0. \tag{4.107}$$

From equations (4.68), (4.99), (4.103), and (4.104), we have

$$C_{\varepsilon 2} - C_{\varepsilon 1} = \frac{\kappa^2}{\sqrt{C_\nu}\sigma_\varepsilon}. \tag{4.108}$$

This relationship among model constants was taken into consideration in equation (4.71). Readers may consult [4.17] for a more detailed analytical study of the $K - \varepsilon$ model.

In channel flow, equation (4.63) gives

$$\langle u'^2 \rangle = \langle v'^2 \rangle = \langle w'^2 \rangle = \tfrac{2}{3}K, \tag{4.109}$$

as three turbulent intensities. Such finding is inconsistent with the observational result (3.69). On the other hand, equation (4.17) gives

$$\langle u'^2 \rangle = \tfrac{2}{3}K + \left(\tfrac{1}{3}\eta_1^{(2)} + 2\eta_2^{(2)}\right)\left(\frac{\mathrm{d}U}{\mathrm{d}y}\right)^2, \tag{4.110a}$$

$$\langle v'^2 \rangle = \tfrac{2}{3}K + \left(\tfrac{1}{3}\eta_1^{(2)} - 2\eta_2^{(2)}\right)\left(\frac{\mathrm{d}U}{\mathrm{d}y}\right)^2, \tag{4.110b}$$

$$\langle w'^2 \rangle = \tfrac{2}{3}K - \tfrac{2}{3}\eta_1^{(2)}\left(\frac{\mathrm{d}U}{\mathrm{d}y}\right)^2, \tag{4.110c}$$

and the above inconsistency may be overcome. In the derivation of the logarithmic-velocity law, we need only R_{xy}, to which the second-order parts in equation (4.17) do not contribute. As a result, no change is necessary for the velocity law.

4.2.5.3 Square-duct flow

In §2.1.2 we remarked on the occurrence of secondary flows in a square duct as an instance of the structure formation in turbulent flows (figure 2.5). In the region far from the entrance, the flow is fully-developed and homogeneous in the x direction. Then the mean velocity is expressed as

$$\mathbf{U} = [U(y,z), V(y,z), W(y,z)]. \tag{4.111}$$

The occurrence of secondary flows at the cross-section is equivalent to the existence of the x component of the mean vorticity

$$\Omega_x = \frac{\partial W}{\partial y} - \frac{\partial V}{\partial z}. \tag{4.112}$$

From equation (4.62), V and W obey

$$V\frac{\partial V}{\partial y} + W\frac{\partial V}{\partial z} = -\frac{\partial P}{\partial y} - \frac{\partial R_{yy}}{\partial y} - \frac{\partial R_{yz}}{\partial z} + \nu\nabla^2 V, \tag{4.113}$$

$$V\frac{\partial W}{\partial y} + W\frac{\partial W}{\partial z} = -\frac{\partial P}{\partial z} - \frac{\partial R_{yz}}{\partial y} - \frac{\partial R_{zz}}{\partial z} + \nu\nabla^2 W, \tag{4.114}$$

which leads to

$$V\frac{\partial \Omega_x}{\partial y} + W\frac{\partial \Omega_x}{\partial z} = \frac{\partial^2}{\partial y\,\partial z}(R_{yy} - R_{zz}) - \left(\frac{\partial^2}{\partial y^2} - \frac{\partial^2}{\partial z^2}\right)R_{yz} + \nu\nabla^2\Omega_x. \tag{4.115}$$

In the turbulent-viscosity representation (4.63), we have

$$R_{yy} = \tfrac{2}{3}K - 2\nu_{\mathrm{T}} \frac{\partial V}{\partial y}, \qquad (4.116a)$$

$$R_{zz} = \tfrac{2}{3}K - 2\nu_{\mathrm{T}} \frac{\partial W}{\partial z}, \qquad (4.116b)$$

$$R_{yz} = -\nu_{\mathrm{T}} \left(\frac{\partial W}{\partial y} + \frac{\partial V}{\partial z} \right), \qquad (4.116c)$$

the first two of which give

$$R_{yy} - R_{zz} = -2\nu_{\mathrm{T}} \left(\frac{\partial V}{\partial y} - \frac{\partial W}{\partial z} \right). \qquad (4.117)$$

A turbulent state with no secondary flows is given by

$$V = W = \Omega_x = 0. \qquad (4.118)$$

Equation (4.118) satisfies equation (4.115) under equation (4.117). That is, secondary flows cannot exist under the turbulent-viscosity representation (4.63). This conclusion is natural since the turbulent viscosity expresses the enhancement of diffusion effects due to velocity fluctuation and contribute to the destruction of distinct mean-flow structures such as secondary flows.

Under equation (4.17), the counterpart of equation (4.117) is given by

$$R_{yy} - R_{zz} = -2\nu_{\mathrm{T}} \left(\frac{\partial V}{\partial y} - \frac{\partial W}{\partial z} \right) + \left(\eta_1^{(2)} - 2\eta_2^{(2)}\right) \left[\left(\frac{\partial U}{\partial y} \right)^2 - \left(\frac{\partial U}{\partial z} \right)^2 \right] + 4\eta_1^{(2)} \left[\left(\frac{\partial V}{\partial y} \right)^2 - \left(\frac{\partial W}{\partial z} \right)^2 \right] + 4\eta_2^{(2)} \left[\left(\frac{\partial V}{\partial z} \right)^2 - \left(\frac{\partial W}{\partial y} \right)^2 \right]. \qquad (4.119)$$

The big difference between equations (4.117) and (4.119) is the explicit dependence of the latter on the axial velocity U through the second term. In order to see the role of the term, we consider the situation with very weak secondary flows and retain only the second term on the right-hand side of equation (4.119). We substitute it into equation (4.115) and may find Ω_x through the balance with the third or viscous term $\nu \nabla^2 \Omega_x$. There the anisotropy of turbulent intensity is the cause of the secondary-flow generation. The occurrence of Ω_x in a general situation may be confirmed by the numerical simulation of the turbulence model with equation (4.119) embedded [4.9, 4.18]. The magnitude of V and W is only a few per cent of U (axial velocity), but the existence of the former changes the U profile drastically, compared with the state with no secondary flows that is obtained from the turbulent-viscosity representation (4.63). For instance, U is accelerated near the corners of the square duct.

4.3 Second-Order Modelling

In second-order modelling based on equation (3.33), modelling the production term P_{Rij} is not necessary, and the remaining three terms, that is, Π_{ij} (pressure–strain term), ε_{ij} (dissipation term), and T_{Rijl} (transport term) are its targets [4.3–4.7].

4.3.1 Modelling of Pressure–Strain Term

As was discussed in §3.3.2.3 on channel turbulence, Π_{ij} plays a critical role of distributing energy among three intensities. We first consider the situation that the initial three turbulent intensities are given by

$$\langle u'^2 \rangle > \langle v'^2 \rangle = \langle w'^2 \rangle, \tag{4.120}$$

in an infinite region with no mean flow. Under the assumption of homogeneity, equation (4.11) is reduced to

$$\frac{\partial B_{ij}}{\partial t} = \Pi_{ij} - [\varepsilon_{ij}]_{\mathrm{D}}. \tag{4.121}$$

From equation (4.120), $\langle v'^2 \rangle$ and $\langle w'^2 \rangle$ drain energy from $\langle u'^2 \rangle$, and the isotropic state is generated eventually. For a positive nondimensional constant $C_{\Pi 1}$ we may express this process by adopting

$$\Pi_{ij} = -C_{\Pi 1} \frac{\varepsilon}{K} B_{ij} = -C_{\Pi 1} \varepsilon b_{ij}, \tag{4.122}$$

where b_{ij} denotes the nondimensionalized anisotropic tensor

$$b_{ij} = \frac{B_{ij}}{K}. \tag{4.123}$$

By substituting equation (4.122) into equation (4.121), we may confirm that the relaxation time of anisotropy has been chosen to be proportional to the time scale of energy-containing eddies, equation (3.144).

In general inhomogeneous turbulent flows, equation (4.42a) shows that Π_{ij} is explicitly associated with $\boldsymbol{U}$. In the current second-order modelling [4.4–4.6, 4.19, 4.20], Π_{ij} is assumed to be a functional of the mean strain and vorticity tensors, S_{ij} and Ω_{ij}, in addition to b_{ij}, K, and ε; namely, we write

$$\Pi_{ij} = \Pi_{ij}\{K, \varepsilon, b_{ij}, S_{ij}, \Omega_{ij}\}. \tag{4.124}$$

Specifically, Π_{ij} is assumed to be first-order in S_{ij} and Ω_{ij}. This assumption is due to the linear dependence of equation (4.42a) on them, but their nonlinear dependence and the dependence on higher-order derivatives of $\boldsymbol{U}$ cannot be ruled out [the latter may be understood from the expansion of $\boldsymbol{U}(\boldsymbol{x}')$ around $\boldsymbol{x}$ in equation (4.42a)].

Under the constraint of the linear dependence on S_{ij} and Ω_{ij}, the simplest model for Π_{ij} is

$$\Pi_{ij} = -C_{\Pi 1}\varepsilon b_{ij} + C_{\Pi 2}KS_{ij} + C_{\Pi 3}K[b_{il}S_{lj} + b_{jl}S_{li}]_{\mathrm{D}} + C_{\Pi 4}K(b_{il}\Omega_{lj} + b_{jl}\Omega_{li}). \tag{4.125}$$

An instance of the choice of model constants $C_{\Pi n}$ $(n = 1\text{–}4)$ is

$$C_{\Pi 1} = 1.8, \qquad C_{\Pi 2} = 0.4, \qquad C_{\Pi 3} = 0.3, \qquad C_{\Pi 4} = 0.1. \tag{4.126}$$

The second-order terms in b_{ij} may be included in equation (4.125). On the other hand, the third-order terms are reduced to the lower-order terms with the aid of the Cayley–Hamilton formula (4.22). This situation is entirely similar to explicit algebraic modelling of R_{ij} in §4.2.1.

4.3.2 Modelling of Dissipation and Transport Terms

After equation (4.124) for Π_{ij}, we write

$$[\varepsilon_{ij}]_{\mathrm{D}} \equiv \varepsilon_{ij} - \frac{1}{3}\varepsilon\delta_{ij} = [\varepsilon_{ij}]_{\mathrm{D}}\{b_{ij}, S_{ij}, \Omega_{ij}, K, \varepsilon\}. \tag{4.127}$$

Its simplest expression is

$$[\varepsilon_{ij}]_{\mathrm{D}} = C_{\mathrm{D}}\varepsilon b_{ij}, \tag{4.128}$$

with C_{D} as a constant.

We consider the lower wall of a channel, as in §4.2.5.2. Near a solid wall, we expand $\boldsymbol{u}'(u', v', w')$ as

$$u' = a_1 Y + a_2 Y^2 + \cdots, \tag{4.129a}$$

$$v' = b_2 Y^2 + \cdots, \tag{4.129b}$$

$$w' = c_1 Y + c_2 Y^2 + \cdots. \tag{4.129c}$$

Then we have

$$\varepsilon_{xy} = 4\nu\langle a_1 b_2\rangle Y + \cdots, \tag{4.130}$$

$$C_{\mathrm{D}}\varepsilon b_{xy} = 2C_{\mathrm{D}}\nu\langle a_1 b_2\rangle Y + \cdots. \tag{4.131}$$

When equation (4.128) is assumed to be applicable to the wall, equations (4.130) and (4.131) give

$$C_{\mathrm{D}} = 2. \tag{4.132}$$

The transport term T_{Rijl} remains to be modelled [equation (3.37)], and is related to the energy-equation counterpart $\boldsymbol{T}_K$ [equation (3.42)] as

$$T_{Rjji} = 2T_{Ki}. \tag{4.133}$$

As was done for T_K, we retain only the first term as

$$T_{Rijl} \cong -\langle u_i' u_j' u_l' \rangle. \tag{4.134}$$

We apply the gradient-diffusion approximation, equation (4.47), to equation (4.134). The product $u_i' u_j' u_k'$ may be regarded as the transport of $u_i' u_j'$ by u_k'. We consider $u_i' u_j'$ is transported from the large- to the small-$\langle u_i' u_j' \rangle$ region, and take the symmetry of suffices into account. As a result, we have

$$T_{Rijl} = \frac{\nu_T}{\sigma_R} \left(\frac{\partial R_{ij}}{\partial x_l} + \frac{\partial R_{jl}}{\partial x_i} + \frac{\partial R_{li}}{\partial x_j} \right), \tag{4.135}$$

where σ_R is a nondimensional coefficient. More complicated models for T_{Rijl} have been proposed through the change of the scalar diffusion coefficient ν_T / σ_R to a tensorial form [4.19].

4.3.3 The Simplest Second-Order Model and its Relationship with a Higher-Order Algebraic Model

In the current study of second-order modelling, attention has been focused on the modelling of Π_{ij}. Equation (4.125) is a starting point towards more elaborate second-order modelling. We summarize the simplest model for capturing the essence of second-order modelling. The mean-velocity equation is equation (4.62), and the anisotropic part of R_{ij}, B_{ij} [equation (4.5)], obeys equation (4.11). We substitute equations (4.12), (4.125) and (4.128) into the latter, and have

$$\begin{aligned} \frac{\mathrm{D}B_{ij}}{\mathrm{D}t} &= -(C_{\Pi 1} + C_{\mathrm{D}}) \frac{\varepsilon}{K} B_{ij} + (C_{\Pi 2} - \tfrac{2}{3}) K S_{ij} \\ &\quad + (C_{\Pi 3} - \tfrac{1}{2}) [B_{il} S_{lj} + B_{jl} S_{li}]_{\mathrm{D}} \\ &\quad + (C_{\Pi 4} - \tfrac{1}{2}) (B_{il} \Omega_{lj} + B_{jl} \Omega_{li}) + \left[\frac{\partial T_{Rijl}}{\partial x_l} \right]_{\mathrm{D}} + \nu \nabla^2 B_{ij}. \end{aligned} \tag{4.136}$$

Here the transport term T_{Rijl} is given by equation (4.135).

The turbulent-energy equation linked with the isotropic part of R_{ij} and the dissipation-rate equation are the same as for the K-ε model and are given by equations (4.66) and (4.67), respectively.

In second-order modelling, it is difficult to perform the analytical discussions such as for a channel flow in §4.2.5.2 and a square-duct flow in §4.2.5.3 since the transport equation for B_{ij} needs to be solved. Here we examine the relationship of second-order modelling with algebraic modelling. We pay attention to the first term on the right-hand side of equation (4.136), and write

$$B_{ij} = -\frac{1}{C_{\Pi 1}+C_{\mathrm{D}}}\left(\tfrac{2}{3}-C_{\Pi 2}\right)\frac{K^2}{\varepsilon}S_{ij} - \frac{1}{C_{\Pi 1}+C_{\mathrm{D}}}\left(\tfrac{1}{2}-C_{\Pi 3}\right)\frac{K}{\varepsilon}[B_{il}S_{lj}+B_{jl}S_{li}]_{\mathrm{D}}$$
$$-\frac{1}{C_{\Pi 1}+C_{\mathrm{D}}}\left(\tfrac{1}{2}-C_{\Pi 4}\right)\frac{K}{\varepsilon}(B_{il}\Omega_{lj}+B_{jl}\Omega_{li}) - \frac{1}{C_{\Pi 1}+C_{\mathrm{D}}}\frac{K}{\varepsilon}\frac{\mathrm{D}B_{ij}}{\mathrm{D}t}$$
$$+\frac{1}{C_{\Pi 1}+C_{\mathrm{D}}}\frac{K}{\varepsilon}\left[\frac{\partial T_{Rijl}}{\partial x_l}\right]_{\mathrm{D}} + \frac{1}{C_{\Pi 1}+C_{\mathrm{D}}}\frac{K}{\varepsilon}\nu\nabla^2 B_{ij}. \tag{4.137}$$

The expression consisting of the first three terms of equation (4.137) is of the same form as equation (4.16), that is, the basis of algebraic modelling. Specifically, the first term is the turbulent-viscosity representation with

$$\nu_{\mathrm{T}} = \frac{1}{C_{\Pi 1}+C_{\mathrm{D}}}\left(\tfrac{2}{3}-C_{\Pi 2}\right)\frac{K^2}{\varepsilon}. \tag{4.138}$$

The application of the iteration procedure to equation (4.137) generates the terms nonlinear in S_{ij} and Ω_{ij} in equation (4.17), but new effects occur from the fourth term. One of them is the advection effect on ν_{T}, and its inclusion results in the replacement of equation (4.138) with

$$\nu_{\mathrm{T}} = \frac{1}{C_{\Pi 1}+C_{\mathrm{D}}}\left(\tfrac{2}{3}-C_{\Pi 2}\right)\frac{K^2}{\varepsilon}\left(1-\frac{1}{C_{\Pi 1}+C_{\mathrm{D}}}\frac{1}{K}\frac{\mathrm{D}}{\mathrm{D}t}\frac{K^2}{\varepsilon}\right). \tag{4.139}$$

This expression will be shown to be useful in §5.3 in relation to modelling Mach-number effects on turbulence.

From the foregoing relationship between algebraic and second-order models, the latter may be viewed as an approximate analytic solution of the former. On the contrary, a second-order model may be constructed with the aid of an algebraic model. For instance, we apply the renormalization [4.21]

$$S_{ij} \to \frac{B_{ij}}{\nu_{\mathrm{T}}} \tag{4.140}$$

to an algebraic model with its coefficients functionalized using expression (4.26). The resulting model corresponds to the partial functionalization of coefficients in second-order modelling. In the current second-order modelling, we have no clear mathematical recipe for the functionalization of coefficients. The foregoing relationship between these two types of modelling indicates that they are supplementary to each other in the construction of turbulence models.

4.4 A Variational-Method Model

As an instance of the formation of distinct mean-velocity structures in turbulent flow, we referred to a swirling flow in a circular pipe in §2.1.2

[2.3]. In the flow, swirl motion is imposed on the axial flow at the entrance of a pipe. Far downstream, this flow relaxes to a usual pipe flow with its axial component as the sole mean velocity. In the intermediate region subject to swirl effect, a characteristic feature of the flow is the retardation of mean axial velocity near the center-line. With the increase in the entrance swirl intensity, the velocity near the axis is further retarded, eventually resulting in the reversal of flow direction.

The occurrence of such a retarded or reversed mean-flow profile leads to large mean velocity gradients, compared with the nonswirling case. This fact suggests the existence of a mechanism under which such a profile may survive against the turbulent mixing effect represented by the turbulent viscosity. The study of a swirling pipe flow is an interesting theme from the viewpoint of elucidating a coexistence mechanism of a distinct mean-flow profile and turbulent fluctuations.

A swirling pipe flow is also a challenging subject in the study of turbulence modelling. The model based on the simple turbulent-viscosity representation for the Reynolds stress gives rise to the rapid loss of a distinct mean-flow profile possessing the retardation or reversal of axial velocity. Higher-order algebraic and second-order models are still insufficient for the analysis of this flow.

In what follows, we shall adopt an approach different from algebraic and second-order modelling, and examine a mechanism of keeping a distinct mean-velocity profile in a turbulent swirling pipe flow [4.22]. We first obtain a mean flow expressing retarded or reversed axial velocity by the variational method. Next, we seek an expression for the Reynolds stress that is consistent with it.

4.4.1 Helicity and Vortical-Structure Persistence

From equation (3.16), the vorticity $\boldsymbol{\omega}$ obeys

$$\frac{\partial \boldsymbol{\omega}}{\partial t} = \nabla \times (\boldsymbol{u} \times \boldsymbol{\omega}) + \nu \nabla^2 \boldsymbol{\omega}. \tag{4.141}$$

In relation to the first term on the right-hand side, we have the relationship

$$\frac{|\boldsymbol{u} \times \boldsymbol{\omega}|^2}{|\boldsymbol{u}|^2 |\boldsymbol{\omega}|^2} + \frac{|\boldsymbol{u} \cdot \boldsymbol{\omega}|^2}{|\boldsymbol{u}|^2 |\boldsymbol{\omega}|^2} = 1. \tag{4.142}$$

Equation (4.142) indicates that larger $|\boldsymbol{u} \cdot \boldsymbol{\omega}|$ tends to give smaller $|\boldsymbol{u} \times \boldsymbol{\omega}|$. The decrease in $\nabla \times (\boldsymbol{u} \times \boldsymbol{\omega})$ signifies weakening of the energy cascade. This fact suggests that the helicity $\boldsymbol{u} \cdot \boldsymbol{\omega}$ is a candidate for a parameter controlling the sustainment of a distinct mean-flow structure in turbulence.

The integration of equation (3.82) in a whole fluid region V results in

$$\frac{\partial}{\partial t}\int_V \boldsymbol{u}\cdot\boldsymbol{\omega}\,\mathrm{d}V = -\int_V 2\nu\frac{\partial u_j}{\partial x_i}\frac{\partial \omega_j}{\partial x_i}\,\mathrm{d}V$$
$$+\int_S \boldsymbol{n}\cdot\left[-(\boldsymbol{u}\cdot\boldsymbol{\omega})\boldsymbol{u}+\left(-p+\frac{\boldsymbol{u}^2}{2}\right)\boldsymbol{\omega}+\nu\nabla(\boldsymbol{u}\cdot\boldsymbol{\omega})\right]\mathrm{d}S, \quad (4.143)$$

from the Gauss integral theorem. The vorticity arising from the swirl motion is in the axial direction, which is combined with the axial velocity to generate helicity. The imposition of swirling at the entrance is equivalent to the injection of helicity into fluid motion. The injection is made through the $\boldsymbol{u}\cdot\boldsymbol{\omega}$-related part in the second term on the right-hand side of equation (4.143).

Here we should stress that $\boldsymbol{u}\cdot\boldsymbol{\omega}$ is not Galilean-invariant; namely, its magnitude is dependent on the choice of a coordinate system. Throughout the remainder of §4.4 we shall choose the frame fixed at a pipe. A swirl motion imposed at the entrance of a pipe is a source of the helicity of mean flow. In this frame, the flow far downstream is reduced to a usual state with no mean-flow helicity.

4.4.2 Derivation of the Vorticity Equation Using the Variational Method

From the viewpoint of turbulence modelling, the following analysis pays special attention to the mean velocity of a swirling pipe flow whose axial component is restarted or reversed according to the strength of swirl. Such an axial velocity is characterized by a large mean velocity gradient. As an indicator of the strength of spatial variation of mean velocity in a region V, we introduce

$$\Phi = \int_V \left(\frac{\partial U_j}{\partial x_i}+\frac{\partial U_i}{\partial x_j}\right)^2 \mathrm{d}V. \quad (4.144)$$

A mean axial vorticity arising from a mean swirl velocity is combined with a mean axial velocity and results in nonvanishing mean-flow helicity. Then, as an important quantity distinguishing between swirling and nonswirling flows, we choose a total amount of mean-flow helicity, which is defined by

$$\Psi = \int_V \boldsymbol{U}\cdot\boldsymbol{\Omega}\,\mathrm{d}V. \quad (4.145)$$

We use Ψ as a constraint on Φ and seek the state

$$\Phi = \text{minimum} \quad \text{under} \quad \Psi = \text{constant}, \quad (4.146)$$

by the standard variational method. In mathematical terms, equation (4.146) is given by

$$\delta(\Phi + \lambda\Psi) = 0, \quad (4.147)$$

where λ is a Lagrange multiplier, and $\boldsymbol{U}$ is fixed at the boundary S surrounding a flow region V, as

$$\delta \boldsymbol{U} = 0 \quad \text{at } S. \tag{4.148}$$

From equations (4.144), (4.145), and (4.147), we have

$$\int_V (\nabla \times \boldsymbol{\Omega} + \lambda \boldsymbol{\Omega}) \cdot \delta \boldsymbol{U} \, \mathrm{d}V = 0, \tag{4.149}$$

under the constraint (4.148), which gives

$$\boldsymbol{\Omega} = -\frac{1}{\lambda} \nabla \times \boldsymbol{\Omega} \tag{4.150}$$

for arbitrary $\delta \boldsymbol{U}$. In equation (4.150), $\boldsymbol{\Omega}$ is an axial vector (a vector whose sign does not change under the reflection of coordinates), whereas $\nabla \times \boldsymbol{\Omega}$ is a polar vector. Then λ needs to be a pseudoscalar (a scalar whose sign changes under the reflection). Equation (4.150) with $\boldsymbol{\Omega}$ replaced with a magnetic field vector $\boldsymbol{B}$ corresponds to the so-called force-free state in plasma phenomena [4.23], which will be discussed in chapter 6.

We integrate equation (4.150), and have

$$\boldsymbol{U} = -\frac{1}{\lambda} \boldsymbol{\Omega} + \nabla \phi, \tag{4.151}$$

where ϕ is a harmonic function; namely, it obeys $\nabla^2 \phi = 0$ from the solenoidal condition $\nabla \cdot \boldsymbol{U} = 0$. Here we should stress the following two points. First, the essence of equation (4.150) manifests itself in equation (4.151). The first term on the right-hand side signifies that the mean velocity and vorticity are aligned with each other, leading to nonvanishing mean-flow helicity, as is consistent with the requirement for nonvanishing Ψ in equation (4.145). Second, the mean velocity may be expressed in a simple form, with the aid of the mean vorticity. From the curl of equation (4.150), $\boldsymbol{\Omega}$ obeys

$$(\nabla^2 + \lambda^2) \boldsymbol{\Omega} = 0. \tag{4.152}$$

4.4.3 Analysis of Swirling Pipe Flow

4.4.3.1 Cylindrical-coordinate solution

We adopt the cylindrical coordinates (r, θ, z), and z is along the central axis of the cylinder. The velocity $\boldsymbol{U}$ is written as

$$\boldsymbol{U} = [0, U_\theta(r), U_z(r)]. \tag{4.153}$$

In the case of swirling pipe flow, the variation of $\boldsymbol{U}$ along the z axis is important in relation to the relaxation of a retarded or reversed axial flow to a usual pipe flow. In the present variational approach, this relaxation is treated implicitly through the change of the parameter λ.

The mean vorticity $\boldsymbol{\Omega}$, whose component is given by

$$\boldsymbol{\Omega} = [0, \Omega_\theta(r), \Omega_z(r)], \tag{4.154}$$

is written in terms of $\boldsymbol{U}$ as

$$\Omega_\theta = -\frac{\mathrm{d}U_z}{\mathrm{d}r}, \tag{4.155a}$$

$$\Omega_z = \frac{1}{r}\frac{\mathrm{d}}{\mathrm{d}r}(rU_\theta). \tag{4.155b}$$

From equation (4.151), $\boldsymbol{U}$ is related to $\boldsymbol{\Omega}$ as

$$U_\theta = -\frac{1}{\lambda}\Omega_\theta + \frac{\partial\phi}{r\partial\theta}, \tag{4.156a}$$

$$U_z = -\frac{1}{\lambda}\Omega_z + \frac{\partial\phi}{\partial z}. \tag{4.156b}$$

Moreover, we have the relation

$$\Omega_\theta = \frac{1}{\lambda}\frac{\mathrm{d}\Omega_z}{\mathrm{d}r}, \tag{4.157}$$

from equation (4.150).

The z component of equation (4.152) is

$$\frac{\mathrm{d}^2\Omega_z}{\mathrm{d}r^2} + \frac{1}{r}\frac{\mathrm{d}\Omega_z}{\mathrm{d}r} + \lambda^2\Omega_z = 0. \tag{4.158}$$

The solution of equation (4.158), which is regular at the center-line ($r = 0$), is given by

$$\Omega_z = \Omega_z^{(\mathrm{C})} J_0(\lambda r), \tag{4.159}$$

where J_n is the first-kind Bessel function of the nth order, and

$$\Omega_z^{(\mathrm{C})} = \Omega_z(0). \tag{4.160}$$

From equations (4.157) and (4.159), we have

$$\Omega_\theta = -\Omega_z^{(\mathrm{C})} J_1(\lambda r). \tag{4.161}$$

Through the combination of equations (4.156a,b) with equations (4.159) and (4.160), $\boldsymbol{U}$ has been expressed in terms of the Bessel function J_n.

4.4.3.2 Retarded or reversed center-line velocity

A retarded axial flow U_z in a pipe is depicted schematically in figure 4.2 (left; a is the radius of the pipe). From equation (4.155a), the maximum-velocity point of U_z, r_M, corresponds to the zero point of Ω_θ, as is schematically shown in figure 4.2 (right); namely, we have

$$\Omega_\theta(r_\mathrm{M}) = 0. \tag{4.162}$$

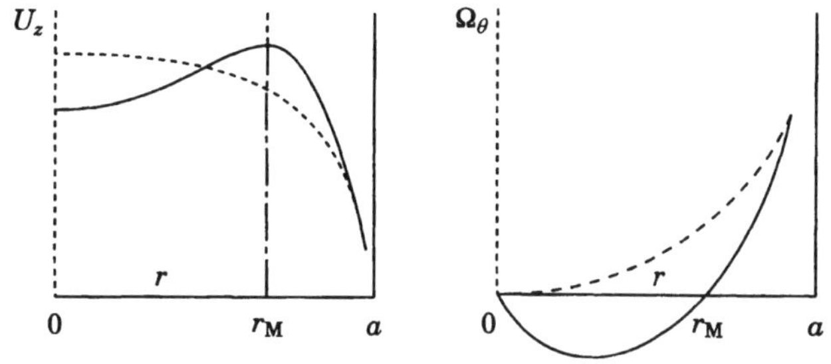

Figure 4.2. Swirling pipe flow. Axial velocity (left); tangential vorticity (right).

The first zero of $J_1(\lambda r_M)$ gives

$$|\lambda| r_M = 3.8. \tag{4.163}$$

Here we should note that λ may take both signs. We first choose positive λ and refer to negative λ later.

From $r_M \leq a$, we have

$$\lambda \geq \frac{3.8}{a}. \tag{4.164}$$

With λ increasing from $3.8/a$, r_M approaches the central axis, and the negative region of Ω_θ between the central axis and r_M shrinks. As a result, equation (4.156b) is reduced to the usual axial velocity in a nonswirling pipe flow; namely, we have

$$U_z \to U_\infty \quad \text{as } \lambda \to \infty. \tag{4.165}$$

Here U_∞ denotes the axial velocity of a pipe flow in the limit of a large Reynolds number and is uniform across the cross-section, except the near-wall region. Equation (4.165) corresponds to the choice

$$\phi = U_\infty z + \phi_0, \tag{4.166}$$

with a constant ϕ_0, resulting in

$$U_z = U_\infty - \frac{1}{\lambda}\Omega_z, \tag{4.167a}$$

$$U_\theta = -\frac{1}{\lambda}\Omega_\theta, \tag{4.167b}$$

for finite λ. Here equation (4.167b) fulfills the symmetry condition on the center-line, namely, $U_\theta(0) = 0$.

On the center-line ($r = 0$), equation (4.167a) is given by

$$U_C^{(S)} \equiv U_z(0) = U_\infty - \frac{\Omega_z^{(C)}}{\lambda}, \tag{4.168}$$

which is written as

$$\frac{U_{\mathrm{C}}^{(\mathrm{S})}}{U_\infty} = 1 - \frac{\Omega_z^{(\mathrm{C})}}{\lambda U_\infty}, \tag{4.169}$$

in nondimensional form. We use equation (4.163) for positive λ and express equation (4.169) as

$$\frac{U_{\mathrm{C}}^{(\mathrm{S})}}{U_\infty} = 1 - 0.26 \frac{r_{\mathrm{M}}\Omega_z^{(\mathrm{C})}}{U_\infty}, \tag{4.170}$$

in terms of measurable quantities.

The second term on the right-hand side of equation (4.170) represents the degree of retardation or reversal of the mean axial velocity on the center-line. It is proportional to the mean vorticity on the axis, $\Omega_z^{(\mathrm{C})}$. The degree of retardation or reversal is also proportional to r_{M} (the point of maximum mean axial velocity). The observation indicates that increasing $\Omega_z^{(\mathrm{C})}$ leads to larger r_{M}, resulting in the larger retardation of the center-line velocity or its reversal [4.3]. This tendency is consistent with equation (4.170). As a result, we may conclude that the retardation or reversal of the mean center-line velocity is tightly linked with positive $\Omega_z^{(\mathrm{C})}$; namely, $\Omega_z^{(\mathrm{C})}$ is a primary controlling parameter in a swirling pipe flow.

In the above discussion, we chose positive λ. In the case of negative λ, Ω_θ becomes positive for $r < r_{\mathrm{M}}$ from equation (4.161) with positive $\Omega_z^{(\mathrm{C})}$. From equation (4.155a), such Ω_θ arises from a retarded axial flow pointing to the negative z direction, while keeping the same tangential velocity U_θ that is written as

$$U_\theta = \frac{\Omega_z^{(\mathrm{C})}}{\lambda} J_1(\lambda r) \tag{4.171}$$

[note that U_θ is unchanged under the change of the sign of λ since $J_1(\lambda r)/\lambda$ is an even function of λ].

4.4.4 Swirl Effect on Reynolds Stress

From equation (4.141), the mean-vorticity equation is given by

$$\frac{\partial \boldsymbol{\Omega}}{\partial t} = \nabla \times (\boldsymbol{U} \times \boldsymbol{\Omega} + \boldsymbol{V}_{\mathrm{M}}) + \nu \nabla^2 \boldsymbol{\Omega}, \tag{4.172}$$

where $\boldsymbol{V}_{\mathrm{M}}$ is named the vortex-motive force and is defined as

$$\boldsymbol{V}_{\mathrm{M}} = \langle \boldsymbol{u}' \times \boldsymbol{\omega}' \rangle. \tag{4.173}$$

We use equation (3.88), and have the relationship between $\boldsymbol{V}_{\mathrm{M}}$ and R_{ij} as

$$V_{\mathrm{M}i} = -\frac{\partial R_{ij}}{\partial x_j} + \frac{\partial K}{\partial x_i}. \tag{4.174}$$

In the light of equation (4.172), a stationary state at a high Reynolds number may arise from the condition

$$\boldsymbol{U} \times \boldsymbol{\Omega} + \boldsymbol{V}_{\mathrm{M}} = 0. \tag{4.175}$$

From equations (4.151) and (4.175), we have

$$\boldsymbol{V}_{\mathrm{M}} = -\nabla\phi \times \boldsymbol{\Omega}. \tag{4.176a}$$

We substitute equation (4.150) into equation (4.176a) and write

$$\boldsymbol{V}_{\mathrm{M}} = \nabla\chi \times (\nabla \times \boldsymbol{\Omega}) = -\nabla\chi \times \nabla^2 \boldsymbol{U}, \tag{4.176b}$$

where a pseudoscalar χ is defined by

$$\chi = \frac{\phi}{\lambda}. \tag{4.177}$$

The i component of equation (4.176b) is rewritten in the form

$$\begin{aligned} V_{\mathrm{M}i} &= -\varepsilon_{ilm} \frac{\partial \chi}{\partial x_l} \frac{\partial^2 U_m}{\partial x_j^2} \\ &= \frac{\partial}{\partial x_j}\left(-\varepsilon_{ilm} \frac{\partial \chi}{\partial x_l} \frac{\partial U_m}{\partial x_j}\right) + \varepsilon_{ilm} \frac{\partial^2 \chi}{\partial x_j \partial x_l} \frac{\partial U_m}{\partial x_j}. \end{aligned} \tag{4.178}$$

We pay attention to the first term in the second relation of equation (4.178). Noting equation (4.174), we may see that the contribution of the first term to R_{ij} is modelled as

$$\varepsilon_{ilm} \frac{\partial U_m}{\partial x_j} \frac{\partial \chi}{\partial x_l} + \varepsilon_{jlm} \frac{\partial U_m}{\partial x_i} \frac{\partial \chi}{\partial x_l}. \tag{4.179}$$

From the viewpoint of turbulence modelling, it is sufficient to model the pseudoscalar χ, not each of ϕ and λ. A pseudoscalar characterizing a helical property of turbulence properties is H_{K} [equation (3.83)]. We use it and write

$$\chi = C_V \frac{K^4}{\varepsilon^3} H_{\mathrm{K}}, \tag{4.180}$$

from dimensional analysis, where C_V is a constant. Equations (4.177) and (4.180) signify that λ is proportional to the inverse of H_{K}. This modelling is consistent with the fact that large $|\lambda|$ corresponds to a weak-helicity case.

Equation (4.179) is rewritten as

$$\frac{1}{2}\left(\varepsilon_{ilm} S_{jm} \frac{\partial \chi}{\partial x_l} + \varepsilon_{jlm} S_{im} \frac{\partial \chi}{\partial x_l}\right) + \frac{1}{2}\left(\Omega_i \frac{\partial \chi}{\partial x_j} + \Omega_j \frac{\partial \chi}{\partial x_i}\right). \tag{4.181}$$

We combine equation (4.181) with the turbulent-viscosity representation, and have

$$R_{ij} = \frac{2}{3}K\delta_{ij} - C_\nu \frac{K^2}{\varepsilon} S_{ij} + \left[\Omega_i \frac{\partial \chi}{\partial x_j} + \Omega_j \frac{\partial \chi}{\partial x_i}\right]_{\mathrm{D}}$$

$$+ C_S \left[\varepsilon_{ilm} S_{jm} \frac{\partial \chi}{\partial x_l} + \varepsilon_{jlm} S_{im} \frac{\partial \chi}{\partial x_l}\right]_{\mathrm{D}} \tag{4.182}$$

(the numerical factor $\frac{1}{2}$ is not essential and may be absorbed into χ). Here the helicity-related part of equation (4.182) with $C_S = 1$ is reduced to equation (4.181).

In equation (4.182), the $\mathbf{\Omega}$-related term was first derived using the statistical theory explained in chapter 5. The H_{K} effect combined with $\mathbf{\Omega}$ was confirmed to contribute to the retardation of a mean axial flow. The H_{K} effect dependent on S_{ij} is noteworthy since the S_{ij} effect in the turbulent-viscosity term contributes only to the destruction of distinct mean-flow profiles such as a retarded profile.

4.5 Subgrid-Scale Modelling

A representative method making full use of advanced computer capability in analyzing turbulent flow is large eddy simulation (LES) with subgrid-scale (SGS) models embedded [2.16, 4.24, 4.25]. In this method, the components whose spatial scales are smaller than grid intervals are eliminated with the aid of a filtering procedure, and effects of filtered-out components are supplemented by a model. This mathematical procedure is called SGS modelling.

4.5.1 Filtering Procedure

We apply a filter G_{F} to a function $f(x)$ as

$$\overline{f}^{\mathrm{F}}(x) = \int_{-\infty}^{\infty} G_{\mathrm{F}}(\xi - x, \Delta_{\mathrm{F}}) f(\xi)\, \mathrm{d}\xi, \tag{4.183}$$

where G_{F} obeys

$$\int_{-\infty}^{\infty} G_{\mathrm{F}}(\xi - x, \Delta_{\mathrm{F}})\, \mathrm{d}\xi = 1, \tag{4.184}$$

and Δ_{F} is the filter width that is related to a computational grid in LES. We denote the deviation of f around $\overline{f}^{\mathrm{F}}$ by f'', that is,

$$f'' = f - \overline{f}^{\mathrm{F}}. \tag{4.185}$$

In LES, $\overline{f}^{\mathrm{F}}$ and f'' are called the grid-scale (GS) and SGS components of f, respectively.

The filtering of $\mathrm{d}f/\mathrm{d}x$ leads to

$$\overline{\frac{\mathrm{d}f(x)}{\mathrm{d}x}}^{\mathrm{F}} = [G_{\mathrm{F}}(\xi - x, \Delta_{\mathrm{F}})f(\xi)]_{-\infty}^{\infty} - \int_{-\infty}^{\infty} \frac{\mathrm{d}G_{\mathrm{F}}(\xi - x, \Delta_{\mathrm{F}})}{\mathrm{d}\xi} f(\xi)\,\mathrm{d}\xi. \tag{4.186}$$

In the case that G_{F} is subject to

$$G_{\mathrm{F}}(\xi - x, \Delta_{\mathrm{F}})f(\xi) \to 0 \qquad (|\xi| \to \infty), \tag{4.187a}$$

$$\frac{\mathrm{d}G_{\mathrm{F}}(\xi - x, \Delta_{\mathrm{F}})}{\mathrm{d}\xi} = -\frac{\mathrm{d}G_{\mathrm{F}}(\xi - x, \Delta_{\mathrm{F}})}{\mathrm{d}x}, \tag{4.187b}$$

we have

$$\begin{aligned}\overline{\frac{\mathrm{d}f}{\mathrm{d}x}}^{\mathrm{F}} &= \int_{-\infty}^{\infty} \frac{\mathrm{d}G_{\mathrm{F}}(\xi - x, \Delta_{\mathrm{F}})}{\mathrm{d}x} f(\xi)\,\mathrm{d}\xi \\ &= \frac{\mathrm{d}}{\mathrm{d}x}\int_{-\infty}^{\infty} G_{\mathrm{F}}(\xi - x, \Delta_{\mathrm{F}})f(\xi)\,\mathrm{d}\xi = \frac{\mathrm{d}\overline{f}^{\mathrm{F}}}{\mathrm{d}x},\end{aligned} \tag{4.188}$$

namely, filtering commutes differentiation.

Filters adopted usually in LES are the top-hat filter

$$G_{\mathrm{T}}(x) = \frac{1}{\Delta_{\mathrm{F}}} \quad \left(|x| < \frac{\Delta_{\mathrm{F}}}{2}\right), \qquad 0 \quad \left(|x| > \frac{\Delta_{\mathrm{F}}}{2}\right), \tag{4.189}$$

and the Gaussian filter

$$G_{\mathrm{G}}(x) = \sqrt{\frac{C_{\mathrm{G}}}{\pi}} \exp(-C_{\mathrm{G}}x^2), \tag{4.190}$$

with C_{G} as a constant related to the filter width (subscripts T and G denote top-hat and Gaussian, respectively). Equation (4.189) obeys equation (4.187a), and equation (4.190) is consistent with it as long as $f(x)$ increases at highest algebraically with x.

For Δ_{F} independent of location, both equations (4.189) and (4.190) obey equation (4.187b). In LES, however, smaller Δ_{F} is adopted near a solid boundary. Then equation (4.187b) is not satisfied strictly, but the commutation between filtering and differentiation is usually assumed. For $f(\boldsymbol{x})$ in three-dimensional space, we introduce a filter

$$\overline{f}^{\mathrm{F}}(\boldsymbol{x}) = \int_{-\infty}^{\infty} G_{\mathrm{F}}(\boldsymbol{\xi} - \boldsymbol{x}, \Delta_x, \Delta_y, \Delta_z)f(\boldsymbol{\xi})\,\mathrm{d}\boldsymbol{\xi}, \tag{4.191}$$

with three filter widths Δ_x, Δ_y, and Δ_z.

In order to see filtering effects on f, we apply the top-hat filter G_{T}, and have

$$\begin{aligned}\overline{f}^{\mathrm{T}} &= \int_{x-(\Delta_{\mathrm{F}}/2)}^{x+(\Delta_{\mathrm{F}}/2)} G_{\mathrm{T}}(\xi - x)f(\xi)\,\mathrm{d}\xi = \int_{-\Delta_{\mathrm{F}}/2}^{\Delta_{\mathrm{F}}/2} G_{\mathrm{T}}(\eta)f(x+\eta)\,\mathrm{d}\eta \\ &= f(x) + \frac{\Delta_{\mathrm{F}}^2}{24}\frac{\partial^2 f(x)}{\partial x^2} + O(\Delta_{\mathrm{F}}^4),\end{aligned} \tag{4.192}$$

by the Taylor expansion. Entirely similarly, the Gaussian filter G_G gives

$$\bar{f}^G = \int_{-\infty}^{\infty} G_G(\xi - x) f(\xi)\, d\xi$$
$$= \sum_{n=0}^{\infty} \frac{1}{n!} \sqrt{\frac{C_G}{\pi}} \frac{\partial^n f(x)}{\partial x^n} \int_{-\infty}^{\infty} \eta^n \exp(-C_G \eta^2)\, d\eta. \qquad (4.193)$$

The retention of the terms up to $n = 2$ results in

$$\bar{f}^G = f(x) + \frac{1}{4C_G} \frac{\partial^2 f(x)}{\partial x^2}. \qquad (4.194)$$

From the comparison between equations (4.192) and (4.194), we see that the Gaussian filter with

$$C_G = \frac{6}{\Delta_F^2} \qquad (4.195)$$

is equivalent to the top-hat filter with the width Δ_F.

From equations (4.192) and (4.193), we have

$$\overline{\bar{f}^F}^F \neq \bar{f}^F, \qquad \overline{f''}^F \neq 0, \qquad \overline{\bar{f}^F f''}^F \neq 0 \qquad (4.196)$$

for both the top-hat and Gaussian filters, in contrast to the ensemble-averaging counterparts

$$\langle\langle f \rangle\rangle = \langle f \rangle, \qquad \langle f' \rangle = 0, \qquad \langle\langle f \rangle f' \rangle = 0. \qquad (4.197)$$

4.5.2 Filtered Equations

4.5.2.1 Grid-scale equations

We apply the three-dimensional filtering, equation (4.191), to equations (3.15) and (3.16), and have

$$\nabla \cdot \bar{\boldsymbol{u}}^F = 0, \qquad (4.198)$$

$$\frac{\partial \bar{u}_i^F}{\partial t} + \frac{\partial}{\partial x_j} \bar{u}_i^F \bar{u}_j^F = -\frac{\partial \bar{p}^F}{\partial x_i} + \frac{\partial}{\partial x_j}(-\tau_{ij}) + \nu \nabla^2 \bar{u}_i^F. \qquad (4.199)$$

Here τ_{ij} is defined as

$$\tau_{ij} = \overline{u_i u_j}^F - \bar{u}_i^F \bar{u}_j^F, \qquad (4.200)$$

and is called the SGS stress. Equation (4.200) is divided into three parts

$$\tau_{ij} = \tau_{ij}^{(I)} + \tau_{ij}^{(II)} + \tau_{ij}^{(III)}, \qquad (4.201)$$

where

$$\tau_{ij}^{(\mathrm{I})} = \overline{\bar{u}_i^{\mathrm{F}} \bar{u}_j^{\mathrm{F}}}^{\mathrm{F}} - \bar{u}_i^{\mathrm{F}} \bar{u}_j^{\mathrm{F}}, \tag{4.202}$$

$$\tau_{ij}^{(\mathrm{II})} = \overline{\bar{u}_i^{\mathrm{F}} u_j''}^{\mathrm{F}} + \overline{u_i'' \bar{u}_j^{\mathrm{F}}}^{\mathrm{F}}, \tag{4.203}$$

$$\tau_{ij}^{(\mathrm{III})} = \overline{u_i'' u_j''}^{\mathrm{F}}. \tag{4.204}$$

Equation (4.202) is composed of the GS velocity only, but (4.203) and (4.204) are dependent on the SGS velocity and need to be modelled in LES. In ensemble-mean modelling, we have no terms corresponding to $\tau_{ij}^{(\mathrm{I})}$ and $\tau_{ij}^{(\mathrm{II})}$. In the sense of small-scale effects on larger-scale motion, $\tau_{ij}^{(\mathrm{III})}$ is similar to the Reynolds stress R_{ij} [equation (3.27)] and is specifically called the SGS Reynolds-stress.

We refer to the relationship of the Galilean invariance with τ_{ij} [4.26]. The Navier–Stokes equation (3.16) is Galilean-invariant, that is, its mathematical form is unchanged under the translation

$$t \to t, \qquad \boldsymbol{x} \to \boldsymbol{x} - \boldsymbol{V}t, \qquad \boldsymbol{u} \to \boldsymbol{u} - \boldsymbol{V}. \tag{4.205}$$

In the light of filtering, the velocity is subject to the translation

$$\bar{\boldsymbol{u}}^{\mathrm{F}} \to \bar{\boldsymbol{u}}^{\mathrm{F}} - \boldsymbol{V}, \qquad \boldsymbol{u}'' \to \boldsymbol{u}''. \tag{4.206}$$

As a result, we have

$$\tau_{ij}^{(\mathrm{I})} \to \overline{\bar{u}_i^{\mathrm{F}} \bar{u}_j^{\mathrm{F}}}^{\mathrm{F}} - \bar{u}_i^{\mathrm{F}} \bar{u}_j^{\mathrm{F}} + V_i \left(\bar{u}_j^{\mathrm{F}} - \overline{\bar{u}_j^{\mathrm{F}}}^{\mathrm{F}} \right) + \left(\bar{u}_i^{\mathrm{F}} - \overline{\bar{u}_i^{\mathrm{F}}}^{\mathrm{F}} \right) V_j, \tag{4.207}$$

$$\tau_{ij}^{(\mathrm{II})} \to \overline{\bar{u}_i^{\mathrm{F}} u_j''}^{\mathrm{F}} + \overline{u_i'' \bar{u}_j^{\mathrm{F}}}^{\mathrm{F}} - V_i \overline{u_j''}^{\mathrm{F}} - \overline{u_i''}^{\mathrm{F}} V_j. \tag{4.208}$$

The $\boldsymbol{V}$-related parts in equations (4.207) and (4.208) cancel each other, and the Galilean invariance of the filtered Navier–Stokes equation is guaranteed. In other words, use of either $\tau_{ij}^{(\mathrm{I})}$ or $\tau_{ij}^{(\mathrm{II})}$ is prohibited from the Galilean-invariance principle.

4.5.2.2 Subgrid-scale-energy equation

As a quantity expressing the strength of the SGS velocity, we introduce the SGS energy

$$K_{\mathrm{S}} \equiv \frac{1}{2} u_{\mathrm{S}}^2 = \frac{1}{2} \overline{\boldsymbol{u}''^2}^{\mathrm{F}}, \tag{4.209}$$

where subscript S denotes subgrid, and u_{S} represents the magnitude of SGS velocity. The SGS energy obeys

$$\frac{\partial K_{\mathrm{S}}}{\partial t} + \overline{(\bar{\boldsymbol{u}}^{\mathrm{F}} \cdot \nabla) \frac{1}{2} \boldsymbol{u}''^2}^{\mathrm{F}} = P_{\mathrm{S}} - \varepsilon_{\mathrm{S}} + \nabla \cdot \boldsymbol{T}_{\mathrm{S}} + \Xi_{\mathrm{S}}. \tag{4.210}$$

Here each term on the right-hand side is given by

$$P_S = -\overline{u_i'' u_j'' \frac{\partial \bar{u}_j^F}{\partial x_i}}^F, \tag{4.211}$$

$$\varepsilon_S = \nu \overline{\left(\frac{\partial u_j''}{\partial x_i} \right)^2}^F, \tag{4.212}$$

$$\boldsymbol{T}_S = -\overline{\left(\tfrac{1}{2}\boldsymbol{u}''^2 + p''\right)\boldsymbol{u}''}^F + \nu \nabla K_S, \tag{4.213}$$

$$\Xi_S = \overline{u_i'' \frac{\partial \tau_{ij}}{\partial x_j}}^F, \tag{4.214}$$

which correspond to equations (3.40)–(3.42) in the ensemble averaging and are called the SGS production, dissipation, and transport terms, respectively. Equation (4.214) has no ensemble-averaging counterpart.

4.5.3 Fixed-Parameter Modelling

4.5.3.1 Smagorinsky model

The prototype of ensemble-mean models is the K-ε model of turbulent-viscosity type in §4.2.4. We focus attention on $\tau_{ij}^{(\mathrm{III})}$ in τ_{ij} and construct its model with the aid of the similar concept. We first introduce the SGS viscosity ν_S and write it as

$$[\tau_{ij}^{(\mathrm{III})}]_D \equiv \tau_{ij}^{(\mathrm{III})} - \tfrac{2}{3} K_S \delta_{ij} = -\nu_S \bar{s}_{ij}^F, \tag{4.215}$$

where $\bar{s}_{ij}^F$ is the GS velocity strain tensor

$$\bar{s}_{ij}^F = \frac{\partial \bar{u}_j^F}{\partial x_i} + \frac{\partial \bar{u}_i^F}{\partial x_j}. \tag{4.216}$$

In the K-ε model, the turbulent viscosity ν_T is written in terms of two turbulence quantities, that is, the turbulent energy K and its dissipation rate ε. In the SGS modelling, the filter width Δ_F is an important quantity characterizing the SGS velocity. Then we may express ν_S by introducing one more SGS quantity. We adopt u_S in equation (4.209), and model ν_S as

$$\nu_S = C_{S\nu} \Delta_F u_S, \tag{4.217}$$

with $C_{S\nu}$ as a numerical constant. In actual LES, the filter width usually differs in three directions. For three different filter widths $(\Delta_x, \Delta_y, \Delta_z)$, we define one characteristic width Δ_F as

$$\Delta_F = (\Delta_x \Delta_y \Delta_z)^{1/3} \tag{4.218a}$$

or

$$\Delta_{\rm F} = \left(\frac{\Delta_x^2 + \Delta_y^2 + \Delta_z^2}{3} \right)^{1/2}. \tag{4.218b}$$

A method of estimating $u_{\rm S}$ in equation (4.217) is the use of equation (4.210) for $K_{\rm S}$ and is called one-equation SGS modelling [4.27–4.29]. In current LESs, a simpler method is often adopted, leading to the Smagorinsky model that is the prototype of all SGS models. In what follows, we shall explain the method.

In §3.5 we showed that the production and dissipation terms are always important in equation (3.175) for K, although the role of the other terms becomes less important in some flows such as channel turbulence. This finding is not simply applicable to the SGS counterpart (4.210), but the production and dissipation terms are still important. Then we pick up these terms and assume

$$P_{\rm S} \cong \varepsilon_{\rm S}. \tag{4.219}$$

In order to simplify equation (4.211), we adopt the approximation [4.30]

$$\overline{\phi\psi}^{\rm F} \cong \bar{\phi}^{\rm F} \bar{\psi}^{\rm F}. \tag{4.220}$$

It is often used in various types of SGS modelling, but we should note that its straightforward application to equation (4.200) leads to the meaningless result of vanishing τ_{ij}. We use equation (4.220) in a limited manner, as will be done below. We apply it and approximate

$$P_{\rm S} = -\overline{u_i'' u_j'' \frac{\partial \bar{u}_i^{\rm F}}{\partial x_j}}^{\rm F} \cong -\tau_{ij}^{\rm (III)} \overline{\frac{\partial \bar{u}_i^{\rm F}}{\partial x_j}}^{\rm F} \quad \text{or} \quad -\tau_{ij}^{\rm (III)} \frac{\partial \bar{u}_i^{\rm F}}{\partial x_j}. \tag{4.221}$$

Next, we use $\Delta_{\rm F}$ and $u_{\rm S}$, and model $\varepsilon_{\rm S}$ as

$$\varepsilon_{\rm S} = C_{{\rm S}\varepsilon} \frac{u_{\rm S}^3}{\Delta_{\rm F}} \tag{4.222}$$

($C_{{\rm S}\varepsilon}$ is a numerical constant). We substitute equations (4.221) and (4.222) into equation (4.219), and use equations (4.215) and (4.217). As a result, we have

$$u_{\rm S} = \sqrt{\frac{C_{{\rm S}\nu}}{2C_{{\rm S}\varepsilon}}} \Delta_{\rm F} \bar{s}^{\rm F}, \tag{4.223}$$

with the magnitude of SGS velocity-strain tensor

$$\bar{s}^{\rm F} = \sqrt{\bar{s}_{ij}^{\rm F} \bar{s}_{ij}^{\rm F}}. \tag{4.224}$$

From equations (4.217) and (4.223), we have

$$\nu_{\rm S} = (C_{\rm S}\Delta_{\rm F})^2 \frac{\bar{s}^{\rm F}}{\sqrt{2}} = (C_{\rm S}\Delta_{\rm F})^2 \sqrt{\tfrac{1}{2}\bar{s}_{ij}^{\rm F}\bar{s}_{ij}^{\rm F}}, \tag{4.225}$$

with the model constant

$$C_{\rm S} = \left(\frac{C_{{\rm S}\nu}^3}{C_{{\rm S}\varepsilon}}\right)^{1/4}. \tag{4.226}$$

Under the approximation $\tau_{ij} \cong \tau_{ij}^{\rm (III)}$, the GS-velocity equation (4.199) may be closed in the combination with equations (4.215) and (4.225). This model is called the Smagorinsky model [4.31, 4.32] whose greatest merit is its simplicity. The model is a mathematical basis in developing more elaborate SGS models.

In LES of turbulent flows bounded by a solid wall such as channel turbulence, the Smagorinsky model is not applicable close to the wall. Molecular viscous effects are dominant there, but the effects are not taken into account in the model. For expressing molecular effects near a wall, we often introduce Van Driest's wall-damping function

$$F_{\rm W}^{\rm (V)} = 1 - \exp\left(-\frac{y_{\rm WF}}{A}\right), \tag{4.227}$$

and replace $\Delta_{\rm F}$ in equation (4.225) with

$$\Delta_{\rm F} \to \Delta_{\rm F} F_{\rm W}^{\rm (V)}. \tag{4.228}$$

In equation (4.227), A is a numerical constant, and $y_{\rm WF}$ is the filtering-related wall-unit coordinate that is defined as

$$y_{\rm WF} = \frac{u_{\rm WF}\, y}{\nu}, \tag{4.229}$$

in terms of the distance from the wall, y and the friction velocity

$$u_{\rm WF} = \sqrt{\nu\left(\frac{\partial \bar{u}^{\rm F}}{\partial y}\right)_{y=0}} \tag{4.230}$$

($\bar{u}^{\rm F}$ is the velocity component along the wall).

The Smagorinsky model with the choice of model constants

$$C_{\rm S} = 0.1, \qquad A = 25 \tag{4.231}$$

can reproduce the logarithmic velocity profile. The computed profile is shown in figure 4.3, and the magnitude of the turbulence intensity in the x direction, $\tilde{u}$, that is defined by

$$\tilde{u} = \sqrt{\langle(\bar{u}^{\rm F} - \langle\bar{u}^{\rm F}\rangle)^2\rangle}, \tag{4.232}$$

is given in figure 4.4 [R_τ is the Reynolds number based on the friction velocity

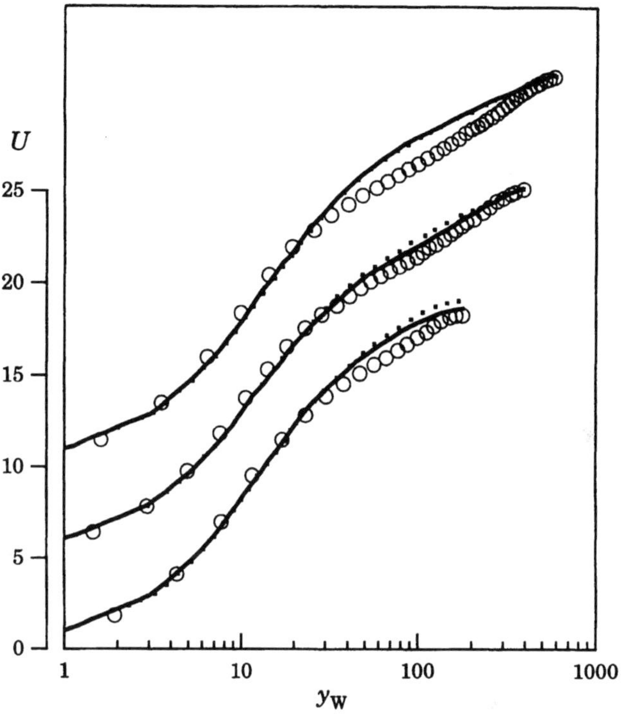

Figure 4.3. Mean velocity profiles at $R_\tau = 180$, 395, and 590 (bottom to top) based on the normalization by the friction velocity: O, DNS; dotted line, Smagorinsky/Van Driest model; solid line, nonequilibrium model.

that is given by equation (4.93)]. In order to enhance the accuracy of the model, some attempts at keeping $\tau_{ij}^{(\mathrm{I})}$ and $\tau_{ij}^{(\mathrm{II})}$ have been made. The use of either of them is prohibited from the Galilean-invariance principle, as was noted in §4.5.2.1.

4.5.3.2 Nonequilibrium model

The Smagorinsky model combined with Van Driest's wall-damping function is dependent on y (the distance from a wall) and u_{WF}. Near a sharp corner and a highly curved boundary, it is not easy to uniquely define y. In the presence of flow separation, there is the location at which u_{WF} vanishes or becomes very small. There y_{WF} based on u_{WF} loses its physical meaning. In the following, we shall show a method of deriving an SGS model free from u_{WF} [4.33].

Near a wall, we write

$$\bar{\boldsymbol{u}}^{\mathrm{F}} = (\bar{u}^{\mathrm{F}}, \bar{v}^{\mathrm{F}}, \bar{w}^{\mathrm{F}}), \qquad \boldsymbol{u}'' = (u'', v'', w''). \tag{4.233}$$

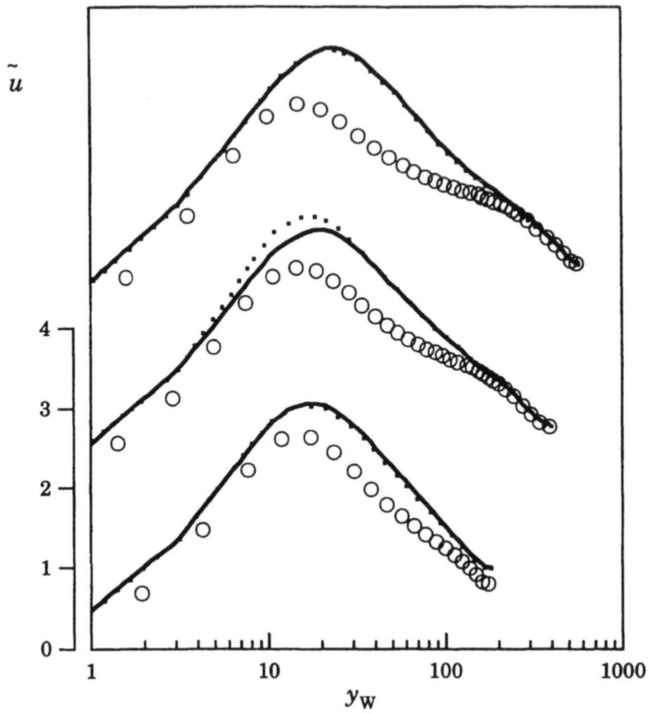

Figure 4.4. Streamwise turbulence intensities at $R_\tau = 180$, 395, and 590 (bottom to top) based on the normalization by the friction velocity: ○, DNS; dotted line, Smagorinsky/Van Driest model; solid line, nonequilibrium model.

From the near-wall velocity asymptotics (4.88) and

$$[\tau_{xy}^{(\mathrm{III})}]_{\mathrm{D}} \cong -\nu_{\mathrm{S}} \frac{\partial \bar{u}^{\mathrm{F}}}{\partial y}, \tag{4.234}$$

we have

$$u_{\mathrm{S}} = O(y), \qquad \nu_{\mathrm{S}} = O(y^3). \tag{4.235}$$

Considering that $u_{\mathrm{S}}/(\bar{s}^{\mathrm{F}}\Delta_{\mathrm{F}})$ is nondimensional, we introduce

$$\nu_{\mathrm{S}} \propto u_{\mathrm{S}}\Delta_{\mathrm{F}}\left(\frac{u_{\mathrm{S}}}{\bar{s} \to \bar{s}^{\mathrm{F}}\Delta_{\mathrm{F}}}\right)^2 = \frac{u_{\mathrm{S}}^3}{\bar{s} \to \bar{s}^{\mathrm{F}}\Delta_{\mathrm{F}}}, \tag{4.236}$$

which obeys the latter of equation (4.235) near a solid wall.

We introduce a wall-damping function

$$F_{\mathrm{W}} = 1 - \exp\left[-\left(C_{\mathrm{W}} \frac{u_{\mathrm{S}}}{\bar{s}^{\mathrm{F}}\Delta_{\mathrm{F}}}\right)^2\right] \tag{4.237}$$

(C_W is a model constant). With $C_\nu^{(S)}$ as a constant, we write

$$\nu_S = C_\nu^{(S)} \Delta_F u_S F_W. \tag{4.238}$$

Since $u_S/(\bar{s}^F \Delta_F)$ increases with the distance from a wall, equation (4.238) tends to equation (4.217) far from the wall. In this stage, u_S is not still related to the GS velocity $\bar{\boldsymbol{u}}^F$. For its estimate, we use equation (4.220) and approximate u_S as

$$u_S \equiv \sqrt{\overline{\boldsymbol{u}''^2}^F} = \sqrt{\overline{(\boldsymbol{u} - \bar{\boldsymbol{u}}^F)^2}} \cong \sqrt{\left(\bar{\boldsymbol{u}}^F - \overline{\bar{\boldsymbol{u}}^F}^F\right)^2}. \tag{4.239}$$

We summarize equations (4.215) and (4.237)–(4.239), and may close the equation for $\bar{\boldsymbol{u}}^F$. The choice of model constants

$$C_\nu^{(S)} = 0.03, \qquad C_W = 21 \tag{4.240}$$

may reproduce the logarithmic velocity profile. The profile and the turbulence intensity $\tilde{u}$ are given in figures 4.3 and 4.4 in the comparison with the counterparts by the Smagorinsky model. This model may be called a nonequilibrium model since it is free from the equilibrium assumption or equation (4.219).

4.5.4 Dynamic Model

The Smagorinsky model with Van Driest's wall-damping function and the nonequilibrium model are based on fixed nondimensional parameters. As the other type of SGS modelling, we may mention the dynamic SGS modelling in which such parameters are determined dynamically in the course of LES [4.34, 4.35].

4.5.4.1 Germano's identity

A great feature of dynamic modelling is the introduction of two different filters in contrast to fixed-parameter modelling. We denote those filters by G_A and G_B, and the filter widths by Δ_A and Δ_B ($>\Delta_A$) (we usually adopt $\Delta_B = 2\Delta_A$). We write

$$\bar{f}^A = \int G_A(\boldsymbol{\xi} - \boldsymbol{x}) f(\boldsymbol{\xi})\, d\boldsymbol{\xi}, \tag{4.241}$$

$$\bar{f}^{AB} \equiv \overline{\bar{f}^A}^B = \int G_B(\boldsymbol{\xi} - \boldsymbol{x}) \bar{f}^A(\boldsymbol{\xi})\, d\boldsymbol{\xi}. \tag{4.242}$$

In the consecutive use of G_A and G_B, we write the effective filter width as

$$\Delta_{AB} = \Delta_{AB}(\Delta_A, \Delta_B). \tag{4.243}$$

For Gaussian filters G_A and G_B, we have

$$\Delta_{AB} = \sqrt{\Delta_A^2 + \Delta_B^2}. \tag{4.244}$$

We apply equation (4.241) to equation (3.16), and write

$$\frac{\partial \bar{u}_i^{\mathrm{A}}}{\partial t} + \frac{\partial}{\partial x_j} \bar{u}_i^{\mathrm{A}} \bar{u}_j^{\mathrm{A}} = -\frac{\partial \bar{p}^{\mathrm{A}}}{\partial x_i} + \frac{\partial}{\partial x_j}(-\tau_{ij}^{(\mathrm{A})}) + \nu \nabla^2 \bar{u}_i^{\mathrm{A}}, \tag{4.245}$$

where the SGS stress $\tau_{ij}^{(\mathrm{A})}$ is given by

$$\tau_{ij}^{(\mathrm{A})} = \overline{u_i u_j}^{\mathrm{A}} - \bar{u}_i^{\mathrm{A}} \bar{u}_j^{\mathrm{A}}. \tag{4.246}$$

The consecutive-filtering counterpart of equation (4.245) are

$$\frac{\partial \bar{u}_i^{\mathrm{AB}}}{\partial t} + \frac{\partial}{\partial x_j} \bar{u}_i^{\mathrm{AB}} \bar{u}_j^{\mathrm{AB}} = -\frac{\partial \bar{p}^{\mathrm{AB}}}{\partial x_i} + \frac{\partial}{\partial x_j}(-\tau_{ij}^{(\mathrm{AB})}) + \nu \nabla^2 \bar{u}_i^{\mathrm{AB}}, \tag{4.247}$$

with

$$\tau_{ij}^{(\mathrm{AB})} = \overline{u_i u_j}^{\mathrm{AB}} - \bar{u}_i^{\mathrm{AB}} \bar{u}_j^{\mathrm{AB}}. \tag{4.248}$$

Between equations (4.246) and (4.248), we have the identity

$$\tau_{ij}^{(\mathrm{AB})} - \overline{\tau_{ij}^{(\mathrm{A})}}^{\mathrm{B}} = L_{ij} \equiv \overline{\bar{u}_i^{\mathrm{A}} \bar{u}_j^{\mathrm{A}}}^{\mathrm{B}} - \bar{u}_i^{\mathrm{AB}} \bar{u}_j^{\mathrm{AB}}, \tag{4.249}$$

which is called Germano's identity. In equations (4.245) and (4.247), both $\tau_{ij}^{(\mathrm{AB})}$ and $\overline{\tau_{ij}^{(\mathrm{A})}}^{\mathrm{B}}$ are unknown, but their difference is expressed in terms of the GS velocity and is calculated in the course of LES. Then there appears the possibility that the model parameters in $\tau_{ij}^{(\mathrm{AB})}$ and $\overline{\tau_{ij}^{(A)}}^{\mathrm{B}}$ may be estimated with the aid of this identity.

4.5.4.2 Subgrid-scale-viscosity model

Germano's identity itself does not suggest what models are proper for $\tau_{ij}^{(\mathrm{AB})}$ and $\overline{\tau_{ij}^{(\mathrm{A})}}^{\mathrm{B}}$. The model adopted widely in current LESs is the SGS-viscosity model given by equation (4.225), where a model parameter is estimated from the identity.

Following equations (4.215) and (4.225), we write

$$[\tau_{ij}^{\mathrm{A}}]_{\mathrm{D}} = -C_{\mathrm{S}}^{(\mathrm{A})} \Delta_{\mathrm{A}}^2 \bar{s}^{(\mathrm{A})} \bar{s}_{ij}^{\mathrm{A}}, \tag{4.250}$$

with

$$\bar{s}_{ij}^{\mathrm{A}} = \frac{\partial \bar{u}_j^{\mathrm{A}}}{\partial x_i} + \frac{\partial \bar{u}_i^{\mathrm{A}}}{\partial x_j}, \tag{4.251}$$

$$\bar{s}^{(\mathrm{A})} = \sqrt{\bar{s}_{ij}^{\mathrm{A}} \bar{s}_{ij}^{\mathrm{A}}}. \tag{4.252}$$

Here $C_{\mathrm{S}}^{(\mathrm{A})}$ is a function of time and location, that is,

$$C_{\mathrm{S}}^{(\mathrm{A})} = C_{\mathrm{S}}^{(\mathrm{A})}(\boldsymbol{x}, t). \tag{4.253}$$

Entirely similarly, we express $\tau_{ij}^{(AB)}$ as

$$[\tau_{ij}^{AB}]_D = -C_S^{(AB)} \Delta_{AB}^2 \bar{S}^{(AB)} \bar{s}_{ij}^{AB}. \qquad (4.254)$$

We substitute equations (4.250) and (4.254) into equation (4.249), and have

$$\overline{C_S^{(A)} \Delta_A^2 \bar{S}^{(A)} \bar{s}_{ij}^{A}}^{B} - C_S^{(AB)} \Delta_{AB}^2 \bar{S}^{(AB)} \bar{s}_{ij}^{AB} = [L_{ij}]_D. \qquad (4.255)$$

As may be seen from equation (4.249), the right-hand side of equation (4.255) is expressed in terms of the GS velocity only. In order to estimate the model coefficients on the left-hand side, we approximate

$$C_S^{(A)} \cong C_S^{(AB)}, \qquad \overline{C_S^{(A)} \Delta_A^2 \bar{S}^{(A)} \bar{s}_{ij}^{A}}^{B} \cong C_S^{(A)} \overline{\Delta_A^2 \bar{S}^{(A)} \bar{s}_{ij}^{A}}^{B}. \qquad (4.256)$$

As a result, we have

$$C_S^{(A)} M_{ij} = [L_{ij}]_D, \qquad (4.257)$$

where

$$M_{ij} = \overline{\Delta_A^2 \bar{S}^{(A)} \bar{s}_{ij}^{A}}^{B} - \Delta_{AB}^2 \bar{S}^{(AB)} \bar{s}_{ij}^{AB}. \qquad (4.258)$$

Equation (4.257) is a six-component equation, but the unknown is $C_S^{(A)}$ only. A method of estimating $C_S^{(A)}$ in a rational way is Lilly's least-square method [4.36]. There we introduce

$$D_R = ([L_{ij}]_D - C_S^{(A)} M_{ij})^2, \qquad (4.259)$$

which represents the degree of the error arising from an approximate solution of equation (4.257). In order to make D_R minimum, we impose

$$\frac{dD_R}{dC_S^{(A)}} = 0, \qquad (4.260)$$

which results in

$$C_S^{(A)} = \frac{M_{ij}[L_{ij}]_D}{M_{ij}^2}. \qquad (4.261)$$

This estimate does not always guarantee the positiveness of $C_S^{(A)}$. Negative $C_S^{(A)}$ gives rise to numerical instability in LES, and various mathematical devices for avoiding it have been studied [4.37–4.39].

The merit of the dynamic modelling lies in the following two points.

1. Nondimensional coefficients in SGS models may be estimated automatically in the course of LES. Properties intrinsic to each turbulent flow are expected to be reflected on the coefficients.
2. The introduction of wall-damping functions represented by Van Driest's function is not necessary. This point is important in dealing with a sharp

corner and a highly curved boundary, as was noted in relation to a non-equilibrium model in §4.5.3.2.

From the viewpoint of the accuracy of computed results, there is still room for the improvcment of dynamic modelling. For instance, in channel turbulence the accuracy of the mean velocity profile and turbulent intensities computed by the dynamic SGS-viscosity model (4.250) is lower than the counterparts by the Smagorinsky and nonequilibrium models. Mathematical devices for rectifying such shortfalls and properly treating frame-rotation effects are under intensive study, such as the inclusion of $\tau_{ij}^{(\mathrm{I})}$ and $\tau_{ij}^{(\mathrm{II})}$, the choice of the SGS viscosity different from equation (4.250), etc. [4.40–4.45]. Readers may consult [4.46, 4.47] for novel SGS modelling.

References

[4.1] Kasagi N and Shikazono N 1995 *Proc. Roy. Soc. Lond. A* **451** 257

[4.2] Moin P and Mahesh K 1998 *Ann. Rev. Fluid Mech.* **30** 539

[4.3] Lumley J L 1978 *Adv. Appl. Mech.* **18** 123

[4.4] Halläback M, Johansson A V and Burden A D 1996 *Turbulence and Transition Modelling* eds M Halläback, D S Henningson, A V Johansson and P H Alfredson (Dordrecht: Kluwer) p 81

[4.5] Shih T-H 1996 *Turbulence and Transition Modelling* eds M Halläback, D S Henningson, A V Johansson and P H Alfredson (Dordrecht: Kluwer) p 155

[4.6] Piquet J 1999 *Turbulent Flows* (Berlin: Springer)

[4.7] Pope S B 2000 *Turbulent Flows* (Cambridge: Cambridge University Press)

[4.8] Yoshizawa A 1984 *Phys. Fluids* **27** 1377

[4.9] Speziale C G 1987 *J. Fluid Mech.* **178** 459

[4.10] Speziale C G 1991 *Ann. Rev. Fluid Mech.* **23** 107

[4.11] Taulbee D B 1992 *Phys. Fluids A* **4** 2555

[4.12] Gatski T B and Speziale C G 1993 *J. Fluid Mech.* **254** 59

[4.13] Craft T J, Launder B E and Suga K 1997 *Intl. J. Heat Fluid Flow* **18** 15

[4.14] Abe K, Kondoh T and Nagano Y 1997 *Intl. J. Heat Fluid Flow* **18** 266

[4.15] Launder B E and Sharma B I 1974 *Lett. Heat Mass Transf.* **1** 131

[4.16] Speziale C G and Gatski T B 1997 *J. Fluid Mech.* **344** 155

[4.17] Takemitsu N 1990 *J. Fluids Eng.* **112** 193

[4.18] Nisizima S 1990 *Theor. Comput. Fluid Dyn.* **2** 61

[4.19] Launder B E, Reece G and Rodi W 1975 *J. Fluid Mech.* **68** 537

[4.20] Speziale C G, Sarkar S and Gatski T B 1991 *J. Fluid Mech.* **227** 245

[4.21] Yoshizawa A 1993 *Phys. Rev. E* **48** 273

[4.22] Yoshizawa A, Yokoi N, Nisizima S, Itoh S-I and Itoh K 2001 *Phys. Fluids* **13** 2309

[4.23] Taylor J B 1974 *Phys. Rev. Lett.* **33** 1139

[4.24] Germano M 2000 *Advanced Turbulent Flow Computation* eds R Peyret and E Krause (New York: Springer) p 81

[4.25] Sagaut P 2001 *Large Eddy Simulation for Incompressible Flows* (New York: Springer)

[4.26] Speziale C G 1985 *J. Fluid Mech.* **156** 55

[4.27] Schumann U 1975 *J. Comput. Phys.* **18** 376
[4.28] Yoshizawa A and Horiuti K 1985 *J. Phys. Soc. Jpn.* **54** 2834
[4.29] Okamoto M and Shima N 1999 *JSME Intl. J. B* **42** 154
[4.30] Bardina J 1983 PhD Dissertation Stanford University
[4.31] Smagorinsky J S 1963 *Mon. Weather Rev.* **91** 99
[4.32] Deardorff J W 1970 *J. Fluid Mech.* **41** 453
[4.33] Yoshizawa A, Kobayashi K, Kobayashi T and Taniguchi N 2000 *Phys. Fluids* **12** 2338
[4.34] Germano M, Piomelli U, Moin P and Cabot W 1991 *Phys. Fluids A* **3** 1760
[4.35] Germano M 1992 *J. Fluid Mech.* **238** 325
[4.36] Lilly D 1992 *Phys. Fluids A* **4** 633
[4.37] Ghosal S, Lund T S, Moin P and Akselvoll K 1995 *J. Fluid Mech.* **286** 229
[4.38] Piomelli U and Liu J 1995 *Phys. Fluids* **7** 839
[4.39] Meneveau C, Lund T S and Cabot W H 1996 *J. Fluid Mech.* **319** 353
[4.40] Zang Y, Street R L and Koseff J R 1993 *Phys. Fluids A* **5** 3186
[4.41] Vreman B, Geurts B and Kuerten H 1994 *Phys. Fluids A* **6** 4057
[4.42] Horiuti K 1997 *Phys. Fluids* **9** 3443
[4.43] Tsubokura M 2001 *Phys. Fluids* **13** 500
[4.44] Kobayashi H and Shimomura Y 2001 *Phys. Fluids* **13** 2350
[4.45] Morinishi Y and Vasilyev O V 2001 *Phys. Fluids* **13** 2912
[4.46] Stolz S and Adams N A 1999 *Phys. Fluids* **11** 1699
[4.47] Domaradzki J A and Loh K-C 1999 *Phys. Fluids* **11** 2330

Chapter 5

Statistical Theory of Fluid Turbulence

5.1 Mathematical Methods Necessary for Turbulence Theory

In the investigation into turbulence in wavenumber space, we need to deal with complicated interactions among waves or eddies that are governed by equation (3.120). For abstracting some of their characteristics, we often resort to the approximations whose range of validity is not so evident. In this section we shall refer to some approximate methods that are utilized in the study of statistical turbulence theory.

5.1.1 Partial Summation of Infinite Series

In order to understand the method of partial summation of infinite series, we consider an ordinary differential equation

$$\frac{\mathrm{d}f}{\mathrm{d}x} = f^2, \qquad f(0) = 1. \tag{5.1}$$

Its solution is simply obtained as

$$f = \frac{1}{1-x}. \tag{5.2}$$

We seek the solution of equation (5.1) by the perturbation method. For small x, we have $f \cong 1$ from the condition, which is substituted into the right-hand side of equation (5.1). Then we have

$$f = 1 + x. \tag{5.3}$$

We substitute equation (5.3) into the right-hand side of equation (5.1) and retain the terms up to $O(x)$ since equation (5.3) is correct to the order. Then we have

$$f = 1 + x + x^2. \tag{5.4}$$

Repetition of the foregoing procedure results in

$$f = 1 + x + x^2 + x^3 + \cdots = \sum_{n=0}^{\infty} x^n, \tag{5.5}$$

which is the Taylor expansion of equation (5.2) for $x < 1$.

Equation (5.2) has a singular point at $x = 1$, and equation (5.5) is not applicable beyond $x = 1$. In order to obtain the information about f for $x > 1$, it is necessary to derive equation (5.2) from equation (5.5). In the case of a simple equation such as equation (5.1), we can obtain the exact solution by constructing an infinite-series solution and summing up the series. For a complicated system like equation (3.120), it is difficult to derive its exact solution from such a perturbational solution. As a result, we cannot but infer an approximate solution close to the true solution, with the aid of a limited number of terms in the perturbational expansion.

In equation (5.5), we retain the terms up to $O(x^2)$ and write

$$f = 1 + (1 + x)x. \tag{5.6}$$

Noting that the coefficient of the second term, $1 + x$, is an approximation to f for small x, we rewrite equation (5.6) as

$$f = 1 + fx, \tag{5.7}$$

resulting in the exact solution (5.2). This result is accidental and cannot be expected in a general case. Here the noteworthy point is that the terms up to an infinite order of x can be included through the replacement such as equation (5.7) even if the summation is partial in general. This procedure may be called the partial summation or the renormalization of infinite series.

5.1.2 Gaussian Distribution Function

We denote the probability of x being between x and $x + \mathrm{d}x$ by $P(x)\,\mathrm{d}x$ and call $P(x)$ the probability distribution density or, simply, the probability distribution. In the case that $P(x)$ is given by

$$P(x) = \frac{1}{\sqrt{2\pi}\sigma} \exp\left(-\frac{1}{2\sigma^2} x^2\right), \tag{5.8}$$

with $(-\infty, \infty)$ as the range of x, we say that x obeys the Gaussian distribution. Equation (5.8) represents the probability distribution for x varying randomly around $x = 0$. Under this $P(x)$, we have

$$\langle x^n \rangle = \int_{-\infty}^{\infty} x^n P(x)\,\mathrm{d}x = \begin{cases} (n-1)!!\sigma^n & \text{for even } n \\ 0 & \text{for odd } n, \end{cases} \tag{5.9}$$

where σ and σ^2 are the standard deviation and variance, respectively (the former is an indicator of the degree of variation of x around $x = 0$).

In the calculation of mean values, the characteristic function defined by

$$\Psi(y) \equiv \langle e^{ixy} \rangle = \int_{-\infty}^{\infty} e^{ixy} P(x)\, dx \tag{5.10}$$

is more manageable than $P(x)$. The function $\Psi(y)$ corresponds to the Fourier mode with y as the wavenumber [see equation (3.97)]. The first relation of equation (5.9) is replaced with

$$\langle x^n \rangle = \frac{1}{i^n} \left(\frac{d^n \Psi}{dy^n} \right)_{y=0}. \tag{5.11}$$

From equations (5.8) and (5.10), we have

$$\Psi(y) = \exp\left(-\frac{\sigma^2}{2} y^2 \right), \tag{5.12}$$

and equation (5.9) may be easily reproduced.

In the case of N variables, we write

$$\boldsymbol{x} = (x_1, x_2, \ldots, x_N), \qquad \boldsymbol{y} = (y_1, y_2, \ldots, y_N). \tag{5.13}$$

The Gaussian distribution and its characteristic function corresponding to equations (5.8) and (5.12) are given by

$$P(\boldsymbol{x}) = \prod_{n=1}^{N} \frac{1}{\sqrt{2\pi}\sigma_{(n)}} \exp\left(-\frac{1}{2\sigma_{(n)}^2} x_{(n)}^2 \right), \tag{5.14}$$

$$\Psi(\boldsymbol{y}) = \prod_{n=1}^{N} \exp\left(-\frac{\sigma_{(n)}^2}{2} y_{(n)}^2 \right), \tag{5.15}$$

respectively. Here and hereafter, the summation convention is not applied to parenthesized subscripts. From equation (5.15), we have

$$\begin{aligned} \langle x_i x_j x_k x_l \rangle &= \frac{1}{i^4} \left(\frac{d^4 \Psi}{dy_i\, dy_j\, dy_k\, dy_l} \right)_{y=0} \\ &= \langle x_{(i)} x_{(j)} \rangle \langle x_{(k)} x_{(l)} \rangle \delta_{ij} \delta_{kl} + \langle x_{(i)} x_{(k)} \rangle \langle x_{(j)} x_{(l)} \rangle \delta_{ik} \delta_{jl} \\ &\quad + \langle x_{(i)} x_{(l)} \rangle \langle x_{(j)} x_{(k)} \rangle \delta_{il} \delta_{jk}. \end{aligned} \tag{5.16}$$

In general, the mean value of an nth-order function with even n is divided into the products of variances.

We consider that $\boldsymbol{x}$ in equation (5.13) is a function of time t, that is, $\boldsymbol{x} = \boldsymbol{x}(t)$. The probability distribution $P(\boldsymbol{x}; t)$ obeys the Liouville equation

$$\frac{\partial P(\boldsymbol{x}; t)}{\partial t} + \frac{\partial}{\partial x_i} \left[\frac{dx_i}{dt} P(\boldsymbol{x}; t) \right] = 0. \tag{5.17}$$

This equation corresponds to the equation of continuity in phase space [ρ and u_i in equation (3.5) are replaced with $P(\boldsymbol{x}; t)$ and dx_i/dt, respectively].

We consider a simple system

$$\frac{\mathrm{d}x_i}{\mathrm{d}t} = -\gamma_{(i)} x_i + \zeta_i. \tag{5.18}$$

Here $\gamma_{(i)}$ is the damping rate of mode x_i, and ζ_i subject to

$$\langle \zeta_i \rangle = 0 \tag{5.19}$$

is assumed to be a random white noise, that is, its statistical properties are characterized by the variances with the infinitesimal correlation time

$$\langle \zeta_i(t)\zeta_j(t')\rangle = 2\eta_{(i)}(t)\delta_{ij}\delta(t-t'), \tag{5.20}$$

as in equation (5.16).

We substitute equation (5.18) into equation (5.17), and have

$$\frac{\partial P(\boldsymbol{x};t)}{\partial t} - \frac{\partial}{\partial x_i}[\gamma_{(i)} x_i P(\boldsymbol{x};t)] + \frac{\partial}{\partial x_i}[\zeta_i P(\boldsymbol{x};t)] = 0. \tag{5.21}$$

The average of equation (5.21) with respect to ζ results in

$$\frac{\partial P_\zeta(\boldsymbol{x};t)}{\partial t} - \frac{\partial}{\partial x_i}[\gamma_{(i)} x_i P_\zeta(\boldsymbol{x};t)] + \frac{\partial}{\partial x_i}\langle \zeta_i P(\boldsymbol{x};t)\rangle_\zeta = 0, \tag{5.22}$$

where

$$P_\zeta(\boldsymbol{x};t) \equiv \langle P(\boldsymbol{x})\rangle_\zeta. \tag{5.23}$$

In order to express the last term of equation (5.22) using $P_\zeta(\boldsymbol{x};t)$, we formally integrate equation (5.21) as

$$P(\boldsymbol{x};t) = \int^t \left(\frac{\partial}{\partial x_i}[\gamma_{(i)} x_i P(\boldsymbol{x};\tau)] - \frac{\partial}{\partial x_i}[\zeta_i P(\boldsymbol{x};\tau)]\right) \mathrm{d}\tau. \tag{5.24}$$

We substitute equation (5.22) into the last term, and make use of equations (5.19) and (5.20). Then equation (5.22) is reduced to

$$\frac{\partial P_\zeta(\boldsymbol{x};t)}{\partial t} - \frac{\partial}{\partial x_i}[\gamma_{(i)} x_i P_\zeta(\boldsymbol{x};t)] - \eta_{(i)} \frac{\partial^2 P_\zeta(\boldsymbol{x};t)}{\partial x_i^2} = 0. \tag{5.25}$$

We introduce the characteristic function corresponding to $P_\zeta(\boldsymbol{x};t)$ as

$$\Psi_\zeta(\boldsymbol{y};t) = \int_{-\infty}^{\infty} \cdots \int_{-\infty}^{\infty} \exp\left(\mathrm{i}\sum_{n=1}^{N} x_{(n)} y_{(n)}\right) P_\zeta(\boldsymbol{x};t)\, \mathrm{d}x_1 \cdots \mathrm{d}x_N. \tag{5.26}$$

From equation (5.25), we have

$$\frac{\partial \Psi_\zeta(\boldsymbol{y};t)}{\partial t} + \gamma_{(i)} y_i \frac{\partial \Psi_\zeta(\boldsymbol{y};t)}{\partial y_i} + \eta_{(i)} \boldsymbol{y}^2 \Psi_\zeta(\boldsymbol{y};t) = 0. \tag{5.27}$$

The time-independent solution of equation (5.27) is simply obtained as

$$\Psi_\zeta(\boldsymbol{y};t) = \exp\left(-\sum_{n=1}^{N} \frac{\sigma_{(n)}^2}{2} y_{(n)}^2\right), \tag{5.28}$$

where

$$\sigma_{(i)} = \sqrt{\frac{\eta_{(i)}}{\gamma_{(i)}}}. \tag{5.29}$$

Namely, the variable $\boldsymbol{x}$, whose temporal change is governed by equation (5.18), obeys the Gaussian distribution.

5.1.3 Solution of Differential Equation Using Method of Partial Summation

The partial summation and the Gaussianity of distribution constitute primary ingredients in the formulation of statistical turbulence theory. In order to give its outline, we consider a time-marching equation

$$\frac{\mathrm{d}f}{\mathrm{d}t} + \lambda f = A\{f\}, \qquad f(0) = 1, \tag{5.30}$$

where λ is a positive constant, and A is a functional of f. For A nonlinear in f, it is generally difficult to find the exact solution of equation (5.30). For explaining an approximate approach to such an equation, we choose the linear case

$$A = -\mathrm{i}\alpha f. \tag{5.31}$$

Here α is assumed to vary randomly around $\alpha = 0$, namely, it obeys the Gaussian distribution

$$P(\alpha) = \frac{1}{\sqrt{2\pi}\sigma} \exp\left(-\frac{1}{2\sigma^2}\alpha^2\right) \tag{5.32}$$

(σ is the standard deviation).

We integrate equation (5.30) formally, and have

$$f = \hat{f}(t) - \mathrm{i}\alpha \int_0^t \hat{g}(t - t_1) f(t_1)\, \mathrm{d}t_1. \tag{5.33}$$

Here $\hat{f}(t)$ and $\hat{g}(t)$ are defined as

$$\hat{f}(t) = \mathrm{e}^{-\lambda t}, \tag{5.34}$$

$$\hat{g}(t) = S(t)\,\mathrm{e}^{-\lambda t}, \tag{5.35}$$

respectively, with $S(t) = 1$ $(t > 0)$ and 0 $(t < 0)$ (the unit step function). We should note that $\hat{g}(t)$ is the Green's function obeying equation (5.30) with A replaced with $\delta(t)$.

For equation (5.31), the solution of equation (5.30) is easily obtained as

$$f = \exp[-(\lambda + \mathrm{i}\alpha)t]. \tag{5.36}$$

Under the Gaussian distribution (5.32), we have

$$F = \langle f \rangle = \mathrm{e}^{-\lambda t} \exp\left(-\frac{\sigma^2}{2} t^2\right). \tag{5.37}$$

By an iteration method, we solve equation (5.33). In the first iteration, we substitute $f = \hat{f}(t)$ into the second part, and have

$$f = \hat{f}(t) - \mathrm{i}\alpha \int_0^t \hat{g}(t - t_1)\hat{f}(t_1)\,\mathrm{d}t_1. \tag{5.38}$$

The average with respect to α leads to vanishing of the α-related part. Through the second iteration based on equation (5.38), we have

$$F_{\mathrm{A}} = \hat{f}(t) - \sigma^2 \int_0^t \hat{g}(t - t_1)\,\mathrm{d}t_1 \int_0^{t_1} \hat{g}(t_1 - t_2)\hat{f}(t_2)\,\mathrm{d}t_2, \tag{5.39}$$

from equation (5.32). Here we denote the approximate mean value of f by F_{A} to distinguish the exact counterpart, equation (5.37). The repetition of this iteration generates an expression of infinite series in σ^2, as

$$F_{\mathrm{A}} = \sum_{n=0}^{\infty} C_n(t)\sigma^{2n}. \tag{5.40}$$

It is highly probable that equation (5.40) is not applicable for large σ, as in equation (5.5). In order to enlarge the range of applicability concerning σ, we follow equations (5.6) and (5.7), and make the replacement

$$\hat{f} \to F_{\mathrm{A}}, \tag{5.41}$$

in the second term of equation (5.39), resulting in

$$F_{\mathrm{A}} = \hat{f}(t) - \sigma^2 \int_0^t \hat{g}(t - t_1)\,\mathrm{d}t_1 \int_0^{t_1} \hat{g}(t_1 - t_2)F_{\mathrm{A}}(t_2)\,\mathrm{d}t_2. \tag{5.42}$$

The expression that is obtained from equation (5.42) by the iteration procedure does not coincide with equation (5.40) in general, but equation (5.42) contains the σ effects up to an infinite order and is free from the difficulty of a limited range of σ in equation (5.40).

Equation (5.42) may be rewritten in the form

$$\frac{\mathrm{d}F_{\mathrm{A}}}{\mathrm{d}t} + \lambda F_{\mathrm{A}} = -\sigma^2 \int_0^t \hat{g}(t - t_1)F_{\mathrm{A}}(t_1)\,\mathrm{d}t_1. \tag{5.43}$$

In the context of the Navier–Stokes equation (3.120), λ corresponds to the molecular viscosity ν. In the limit $\lambda \to 0$ signifying a state at a high Reynolds number, equation (5.43) is reduced to

$$\frac{\mathrm{d}F_{\mathrm{A}}}{\mathrm{d}t} = -\sigma^2 \int_0^t F_{\mathrm{A}}(t_1)\,\mathrm{d}t_1. \tag{5.44}$$

The solution subject to the initial condition in equation (5.30) is given by

$$F_{\mathrm{A}}(t) = \cos \sigma t. \tag{5.45}$$

This approximate solution has an oscillating behavior in time, but its exact counterpart (5.37) decays monotonically with time. There is a big difference between these two solutions.

We now seek the cause of the occurrence of the foregoing discrepancy. The exact solution obeys

$$\frac{\mathrm{d}F}{\mathrm{d}t} = -\sigma^2 tF \tag{5.46}$$

in the limit $\lambda \to 0$. A crucial difference between equations (5.44) and (5.46) lies in effects of past memory. In equation (5.46), the time evolution of F is determined by the quantities at present, that is, F is subject to the Markovian process. On the other hand, the time evolution of F_{A} is dependent uniformly on the events in the past. Such dependence is not proper since events in a nearer past have a bigger influence on the present event. This shortfall of F_{A} originates from $\hat{g}$ in equation (5.43), which tends to one in the limit $\lambda \to 0$ and cannot put more weight on events in a nearer past. From the viewpoint of putting emphasis on events at the time close to the present, we take out F_{A} from the integral of equation (5.44), namely, we perform the Markovianization, which leads to equation (5.46).

The foregoing procedures from equation (5.30) to equation (5.43) may be summarized as follows.

(i) We obtain the perturbational solution by an iteration method based on a linear solution.
(ii) We calculate the mean value with the aid of the Gaussianity of distribution.
(iii) We perform the renormalization of the resulting expression.

These procedures constitute primary mathematical ingredients in various types of statistical turbulence theories, whose prototype is the direct-interaction approximation (DIA) [5.1, 5.2].

5.2 Theoretical Approach to Inhomogeneous Turbulence

In the study of statistical turbulence theories of homogeneous isotropic turbulence, attention has been focused on the inertial-range and smaller-scale properties of turbulence [3.4–3.6]. A characteristic time scale of energy-containing eddies that is represented by equation (4.15) is one of the important factors in turbulence modelling. It is linked with the turbulent-energy production mechanism, and its examination is beyond the scope of such theories. Then there is a wide gap between past studies of turbulence theories and modelling.

In this section we give an account of turbulence theories from a slightly different viewpoint. Namely, we first show a theoretical framework of inhomogeneous turbulence, specifically, a two-scale direct-interaction approximation (TSDIA) [2.16, 4.8] and clarify what is essentially important in both the studies of homogeneous and inhomogeneous turbulence.

Through this exposition we shall be able to understand what development of homogeneous turbulence theories is helpful to the development of turbulence modelling necessary for the analysis of real-world flows.

5.2.1 Perturbational Method to Turbulence

5.2.1.1 Introduction of two-scale variables

In the decomposition of f by equation (3.22), the spatial or temporal variations of the mean F and the fluctuation f' are depicted schematically in figure 3.2. In general, the variation of F is much slower, compared with f'. A method useful for describing such different properties is the introduction of multiple scales.

In order to distinguish between spatial and temporal variations of F and f', we introduce a positive parameter

$$\delta_S \ll 1. \tag{5.47}$$

Using it, we construct two spatial and temporal variables

$$\boldsymbol{\xi}\ (=\boldsymbol{x}), \quad \tau\ (=t); \qquad \boldsymbol{X}\ (=\delta_S \boldsymbol{x}), \quad T\ (=\delta_S t). \tag{5.48}$$

Here δ_S is not a real nondimensional parameter such as a Reynolds number and is called the scale parameter hereafter. Its mathematical meaning will become clear in the following discussions.

For the change of the original coordinate

$$\boldsymbol{x} \to \boldsymbol{x} + \boldsymbol{r}, \tag{5.49}$$

we have

$$\boldsymbol{\xi} \to \boldsymbol{\xi} + \boldsymbol{r}, \qquad \boldsymbol{X} \to \boldsymbol{X} + \delta_S \boldsymbol{r}, \tag{5.50}$$

which indicates that the change of $\boldsymbol{X}$ is small under equation (5.47). We write equation (3.22) as

$$f = F(\boldsymbol{X}, T) + f'(\boldsymbol{\xi}, \tau; \boldsymbol{X}, T). \tag{5.51}$$

From equation (5.50), we have

$$F(\boldsymbol{X}, T) \to F(\boldsymbol{X}, T) + \delta_S r_i \frac{\partial F(\boldsymbol{X}, T)}{\partial X_i} + O[(\delta_S r)^2], \tag{5.52a}$$

$$f'(\boldsymbol{\xi}, \tau; \boldsymbol{X}, T) \to f'(\boldsymbol{\xi}, \tau; \boldsymbol{X}, T) + r_i \frac{\partial}{\partial \xi_i} f'(\boldsymbol{\xi}, \tau; \boldsymbol{X}, T) + O(\delta_S r), \tag{5.52b}$$

which guarantee that the variation of F is much slower than that of f'. Hereafter $(\boldsymbol{X}, T)$ and $(\boldsymbol{\xi}, \tau)$ are called slow and fast variables, respectively. We should note the dependence of f' on $(\boldsymbol{X}, T)$. By adopting $\boldsymbol{u}'$ as f', we may understand it since $\boldsymbol{u}'$ is connected with $\boldsymbol{U}$ through equation (3.31).

We apply equations (5.48) and (5.51) to equations (3.30) and (3.31). We have

$$\frac{\partial u_i'}{\partial \xi_i} + \delta_S \frac{\partial u_i'}{\partial X_i} = 0, \tag{5.53}$$

$$\frac{\partial u_i'}{\partial \tau} + U_j \frac{\partial u_i'}{\partial \xi_j} + \frac{\partial}{\partial \xi_j} u_j' u_i' + \frac{\partial p'}{\partial \xi_i} - \nu \nabla_\xi^2 u_i'$$
$$= \delta_S \left(-u_j' \frac{\partial U_i}{\partial X_j} - \frac{\mathrm{D} u_i'}{\mathrm{D} T} - \frac{\partial p'}{\partial X_i} - \frac{\partial}{\partial X_j}(u_j' u_i' - R_{ij}) + 2\nu \frac{\partial u_i'}{\partial X_j \partial \xi_j} \right)$$
$$+ \delta_S^2 (\nu \nabla_X^2 u_i'), \tag{5.54}$$

where

$$\nabla_\xi = \left(\frac{\partial}{\partial \xi_i} \right), \qquad \nabla_X = \left(\frac{\partial}{\partial X_i} \right), \tag{5.55}$$

$$\frac{\mathrm{D}}{\mathrm{D}T} = \frac{\partial}{\partial T} + \mathbf{U} \cdot \nabla_X. \tag{5.56}$$

5.2.1.2 Fourier representation of fast-varying modes

In §3.4 we discussed homogeneous turbulence with no mean flow, specifically, isotropic turbulence. There $\boldsymbol{u}'$ is written in the Fourier representation as equation (3.97a), and $\boldsymbol{u}'(\boldsymbol{k})$ corresponds to a wave or an eddy whose spatial scale is $2\pi/k$. The second term on the left-hand side of equation (5.54) shows that $\boldsymbol{u}'$ is swept away by the slowly-varying mean flow $\boldsymbol{U}$. In order to properly express this situation, we introduce the Fourier representation in the frame moving with the velocity $\boldsymbol{U}$; namely, we write

$$f'(\boldsymbol{\xi}, \boldsymbol{X}; \tau, T) = \int f'(\boldsymbol{k}, \boldsymbol{X}; \tau, T) \exp[-\mathrm{i}\boldsymbol{k} \cdot (\boldsymbol{\xi} - \boldsymbol{U}\tau)] \, \mathrm{d}\boldsymbol{\xi}. \tag{5.57}$$

We apply equation (5.57) to equations (5.53) and (5.54), and have

$$\boldsymbol{k} \cdot \boldsymbol{u}'(\boldsymbol{k}; \tau) = \delta_S [-\mathrm{i} \nabla_X^* \cdot \boldsymbol{u}'(\boldsymbol{k}; \tau)], \tag{5.58}$$

$$\frac{\partial u_i'(\boldsymbol{k}; \tau)}{\partial \tau} + \nu k^2 u_i'(\boldsymbol{k}; \tau) - \mathrm{i} k_i p'(\boldsymbol{k}; \tau)$$
$$- \mathrm{i} k_j \iint u_i'(\boldsymbol{p}; \tau) u_j'(\boldsymbol{q}; \tau) \delta(\boldsymbol{k} - \boldsymbol{p} - \boldsymbol{q}) \, \mathrm{d}\boldsymbol{p} \, \mathrm{d}\boldsymbol{q}$$
$$= \delta_S \left(-u_j'(\boldsymbol{k}; \tau) \frac{\partial U_i}{\partial X_j} - \frac{\mathrm{D}^* u_i'(\boldsymbol{k}; \tau)}{\mathrm{D} T^*} - \frac{\partial^* p'(\boldsymbol{k}; \tau)}{\partial X_i^*} \right.$$
$$\left. - \iint \frac{\partial^*}{\partial X_j^*} [u_i'(\boldsymbol{p}; \tau) u_i'(\boldsymbol{q}; \tau)] \delta(\boldsymbol{k} - \boldsymbol{p} - \boldsymbol{q}) \, \mathrm{d}\boldsymbol{p} \, \mathrm{d}\boldsymbol{q} + \delta(\boldsymbol{k}) \frac{\partial R_{ij}}{\partial X_j} \right), \tag{5.59}$$

with

$$\left(\frac{\mathrm{D}^*}{\mathrm{D}T^*}, \nabla_X^*\right) = \exp(-\mathrm{i}\boldsymbol{k}\cdot\boldsymbol{U}\tau)\left(\frac{\mathrm{D}}{\mathrm{D}T}, \nabla_X\right)\exp(\mathrm{i}\boldsymbol{k}\cdot\boldsymbol{U}\tau). \tag{5.60}$$

Here and hereafter, the dependency of f' on slow variables $\boldsymbol{X}$ and T is not written explicitly, except when necessary. The new differential operators given by equation (5.60) arise from the use of the Fourier representation in the moving frame. On the right-hand side of equation (5.59), the R_{ij}-related term is neglected owing to their minor importance in later analyses.

In equation (5.58), we should note that $\boldsymbol{u}'(\boldsymbol{k};\tau)$ does not obey the usual solenoidal condition concerning $\boldsymbol{k}$, equation (3.106). This situation originates from the fact that the Fourier representation of $\boldsymbol{u}'$ was made for the fast-varying components of motion only. In order to properly deal with equation (5.58), we introduce the transformation [5.3]

$$\boldsymbol{u}'(\boldsymbol{k};\tau) = \boldsymbol{v}'(\boldsymbol{k};\tau) + \delta_{\mathrm{S}}\left(-\mathrm{i}\,\frac{\boldsymbol{k}}{k^2}\nabla_X^*\cdot\boldsymbol{u}'(\boldsymbol{k};\tau)\right). \tag{5.61}$$

As a result, $\boldsymbol{v}'(\boldsymbol{k};\tau)$ is subject to the solenoidal condition concerning $\boldsymbol{k}$, that is,

$$\boldsymbol{k}\cdot\boldsymbol{v}'(\boldsymbol{k};\tau) = 0, \tag{5.62}$$

and various mathematical tools developed in the study of homogeneous turbulence are available for inhomogeneous turbulence.

5.2.1.3 *Scale-parameter expansion*

Once the δ_{S}-related terms in equations (5.58) and (5.59) are dropped, we have the same system of equations as for homogeneous turbulence, equations (3.106) and (3.118), except the implicit influence of slowly-varying properties of $\boldsymbol{U}$ through $\boldsymbol{X}$ and T. This mathematical situation suggests that inhomogeneous turbulence may be investigated by the perturbational method based on δ_{S}.

We expand

$$\boldsymbol{u}'(\boldsymbol{k};\tau) = \sum_{n=0}^{\infty}\delta_{\mathrm{S}}^n\boldsymbol{u}'_n(\boldsymbol{k};\tau), \qquad \boldsymbol{v}'(\boldsymbol{k};\tau) = \sum_{n=0}^{\infty}\delta_{\mathrm{S}}^n\boldsymbol{v}'_n(\boldsymbol{k};\tau), \tag{5.63}$$

$$p'(\boldsymbol{k};\tau) = \sum_{n=0}^{\infty}\delta_{\mathrm{S}}^n p'_n(\boldsymbol{k};\tau). \tag{5.64}$$

We substitute equations (5.63) and (5.64) into equations (5.59), (5.61), and (5.62), and have

$$\boldsymbol{u}'_n(\boldsymbol{k};\tau) = \boldsymbol{v}'_n(\boldsymbol{k};\tau) - \mathrm{i}\,\frac{\boldsymbol{k}}{k^2}\nabla_X^*\cdot\boldsymbol{u}'_{n-1}(\boldsymbol{k};\tau), \tag{5.65}$$

$$\frac{\partial u'_{ni}(\boldsymbol{k};\tau)}{\partial \tau} + \nu k^2 u'_{ni}(\boldsymbol{k};\tau) - \mathrm{i}k_i p'_n(\boldsymbol{k};\tau)$$
$$- \mathrm{i}k_j \iint [u'_{ni}(\boldsymbol{p};\tau)u'_{0j}(\boldsymbol{q};\tau) + u'_{0i}(\boldsymbol{p};\tau)u'_{nj}(\boldsymbol{q};\tau)]\delta(\boldsymbol{k}-\boldsymbol{p}-\boldsymbol{q})\,\mathrm{d}\boldsymbol{p}\,\mathrm{d}\boldsymbol{q}$$
$$= -u'_{n-1j}(\boldsymbol{k};\tau)\frac{\partial U_i}{\partial X_j} - \frac{\mathrm{D}^* u'_{n-1i}(\boldsymbol{k};\tau)}{\mathrm{D}T^*} - \frac{\partial^* p'_{n-1}(\boldsymbol{k};\tau)}{\partial X_i^*}$$
$$- \sum_{m=0}^{n-1} \iint \frac{\partial^*}{\partial X_j^*}[u'_{mi}(\boldsymbol{p};\tau)u'_{n-m-1i}(\boldsymbol{q};\tau)]\delta(\boldsymbol{k}-\boldsymbol{p}-\boldsymbol{q})\,\mathrm{d}\boldsymbol{p}\,\mathrm{d}\boldsymbol{q}, \tag{5.66}$$

where $\boldsymbol{v}'_n(\boldsymbol{k};\tau)$ obeys the solenoidal condition

$$\boldsymbol{k}\cdot\boldsymbol{v}'_n(\boldsymbol{k};\tau) = 0. \tag{5.67}$$

Lowest-order equations

As has already been noted, the lowest-order equations are of the same form as for homogeneous turbulence. Then we have

$$\frac{\partial u'_{0i}(\boldsymbol{k};\tau)}{\partial \tau} + \nu k^2 u'_{0i}(\boldsymbol{k};\tau)$$
$$- \mathrm{i}M_{ijl}(\boldsymbol{k}) \iint u'_{0j}(\boldsymbol{p};\tau)u'_{0l}(\boldsymbol{q};\tau)\delta(\boldsymbol{k}-\boldsymbol{p}-\boldsymbol{q})\,\mathrm{d}\boldsymbol{p}\,\mathrm{d}\boldsymbol{q} = 0, \tag{5.68}$$

$$p'_0(\boldsymbol{k};\tau) = -\frac{k_i k_j}{k^2}\iint u'_{0i}(\boldsymbol{p};\tau)u'_{0j}(\boldsymbol{q};\tau)\delta(\boldsymbol{k}-\boldsymbol{p}-\boldsymbol{q})\,\mathrm{d}\boldsymbol{p}\,\mathrm{d}\boldsymbol{q}, \tag{5.69}$$

from equations (3.119) and (3.120), where we should note

$$\boldsymbol{u}'_0(\boldsymbol{k};\tau) = \boldsymbol{v}'_0(\boldsymbol{k};\tau). \tag{5.70}$$

First-order equations

Explicit effects of mean flow occur in the first-order equations. From equations (5.65)–(5.67), these equations are written as

$$\boldsymbol{u}'_1(\boldsymbol{k};\tau) = \boldsymbol{v}'_1(\boldsymbol{k};\tau) - \mathrm{i}\frac{\boldsymbol{k}}{k^2}\nabla_X^*\cdot\boldsymbol{u}'_0(\boldsymbol{k};\tau), \tag{5.71}$$

$$\frac{\partial u'_{1i}(\boldsymbol{k};\tau)}{\partial \tau} + \nu k^2 u'_{1i}(\boldsymbol{k};\tau) - \mathrm{i}k_i p'_1(\boldsymbol{k};\tau)$$
$$= -u'_{0j}(\boldsymbol{k};\tau)\frac{\partial U_i}{\partial X_j} - \frac{\mathrm{D}^* u'_{0i}(\boldsymbol{k};\tau)}{\mathrm{D}T^*} - \frac{\partial^* p'_0(\boldsymbol{k};\tau)}{\partial X_i^*}$$
$$- \iint \frac{\partial^*}{\partial X_j^*}[u'_{0i}(\boldsymbol{p};\tau)u'_{0j}(\boldsymbol{q};\tau)]\delta(\boldsymbol{k}-\boldsymbol{p}-\boldsymbol{q})\,\mathrm{d}\boldsymbol{p}\,\mathrm{d}\boldsymbol{q}, \tag{5.72}$$

with

$$\boldsymbol{k}\cdot\boldsymbol{v}'_1(\boldsymbol{k};\tau) = 0. \tag{5.73}$$

We substitute equation (5.71) into equation (5.72), and apply equation (5.73). Then we have

$$p'_1(\boldsymbol{k};\tau) = -2\mathrm{i}\frac{k_j}{k^2}\frac{\partial U_j}{\partial X_i}u'_{0i}(\boldsymbol{k};\tau)$$
$$-2\frac{k_i k_j}{k^2}\iint u'_{0i}(\boldsymbol{p};\tau)u'_{1j}(\boldsymbol{q};\tau)\delta(\boldsymbol{k}-\boldsymbol{p}-\boldsymbol{q})\,\mathrm{d}\boldsymbol{p}\,\mathrm{d}\boldsymbol{q}$$
$$-2\mathrm{i}\frac{1}{k^2}M_{lij}(\boldsymbol{k})\iint \frac{\partial^*}{\partial X^*_l}[u'_{0i}(\boldsymbol{p};\tau)u'_{0j}(\boldsymbol{q};\tau)]\delta(\boldsymbol{k}-\boldsymbol{p}-\boldsymbol{q})\,\mathrm{d}\boldsymbol{p}\,\mathrm{d}\boldsymbol{q}. \quad (5.74)$$

We use equations (5.71) and (5.74), and eliminate p'_1 from equation (5.72), obtaining

$$\frac{\partial v'_{1i}(\boldsymbol{k};\tau)}{\partial\tau} + \nu k^2 v'_{1i}(\boldsymbol{k};\tau) - 2\mathrm{i}M_{ijl}(\boldsymbol{k})\iint u'_{0j}(\boldsymbol{p};\tau)v'_{1l}(\boldsymbol{q};\tau)\delta(\boldsymbol{k}-\boldsymbol{p}-\boldsymbol{q})\,\mathrm{d}\boldsymbol{p}\,\mathrm{d}\boldsymbol{q}$$
$$= I_{1i}(\boldsymbol{k};\tau) \equiv -D_{il}(\boldsymbol{k})u'_{0j}(\boldsymbol{k};\tau)\frac{\partial U_l}{\partial X_j} - D_{ij}(\boldsymbol{k})\frac{\mathrm{D}^* u'_{0j}(\boldsymbol{k};\tau)}{\mathrm{D}T^*}$$
$$+ 2M_{ijl}(\boldsymbol{k})\iint \frac{q_l}{q^2}u'_{0j}(\boldsymbol{p};\tau)\nabla^*_X\cdot\boldsymbol{u}'_0(\boldsymbol{q};\tau)\delta(\boldsymbol{k}-\boldsymbol{p}-\boldsymbol{q})\,\mathrm{d}\boldsymbol{p}\,\mathrm{d}\boldsymbol{q}$$
$$- D_{in}(\boldsymbol{k})M_{jlmn}(\boldsymbol{k})\iint \frac{\partial^*}{\partial X^*_m}[u'_{0j}(\boldsymbol{p};\tau)u'_{0l}(\boldsymbol{q};\tau)]\delta(\boldsymbol{k}-\boldsymbol{p}-\boldsymbol{q})\,\mathrm{d}\boldsymbol{p}\,\mathrm{d}\boldsymbol{q}, \quad (5.75)$$

where

$$M_{ijlm}(\boldsymbol{k}) = \tfrac{1}{2}\delta_{il}\delta_{jm} + \tfrac{1}{2}\delta_{im}\delta_{jl} - \frac{k_i k_j}{k^2}\delta_{lm}. \quad (5.76)$$

5.2.2 Introduction of Green's Function

The lowest-order equation (5.68) is not dependent explicitly on $\boldsymbol{U}$, but it is a nonlinear equation. The first-order equation (5.75) depends on both $\boldsymbol{U}$ and $\boldsymbol{u}'_0$ in a complicated manner, but it is a linear equation. The latter equation may be integrated formally with the aid of a Green's function.

5.2.2.1 Green's function in inhomogeneous turbulence

We rewrite equation (5.75) as

$$\frac{\partial v'_{1i}(\boldsymbol{k};\tau)}{\partial\tau} + \nu k^2 v'_{1i}(\boldsymbol{k};\tau) - 2\mathrm{i}M_{ijl}(\boldsymbol{k})\iint u'_{0j}(\boldsymbol{p};\tau)v'_{1l}(\boldsymbol{q};\tau)\delta(\boldsymbol{k}-\boldsymbol{p}-\boldsymbol{q})\,\mathrm{d}\boldsymbol{p}\,\mathrm{d}\boldsymbol{q}$$
$$= \int \delta(\boldsymbol{k}-\boldsymbol{k}_1)\,\mathrm{d}\boldsymbol{k}_1\int_{-\infty}^{\tau} I_{1i}(\boldsymbol{k}_1;\tau_1)\delta(\tau-\tau_1)\,\mathrm{d}\tau_1. \quad (5.77)$$

In correspondence to the left-hand side of equation (5.77), we introduce the Green's function $G'_{ij}(\boldsymbol{k},\boldsymbol{k}';\tau,\tau')$ obeying

$$\frac{\partial G'_{ij}(\boldsymbol{k},\boldsymbol{k}';\tau,\tau')}{\partial\tau} + \nu k^2 G'_{ij}(\boldsymbol{k},\boldsymbol{k}';\tau,\tau') - 2\mathrm{i}M_{ilm}(\boldsymbol{k})\iint u'_{0l}(\boldsymbol{p},\tau)G'_{mj}(\boldsymbol{q},\boldsymbol{k}';\tau,\tau')\delta(\boldsymbol{k}-\boldsymbol{p}-\boldsymbol{q})\,\mathrm{d}\boldsymbol{p}\,\mathrm{d}\boldsymbol{q} = D_{ij}(\boldsymbol{k}')\delta(\boldsymbol{k}-\boldsymbol{k}')\delta(\tau-\tau'). \tag{5.78}$$

We use G'_{ij} and integrate equation (5.77) as

$$v'_{1i}(\boldsymbol{k};\tau) = \int \mathrm{d}\boldsymbol{k}_1 \int_{-\infty}^{\tau} G'_{ij}(\boldsymbol{k},\boldsymbol{k}_1;\tau,\tau_1)I_{1j}(\boldsymbol{k}_1;\tau_1)\,\mathrm{d}\tau_1. \tag{5.79}$$

Here we chose $-\infty$ as the lower limit of time integral. This is due to the premise that our interest lies in a turbulence state independent of initial conditions. Considering the dependence of the right-hand side of equation (5.78) on $\delta(\boldsymbol{k}-\boldsymbol{k}')$, we put

$$G'_{ij}(\boldsymbol{k},\boldsymbol{k}';\tau,\tau') = G'_{ij}(\boldsymbol{k}';\tau,\tau')\delta(\boldsymbol{k}-\boldsymbol{k}'). \tag{5.80}$$

We substitute equation (5.80) into equation (5.78), and perform the integration with respect to $\boldsymbol{k}'$. Then we have

$$\frac{\partial G'_{ij}(\boldsymbol{k};\tau,\tau')}{\partial\tau} + \nu k^2 G'_{ij}(\boldsymbol{k};\tau,\tau') - 2\mathrm{i}M_{ilm}(\boldsymbol{k})\iint u'_{0l}(\boldsymbol{p};\tau)G'_{mj}(\boldsymbol{q};\tau,\tau')\delta(\boldsymbol{k}-\boldsymbol{p}-\boldsymbol{q})\,\mathrm{d}\boldsymbol{p}\,\mathrm{d}\boldsymbol{q} = D_{ij}(\boldsymbol{k})\delta(\tau-\tau'). \tag{5.81}$$

In this context, equation (5.79) is reduced to

$$v'_{1i}(\boldsymbol{k};\tau) = \int_{-\infty}^{\tau} G'_{ij}(\boldsymbol{k};\tau,\tau_1)I_{1j}(\boldsymbol{k};\tau_1)\,\mathrm{d}\tau_1. \tag{5.82}$$

From equations (5.71) and (5.82), the first-order solution $\boldsymbol{u}'_1$ is given by

$$u'_{1i}(\boldsymbol{k};\tau) = \int_{-\infty}^{\tau} G'_{ij}(\boldsymbol{k};\tau,\tau_1)I_{1j}(\boldsymbol{k};\tau_1)\,\mathrm{d}\tau_1 - \mathrm{i}\frac{k_i}{k^2}\nabla^*_X\cdot\boldsymbol{u}'_0(\boldsymbol{k};\tau). \tag{5.83}$$

The second-order solution $\boldsymbol{u}'_2$ may be calculated in an entirely similar manner [2.16, 4.8, 5.4]. The manipulation, however, is very complicated owing to the occurrence of many terms. To alleviate the complexity, we focus attention on the effects linear in $\boldsymbol{u}'_0$. After this simplification, we have

$$\boldsymbol{u}'_2(\boldsymbol{k};\tau) = \boldsymbol{v}'_2(\boldsymbol{k};\tau) - \mathrm{i}\frac{\boldsymbol{k}}{k^2}\nabla^*_X\cdot\boldsymbol{v}'_1(\boldsymbol{k};\tau) - \frac{\boldsymbol{k}k_i}{k^4}\frac{\partial^*}{\partial X^*_i}\nabla^*_X\cdot\boldsymbol{u}'_0(\boldsymbol{k};\tau). \tag{5.84}$$

Here $\boldsymbol{v}_2'$ is given by

$$v_{2i}'(\boldsymbol{k};\tau) = \int_{-\infty}^{T} G_{ij}'(\boldsymbol{k};\tau,\tau_1) I_{2j}(\boldsymbol{k};\tau_1)\, d\tau_1, \tag{5.85}$$

with $\boldsymbol{I}_2$ defined by

$$\begin{aligned} I_{2i}(\boldsymbol{k};\tau) &= \mathrm{i}\frac{k_m}{k^2} D_{ij}(\boldsymbol{k}) \frac{\partial^2 U_m}{\partial X_j \partial X_n} u_{0n}'(\boldsymbol{k};\tau) - D_{im}(\boldsymbol{k}) \frac{\partial U_m}{\partial X_j} v_{1j}'(\boldsymbol{k};\tau) \\ &\quad + \mathrm{i}\frac{k_j}{k^2} D_{im}(\boldsymbol{k}) \frac{\partial U_j}{\partial X_n} \frac{\partial^* u_{0n}'(\boldsymbol{k};\tau)}{\partial X_m^*} + \mathrm{i}\frac{k_m}{k^2} D_{ij}(\boldsymbol{k}) \frac{\partial U_j}{\partial x_m} \nabla_X^* \cdot \boldsymbol{u}_0'(\boldsymbol{k};\tau) \\ &\quad - \frac{\mathrm{D}^* v_{1i}'(\boldsymbol{k};\tau)}{\mathrm{D}T^*} + \frac{1}{k^2} D_{ij}(\boldsymbol{k}) \frac{\partial}{\partial \tau} \frac{\partial^*}{\partial X_j^*} \nabla_X^* \cdot \boldsymbol{u}_0'(\boldsymbol{k};\tau). \end{aligned} \tag{5.86}$$

5.2.2.2 Green's function in homogeneous turbulence and isotropic turbulence theory

In §5.2.2.1 the Green's function G_{ij}' was introduced in relation to the perturbational solution of a system of equations for inhomogeneous turbulence. There the lowest-order solution $\boldsymbol{u}_0'$ was treated as known. For calculating various statistical quantities with the aid of the perturbational solution, it is indispensable to know the statistical properties of $\boldsymbol{u}_0'$. No explicit effects of $\boldsymbol{U}$ enter equation (5.68) for $\boldsymbol{u}_0'$, and such effects arise through the implicit dependence on slow variables $\boldsymbol{X}$ and T. Then the statistical properties of $\boldsymbol{u}_0'$ may be examined by homogeneous turbulence theories.

Direct-interaction approximation

The prototype of homogeneous turbulence theories is the direct-interaction approximation (DIA) by Kraichnan [3.4, 3.5, 5.1, 5.2]. Its understanding is helpful to understanding not only other homogeneous turbulence theories, but also the mathematical procedures for evaluating the Reynolds stress, etc., in inhomogeneous turbulence.

Following the method of solving equation (5.30) in a perturbational manner, we regard the third term on the left-hand side of equation (5.68) as a perturbation, and formally integrate

$$\begin{aligned} u_{0i}'(\boldsymbol{k},\tau) &= w_i(\boldsymbol{k},\tau) + \mathrm{i}M_{mjl}(\boldsymbol{k}) \iint \delta(\boldsymbol{k}-\boldsymbol{p}-\boldsymbol{q})\, d\boldsymbol{p}\, d\boldsymbol{q} \\ &\quad \times \int_{-\infty}^{T} \hat{G}_{im}(\boldsymbol{k};\tau,\tau_1) u_{0j}'(\boldsymbol{p};\tau_1) u_{0l}'(\boldsymbol{q};\tau_1)\, d\tau_1. \end{aligned} \tag{5.87}$$

Here the first term $\boldsymbol{w}$, which is assumed to be a random flow obeying a Gaussian distribution, denotes $\boldsymbol{u}_0'$ at $t = -\infty$. The second term expresses the effect arising from the nonlinear interaction thereafter. Moreover $\hat{G}_{ij}$ obeys

$$\frac{\partial \hat{G}_{ij}(\boldsymbol{k};\tau,\tau')}{\partial \tau} + \nu k^2 \hat{G}_{ij}(\boldsymbol{k};\tau,\tau') = D_{ij}(\boldsymbol{k})\delta(\tau-\tau'), \tag{5.88}$$

and is given by

$$\hat{G}_{ij}(\boldsymbol{k};\tau,\tau') = D_{ij}(\boldsymbol{k})S(\tau-\tau')\exp[-\nu k^2(\tau-\tau')]. \tag{5.89}$$

Equation (5.88) is reduced from equation (5.78) with the nonlinear term dropped, and corresponds to the Green's or response equation in the low-Reynolds-number limit. We should note that $\hat{G}_{ij}$ is a deterministic quantity.

Similar to equation (5.87), we rewrite equation (5.78) as

$$G'_{ij}(\boldsymbol{k};\tau,\tau') = \hat{G}_{ij}(\boldsymbol{k};\tau,\tau') + 2\mathrm{i}M_{nlm}(\boldsymbol{k})\iint \delta(\boldsymbol{k}-\boldsymbol{p}-\boldsymbol{q})\,\mathrm{d}\boldsymbol{p}\,\mathrm{d}\boldsymbol{q}$$
$$\times \int_{\tau'}^{\tau} \hat{G}_{in}(\boldsymbol{k};\tau,\tau_1)u'_{0l}(\boldsymbol{p};\tau_1)G'_{mj}(\boldsymbol{q};\tau_1,\tau')\,\mathrm{d}\tau_1. \tag{5.90}$$

Here the lower limit of time integral, τ', means that a disturbance is added at τ' and its effect occurs for $\tau > \tau'$.

We solve equations (5.87) and (5.90) in an iterative manner with each first term as the leading part. As a result, we have

$$u'_{0i}(\boldsymbol{k};\tau) = w_i(\boldsymbol{k};\tau) + \mathrm{i}M_{mjl}(\boldsymbol{k})\iint \delta(\boldsymbol{k}-\boldsymbol{p}-\boldsymbol{q})\,\mathrm{d}\boldsymbol{p}\,\mathrm{d}\boldsymbol{q}$$
$$\times \int_{-\infty}^{\tau} \hat{G}_{im}(\boldsymbol{k};\tau,\tau_1)w_j(\boldsymbol{p};\tau_1)w_l(\boldsymbol{q};\tau_1)\,\mathrm{d}\tau_1 + \cdots, \tag{5.91}$$

$$G'_{ij}(\boldsymbol{k};\tau,\tau') = \hat{G}_{ij}(\boldsymbol{k};\tau,\tau') + 2\mathrm{i}M_{nlm}(\boldsymbol{k})\iint \delta(\boldsymbol{k}-\boldsymbol{p}-\boldsymbol{q})\,\mathrm{d}\boldsymbol{p}\,\mathrm{d}\boldsymbol{q}$$
$$\times \int_{\tau'}^{\tau} \hat{G}_{in}(\boldsymbol{k};\tau,\tau_1)w_l(\boldsymbol{p};\tau_1)\hat{G}_{mj}(\boldsymbol{q};\tau_1,\tau')\,\mathrm{d}\tau_1 + \cdots. \tag{5.92}$$

In the DIA, the two-time covariance

$$Q_{ij}(\boldsymbol{k};\tau,\tau') = \frac{\langle u'_{0i}(\boldsymbol{k};\tau)u'_{0j}(\boldsymbol{k}';\tau')\rangle}{\delta(\boldsymbol{k}+\boldsymbol{k}')} \tag{5.93}$$

and the mean of the Green's function

$$G_{ij}(\boldsymbol{k};\tau,\tau') = \langle G'_{ij}(\boldsymbol{k};\tau,\tau')\rangle \tag{5.94}$$

are chosen as the fundamental statistical quantities of turbulence. We construct the equations governing these two quantities.

From equation (5.68), equation (5.93) obeys

$$LQ_{ij}(\boldsymbol{k};\tau,\tau') \equiv \left(\frac{\partial}{\partial\tau} + \nu k^2\right) Q_{ij}(\boldsymbol{k};\tau,\tau')$$
$$= \mathrm{i}M_{ilm}(\boldsymbol{k}) \iint \frac{\langle u'_{0l}(\boldsymbol{p};\tau) u'_{0m}(\boldsymbol{q};\tau) u'_{0j}(\boldsymbol{k}';\tau')\rangle}{\delta(\boldsymbol{k}+\boldsymbol{k}')}$$
$$\times \delta(\boldsymbol{k}-\boldsymbol{p}-\boldsymbol{q})\,\mathrm{d}\boldsymbol{p}\,\mathrm{d}\boldsymbol{q}. \tag{5.95}$$

We substitute the perturbational solution (5.91) into the right-hand side of equation (5.95), and make use of the Gaussianity of $\boldsymbol{w}$. Then the right-hand side is expressed in terms of the second-order correlation concerning $\boldsymbol{w}$, that is,

$$\hat{Q}_{ij}(\boldsymbol{k};\tau,\tau') = \frac{\langle w_i(\boldsymbol{k};\tau) w_j(\boldsymbol{k}';\tau')\rangle}{\delta(\boldsymbol{k}+\boldsymbol{k}')}. \tag{5.96}$$

Here we retain the contributions of the lowest order in $\hat{Q}_{ij}$, and have

$$LQ_{ij}(\boldsymbol{k};\tau,\tau') = 2\iint M_{iab}(\boldsymbol{k}) M_{ecd}(\boldsymbol{k}) \delta(\boldsymbol{k}-\boldsymbol{p}-\boldsymbol{q})\,\mathrm{d}\boldsymbol{p}\,\mathrm{d}\boldsymbol{q}$$
$$\times \int_{-\infty}^{\tau'} \hat{G}_{je}(\boldsymbol{k}';\tau',\tau_1) \hat{Q}_{ac}(\boldsymbol{p};\tau,\tau_1) \hat{Q}_{bd}(\boldsymbol{q};\tau,\tau_1)\,\mathrm{d}\tau_1$$
$$- 4\iint M_{iab}(\boldsymbol{k}) M_{dce}(\boldsymbol{q}) \delta(\boldsymbol{k}-\boldsymbol{p}-\boldsymbol{q})\,\mathrm{d}\boldsymbol{p}\,\mathrm{d}\boldsymbol{q}$$
$$\times \int_{-\infty}^{\tau} \hat{G}_{bd}(\boldsymbol{q};\tau,\tau_1) \hat{Q}_{ac}(\boldsymbol{p};\tau,\tau_1) \hat{Q}_{ej}(-\boldsymbol{k};\tau_1,\tau')\,\mathrm{d}\tau_1. \tag{5.97}$$

Following the renormalization procedures explained in §5.1.3, we make the replacement

$$\hat{Q}_{ij}(\boldsymbol{k};\tau,\tau') \to Q_{ij}(\boldsymbol{k};\tau,\tau'), \qquad \hat{G}_{ij}(\boldsymbol{k};\tau,\tau') \to G_{ij}(\boldsymbol{k};\tau,\tau'), \tag{5.98}$$

which gives

$$LQ_{ij}(\boldsymbol{k};\tau,\tau') = 2\iint M_{iab}(\boldsymbol{k}) M_{ecd}(\boldsymbol{k}) \delta(\boldsymbol{k}-\boldsymbol{p}-\boldsymbol{q})\,\mathrm{d}\boldsymbol{p}\,\mathrm{d}\boldsymbol{q}$$
$$\times \int_{-\infty}^{\tau'} G_{je}(\boldsymbol{k}';\tau',\tau_1) Q_{ac}(\boldsymbol{p};\tau,\tau_1) Q_{bd}(\boldsymbol{q};\tau,\tau_1)\,\mathrm{d}\tau_1$$
$$- 4\iint M_{iab}(\boldsymbol{k}) M_{dce}(\boldsymbol{q}) \delta(\boldsymbol{k}-\boldsymbol{p}-\boldsymbol{q})\,\mathrm{d}\boldsymbol{p}\,\mathrm{d}\boldsymbol{q}$$
$$\times \int_{-\infty}^{\tau} G_{bd}(\boldsymbol{q};\tau,\tau_1) Q_{ac}(\boldsymbol{p};\tau,\tau_1) Q_{ej}(-\boldsymbol{k};\tau_1,\tau')\,\mathrm{d}\tau_1. \tag{5.99}$$

If the latter half is not made in equation (5.98), the resulting equation for $\tau = \tau'$ is coincident with the so-called quasi-normal approximation [5.5]. In the formalism, the past events are overestimated, as was discussed in §5.1.3. Its typical shortfall is the occurrence of the negative energy spectrum.

We apply entirely the same procedure to G_{ij}, and have

$$LG_{ij}(\boldsymbol{k};\tau,\tau') = D_{ij}(\boldsymbol{k})\delta(\tau-\tau') - 4\iint M_{iab}(\boldsymbol{k})M_{dce}(\boldsymbol{q})\delta(\boldsymbol{k}-\boldsymbol{p}-\boldsymbol{q})\,\mathrm{d}\boldsymbol{p}\,\mathrm{d}\boldsymbol{q}$$
$$\times \int_{\tau'}^{\tau} G_{bd}(\boldsymbol{q};\tau,\tau_1)G_{ej}(-\boldsymbol{k};\tau_1,\tau')Q_{ac}(\boldsymbol{p};\tau,\tau_1)\,\mathrm{d}\tau_1. \qquad (5.100)$$

This is combined with equation (5.99) to constitute the DIA system of equations. The derivation of the DIA system may be more systematic with the aid of a diagrammatic representation [2.16, 3.5, 5.6].

Difficulty about Green's function

In order to see the relationship of the foregoing DIA system with the Kolmogorov spectrum (3.138), we assume the isotropy of turbulence and write

$$Q_{ij}(\boldsymbol{k};\tau,\tau') = D_{ij}(\boldsymbol{k})Q(k;\tau,\tau'), \qquad (5.101)$$
$$G_{ij}(\boldsymbol{k};\tau,\tau') = D_{ij}(\boldsymbol{k})G(k;\tau,\tau'), \qquad (5.102)$$

from equation (3.108) with the helicity part dropped.

We substitute equations (5.101) and (5.102) into equations (5.99) and (5.100), and have

$$LQ(k;\tau,\tau') = k^2\iint\Bigg(N_{Q1}(k,p,q)\int_{-\infty}^{\tau'} G(k;\tau',\tau_1)Q(p;\tau,\tau_1)Q(q;\tau,\tau_1)\,\mathrm{d}\tau_1$$
$$- N_{Q2}(k,p,q)\int_{-\infty}^{\tau} G(q;\tau,\tau_1)Q(p;\tau,\tau_1)Q(k;\tau_1,\tau')\,\mathrm{d}\tau_1\Bigg)$$
$$\times\,\delta(\boldsymbol{k}-\boldsymbol{p}-\boldsymbol{q})\,\mathrm{d}\boldsymbol{p}\,\mathrm{d}\boldsymbol{q}, \qquad (5.103)$$

$$LG(k;\tau,\tau') = \delta(\tau-\tau') - k^2\iint N_G(k,p,q)\delta(\boldsymbol{k}-\boldsymbol{p}-\boldsymbol{q})\,\mathrm{d}\boldsymbol{p}\,\mathrm{d}\boldsymbol{q}$$
$$\times\int_{\tau'}^{\tau} G(q;\tau,\tau_1)G(k;\tau_1,\tau')Q(p;\tau,\tau_1)\,\mathrm{d}\tau_1. \qquad (5.104)$$

For geometrical factors, we have the relations

$$N_{Q1}(k,p,q) = M_{eab}(\boldsymbol{k})M_{ecd}(\boldsymbol{k})D_{ac}(\boldsymbol{p})D_{bd}(\boldsymbol{q})$$
$$= N(k,p,q) \equiv \frac{q}{k}(xz+y^3), \qquad (5.105)$$

$$N_{Q2}(k,p,q) = N_G(k,p,q)$$
$$= 2M_{dab}(\boldsymbol{k})M_{bcd}(\boldsymbol{q})D_{ac}(\boldsymbol{p}) = N(k,p,q), \qquad (5.106)$$

where x, y, and z are the cosines of the angles opposite to sides k, p, and q that constitute a triangle. Specifically, the one-time covariance $Q(k;\tau,\tau)$ obeys

$$\left(\frac{\partial}{\partial\tau}+2\nu k^2\right)Q(k;\tau,\tau)=2k^2\iint N(k,p,q)$$
$$\times\left(\int_{-\infty}^{\tau}\mathrm{d}\tau_1 G(k;\tau,\tau_1)Q(p;\tau,\tau_1)Q(q;\tau,\tau_1)\right.$$
$$\left.-\int_{-\infty}^{\tau}\mathrm{d}\tau_1 G(q;\tau,\tau_1)Q(p;\tau,\tau_1)Q(k;\tau_1,\tau)\right)$$
$$\times\,\delta(\boldsymbol{k}-\boldsymbol{p}-\boldsymbol{q})\,\mathrm{d}\boldsymbol{p}\,\mathrm{d}\boldsymbol{q}. \tag{5.107}$$

For $Q(k;\tau,\tau')$ and $G(k;\tau,\tau')$, we assume the simplest stationary expressions

$$Q(k;\tau,\tau')=\sigma(k)\exp(-\omega(k)|\tau-\tau'|), \tag{5.108}$$

$$G(k;\tau,\tau')=S(\tau-\tau')\exp(-\omega(k)(\tau-\tau')), \tag{5.109}$$

where $\sigma(k)$ is related to the energy spectrum $E(k)$ as

$$E(k)=4\pi k^2\sigma(k). \tag{5.110}$$

From the discussions on the Kolmogorov spectrum in §3.4.2.1, we may write

$$\sigma(k)=\frac{K_O}{4\pi}\varepsilon^{2/3}k^{-11/3}, \tag{5.111}$$

$$\omega(k)=\frac{1}{C_\tau}\varepsilon^{1/3}k^{2/3} \tag{5.112}$$

[see equations (3.138) and (3.139)].

We substitute equations (5.108) and (5.109) into equation (5.107), and have

$$2\nu k^2\sigma(k)=2k^2\iint N(k,p,q)\frac{\sigma(p)[\sigma(q)-\sigma(k)]}{\omega(k)+\omega(p)+\omega(q)}$$
$$\times\,\delta(\boldsymbol{k}-\boldsymbol{p}-\boldsymbol{q})\,\mathrm{d}\boldsymbol{p}\,\mathrm{d}\boldsymbol{q}. \tag{5.113}$$

This type of equation may be also derived by the method based on the Liouville equation for the probability distribution function that was discussed in §5.1.2 [5.7, 5.8]. We choose the wavevector $\boldsymbol{r}$ obeying

$$k_\mathrm{E}\ll r\ll k_\mathrm{D}, \tag{5.114}$$

where k_E and k_D are the wavenumbers characterizing the energy-containing and -dissipation ranges, respectively, which are given by equations (3.142a) and (3.146a). We integrate equation (5.113) with respect to the wavevector $\boldsymbol{k}$, as

$$2\nu \int_{k>r} r^2 \sigma(k)\, \mathrm{d}\boldsymbol{k} = 2\int_{k>r} k^2\, \mathrm{d}\boldsymbol{k} \iint N(k,p,q) \frac{\sigma(p)[\sigma(q)-\sigma(k)]}{\omega(k)+\omega(p)+\omega(q)} \times \delta(\boldsymbol{k}-\boldsymbol{p}-\boldsymbol{q})\, \mathrm{d}\boldsymbol{p}\, \mathrm{d}\boldsymbol{q}. \quad (5.115)$$

We use equations (3.129) and (3.130), and have

$$2\nu \int_{k>r} k^2 \sigma(k)\, \mathrm{d}\boldsymbol{k} = 2\nu \int_{k>r} k^2 E(k)\, \mathrm{d}k = \varepsilon, \quad (5.116)$$

where we should note that little energy is dissipated in the inertial range. We combined equations (5.111) and (5.112) with equations (5.115) and (5.116), obtaining

$$K_O^2 C_\tau = 5.2. \quad (5.117)$$

The simple expressions such as equations (5.108) and (5.109) cannot satisfy the response equation (5.104) exactly. Then we seek its weak solution; namely, we integrate equation (5.104) as

$$\int_{\tau'+0}^{\infty} \left(LG(k;\tau,\tau') + k^2 \iint N_G(k,p,q) \delta(\boldsymbol{k}-\boldsymbol{p}-\boldsymbol{q})\, \mathrm{d}\boldsymbol{p}\, \mathrm{d}\boldsymbol{q} \times \int_{\tau'}^{\tau} G(q;\tau,\tau_1) G(k;\tau_1,\tau') Q(p;\tau,\tau_1)\, \mathrm{d}\tau_1 \right) \mathrm{d}\tau = 0. \quad (5.118)$$

Then we have

$$\omega(k) = \nu k^2 + k^2 \iint N(k,p,q) \frac{\sigma(p)}{\omega(p)+\omega(q)} \delta(\boldsymbol{k}-\boldsymbol{p}-\boldsymbol{q})\, \mathrm{d}\boldsymbol{p}\, \mathrm{d}\boldsymbol{q}, \quad (5.119)$$

from equations (5.108) and (5.109).

On substituting equations (5.111) and (5.112) into equation (5.119), we encounter the difficulty that the resulting integral does not converge in the lower limit of $\boldsymbol{p} \to 0$. Such divergence of the integral is called the infrared divergence [5.9–5.11]. Its cause is quite similar to the effect of $\boldsymbol{U}$ on $\boldsymbol{u}'$ in equation (5.57) with $\boldsymbol{u}'$ adopted as f'. As may be seen from the second term on the left-hand side of equation (5.54), the small eddies expressed by $\boldsymbol{u}'$ are swept away by the large-scale motion $\boldsymbol{U}$. The time-scale due to this sweeping-away effect, τ_S, is

$$\tau_S \propto (\boldsymbol{k} \cdot \boldsymbol{U})^{-1}. \quad (5.120)$$

It is different from equation (3.139) that expresses the lifetime of eddies whose size is $2\pi/k$.

In the context of inhomogeneous turbulence, the explicit sweeping-away effect may be removed through the moving-frame Fourier representation, equation (5.57). In the investigation of homogeneous turbulence, however, the sweeping-away effect still survives in the two-time equations (5.103) and (5.104), resulting in the above difficulty. The construction of a formalism free from the difficulty has long been a central theme in the study of homogeneous turbulence. At present, the infrared divergence was successfully removed with the aid of the Lagrangian description of turbulence field that will be referred to below.

In the following explanation of inhomogeneous-turbulence theory, we make use of equations (5.108) and (5.109). This is solely for the estimate of anisotropy induced by mean-velocity gradients. We should stress that the Lagrangian formalism is necessary for the analysis of homogeneous-turbulence statistics themselves. With this point in mind, we adopt

$$K_O \cong 1.5 \tag{5.121}$$

as a typical observational value, which leads to

$$C_\tau = 2.3, \tag{5.122}$$

from equation (5.117).

Eddy-damped quasi-normal Markovianized approximation

The difficulty encountered by the DIA arises from the two-point quantities such as $Q(k;\tau,\tau')$ and $G(k;\tau,\tau')$, whereas $Q(k;\tau,\tau)$ is free from it, as is seen from equation (5.107). The method using only $Q(k;\tau,\tau)$ is the eddy-damped quasi-normal Markovianized approximation (EDQNMA) [3.6]. There we approximate

$$Q(k;\tau,\tau') = G(k;\tau,\tau')Q(k;\tau,\tau), \tag{5.123}$$

for $\tau > \tau'$. Instead of solving the equation for $G(k;\tau,\tau')$, we model $\omega(k)$ in equation (5.108) as

$$\omega(k) = C'_\omega \left(\int_0^k r^4 Q(r;\tau,\tau)\,\mathrm{d}r \right)^{1/2} + \nu k^2, \tag{5.124}$$

from dimensional consideration. Here the numerical factor C'_ω is determined so that equation (5.107) may give the Kolmogorov constant K_O around 1.5. The retention of only the molecular-viscosity effect leads to the so-called quasi-normal approximation.

The choice of $\omega(k)$ is not unique, and the EDQNMA is subject to some theoretical ambiguity. Its merit, however, lies in its simplicity and applicability. This method has already been applied to the study of homogeneous-shear turbulence, effects of frame rotation on turbulence properties, etc. [5.12, 5.13]. In the latter case, the rotation effect on $\omega(k)$ needs to be included.

Lagrangian formalism

In order to properly treat the effect of energy-containing eddies sweeping away smaller eddies, we need to examine the evolution of a fluid blob or an eddy in a Lagrangian sense [5.2, 5.14, 5.15]. We consider the blob at time s that was located at $\boldsymbol{x}$ at a previous time τ. Its velocity is denoted by $\boldsymbol{u}'(\boldsymbol{x}, \tau | s)$. The Eulerian velocity $\boldsymbol{u}'(\boldsymbol{x}, \tau)$ is given by $\boldsymbol{u}'(\boldsymbol{x}, \tau | \tau)$. The condition that this blob also passes the location $\boldsymbol{x}'$ at time τ' $(\tau < \tau' < s)$ gives

$$\boldsymbol{x}' = \boldsymbol{x} + \int_{\tau}^{\tau'} \boldsymbol{u}'(\boldsymbol{x}, \tau | \tau_1)\, \mathrm{d}\tau_1, \tag{5.125}$$

$$\boldsymbol{u}'(\boldsymbol{x}', \tau' | s) = \boldsymbol{u}'(\boldsymbol{x}, \tau | s). \tag{5.126}$$

For infinitesimal $|\tau - \tau'|$, equation (5.126) is reduced to

$$\frac{\partial \boldsymbol{u}'(\boldsymbol{x}, \tau | s)}{\partial \tau} + [\boldsymbol{u}'(\boldsymbol{x}, \tau) \cdot \nabla]\boldsymbol{u}'(\boldsymbol{x}, \tau | s) = 0. \tag{5.127}$$

In the Eulerian frame, $\langle u_i'(\boldsymbol{x}, \tau) u_j'(\boldsymbol{x}, \tau') \rangle$ for $\tau > \tau'$ corresponds to the correlation between one blob located at $\boldsymbol{x}$ at time τ' and the other blob that was located at $\boldsymbol{x} - \int_{\tau'}^{\tau} \boldsymbol{u}'\, \mathrm{d}\tau_1$ at time τ' and newly occupies the location $\boldsymbol{x}$ at time τ. The Kolmogorov energy spectrum is linked with the energy cascade process in which one eddy is split to a number of smaller ones, and our concern is the historical development of those eddies. This process may be pursued properly with the aid of the Lagrangian correlation function

$$Q_{ij}^L(\boldsymbol{x}, \tau; \boldsymbol{x} | \rightarrow x, \tau') = \langle u_i'(\boldsymbol{x}, \tau' | \tau) u_j'(\boldsymbol{x}', \tau') \rangle. \tag{5.128}$$

Here $\boldsymbol{u}'(\boldsymbol{x}, \tau' | \tau)$ expresses such an historical evolution of eddies.

In correspondence to $Q_{ij}^L(\boldsymbol{x}, \tau; \boldsymbol{x}', \tau')$, we introduce the response function for $\boldsymbol{u}'(\boldsymbol{x}, \tau' | \tau)$, $G_{ij}^L(\boldsymbol{x}, \tau; \boldsymbol{x}', \tau')$, and apply the renormalization procedures that are used in the DIA. The Lagrangian method may reproduce the Kolmogorov spectrum with a reasonable estimate of the numerical constant.

Renormalization-group method

The renormalization-group (RNG) method was originally developed in the study of critical phenomena such as the phase transition of many-body systems [5.16]. The basic concept of the method may be summarized as follows. We consider the wavenumber range $0 < k < \Lambda$, and divide it into two ranges as

$$\text{Range I:} \quad 0 < k < b\Lambda, \qquad \text{Range II:} \quad b\Lambda < k < 1, \tag{5.129}$$

where $b = \exp(-l)$ $(l > 0)$. In correspondence to Range I and Range II, the velocity $\boldsymbol{u}'(\boldsymbol{k})$ is expressed as

$$\boldsymbol{u}'(\boldsymbol{k}) = \boldsymbol{u}_{\mathrm{I}}'(\boldsymbol{k}) + \boldsymbol{u}_{\mathrm{II}}'(\boldsymbol{k}), \tag{5.130}$$

where

$$\boldsymbol{u}'_{\mathrm{I}}(\boldsymbol{k}) = \boldsymbol{u}'(\boldsymbol{k})S(b\Lambda - k), \tag{5.131a}$$

$$\boldsymbol{u}'_{\mathrm{II}}(\boldsymbol{k}) = \boldsymbol{u}'(\boldsymbol{k})S(k - b\Lambda). \tag{5.131b}$$

Following equation (3.120b), we write

$$\left(\frac{\partial}{\partial \tau} + \nu_0 k^2\right) u'_i(\boldsymbol{k}) = f_i(\boldsymbol{k}) + \mathrm{i}M_{ijl}(\boldsymbol{k}) \iint u'_j(\boldsymbol{p}) u'_l(\boldsymbol{q}) \delta(\boldsymbol{k} - \boldsymbol{p} - \boldsymbol{q})\, \mathrm{d}\boldsymbol{p}\, \mathrm{d}\boldsymbol{q}. \tag{5.132}$$

In the original equation (3.120b), ν_0 is equal to the molecular viscosity ν, and a random force $\boldsymbol{f}(\boldsymbol{k})$ has been inserted newly.

We substitute equation (5.130) into equation (5.132), and have

$$\begin{aligned}\left(\frac{\partial}{\partial t} + \nu_0 k^2\right) (u'_{\mathrm{I}i}(\boldsymbol{k}) + u'_{\mathrm{II}i}(\boldsymbol{k})) = f_i(\boldsymbol{k}) + \mathrm{i}M_{ijl}(\boldsymbol{k}) \iint & [u'_{\mathrm{I}j}(\boldsymbol{p}) u'_{\mathrm{I}l}(\boldsymbol{q}) \\ & + 2u'_{\mathrm{I}j}(\boldsymbol{p}) u'_{\mathrm{II}l}(\boldsymbol{q}) + u'_{\mathrm{II}j}(\boldsymbol{p}) u'_{\mathrm{II}l}(\boldsymbol{q})] \\ & \times \delta(\boldsymbol{k} - \boldsymbol{p} - \boldsymbol{q})\, \mathrm{d}\boldsymbol{p}\, \mathrm{d}\boldsymbol{q}. \end{aligned} \tag{5.133}$$

We focus attention on the interaction among $\boldsymbol{u}'_{\mathrm{II}}$, dropping the $\boldsymbol{u}'_{\mathrm{I}}$ and $\boldsymbol{u}'_{\mathrm{II}}$ interaction. Using the perturbational solution of $\boldsymbol{u}'_{\mathrm{II}}$ of the type (5.91), we evaluate the interaction among $\boldsymbol{u}'_{\mathrm{II}}$, and incorporate the effect into the ν_0-related part. In this evaluation, the dimension of the $\boldsymbol{k}$ space is extended to d ($\neq 3$). Next, we extend Range I into the full one, that is, $0 < k < \Lambda$, through the rescaling $k \to \exp(l)k$. As a result, we have

$$\nu_0 \to \nu_0 + \Delta\nu(l). \tag{5.134}$$

From equation (5.134), we derive a differential equation for ν_0 and seek its fixed point.

The RNG method was originally devised to abstract the power laws concerning k, as is suggested by the rescaling $k \to \exp(l)k$. The deduction of the Kolmogorov $-5/3$ power spectrum is within the scope of this method, but the estimate of the proportional constant, that is, K_O in equation (3.138), is not so. In the first RNG analysis [5.17] with emphasis on the estimate of K_O, the rescaling was skipped, and the energy equation in the DIA system, equation (5.113), was used as a supplementary equation. This RNG approach to turbulence has been subject to many discussions [3.5, 5.18–5.21]. The RNG method has also been applied to the study of inhomogeneous turbulence, specifically, the analysis of the Reynolds stress R_{ij} [5.22, 5.23].

5.2.3 Statistical Evaluation of Reynolds Stress

5.2.3.1 Wavenumber-space representation

Under the two-scale description based on equations (5.48) and (5.51), the

Reynolds stress

$$R_{ij} = \langle u'_i(\boldsymbol{x}) u'_j(\boldsymbol{x}) \rangle \tag{5.135}$$

is expressed in the form

$$R_{ij} = \langle u'_i(\boldsymbol{\xi}, \boldsymbol{X}; \tau, T) u'_j(\boldsymbol{\xi}, \boldsymbol{X}; \tau, T) \rangle. \tag{5.136}$$

We assume the homogeneity concerning $\boldsymbol{\xi}$, and define

$$R_{ij}(\boldsymbol{k}, \boldsymbol{X}; \tau, T) = \frac{\langle u'_i(\boldsymbol{k}, \boldsymbol{X}; \tau, T) u'_j(\boldsymbol{k}', \boldsymbol{X}; \tau, T) \rangle}{\delta(\boldsymbol{k} + \boldsymbol{k}')}. \tag{5.137}$$

Using equation (5.137), we may write

$$R_{ij} = \int R_{ij}(\boldsymbol{k}, \boldsymbol{X}; \tau, T) \, \mathrm{d}\boldsymbol{k}. \tag{5.138}$$

We substitute the scale-parameter expansion (5.63) into equation (5.137), and retain the terms up to $O(\delta_S)$. Then we have

$$R_{ij}(\boldsymbol{k}; \tau) = \frac{\langle u'_{0i}(\boldsymbol{k}; \tau) u'_{0j}(\boldsymbol{k}'; \tau) \rangle}{\delta(\boldsymbol{k} + \boldsymbol{k}')} + \delta_S \left(\frac{\langle u'_{1i}(\boldsymbol{k}; \tau) u'_{0j}(\boldsymbol{k}'; \tau) \rangle}{\delta(\boldsymbol{k} + \boldsymbol{k}')} + \frac{\langle u'_{0i}(\boldsymbol{k}; \tau) u'_{1j}(\boldsymbol{k}'; \tau) \rangle}{\delta(\boldsymbol{k} + \boldsymbol{k}')} \right). \tag{5.139}$$

The $O(1)$ term has already been given by equations (5.93) and (5.101), and are written as

$$\frac{\langle u'_{0i}(\boldsymbol{k}; \tau) u'_{0j}(\boldsymbol{k}'; \tau) \rangle}{\delta(\boldsymbol{k} + \boldsymbol{k}')} = D_{ij}(\boldsymbol{k}) Q(k; \tau, \tau). \tag{5.140}$$

In relation to the first part of the $O(\delta_S)$ term, we define

$$R^{(1)}_{ij} = -\frac{\partial U_l}{\partial X_m} \int_{-\infty}^{\tau} \frac{\langle G'_{il}(\boldsymbol{k}; \tau, \tau_1) u'_{0m}(\boldsymbol{k}; \tau_1) u'_{0j}(\boldsymbol{k}'; \tau) \rangle}{\delta(\boldsymbol{k} + \boldsymbol{k}')} \, \mathrm{d}\tau_1, \tag{5.141}$$

$$R^{(2)}_{ij} = -\int_{-\infty}^{\tau} \frac{\left\langle G'_{il}(\boldsymbol{k}; \tau, \tau_1) \dfrac{\mathrm{D}^* u'_{0l}(\boldsymbol{k}; \tau_1)}{\mathrm{D}T^*} u'_{0j}(\boldsymbol{k}'; \tau) \right\rangle}{\delta(\boldsymbol{k} + \boldsymbol{k}')} \, \mathrm{d}\tau_1. \tag{5.142}$$

Then we may write

$$\frac{\langle u'_{1i}(\boldsymbol{k}; \tau) u'_{0j}(\boldsymbol{k}'; \tau) \rangle}{\delta(\boldsymbol{k} + \boldsymbol{k}')} = R^{(1)}_{ij} + R^{(2)}_{ij}, \tag{5.143}$$

from equation (5.83) for $\boldsymbol{u}'_1$ [we may confirm that the third and fourth terms of $\boldsymbol{I}_1$ in equation (5.75) make no contribution since it is an odd function of $\boldsymbol{k}$]. We substitute the perturbational solution (5.91) and (5.92), and retain the lowest-order contribution in $\hat{Q}$. As a result, we have

$$R^{(1)}_{ij} = -\frac{\partial U_l}{\partial X_m} \int_{-\infty}^{\tau} \hat{G}_{il}(\boldsymbol{k}; \tau, \tau_1) \hat{Q}_{mj}(\boldsymbol{k}; \tau, \tau_1) \, \mathrm{d}\tau_1. \tag{5.144}$$

The application of the renormalization (5.98) to equation (5.144) leads to

$$R_{ij}^{(1)} = -D_{il}(\boldsymbol{k})D_{mj}(\boldsymbol{k})\left(\int_{-\infty}^{T} G(k;\tau,\tau_1)Q(k;\tau,\tau_1)\,\mathrm{d}\tau_1\right)\frac{\partial U_l}{\partial X_m}, \tag{5.145}$$

under the isotropic assumption (5.101) and (5.102). We combine equation (5.145) with its counterpart of the second part of the $O(\delta_S)$ term in equation (5.139), and obtain

$$R_{ij}^{(1)} + R_{ji}^{(1)} = -[D_{il}(\boldsymbol{k})D_{mj}(\boldsymbol{k}) + D_{jl}(\boldsymbol{k})D_{mi}(\boldsymbol{k})]\frac{\partial U_l}{\partial X_m} \times \left(\int_{-\infty}^{T} G(k;\tau,\tau_1)Q(k;\tau,\tau_1)\,\mathrm{d}\tau_1\right). \tag{5.146}$$

Equation (5.142) may be evaluated in entirely the same manner. In correspondence to equation (5.144), we have

$$R_{ij}^{(2)} = -\int_{-\infty}^{T} \hat{G}_{il}(\boldsymbol{k};\tau,\tau_1)\frac{\left\langle \dfrac{\mathrm{D}^* w_l(\boldsymbol{k};\tau_1)}{\mathrm{D}T^*} w_j(\boldsymbol{k}';\tau)\right\rangle}{\delta(\boldsymbol{k}+\boldsymbol{k}')}\,\mathrm{d}\tau_1. \tag{5.147}$$

After equation (5.102), we write the lowest-order part of G_{ij}, $\hat{G}_{ij}$, as

$$\hat{G}_{ij}(\boldsymbol{k};\tau,\tau') = D_{ij}(\boldsymbol{k})\hat{G}(k;\tau,\tau'), \tag{5.148}$$

and note

$$D_{ij}(\boldsymbol{k})w_j(\boldsymbol{k};\tau) = w_i(\boldsymbol{k};\tau). \tag{5.149}$$

Then we have

$$R_{ij}^{(2)} = -\int_{-\infty}^{T} \hat{G}(k;\tau,\tau_1)\frac{\left\langle \dfrac{\mathrm{D}^* w_i(\boldsymbol{k};\tau_1)}{\mathrm{D}T^*} w_j(\boldsymbol{k}';\tau)\right\rangle}{\delta(\boldsymbol{k}+\boldsymbol{k}')}\,\mathrm{d}\tau_1, \tag{5.150}$$

which gives

$$\begin{aligned} R_{ij}^{(2)} + R_{ji}^{(2)} &= -\int_{-\infty}^{T} \hat{G}(k;\tau,\tau_1)\frac{\mathrm{D}^*}{\mathrm{D}T^*}\frac{\langle w_i(\boldsymbol{k};\tau_1)w_j(\boldsymbol{k}';\tau)\rangle}{\delta(\boldsymbol{k}+\boldsymbol{k}')}\,\mathrm{d}\tau_1 \\ &= -\int_{-\infty}^{T} \hat{G}(k;\tau,\tau_1)\frac{\mathrm{D}^*\hat{Q}_{ij}(\boldsymbol{k};\tau,\tau_1)}{\mathrm{D}T^*}\,\mathrm{d}\tau_1. \end{aligned} \tag{5.151}$$

Under the renormalization (5.98), equation (5.151) results in

$$R_{ij}^{(2)} + R_{ji}^{(2)} = -D_{ij}(\boldsymbol{k})\int_{-\infty}^{T} G(k,\tau,\tau_1)\frac{\mathrm{D}Q(k,\tau,\tau_1)}{\mathrm{D}T}\,\mathrm{d}\tau_1, \tag{5.152}$$

where use has been made of the replacement

$$\frac{\mathrm{D}^*}{\mathrm{D}T^*} \rightarrow \frac{\mathrm{D}}{\mathrm{D}T} \tag{5.153}$$

since the neglected part is odd in $\boldsymbol{k}$ and gives no contribution to the final result.

We substitute equations (5.140), (5.143), (5.146), and (5.152) into equation (5.139), and make the replacement

$$\boldsymbol{X} \to \delta_S \boldsymbol{x}, \qquad T \to \delta_S t. \tag{5.154}$$

Then we have

$$\begin{aligned} R_{ij}(\boldsymbol{k}, \boldsymbol{x}; \tau, t) &= D_{ij}(\boldsymbol{k})\left(Q(k;\tau,\tau) - \int_{-\infty}^{\tau} G(k;\tau,\tau_1)\frac{\mathrm{D}Q(k;\tau,\tau_1)}{\mathrm{D}t}\,\mathrm{d}\tau_1 \right) \\ &\quad - [D_{il}(\boldsymbol{k})D_{mj}(\boldsymbol{k}) + D_{jl}(\boldsymbol{k})D_{mi}(\boldsymbol{k})] \\ &\quad \times \left(\int_{-\infty}^{\tau} G(k;\tau,\tau_1)Q(k;\tau,\tau_1)\,\mathrm{d}\tau_1 \right)\frac{\partial U_l}{\partial x_m}, \end{aligned} \tag{5.155}$$

with $\mathrm{D}/\mathrm{D}t$ defined in equation (3.26) [2.16, 5.3]. Here we should note that the scale parameter δ_S has disappeared automatically.

We rewrite equation (5.138) as

$$R_{ij} = \int_0^{\infty} \mathrm{d}k \int_{S(k)} R_{ij}(\boldsymbol{k}, \boldsymbol{x}; \tau, t)\,\mathrm{d}S, \tag{5.156}$$

where $S(k)$ denotes the surface with k as the radius. From the formulae

$$\int \frac{k_i k_j}{k^2}\,\mathrm{d}\boldsymbol{k} = \frac{1}{3}\delta_{ij}\int \mathrm{d}\boldsymbol{k}, \tag{5.157a}$$

$$\int \frac{k_i k_j k_l k_m}{k^4}\,\mathrm{d}\boldsymbol{k} = \frac{1}{15}(\delta_{ij}\delta_{lm} + \delta_{il}\delta_{jm} + \delta_{im}\delta_{jl})\int \mathrm{d}\boldsymbol{k}, \tag{5.157b}$$

equation (5.156) with equation (5.155) is reduced to

$$B_{ij} \equiv R_{ij} - \tfrac{2}{3}K\delta_{ij} = -\nu_T S_{ij}. \tag{5.158}$$

Here the turbulent energy K and the turbulent viscosity ν_T are given by

$$K = \int_0^{\infty} Q(k;\tau,\tau)\,\mathrm{d}\boldsymbol{k} - \int_0^{\infty} \mathrm{d}\boldsymbol{k}\int_{-\infty}^{\tau} G(k;\tau,\tau_1)\frac{\mathrm{D}Q(k;\tau,\tau_1)}{\mathrm{D}t}\,\mathrm{d}\tau_1, \tag{5.159}$$

$$\nu_T = \frac{7}{15}\int_0^{\infty} \mathrm{d}\boldsymbol{k}\int_{-\infty}^{\tau} G(k;\tau,\tau_1)Q(k;\tau,\tau_1)\,\mathrm{d}\tau_1, \tag{5.160}$$

and the mean velocity-strain tensor S_{ij} is defined by equation (4.6). From equation (5.159), the energy spectrum $E(k)$ is

$$E(k) = 4\pi k^2\left(Q(k;\tau,\tau) - \int_{-\infty}^{\tau} G(k;\tau,\tau_1)\frac{\mathrm{D}(k;\tau,\tau_1)}{\mathrm{D}t}\,\mathrm{d}\tau_1 \right). \tag{5.161}$$

Equation (5.158) is the turbulent-viscosity representation for the Reynolds stress and corresponds to the first term in equation (4.17) that was constructed in a heuristic manner. The difference between these two expressions lies in the choice of basic statistical quantities. In the former, two-time quantities in wavenumber space are adopted as such quantities, compared with the one-point quantities in physical space in the latter. To derive the latter from the former is important for giving a theoretical basis to heuristic turbulence modelling and improving existing turbulence models from the standpoint of statistical turbulence theory.

5.2.3.2 Physical-space representation

Equations (5.159) and (5.160) are written in terms of the velocity correlation $Q(k;\tau,\tau')$ and the Green's function $G(k;\tau,\tau')$ of the $O(1)$ part in the δ_S expansion. The correlation $Q(k;\tau,\tau')$ is related to the energy spectrum $E(k)$ as equation (3.114), and its low-wavenumber components play the role of reserving the energy supplied from the mean flow. On the other hand, the high-wavenumber components are connected with the energy dissipation process through equations (3.129) and (3.130). It is generally difficult to express $Q(k;\tau,\tau')$ possessing such a broad role in a compact mathematical form. As was explained in §3.4.2.2, however, the inertial range occurring at high Reynolds numbers partially shares some properties with the energy-containing and -dissipation ranges, through the Kolmogorov spectrum (3.138). We shall make full use of this fortunate situation and reduce equations (5.159) and (5.160) to one-point expressions in physical space.

We approximate $Q(k;\tau,\tau')$ with the aid of the Kolmogorov spectrum (3.138). The simplest expressions for $Q(k;\tau,\tau')$ and $G(k;\tau,\tau')$ leading to the spectrum (3.138) are equations (5.108) and (5.109) with equations (5.111) and (5.112). As the numerical constants in them, we adopt equations (5.121) and (5.122). We denote the wavenumber characterizing the energy-containing range by k_E. We approximate the integral in wavenumber space by

$$\int_0^\infty \mathrm{d}k \to \int_{k_E}^\infty \mathrm{d}k, \tag{5.162}$$

as in figure 5.1. In correspondence to k_E, we introduce the characteristic length in the energy-containing region, l_E, through

$$l_E(\boldsymbol{x},t) = \frac{2\pi}{k_E(\boldsymbol{x},t)}. \tag{5.163}$$

Here the important point is that these characteristic quantities depend on location and time. In this context, the energy dissipation rate occurring in equations (5.111) and (5.112), ε, also changes spatially and temporally, and may be written as

$$\varepsilon = \varepsilon(\boldsymbol{x},t). \tag{5.164}$$

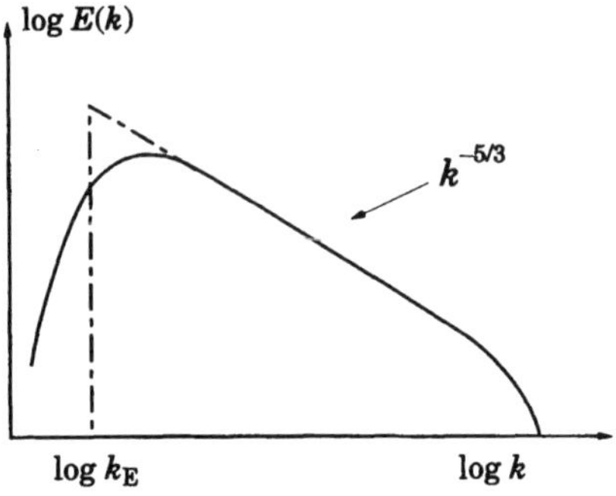

Figure 5.1. Inertial-range approximation to the energy spectrum.

We substitute equations (5.108) and (5.109) into equation (5.159), and have

$$K = K_O \varepsilon(\boldsymbol{x}, t)^{2/3} \int_{k \geq k_E} k^{-5/3}\, \mathrm{d}k - \frac{K_O C_\tau}{4} \int_{k \geq k_E} k^2 \left(2\varepsilon(\boldsymbol{x}, t)^{-1/3} k^{-2/3} \frac{\mathrm{D}}{\mathrm{D}t}[\varepsilon(\boldsymbol{x}, t)^{2/3} k^{-11/3}] - k^{-5} \frac{\mathrm{D}}{\mathrm{D}t}[\varepsilon(\boldsymbol{x}, t)^{1/3} k^{2/3}] \right) \mathrm{d}k. \tag{5.165}$$

We make the transformation

$$s = k/k_\mathrm{E}, \tag{5.166}$$

and note the dependence of k_E on $(\boldsymbol{x}, t)$. We substitute equation (5.163) into the resulting expression, and adopt equations (5.121) and (5.122) as K_O and C_τ in equation (5.165). As a result, we have

$$K = K\{l_\mathrm{E}, \varepsilon\} = C_{K1} \varepsilon^{2/3} l_\mathrm{E}^{2/3} - C_{K2} \varepsilon^{-2/3} l_\mathrm{E}^{4/3} \frac{\mathrm{D}\varepsilon}{\mathrm{D}t} - C_{K3} \varepsilon^{1/3} l_\mathrm{E}^{1/3} \frac{\mathrm{D} l_\mathrm{E}}{\mathrm{D}t}, \tag{5.167}$$

where coefficients C_{Kn} ($n = 1$–3) are evaluated as

$$C_{K1} = 0.67, \qquad C_{K2} = 0.058, \qquad C_{K3} = 0.47. \tag{5.168}$$

From the viewpoint of the scale-parameter expansion (5.63), the first term of equation (5.167) is of $O(1)$, whereas the remaining two are of $O(\delta_\mathrm{S})$. Then we solve equation (5.167) in the perturbational manner based on the first term, and have

$$l_\mathrm{E} = l_\mathrm{E}\{K, \varepsilon\} = C_{l1} K^{3/2} \varepsilon^{-1} + C_{l2} K^{3/2} \varepsilon^{-2} \frac{\mathrm{D}K}{\mathrm{D}t} - C_{l3} K^{5/2} \varepsilon^{-3} \frac{\mathrm{D}\varepsilon}{\mathrm{D}t}, \tag{5.169}$$

with

$$C_{l1} = 1.8, \qquad C_{l2} = 4.4, \qquad C_{l3} = 2.6. \tag{5.170}$$

In the stage of equation (5.162), $k_{\rm E}$ was introduced as the unknown lower limit of integral, but it has been related to the observable physical quantities K and ε though equations (5.163) and (5.169).

We evaluate equation (5.160) similarly and combine it with the result from the $O(\delta_{\rm S}^2)$ analysis [2.16, 5.4]. Then we have

$$\nu_{\rm T} = \nu_{\rm T}\{l_{\rm E}, \varepsilon\} = C_{\nu l}\varepsilon^{1/3} l_{\rm E}^{4/3} - C_{\nu l\varepsilon}\frac{l_{\rm E}^2}{\varepsilon}\frac{{\rm D}\varepsilon}{{\rm D}t} - C_{\nu ll} l_{\rm E}\frac{{\rm D}l_{\rm E}}{{\rm D}t}, \tag{5.171}$$

where

$$C_{\nu l} = 0.054, \qquad C_{\nu l\varepsilon} = 0.011, \qquad C_{\nu ll} = 0.11. \tag{5.172}$$

The substitution of equation (5.169) into equation (5.171) leads to

$$\nu_{\rm T} = C_{\nu 1}\left(1 - C_{\nu 2}\frac{1}{\varepsilon}\frac{{\rm D}K}{{\rm D}t} + C_{\nu 3}\frac{K}{\varepsilon^2}\frac{{\rm D}\varepsilon}{{\rm D}t}\right)\frac{K^2}{\varepsilon}, \tag{5.173}$$

with

$$C_{\nu 1} = 0.12, \qquad C_{\nu 2} = 1.2, \qquad C_{\nu 3} = 0.76. \tag{5.174}$$

Here only the first term in equation (5.169) should be used as long as the ${\rm D}/{\rm D}t$-related terms in equation (5.171) are dropped. These ${\rm D}/{\rm D}t$-related terms that are combined with the mean velocity-strain rate S_{ij} are of $O(\delta_{\rm S}^2)$ in the TSDIA analysis. In §4.3.3, the ${\rm D}/{\rm D}t$ or advection effect on $\nu_{\rm T}$ was also derived by a heuristic approach [see equation (4.139)].

Equation (5.173) with the ${\rm D}/{\rm D}t$-related parts dropped reduces to the turbulent viscosity (4.64), although the proportional constant is a little larger than the value optimized in channel turbulence, equation (4.101). As a typical instance of the flows to which equation (4.64) with fixed C_ν is not applicable, we may mention homogeneous-shear turbulence. There the mean velocity is unidirectional, and the mean velocity gradient is constant in the direction normal to the flow. We denote the constant shear rate by S_∞. The spatial-derivative terms in equation (4.66) vanish, resulting in

$$\frac{{\rm d}K}{{\rm d}t} = P_K - \varepsilon = C_\nu\varepsilon\left[\left(\frac{KS_\infty}{\varepsilon}\right)^2 - \frac{1}{C_\nu}\right], \tag{5.175}$$

under the turbulent-viscosity representation (4.63) based on equation (4.64). In the K-ε model explained in §4.2.4, we have

$$\frac{{\rm d}\varepsilon}{{\rm d}t} = C_{\varepsilon 1}\frac{\varepsilon}{K}P_K - C_{\varepsilon 2}\frac{\varepsilon^2}{K} = C_{\varepsilon 1}C_\nu\frac{\varepsilon^2}{K}\left[\left(\frac{KS_\infty}{\varepsilon}\right)^2 - \frac{C_{\varepsilon 2}}{C_{\varepsilon 1}C_\nu}\right]. \tag{5.176}$$

From equations (5.175) and (5.176), the nondimensional shear rate KS_∞/ε obeys

$$\frac{\rm d}{{\rm d}t}\frac{KS_\infty}{\varepsilon} = -C_\nu(C_{\varepsilon 1} - 1)S_\infty\left[\left(\frac{KS_\infty}{\varepsilon}\right)^2 - \frac{C_{\varepsilon 2} - 1}{C_\nu(C_{\varepsilon 1} - 1)}\right]. \tag{5.177}$$

The stationary state of equation (5.177) is

$$\frac{KS_\infty}{\varepsilon} = \sqrt{\frac{C_{\varepsilon 2} - 1}{C_\nu (C_{\varepsilon 1} - 1)}} = 5, \tag{5.178}$$

under equation (4.71).

When we start from the initial nondimensional shear rate

$$\frac{KS_\infty}{\varepsilon} > \sqrt{\frac{C_{\varepsilon 2}}{C_{\varepsilon 1} C_\nu}} = 3.9, \tag{5.179}$$

we eventually reach equation (5.178). During this period, K and ε continue to grow indefinitely, as is seen from equations (5.175) and (5.176). This behavior of the K-ε model is qualitatively correct, but its quantitative accuracy deteriorates with increasing K and ε, as has been confirmed from the comparison with the results of the direct numerical simulation of equation (3.16). The defect may be removed with the aid of equation (5.173) [5.24]. In the light of characteristic time scales, S_∞^{-1} expresses the decorrelation time due to the mean shearing motion. Equation (5.178) signifies that this time is much shorter than the turnover time of energy-containing eddies, K/ε. In such a situation, the equilibrium state of turbulence does not occur. This is the reason why equation (4.64) with fixed C_ν does not work there.

In order to see the relationship of the defect of equation (4.64) with fixed C_ν with the equilibrium property of the Kolmogorov spectrum (3.138), we examine $E(k)$ obtained from the TSDIA, equation (5.161). It is written as

$$E(k) \propto \varepsilon_{\mathrm{eff}}^{2/3} k^{-5/3}, \tag{5.180}$$

from equations (5.108), (5.109), (5.111), (5.112), (5.121), and (5.122). Here $\varepsilon_{\mathrm{eff}}$ signifies the effective energy transfer rate

$$\varepsilon_{\mathrm{eff}} = [1 - C_{\mathrm{EN}} \,\mathrm{sgn}(\dot{\varepsilon})(l_{\mathrm{N}} k)^{-2/3}]^{3/2} \varepsilon, \tag{5.181}$$

where $\dot{\varepsilon} = \mathrm{d}\varepsilon/\mathrm{d}t$, $\mathrm{sgn}(\dot{\varepsilon}) = 1$ $(\dot{\varepsilon} > 0)$, -1 $(\dot{\varepsilon} < 0)$, $C_{\mathrm{EN}} = 0.6$, and the length scale l_{N} is defined by

$$l_{\mathrm{N}} = \frac{\varepsilon^2}{\dot{\varepsilon}^{3/2}}, \tag{5.182}$$

which is the length linked with the nonstationary energy transfer process. The deviation of equation (5.180) from the Kolmogorov spectrum (3.138) becomes important for small k. This point makes a sharp contrast with the intermittency effects arising from the fluctuation of energy dissipation at fine scales (large k) that were discussed in §3.4.3.1.

Equation (5.181) indicates that the energy transfer is suppressed in the phase of positive $\dot{\varepsilon}$ [4.21]. Physically, this fact signifies that the dissipation process cannot catch up with the process of increasing energy supply instantaneously, resulting in the effective decrease in the energy transfer or the time lag of energy transfer. The defect of the standard K-ε model in the foregoing homogeneous-shear turbulence arises from the neglect of such a time lag.

In §4.2.1 we derived nonlinear algebraic expressions for the Reynolds stress in a heuristic manner, as in equations (4.17) and (4.20). The first term in them has been derived theoretically through the discussions in §5.2.3.1. We may obtain the remaining terms by carrying out the analysis of higher order in δ_S. In the extension of the calculation of equation (5.139) to $O(\delta_S^2)$, δ_S occurs in the combination of the mean velocity U, as [2.16, 4.8, 5.4]

$$\delta_S^2 \frac{\partial U_l}{\partial X_i}\frac{\partial U_l}{\partial X_j}, \qquad \delta_S^2 \left(\frac{\partial U_l}{\partial X_i}\frac{\partial U_j}{\partial X_l} + \frac{\partial U_l}{\partial X_j}\frac{\partial U_i}{\partial X_l}\right), \qquad \delta_S^2 \frac{\partial U_i}{\partial X_l}\frac{\partial U_j}{\partial X_l}, \tag{5.183a}$$

$$\delta_S^2 \frac{\mathrm{D}}{\mathrm{D}T}\left(\frac{\partial U_j}{\partial X_i} + \frac{\partial U_i}{\partial X_j}\right). \tag{5.183b}$$

The second-order nonlinear terms in equation (4.17) arise from equation (5.183a). We also have a new term given by equation (5.183b). The $\mathrm{D}/\mathrm{D}t$-related parts in equation (5.173), which was first pointed out by the theoretical method, will be shown to play an important role in modelling Mach-number effects.

5.3 Contributions to Turbulence Modelling

5.3.1 Modelling of the Turbulent-Energy Equation

As an instance of the theoretical suggestions to turbulence modelling, we mention the modelling of the turbulent energy equation (3.39). Under the choice of K and ε as the fundamental quantities characterizing turbulent flow, what is yet to be modelled is the transport rate $\boldsymbol{T}_K$ [equation (3.42)]. We divide it into two parts as

$$\boldsymbol{T}_K = \boldsymbol{T}_{KV} + \boldsymbol{T}_{KP}, \tag{5.184}$$

where

$$\boldsymbol{T}_{KV} = -\left\langle \frac{\boldsymbol{u}'^2}{2}\boldsymbol{u}' \right\rangle, \tag{5.185}$$

$$\boldsymbol{T}_{KP} = -\langle p'\boldsymbol{u}' \rangle. \tag{5.186}$$

Hereafter $\boldsymbol{T}_{KV}$ and $\boldsymbol{T}_{KP}$ are called the velocity and pressure transports, respectively. As is referred to in §4.2.3.1, $\boldsymbol{T}_{KP}$ is usually considered to be negligible, compared with $\boldsymbol{T}_{KV}$, and $\boldsymbol{T}_K$ is modelled as equation (4.50).

We are in a position to refer to the modelling of $\boldsymbol{T}_K$ from the viewpoint of the TSDIA [5.25]. In order to show a brief outline of its calculation procedure, we consider $\boldsymbol{T}_{KP}$. First, we expand the pressure–velocity correlation $\langle p'\boldsymbol{u}'\rangle$ as

$$\langle p'\boldsymbol{u}'\rangle = \langle p'_0\boldsymbol{u}'_0\rangle + \delta(\langle p'_1\boldsymbol{u}'_0\rangle + \langle p'_0\boldsymbol{u}'_1\rangle) + \delta^2(\langle p'_1\boldsymbol{u}'_1\rangle + \langle p'_2\boldsymbol{u}'_0\rangle + \langle p'_0\boldsymbol{u}'_2\rangle) + O(\delta_S^3). \tag{5.187}$$

We substitute equations (5.69), (5.71), and (5.74) into the $O(\delta_S)$ part of the Fourier representation of equation (5.187). In the DIA formalism, we retain the terms of the lowest order in the velocity correlation Q and the Green's function G. The resulting $O(\delta_S)$ expression for $\boldsymbol{T}_{KP}$ is quadratic in Q and linear in G. We perform a similar calculation for $\boldsymbol{T}_{KV}$. We denote the $O(\delta_S)$ contributions to $\boldsymbol{T}_{KP}$ and $\boldsymbol{T}_{KV}$ by $\boldsymbol{T}_{KP1}$ and $\boldsymbol{T}_{KV1}$, respectively, which are given by [5.26]

$$\boldsymbol{T}_{KP1} = -0.014\frac{K^2}{\varepsilon}\nabla K + 0.0023\frac{K^3}{\varepsilon^2}\nabla\varepsilon, \tag{5.188}$$

$$\boldsymbol{T}_{KV1} = 0.16\frac{K^2}{\varepsilon}\nabla K - 0.049\frac{K^3}{\varepsilon^2}\nabla\varepsilon. \tag{5.189}$$

Their combination leads to

$$\boldsymbol{T}_K = 0.15\frac{K^2}{\varepsilon}\nabla K - 0.047\frac{K^3}{\varepsilon^2}\nabla\varepsilon. \tag{5.190}$$

From the comparison between equations (5.188) and (5.189), we may see

$$|\boldsymbol{T}_{KV}| \gg |\boldsymbol{T}_{KP}|, \tag{5.191}$$

which is consistent with the conjecture leading to equation (4.50), apart from the dependence on the inhomogeneity of ε, that is, $\nabla\varepsilon$. In reality, expression (5.191) has already been confirmed by the direct numerical simulation (DNS) of channel turbulence [5.27]. Recently, some findings contradicting expression (5.191) have been presented by the DNSs of turbulent flows around bluff bodies [5.28, 5.29]. Specifically, the situation entirely opposite to inequality (5.191) is confirmed in the recirculation zones arising from flow separation [5.29].

As a clue to the resolution of the foregoing difficulty about $\boldsymbol{T}_{KP}$ and $\boldsymbol{T}_{KV}$, we may mention effects of $\boldsymbol{U}$ on them. Equations (5.188) and (5.189) contain no explicit dependence on $\boldsymbol{U}$. This situation is rather strange when the importance of explicit mean-flows effects on the pressure–strain correlation Π_{ij} is taken into account [see equation (4.125)]. Those effects come from the $O(\delta_S^2)$ in equation (5.187). The contribution of the lowest order in Q, which is linear in Q, occurs in close relation to equation (4.42a). We denote it by $\boldsymbol{T}_{KP2}$, which is given by [5.30]

$$T_{KP2i} = -\zeta_{Kij}\frac{\partial K}{\partial x_j} + \zeta_{\varepsilon ij}\frac{\partial \varepsilon}{\partial x_j} - 0.0057\frac{K^4}{\varepsilon^3}\nabla^2 U_i, \tag{5.192}$$

with

$$\zeta_{Kij} = 0.12\frac{K^2}{\varepsilon}\left(\frac{3}{10}\frac{K}{\varepsilon}S_{ij} - \frac{1}{6}\frac{K}{\varepsilon}\Omega_{ij}\right), \tag{5.193}$$

$$\zeta_{\varepsilon ij} = 0.064\frac{K^3}{\varepsilon^2}\left(\frac{3}{10}\frac{K}{\varepsilon}S_{ij} - \frac{1}{6}\frac{K}{\varepsilon}\Omega_{ij}\right). \tag{5.194}$$

The first two terms in equation (5.192) represent the anisotropic transports due to mean strain and vorticity effects, in contrast to the isotropic transports in equations (5.188) and (5.189). The third term expresses the curvature effect of mean-flow streamline. These effects are supposed to become important in recirculation regions in which flow properties change considerably in the streamwise direction.

5.3.2 Modelling of the Mach-Number Effect

In §2.3 we referred to Mach-number effects on turbulent flows, specifically, the drastic suppression of the growth rate in a free-shear layer flow. Supersonic channel flows may be treated through the incorporation of mean-density variation into current low-Mach-number models, for instance, the K-ε model. In reality, the DNS of those flows shows that the supersonic effects on the mean velocity and temperature may be explained mainly through the mean-density change [5.31]. Such a K-ε model, however, cannot explain the suppression of the growth rate in a free-shear layer flow. Several models reproducing the suppression have already been presented [5.32–5.37]. There the newly added turbulent compressibility effects tend to have a spurious effect on supersonic channel flows.

A turbulence model applicable to high-Mach-number flows is required to obey the following two requirements.

(a) In supersonic channel flows, compressibility effects in a proposed model appear through the change of mean density; namely, newly added turbulent compressibility effects vanish there.
(b) A proposed model may reproduce the observed suppression of the growth rate in free-shear layer flows.

The construction of a model consistent with these requirements has been one of the challenging themes in the study of compressible turbulence modelling.

In modelling high-speed turbulent flows, we usually adopt the mass-weighted ensemble averaging. There the mean of a quantity f and the fluctuation around it are defined by

$$\hat{f} = \{f\}_{\mathrm{M}} \equiv \frac{\langle \rho f \rangle}{\bar{\rho}}, \tag{5.195}$$

$$f' = f - \hat{f}, \tag{5.196}$$

respectively, where $\langle \rangle$ denotes the ensemble mean, and

$$\bar{\rho} = \langle \rho \rangle, \tag{5.197}$$

and subscript M signifies mass-weighted. We adopt

$$f = (\boldsymbol{u}, p, \zeta) \tag{5.198}$$

as f.

We apply equation (5.195) to equations (3.5), (3.9), and (3.12) and have

$$\frac{\partial \bar{\rho}}{\partial t} + \nabla \cdot (\bar{\rho}\hat{\boldsymbol{u}}) = 0, \tag{5.199}$$

$$\frac{\partial}{\partial t}\bar{\rho}\hat{u}_i + \frac{\partial}{\partial x_j}\bar{\rho}\hat{u}_i\hat{u}_j = -\frac{\partial \bar{p}}{\partial x_i} + \frac{\partial}{\partial x_j}(-\bar{\rho}R_{ij}), \tag{5.200}$$

$$\frac{\partial}{\partial t}\bar{\rho}\hat{\zeta} + \nabla \cdot (\bar{\rho}\hat{\boldsymbol{u}}\hat{\zeta}) = -\langle p \nabla \cdot \boldsymbol{u} \rangle + \nabla \cdot (-\bar{\rho}\boldsymbol{H}_\zeta), \tag{5.201}$$

where the mass-weighted Reynolds stress and internal-energy flux, R_{ij} and $\boldsymbol{H}_\zeta$, are denoted by

$$R_{ij} = \{u'_i u'_j\}_{\mathrm{M}}, \tag{5.202}$$

$$\boldsymbol{H}_\zeta = \{\zeta' \boldsymbol{u}'\}_{\mathrm{M}}, \tag{5.203}$$

and the molecular-diffusion terms related to μ and κ_θ were dropped since they do not play an important role in the following discussions. The ensemble-mean pressure $\bar{p}$ in equation (5.200) may be rewritten as

$$\bar{p} = \langle (\gamma - 1)\rho\hat{\zeta} \rangle = (\gamma - 1)\bar{\rho}\hat{\zeta}. \tag{5.204}$$

Modelling $\langle p \nabla \cdot \boldsymbol{u} \rangle$ in equation (5.201) is difficult in general. We simply approximate it as

$$\langle p \nabla \cdot \boldsymbol{u} \rangle \cong \bar{p} \nabla \cdot \hat{\boldsymbol{u}} = (\gamma - 1)\bar{\rho}\hat{\zeta} \nabla \cdot \hat{\boldsymbol{u}}. \tag{5.205}$$

A big difference between the turbulence properties of free-shear layer and channel flows is the inhomogeneity in the direction of primary mean flow, that is, the streamwise inhomogeneity. A Lagrange-derivative/related quantity expressing advection effects is a proper indicator of such a difference. With this point in mind, we pay special attention to equation (5.173) and write

$$\nu_{\mathrm{T}} = 0.12\frac{K^2}{\varepsilon}\left(1 - 0.76\left(1.6\frac{1}{\varepsilon}\frac{\mathrm{D}K}{\mathrm{D}t} - \frac{K}{\varepsilon^2}\frac{\mathrm{D}\varepsilon}{\mathrm{D}t}\right)\right)$$

$$\cong 0.12\frac{K^2}{\varepsilon}\left(1 - 0.8\frac{1}{K}\frac{\mathrm{D}}{\mathrm{D}t}\left(\frac{K^2}{\varepsilon}\right)\right). \tag{5.206}$$

In the TSDIA formalism, the $\mathrm{D}/\mathrm{D}t$-related terms in equation (5.206) were obtained as the correction to the equilibrium turbulent viscosity ν_{TE} [equation (4.64)]. As a method for enlarging its applicability, we resort to the simplest Padé approximation and write

$$\nu_{\mathrm{T}} = \nu_{\mathrm{TN}} \equiv \frac{\nu_{\mathrm{TE}}}{1 + C_{\mathrm{N}}\dfrac{1}{K}\dfrac{\mathrm{D}}{\mathrm{D}t}\left(\dfrac{K^2}{\varepsilon}\right)} \tag{5.207}$$

(subscript N denotes nonequilibrium), where the equilibrium part ν_{TN} is given by equation (4.64) with $C_\nu = 0.09$, and

$$C_{\mathrm{N}} = 0.8 \tag{5.208}$$

(the theoretically estimated value). We should note that this nonequilibrium effect vanishes identically in fully-developed channel flow.

A representative nondimensional parameter characterizing high-speed turbulent flows is the turbulent Mach number

$$M_{\mathrm{T}} = \frac{\sqrt{\langle \mathbf{u}'^2 \rangle}}{\bar{a}} = \frac{\sqrt{2K}}{\bar{a}}. \tag{5.209}$$

Here $\bar{a}$ is the local mean sound velocity, which is given by

$$\bar{a} = \sqrt{\frac{\mathrm{d}\bar{p}}{\mathrm{d}\bar{\rho}}} = \sqrt{\gamma(\gamma - 1)\hat{e}} \tag{5.210}$$

from equation (3.14) and the adiabatic relation

$$\frac{\bar{p}}{\bar{p}_{\mathrm{R}}} = \left(\frac{\bar{\rho}}{\bar{\rho}_{\mathrm{R}}}\right)^{\gamma} \tag{5.211}$$

(subscript R denotes reference).

Supersonic boundary layers are inhomogeneous in the streamwise direction, but its inhomogeneity is much weaker in the region far from the leading edge, compared with free-shear layer flows. In reality, we may examine supersonic boundary layers by the K-ε model with effects of mean-density variation included. This fact indicates that the manifestation of compressibility effects is related to the degree of streamwise inhomogeneity. From these discussions, we incorporate the M_{T} effect into ν_{T} of equation

(5.207) in the combination with the streamwise inhomogeneity as [5.39]

$$\nu_{\mathrm{T}} = \nu_{\mathrm{TNC}} \equiv \frac{\nu_{\mathrm{TE}}}{1 + C_{\mathrm{N}} \frac{1}{K} \frac{\mathrm{D}}{\mathrm{D}t}\left(\frac{K^2}{\varepsilon}\right) + C_{\mathrm{M}} \frac{M_{\mathrm{T}}^2}{K} \frac{\mathrm{D}}{\mathrm{D}t}\left(\frac{K^2}{\varepsilon}\right)}, \tag{5.212}$$

where

$$\nu_{\mathrm{TE}} = C_\nu \frac{K^2}{\varepsilon}, \tag{5.213}$$

and C_{M} is a model constant and will be determined through its test. This compressibility effect may be shown to be closely related to the gradient Mach number pointed out in the DNS [5.40] of homogeneous-shear turbulence. In fully-developed channel flow, the M_{T}-related part in equation (5.212) vanishes identically owing to $\mathrm{D}/\mathrm{D}t = 0$. Then the present model fulfills the foregoing requirement (a).

Using equation (5.212), we model

$$R_{ij} = \tfrac{2}{3} K \delta_{ij} - \nu_{\mathrm{T}} \hat{s}_{ij}, \tag{5.214}$$

$$\boldsymbol{H}_\zeta = -\frac{\nu_{\mathrm{T}}}{\sigma_\zeta} \nabla \hat{\zeta}, \tag{5.215}$$

where

$$\hat{s}_{ij} = \frac{\partial \hat{u}_j}{\partial x_i} + \frac{\partial \hat{u}_i}{\partial x_j} - \frac{2}{3} \nabla \cdot \hat{\boldsymbol{u}} \delta_{ij}. \tag{5.216}$$

and σ_ζ is a model constant.

In the mass-weighted averaging, the turbulent energy K and the dissipation rate ε are defined by

$$K = \{\tfrac{1}{2} \boldsymbol{u}'^2\}_{\mathrm{M}}, \tag{5.217}$$

$$\varepsilon = \bar{\nu} \left\{ \left(\frac{\partial u'_j}{\partial x_i} + \frac{\partial u'_i}{\partial x_j} - \frac{2}{3} \nabla \cdot \boldsymbol{u}' \delta_{ij} \right) \frac{\partial u'_j}{\partial x_i} \right\}_{\mathrm{M}}, \tag{5.218}$$

with $\bar{\nu} = \bar{\mu}/\bar{\rho}$. As their equations, we adopt the straightforward extension of the incompressible equations (4.66) and (4.67) through the inclusion of mean density; namely, we use

$$\bar{\rho} \frac{\mathrm{D}K}{\mathrm{D}t} = -\bar{\rho} R_{ij} \frac{\partial \hat{u}_j}{\partial x_i} - \bar{\rho}\varepsilon + \nabla \cdot \left(\bar{\rho} \frac{\nu_{\mathrm{TE}}}{\sigma_{\mathrm{K}}} \nabla K \right), \tag{5.219}$$

$$\bar{\rho} \frac{\mathrm{D}\varepsilon}{\mathrm{D}t} = -C_{\varepsilon 1} \frac{\varepsilon}{K} \bar{\rho} R_{ij} \frac{\partial \hat{u}_j}{\partial x_i} - C_{\varepsilon 2} \bar{\rho} \frac{\varepsilon^2}{K} + \nabla \cdot \left(\bar{\rho} \frac{\nu_{\mathrm{TE}}}{\sigma_\varepsilon} \nabla \varepsilon \right). \tag{5.220}$$

We should note that the explicit supersonic effects in this model occur through only ν_{T} in R_{ij} and $\boldsymbol{H}_\zeta$.

The present model contains eight model constants:

$$C_\nu = 0.09, \quad \sigma_K = 1, \quad C_{\varepsilon 1} = 1.4, \quad C_{\varepsilon 2} = 1.9, \quad \sigma_\varepsilon = 1.3, \quad \sigma_\zeta = 1, \tag{5.221a}$$

$$C_{\mathrm{N}} = 0.8, \quad C_{\mathrm{M}}. \tag{5.221b}$$

Of these model constants, equation (5.221a) corresponds to the set of model constants widely adopted in the K-ε model, that is, equation (4.71). In equation (5.221b), the first is the theoretically estimated value, and the latter will be determined through the application to free-shear layer flows.

We are in a position to examine the validity of the present model. For this purpose, we apply the model to the fully developed region of a free-shear layer flow (see figure 2.10). We adopt the Cartesian coordinates (x, y, z), where x is along the two free streams, and y is normal to them. The flow quantities of the faster stream are denoted by attaching subscript 1, such as $(\hat{u}_1, \bar{\rho}_1, \hat{\zeta}_1)$, whereas the slower counterparts are denoted by using subscript 2.

The two free streams are characterized by the ratios

$$\gamma_u = \frac{\hat{u}_2}{\hat{u}_1}, \quad \gamma_\rho = \frac{\bar{\rho}_2}{\bar{\rho}_1}, \quad \gamma_\zeta = \frac{\hat{\zeta}_2}{\hat{\zeta}_1}. \tag{5.222}$$

The most important parameter characterizing high-Mach-number effects on free-shear layer flows is the convective Mach number, which is defined by equation (2.2).

We introduce the normalized velocity in the x direction,

$$u_{\mathrm{R}} = \frac{\hat{u} - \hat{u}_2}{\hat{u}_1 - \hat{u}_2}. \tag{5.223}$$

From equation (5.223), we may define the width of the free-shear layer, δ, as

$$\delta = y_{0.9} - y_{0.1}, \tag{5.224}$$

where y_s denotes the coordinate satisfying

$$s = u_{\mathrm{R}}(y_s). \tag{5.225}$$

The growth rate of the layer, G, is defined by

$$G = \frac{\mathrm{d}\delta}{\mathrm{d}x}. \tag{5.226}$$

We normalize the growth rate by the low-Mach-number counterpart as

$$G_{\mathrm{N}} = \frac{G}{G_{\mathrm{L}}}, \tag{5.227}$$

which is equivalent to equation (2.3).

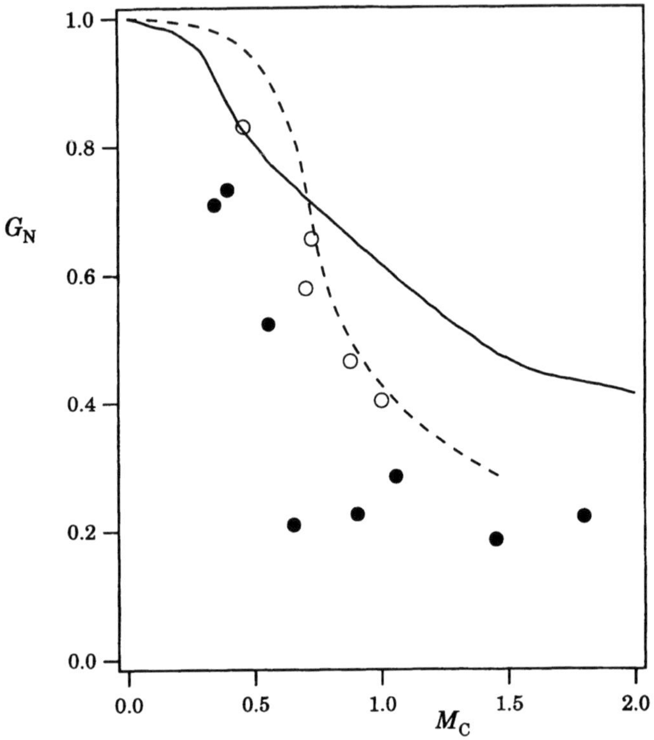

Figure 5.2. Dependence of the normalized growth rate on the convective Mach number. ——, Present model; – – –, Langley curve [5.41]; ●, observations [5.42]; ○, observations [5.43].

In the following, we consider the case

$$\gamma_u = \frac{1}{2}, \qquad \gamma_\rho = 1, \qquad \gamma_\zeta = 1. \tag{5.228}$$

The last two indicate that

$$a_1 = a_2 = a, \tag{5.229}$$

which is used as the reference velocity for normalizing the velocity. The observed values of G_N scatter considerably, but they are around 0.4 at $M_C = 1$ (the data are shown in figure 5.2).

The dependence of G_N on C_M at $M_C = 1$ is examined under equation (5.228). The dependence is gradual; namely, we have $G_N \cong 0.6$ for $C_M = 30$ and $G_N \cong 0.4$ for $C_M = 100$. In figure 5.2, the G_N are plotted for $C_M = 30$ in the comparison with observations [5.41–5.43]. The latter results are limited to M_C lower than 2, and the behavior of G_N for large M_C is not known.

In the context of the behavior of G_N for large M_C, it is significant to refer to the relationship of the collimation of bipolar jets explained in §2.3. As the causes of collimation, we may consider the following two. One is the suppression of the growth rate by supersonic effects, and the other is that due to magnetic forces. The former is related to the behavior of G_N for large M_C, and figure 5.2 suggests that supersonic effects are likely to make an important contribution to the collimation, although we need more work for drawing a more decisive conclusion. The latter magnetic effects will be discussed in §6.7.2.

References

[5.1] Kraichnan R H 1959 *J. Fluid Mech.* **5** 497
[5.2] Kraichnan R H 1977 *J. Fluid Mech.* **83** 349
[5.3] Hamba F 1987 *J. Phys. Soc. Jpn.* **56** 2721
[5.4] Okamoto M 1994 *J. Phys. Soc. Jpn.* **63** 2102
[5.5] Tatsumi T 1980 *Adv. Appl. Mech.* **20** 39
[5.6] Wyld W H 1961 *Ann. Phys. NY* **14** 143
[5.7] Herring J R 1965 *Phys. Fluids* **8** 2219
[5.8] Edwards S F 1964 *J. Fluid Mech.* **18** 239
[5.9] Kraichnan R H 1964 *Phys. Fluids* **7** 1723
[5.10] Kraichnan R H 1965 *Phys. Fluids* **8** 575
[5.11] Nakano T 1972 *Ann. Phys. NY* **73** 326
[5.12] Cambon C, Jeandel D and Mathieu J 1981 *J. Fluid Mech.* **104** 247
[5.13] Cambon C and Jacquin L 1989 *J. Fluid Mech.* **202** 295
[5.14] Kaneda Y 1981 *J. Fluid Mech.* **107** 131
[5.15] Kaneda Y and Gotoh T 1991 *Phys. Fluids A* **3** 1924
[5.16] Wilson K G and Kogut J 1974 *Phys. Rep.* **12** C 75
[5.17] Yakhot V and Orszag S 1986 *J. Sci. Comput.* **1** 3
[5.18] Dannevik W P, Yakhot V and Orszag S 1987 *Phys. Fluids* **30** 2021
[5.19] Smith L M and Reynolds W C 1992 *Phys. Fluids A* **4** 364
[5.20] Nakano T 1992 *J. Phys. Soc. Jpn.* **61** 3994
[5.21] Eyink G L 1994 *Phys. Fluids* **6** 3063
[5.22] Rubinstein R and Barton J M 1990 *Phys. Fluids A* **2** 1472
[5.23] Rubinstein R and Barton J M 1992 *Phys. Fluids A* **4** 759
[5.24] Yoshizawa A and Nisizima S 1993 *Phys. Fluids A* **5** 3302
[5.25] Yoshizawa A 1982 *J. Phys. Soc. Jpn.* **51** 2326
[5.26] Shimomura Y 1998 *Phys. Fluids* **10** 2636
[5.27] Mansour N N, Kim J and Moin P 1989 *AIAA J.* **27** 1068
[5.28] Le H, Moin P and Kim J 1997 *J. Fluid Mech.* **330** 349
[5.29] Yao Y F, Thomas T G and Sandham N D 2001 *Theor. Comput. Fluid Dyn.* **14** 337
[5.30] Yoshizawa A 2002 *Phys. Fluids* **14** 1736
[5.31] Zeman O 1990 *Phys. Fluids A* **2** 178
[5.32] Sarkar S, Erlebacher G E, Hussaini M Y and Kreiss H O 1991 *J. Fluid Mech.* **227** 473
[5.33] Taulbee D and Van Osdol J 1991 *AIAA* Paper No 91-0524
[5.34] Liou W W, Shih T-H and Duncun B S 1995 *Phys. Fluids* **7** 658
[5.35] Yoshizawa A, Liou W W, Yokoi N and Shih T-H 1997 *Phys. Fluids* **9** 3024

[5.36] Adumitroaie V, Ristorcelli J R and Taulbee D B 1999 *Phys. Fluids* **10** 2696
[5.37] Fujiwara H, Matsuo Y and Arakawa C 2000 *Int. J. Heat and Fluid Flow* **21** 354
[5.38] Huang P G, Coleman G N and Bradshaw P 1995 *J. Fluid Mech.* **305** 185
[5.39] Yoshizawa A, Fujiwara H, Hamba F, Nisizima S and Kumagai Y unpublished
[5.40] Sarkar S 1995 *J. Fluid Mech.* **282** 161
[5.41] Kline S J, Cantwell B J and Lilley G M 1982 in *1980–1981 AFOSR-HTTM-Stanford Conference* ed S J Kline, B J Cantwell and G M Lilley (Stanford: Stanford University Press) p 368
[5.42] Papamoschou D and Roshko A 1988 *J. Fluid Mech.* **197** 453
[5.43] Gobble S G and Button J C 1991 *AIAA J.* **29** 538

Part III

Magnetohydrodynamic Turbulence: Dynamo

Nomenclature

$\boldsymbol{a}$	Vector potential
$\boldsymbol{b}$	Magnetic field
$\boldsymbol{e}$	Electric field
e_0	Charge magnitude of electron
$\boldsymbol{E}_{\mathrm{M}}$	Turbulent electromotive force
$D_{ij}(\boldsymbol{k})$	Solenoidal tensor
$\mathrm{D}/\mathrm{D}t$	Lagrange derivative based on mean velocity
$\mathrm{D}/\mathrm{D}T$	Lagrange derivative based on mean velocity and slow variables
$F, \langle f \rangle$	Ensemble mean of f
f'	Fluctuation of f in the ensemble averaging
$\boldsymbol{f}_{\mathrm{p}}, \boldsymbol{f}_{\mathrm{t}}$	Poloidal and toroidal components of $\boldsymbol{f}$
$\boldsymbol{g}$	Gravitational acceleration vector
$G_{ij}(\boldsymbol{k};\tau,\tau')$	Green's or response function in wavenumber space
H	Turbulent residual helicity
$\boldsymbol{H}_{\theta}$	Turbulent heat flux
$\boldsymbol{j}$	Electric current density
$\boldsymbol{k}$	Wavenumber vector
K	Turbulent magnetohydrodynamic (MHD) energy
K_{R}	Turbulent residual energy
$m_{\mathrm{E}}, m_{\mathrm{I}}$	Mass of electron and ion
M_{ij}	Magnetic strain tensor
$n_{\mathrm{E}}, n_{\mathrm{I}}$	Number density of electron and ion
p	Pressure of electrically conducting fluid
$p_{\mathrm{E}}, p_{\mathrm{I}}$	Pressure of electron and ion gases
P_r	Prandtl number
$P_{r\mathrm{M}}$	Magnetic Prandtl number
P_K	Production rate of turbulent MHD energy
P_W	Production rate of turbulent cross helicity
$Q_{ij}(\boldsymbol{k};\tau,\tau')$	Two-time velocity correlation function in wavenumber space
R_a	Rayleigh number
R_e	Reynolds number
$R_{e\mathrm{M}}$	Magnetic Reynolds number
R_{ij}	Reynolds stress
$R_{ij}^{(\mathrm{E})}$	Elsasser's Reynolds stress
(r,θ,ϕ)	Spherical coordinates
S_{ij}	Mean velocity-strain tensor
T_a	Taylor number
T_K	Transport rate of turbulent MHD energy
T_W	Transport rate of turbulent cross helicity
$\boldsymbol{u}$	Velocity of electrically conducting fluid
$\boldsymbol{v}_{\mathrm{E}}, \boldsymbol{v}_{\mathrm{I}}$	Velocity of electron and ion gases
$\boldsymbol{X}, T$	Slow spatial and temporal variables

W	Turbulent cross helicity
α, α_K, α_M	Coefficients of helicity effect
α_T	Thermal expansion coefficient
β, β_K, β_M	Turbulent resistivities
γ, γ_K, γ_M	Coefficients of cross-helicity effect
$\delta(\boldsymbol{k})$	Dirac's delta function
∇, ∇_ξ, ∇_X	Gradient vectors
ε	Dissipation rate of turbulent MHD energy
ε_{ijk}	Alternating tensor
ε_W	Dissipation rate of turbulent cross helicity
θ	Temperature
λ_M	Magnetic diffusivity
λ_θ	Temperature diffusivity
μ	Molecular viscosity
μ_0	Magnetic permeability
ν	Kinematic viscosity
ν_T	Turbulent viscosity
ν_M	Coefficient of magnetic feedback effect
ξ, τ	Fast spatial and temporal variables
ρ	Mass density of electrically conducting fluid
$\rho_E^{(C)}$, $\rho_I^{(C)}$	Charge density of electron and ion
$\rho_E^{(M)}$, $\rho_I^{(M)}$	Mass density of electron and ion
σ_e	Electric conductivity
(σ, ϑ, z)	Cylindrical coordinates
$\boldsymbol{\phi}$, $\boldsymbol{\psi}$	Elsasser's variables
$\boldsymbol{\omega}$	Vorticity
$\boldsymbol{\omega}_F$	Angular velocity vector of frame rotation
Ω_{ij}	Mean vorticity tensor

Chapter 6

Fundamentals of Mean-Field Theory of Dynamo

6.1 One-Fluid Magnetohydrodynamic Approximation

6.1.1 Fundamental Equations

In §2.2 and §2.3 we referred to the generation of large-scale magnetic-field structures by fluid motion and the high collimation of bipolar jets from an accretion disk. A useful tool for investigating these phenomena is the magnetohydrodynamic (MHD) approximation, specifically, the one-fluid approximation [2.24, 6.1]. In what follows, we shall give its brief account and refer to the limitation.

The motions of electron and ion gases are governed by

$$\frac{\partial \rho_{\mathrm{S}}^{(\mathrm{M})}}{\partial t} + \nabla \cdot \left(\rho_{\mathrm{S}}^{(\mathrm{M})} \boldsymbol{v}_{\mathrm{S}} \right) = 0, \tag{6.1}$$

$$\frac{\partial}{\partial t} \rho_{\mathrm{S}}^{(\mathrm{M})} v_{\mathrm{S}i} + \frac{\partial}{\partial x_j} \rho_{\mathrm{S}}^{(\mathrm{M})} v_{\mathrm{S}j} v_{\mathrm{S}i} = -\frac{\partial p_{\mathrm{S}}}{\partial x_i} + \rho_{\mathrm{S}}^{(\mathrm{C})} (\boldsymbol{e} + \boldsymbol{v}_{\mathrm{S}} \times \boldsymbol{b})_i + C_{\mathrm{S}i}, \tag{6.2}$$

with subscript S denoting I and E for electron and ion, respectively. Here $\rho_{\mathrm{S}}^{(\mathrm{M})}$ is the mass density, $\boldsymbol{v}_{\mathrm{S}}$ is the gas velocity, p_{S} is the gas pressure, $\rho_{\mathrm{S}}^{(\mathrm{C})}$ is the charge density, $\boldsymbol{e}$ is the electric field, and $\boldsymbol{b}$ is the magnetic field. On the right-hand side of equation (6.2), the second term is the Lorentz force, and $\boldsymbol{C}_{\mathrm{S}}$ is the force arising from the collision between electron and ion, which is expressed as

$$\boldsymbol{C}_{\mathrm{E}} = -\boldsymbol{C}_{\mathrm{I}} = -\rho_{\mathrm{E}}^{(\mathrm{M})} (\boldsymbol{v}_{\mathrm{E}} - \boldsymbol{v}_{\mathrm{I}}) \nu_{\mathrm{EI}}, \tag{6.3}$$

with ν_{EI} as the collision frequency. The mass density $\rho_{\mathrm{S}}^{(\mathrm{M})}$ and the charge density $\rho_{\mathrm{S}}^{(\mathrm{C})}$ are written as

$$\rho_{\mathrm{S}}^{(\mathrm{M})} = n_{\mathrm{S}} m_{\mathrm{S}}, \tag{6.4}$$

$$\rho_{\mathrm{S}}^{(\mathrm{C})} = n_{\mathrm{S}} e_{\mathrm{S}}, \tag{6.5}$$

with $e_E = -e_0$ and $e_I = Ze_0$, where n_S and m_S are the number density and the mass of ion or electron, respectively, and e_0 is the charge magnitude of an electron.

In the one-fluid MHD approximation, we introduce the total mass density ρ, the associated fluid velocity $\boldsymbol{u}$, the total pressure p, and the electric current density $\boldsymbol{j}$ as

$$\rho = \rho_E^{(M)} + \rho_I^{(M)}, \tag{6.6}$$

$$\boldsymbol{u} = \frac{\rho_E^{(M)} \boldsymbol{v}_E + \rho_I^{(M)} \boldsymbol{v}_I}{\rho}, \tag{6.7}$$

$$p = p_E + p_I, \tag{6.8}$$

$$\boldsymbol{j} = -n_E e_0 \boldsymbol{v}_E + n_I Z e_0 \boldsymbol{v}_I, \tag{6.9}$$

respectively. Here we note the difference between the masses of electron and ion, that is,

$$\frac{m_E}{m_I} = O(10^{-3}), \tag{6.10}$$

which signifies

$$\rho \cong \rho_I^{(M)}, \tag{6.11}$$

$$\boldsymbol{u} \cong \boldsymbol{v}_I. \tag{6.12}$$

The key assumption of the one-fluid MHD approximation is the quasi-neutrality of charge density,

$$\rho_C = \rho_E^{(C)} + \rho_I^{(C)} = -n_E e_0 + Z n_I e_0 \cong 0. \tag{6.13}$$

Under equation (6.13), equation (6.9) is rewritten as

$$\boldsymbol{j} = -n_E e_0 (\boldsymbol{v}_E - \boldsymbol{v}_I). \tag{6.14}$$

We combine equation (6.1) for electron gas with its ion counterpart, and use equations (6.6) and (6.7). Then we have

$$\frac{\partial \rho}{\partial t} + \nabla \cdot (\rho \boldsymbol{u}) = 0. \tag{6.15}$$

Similarly, the combination of equation (6.2) with equations (6.11)–(6.13) results in

$$\frac{\partial}{\partial t} \rho u_i + \frac{\partial}{\partial x_j} \rho u_j u_i = -\frac{\partial p}{\partial x_i} + (\boldsymbol{j} \times \boldsymbol{b})_i + \frac{\partial}{\partial x_j} \left[\mu \left(\frac{\partial u_j}{\partial x_i} + \frac{\partial u_i}{\partial x_j} - \frac{2}{3} \nabla \cdot \boldsymbol{u} \delta_{ij} \right) \right]. \tag{6.16}$$

Here we should note the following two points. One is the disappearance of the electric field $\boldsymbol{e}$, which arises from the quasi-neutrality assumption, equation (6.13). As a result, the study of fusion phenomena closely related to $\boldsymbol{e}$ effects is beyond the scope of the one-fluid MHD approximation. The other is that the diffusion effect arising from the collision among ions themselves is supplemented as the last part of equation (6.16) [see equation (3.9) for an electrically non-conducting fluid].

In this chapter we shall pay special attention to the case of constant mass density and discuss the generation mechanism of magnetic fields as well as their feedback effects. Density-variation effects will be partially taken into account through the buoyancy force. By this approach, we do not intend to say that effects of density variation are not important in those studies, but we aim at abstracting some of essential ingredients in the generation process of magnetic fields.

In the frame rotating with angular velocity ω_F, the constant-density counterparts of equations (6.15) and (6.16) are [6.2, 6.3]

$$\nabla \cdot \boldsymbol{u} = 0, \tag{6.17}$$

$$\frac{\partial u_i}{\partial t} + \frac{\partial}{\partial x_j} u_j u_i = \left(\frac{\partial}{\partial t} + \boldsymbol{u} \cdot \nabla \right) u_i = -\frac{1}{\rho} \frac{\partial p}{\partial x_i} + \nu \nabla^2 u_i$$
$$+ \frac{1}{\rho} (\boldsymbol{j} \times \boldsymbol{b})_i + 2(\boldsymbol{u} \times \boldsymbol{\omega}_F)_i - \alpha_T (\theta - \theta_R) g_i. \tag{6.18}$$

In the second relation of equation (6.18), the fourth term is the Coriolis force, and the fifth term is the buoyancy force based on the Boussinesq approximation (α_T is the thermal-expansion coefficient, θ_R is the reference temperature, and $\boldsymbol{g}$ is the gravitational-acceleration vector). The pressure p denotes the deviation from the static pressure. The temperature θ obeys equation (3.17), that is,

$$\frac{\partial \theta}{\partial t} + \nabla \cdot (\theta \boldsymbol{u}) = \lambda_\theta \nabla^2 \theta. \tag{6.19}$$

The Boussinesq approximation is originally appropriate for the situation that the temperature difference in fluid motion is not so large. Then the approximation needs careful treatment in the application to the phenomena subject to large temperature difference such as the sun. The buoyancy force originating in the density difference of constituents may also become important in geodynamo. In the outer core of the earth its main constituent is melted iron, but silicon is also included there. Their mass difference is a promising candidate for the force driving melted iron, as well as the thermal force.

We consider equation (6.2) for electron gas, and retain the Lorentz and collision forces. Then we have

$$\boldsymbol{e} + \boldsymbol{v}_E \times \boldsymbol{b} + \frac{m_E \nu_{EI}}{e_0} (\boldsymbol{v}_E - \boldsymbol{v}_I) = 0, \tag{6.20}$$

which is reduced to

$$e + v_E \times b - \frac{m_E \nu_{EI}}{n_E e_0^2} j = 0, \tag{6.21}$$

under equation (6.14). From equations (6.12) and (6.14), we have

$$v_E = u - \frac{1}{n_E e_0} j. \tag{6.22}$$

We substitute equation (6.22) into equation (6.21), and have

$$j = \sigma_e (e + u \times b) + \frac{\sigma_e}{\rho_E^{(C)}} j \times b, \tag{6.23}$$

where σ_e is the electric conductivity defined by

$$\sigma_e = \frac{n_E e_0^2}{m_E \nu_{EI}}. \tag{6.24}$$

The retention of the first part in equation (6.23) leads to the simplest Ohm's law.

In the case of constant fluid density, the use of Alfvén-velocity units leads to the concise form of fundamental equations. In the units, $b/\sqrt{\rho\mu_0}$ has the dimension of velocity, where μ_0 is the magnetic permeability. We make use of this fact and make the replacement

$$\frac{b}{\sqrt{\rho\mu_0}} \to b, \qquad \frac{j}{\sqrt{\rho/\mu_0}} \to j, \qquad \frac{e}{\sqrt{\rho\mu_0}} \to e, \qquad \frac{p}{\rho} \to p. \tag{6.25}$$

Under this replacement, equation (6.18) is reduced to

$$\begin{aligned} \frac{\partial u_i}{\partial t} + \frac{\partial}{\partial x_j} u_j u_i &= \left(\frac{\partial}{\partial t} + u \cdot \nabla\right) u_i \\ &= -\frac{\partial p}{\partial x_i} + \nu \nabla^2 u_i + (j \times b)_i + 2(u \times \omega_F)_i - \alpha_T(\theta - \theta_R) g_i \end{aligned} \tag{6.26a}$$

or

$$\begin{aligned} \frac{\partial u_i}{\partial t} + \frac{\partial}{\partial x_j} u_j u_i = &-\frac{\partial}{\partial x_i}\left(p + \frac{b^2}{2}\right) + \nu \nabla^2 u_i \\ &+ \frac{\partial}{\partial x_j} b_j b_i + 2(u \times \omega_F)_i - \alpha_T(\theta - \theta_R) g_i. \end{aligned} \tag{6.26b}$$

In Alfvén-velocity units, the magnetic induction equation, the Ampère's law, and the Ohm's law are written as

$$\frac{\partial b}{\partial t} = -\nabla \times e, \tag{6.27}$$

$$\boldsymbol{j} = \nabla \times \boldsymbol{b}, \tag{6.28}$$

$$\boldsymbol{j} = \frac{1}{\lambda_{\mathrm{M}}} (\boldsymbol{e} + \boldsymbol{u} \times \boldsymbol{b}), \tag{6.29}$$

respectively, where λ_{M} is defined by

$$\lambda_{\mathrm{M}} = \frac{1}{\sigma_e \mu_0}. \tag{6.30}$$

We use equations (6.28) and (6.29), and eliminate $\boldsymbol{e}$ from equation (6.27). As a result, we have

$$\frac{\partial \boldsymbol{b}}{\partial t} = \nabla \times (\boldsymbol{u} \times \boldsymbol{b}) + \lambda_{\mathrm{M}} \nabla^2 \boldsymbol{b} \tag{6.31a}$$

or

$$\frac{\partial \boldsymbol{b}}{\partial t} + (\boldsymbol{u} \cdot \nabla)\boldsymbol{b} - (\boldsymbol{b} \cdot \nabla)\boldsymbol{u} = \lambda_{\mathrm{M}} \nabla^2 \boldsymbol{b}, \tag{6.31b}$$

from which we may see that λ_{M} signifies the magnetic diffusivity. Here we should recall that the disappearance of $\boldsymbol{e}$ effects is closely related to the quasi-neutrality assumption, equation (6.13), leading to equation (6.14).

6.1.2 Nondimensional Parameters Characterizing Flows

The importance of each term in equations (6.26) and (6.31) changes greatly from one phenomenon to another. The most familiar parameter is the Reynolds number R_e, which is defined by

$$R_e = \frac{\{(\boldsymbol{u} \cdot \nabla)\boldsymbol{u}\}_{l_{\mathrm{R}}, u_{\mathrm{R}}}}{\{\nu \nabla^2 \boldsymbol{u}\}_{l_{\mathrm{R}}, u_{\mathrm{R}}}} = \frac{l_{\mathrm{R}} u_{\mathrm{R}}}{\nu}. \tag{6.32}$$

Here l_{R} and u_{R} are the reference length and the reference velocity, respectively, and $\{f\}_{\xi,\eta}$ denotes the magnitude of f that is estimated using quantities ξ and η.

The nondimensional parameter related to the Coriolis force is the Taylor number

$$T_a = \left(\frac{\{\boldsymbol{u} \times \boldsymbol{\omega}_{\mathrm{F}}\}_{u_{\mathrm{R}}}}{\{\nu \nabla^2 \boldsymbol{u}\}_{l_{\mathrm{R}}, u_{\mathrm{R}}}} \right)^2 = \frac{l_{\mathrm{R}}^4 \omega_{\mathrm{F}}^2}{\nu^2}. \tag{6.33}$$

Since $l_{\mathrm{R}} \omega_{\mathrm{F}}$ is the velocity associated with frame rotation, we may see that T_a is the square of the Reynolds number R_e based on velocity $l_{\mathrm{R}} \omega_{\mathrm{F}}$ and length l_{R}.

The magnitude of the buoyancy force due to the Boussinesq approximation is characterized by the Rayleigh number

$$R_a = \left(\frac{\{\alpha_{\mathrm{T}}(\theta - \theta_{\mathrm{R}})\boldsymbol{g}\}_{\Delta\theta_{\mathrm{R}}}}{\{\nu \nabla^2 \boldsymbol{u}\}_{l_{\mathrm{R}}, u_{\mathrm{R}}}} \right)^2 P_r, \tag{6.34}$$

where $\Delta\theta_R$ is the reference temperature difference characterizing $\theta - \theta_R$, and P_r is the Prandtl number denoted by

$$P_r = \frac{\nu}{\lambda_\theta}. \tag{6.35}$$

The reference velocity linked with the buoyancy force is estimated as

$$\{(\boldsymbol{u}\cdot\nabla)\boldsymbol{u}\}_{l_R,u_R} = \{\alpha_T(\theta - \theta_R)\boldsymbol{g}\}_{\Delta\theta_R}, \tag{6.36}$$

which gives

$$u_R = \sqrt{\alpha_T g \Delta\theta_R l_R}. \tag{6.37}$$

We substitute equation (6.37) into equation (6.34), and have

$$R_a = \frac{\alpha_T g \Delta\theta_R l_R^3}{\nu\lambda_\theta} \tag{6.38}$$

as the Rayleigh number for the Boussinesq buoyancy force. As is seen from equations (6.34) and (6.36), R_a may be regarded as the square of R_e in the case of $P_r = O(1)$.

The magnetic-field counterpart of R_e is the magnetic Reynolds number

$$R_{eM} = \frac{\{\nabla\times(\boldsymbol{u}\times\boldsymbol{b})\}_{l_R,u_R}}{\{\lambda_M\nabla^2\boldsymbol{b}\}_{l_R,u_R}} = \frac{l_R u_R}{\lambda_M} = R_e P_{rM}. \tag{6.39}$$

Here P_{rM} is the magnetic Prandtl number given by

$$P_{rM} = \frac{\nu}{\lambda_M}. \tag{6.40}$$

In equation (6.32) large R_e signifies that the advection or inertia effect is dominant at the length scale l_R and its associated velocity scale u_R, compared with the molecular diffusion effect. This fact does not mean that the latter effect is not important at large R_e, but that it may become important at the spatial scale much smaller than l_R, that is, l_D defined by equation (4.53).

In order to see the physical meaning of T_a, we drop the Lorentz and buoyancy forces in equation (6.26), and take its curl. Then we have

$$\frac{\partial\boldsymbol{\omega}}{\partial t} = \nabla\times[\boldsymbol{u}\times(\boldsymbol{\omega}+2\boldsymbol{\omega}_F)] + \nu\nabla^2\boldsymbol{\omega}. \tag{6.41}$$

In the case of large R_e, we pay special attention to the length scale l_R and drop the molecular diffusion term. In the stationary state, we have

$$\nabla\times[\boldsymbol{u}\times(\boldsymbol{\omega}+2\boldsymbol{\omega}_F)] = 0. \tag{6.42}$$

For the case of a strong Coriolis effect or $T_a \gg R_e$, equation (6.42) is reduced to

$$\nabla\times(\boldsymbol{u}\times\boldsymbol{\omega}_F) = 0, \tag{6.43}$$

which is equivalent to

$$(\boldsymbol{\omega}_F \cdot \nabla)\boldsymbol{u} = 0. \tag{6.44}$$

Namely, the fluid motion does not change along the axis of frame rotation. In this sense, the motion becomes two-dimensional along the axis. This finding is called the Taylor–Proudman theorem.

The Taylor–Proudman theorem becomes important in a spherical or spherical-shell region with the buoyancy force in the radial direction as a primary cause of fluid motion. The buoyancy force whose strength is characterized by the Rayleigh number R_a drives fluid from the inner to the outer part of the region. With the increase in the Taylor number T_a, a fluid blob at one location is trapped around an axis along the frame-rotation vector $\boldsymbol{\omega}_F$. As a result, fluid comes up or down along this axis, while rotating; namely, it is subject to helical motion (a similar situation may be seen in the motion of a charged particle around a magnetic field line). Such fluid motion constitutes the so-called convection columns [6.4], as is depicted schematically in figure 6.1. A typical quantity characterizing the columns is the helicity $\boldsymbol{u} \cdot \boldsymbol{\omega}$, which was discussed in §3.3.2.4 and §4.4.1. The properties of convection columns found by computer simulations will be later referred to in the light of geodynamo and solar dynamo.

In the case of the earth's outer core, we have

$$l_R = O(10^6)\ \text{m}, \qquad \nu = O(10^{-6})\ \text{m}^2\ \text{s}^{-1}, \qquad u_R = O(10^{-4})\ \text{m s}^{-1},$$
$$P_r = O(10^{-1}), \qquad P_{rM} = O(10^{-6}). \tag{6.45}$$

Here ν, P_r, and P_{rM} are the values for melted iron, and u_R is the value inferred from geophysical observations. We use $\omega_F = 7 \times 10^{-5}$ rad s^{-1},

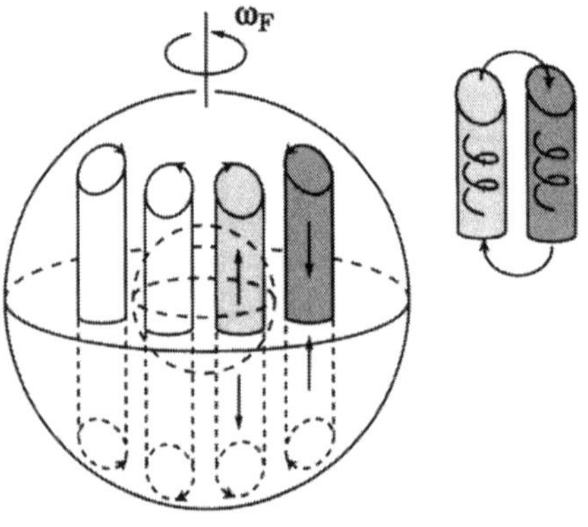

Figure 6.1. Convection columns in a spherical-shell region mimicking the earth's outer core, with attention paid to a specific part of columns in the northern hemisphere (right).

and have

$$R_e = O(10^8), \qquad T_a = O(10^{26}), \qquad R_a = O(10^{15}), \qquad R_{eM} = O(10^2) \tag{6.46}$$

[R_a is estimated from the discussion below equation (6.38), and R_{eM} is found from equation (6.39)]. From equation (6.46) the magnetic field in the outer core is much more diffusive than the momentum. We should also notice the importance of the role of Coriolis-force or helicity effects in the geodynamo owing to $T_a \gg R_e$.

In the case of the solar convective zone, the flow velocity and the spatial scale are much larger, compared with the earth's outer core. Moreover, the convective zone is highly electrically conducting, resulting in lower magnetic diffusivity. As a result, both R_e and R_{eM} are much larger than the values in equation (6.46). On the other hand, T_a in the convective zone is of $O(10^{20})$, which is smaller than the counterpart in equation (6.46). This fact indicates that the role of the convective columns becomes smaller in the solar convective zone. Geometrically, the solar convective zone is a thin spherical shell, and convective columns are limited to the low-latitude region. This situation also suggests a smaller role of helicity effects in the zone, compared with the earth's outer core.

6.1.3 Elsasser's Variables and Conservation Properties

In §3.3.2.2 and §3.3.2.4 we stated that the total amounts of kinetic energy, squared temperature, and helicity are conserved in the absence of molecular effects. For these quantities, their turbulence parts obey the equations whose mathematical and physical bases are firm. It is important to see what quantities may be conserved in MHD flows.

For this purpose, we introduce Elsasser's variables

$$\boldsymbol{\phi} = \boldsymbol{u} + \boldsymbol{b}, \qquad \boldsymbol{\psi} = \boldsymbol{u} - \boldsymbol{b}, \tag{6.47}$$

which lead to

$$\boldsymbol{u} = \frac{\boldsymbol{\phi} + \boldsymbol{\psi}}{2}, \qquad \boldsymbol{b} = \frac{\boldsymbol{\phi} - \boldsymbol{\psi}}{2}. \tag{6.48}$$

In equation (6.26b) we drop the Coriolis and Boussinesq terms, and apply equation (6.47) to the resulting equation. Then we have

$$\frac{\partial \boldsymbol{\phi}}{\partial t} + (\boldsymbol{\psi} \cdot \nabla)\boldsymbol{\phi} = -\nabla\left(p + \frac{\boldsymbol{b}^2}{2}\right) + \frac{\nu + \lambda_M}{2}\nabla^2\boldsymbol{\phi} + \frac{\nu - \lambda_M}{2}\nabla^2\boldsymbol{\psi}, \tag{6.49}$$

$$\frac{\partial \boldsymbol{\psi}}{\partial t} + (\boldsymbol{\phi} \cdot \nabla)\boldsymbol{\psi} = -\nabla\left(p + \frac{\boldsymbol{b}^2}{2}\right) + \frac{\nu + \lambda_M}{2}\nabla^2\boldsymbol{\psi} + \frac{\nu - \lambda_M}{2}\nabla^2\boldsymbol{\phi}. \tag{6.50}$$

In the absence of molecular effects, equations (6.49) and (6.50) give

$$\frac{\partial}{\partial t}\int_V \frac{\phi^2}{2} \mathrm{d}V = -\int_V \nabla \cdot \left[\frac{\phi^2}{2}\psi + \left(p + \frac{b^2}{2}\right)\phi\right] \mathrm{d}V$$
$$= -\int_S n \cdot \left[\frac{\phi^2}{2}\psi + \left(p + \frac{b^2}{2}\right)\phi\right] \mathrm{d}S, \tag{6.51}$$

$$\frac{\partial}{\partial t}\int_V \frac{\psi^2}{2} \mathrm{d}V = -\int_V \nabla \cdot \left[\frac{\psi^2}{2}\phi + \left(p + \frac{b^2}{2}\right)\psi\right] \mathrm{d}V$$
$$= -\int_S n \cdot \left[\frac{\psi^2}{2}\phi + \left(p + \frac{b^2}{2}\right)\psi\right] \mathrm{d}S, \tag{6.52}$$

with the aid of the Gauss integral theorem (3.4) (V and S denote the whole flow region and its surrounding surface, respectively). Then the total amount of each of ϕ^2 and ψ^2 is conserved so long as there are neither their net supply nor loss across the surface.

The conservation of the total amounts of ϕ^2 and ψ^2 is equivalent to that of $\phi^2 \pm \psi^2$. From equation (6.47) we have

$$\frac{1}{4}(\phi^2 + \psi^2) = \frac{u^2}{2} + \frac{b^2}{2}, \tag{6.53}$$

$$\frac{1}{4}(\phi^2 - \psi^2) = u \cdot b, \tag{6.54}$$

which are called the MHD energy and cross helicity, respectively. These two quantities are also conserved in the absence of molecular effects and their net supply or loss across a boundary. The kinetic helicity $u \cdot \omega$, however, is not subject to such a conservation law, unlike the electrically nonconducting case. The foregoing discussions suggest that the MHD energy and the cross helicity are the fundamental quantities in the investigation into MHD flow.

Between the MHD energy and cross helicity, we have the relation

$$\frac{u^2}{2} + \frac{b^2}{2} \geq |u \cdot b|. \tag{6.55}$$

Here the equality holds for

$$u + b = 0 \quad \text{or} \quad u - b = 0, \tag{6.56a}$$

which is equivalent to

$$\phi = 0 \quad \text{or} \quad \psi = 0. \tag{6.56b}$$

6.2 Cowling's Anti-Dynamo Theorem

We introduce the spherical coordinates (r, θ, ϕ) for examining MHD phenomena in a spherical region, as in figure 6.2. We write the toroidal

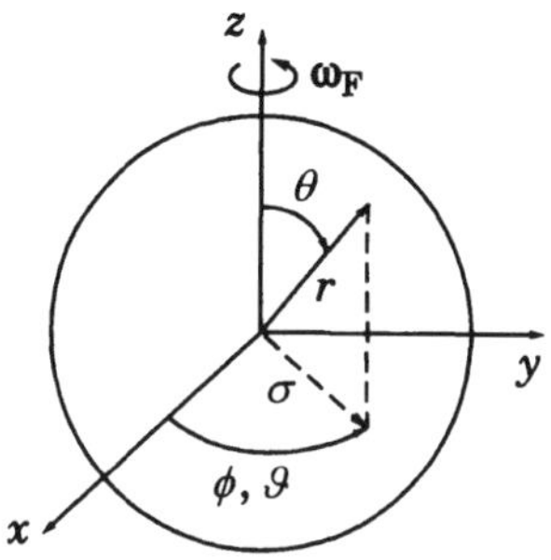

Figure 6.2. Spherical and cylindrical coordinates, (r, θ, ϕ) and (σ, ϑ, z).

and poloidal components of magnetic fields, $\boldsymbol{b}_t$ and $\boldsymbol{b}_p$, as

$$\boldsymbol{b}_t = b_\phi \boldsymbol{e}_\phi, \tag{6.57}$$

$$\boldsymbol{b}_p = b_r \boldsymbol{e}_r + b_\theta \boldsymbol{e}_\theta, \tag{6.58}$$

where $\boldsymbol{e}_r$, $\boldsymbol{e}_\theta$, and $\boldsymbol{e}_\phi$ are the unit vector in each of three directions. In the earth's magnetic fields, the outer core is covered with the electrically non-conducting mantle, and $\boldsymbol{b}_t$ is not observable. Of the poloidal component, the strongest is the dipole field whose present axis is nearly along the axis of earth's rotation.

In relation to the sustainment of axisymmetric magnetic fields, we have Cowling's anti-dynamo theorem [6.5, 6.6]. It says

> "An axisymmetric flow cannot sustain an axisymmetric magnetic field in a stationary sense."

For understanding this theorem, we consider the axisymmetric poloidal field $\boldsymbol{b}_p$, as in figure 6.3. There C expresses the line along which $\boldsymbol{b}_p$ vanishes. On this line, we may rewrite the Ohm's law (6.29) as

$$j_t = \frac{1}{\lambda_M}(\boldsymbol{e} + \boldsymbol{u} \times \boldsymbol{b}_t)_t = \frac{1}{\lambda_M}\boldsymbol{e}_t. \tag{6.59}$$

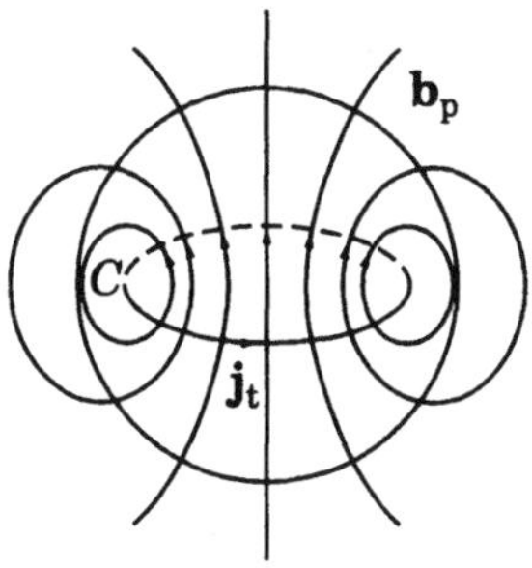

Figure 6.3. Cowling's anti-dynamo.

Here we should note that the toroidal current $\boldsymbol{j}_{\mathrm{t}}$ is axisymmetric owing to the axisymmetry of $\boldsymbol{u}$ and $\boldsymbol{b}$. We integrate $\boldsymbol{j}_{\mathrm{t}}$ along C and have

$$\oint_C \boldsymbol{j}_{\mathrm{t}} \cdot \mathrm{d}\boldsymbol{s} = \frac{1}{\lambda_{\mathrm{M}}} \int_S (\nabla \times \boldsymbol{e}) \cdot \boldsymbol{n}\, \mathrm{d}S = -\frac{1}{\lambda_{\mathrm{M}}} \int_S \frac{\partial \boldsymbol{b}}{\partial t} \cdot \boldsymbol{n}\, \mathrm{d}S, \tag{6.60}$$

from the Stokes integral theorem, where S is an arbitrary surface spanned on C, and $\boldsymbol{n}$ is the unit vector normal to the surface. In a stationary state, equation (6.60) gives

$$\oint_C \boldsymbol{j}_{\mathrm{t}} \cdot \mathrm{d}\boldsymbol{s} = 0, \tag{6.61}$$

which signifies vanishing of $\boldsymbol{j}_{\mathrm{t}}$ itself owing to its axisymmetry. As a result, $\boldsymbol{b}_{\mathrm{P}}$ is not sustainable.

Cowling's anti-dynamo theorem indicates that a nonaxisymmetric motion is indispensable for the stationary sustainment of an axisymmetric component of $\boldsymbol{b}$. From the theorem, the analytical approach to the magnetic-field generation mechanism or dynamo is roughly divided into two categories. In one category, we introduce a small number of asymmetric flow components and study the process through which an axisymmetric component of magnetic field occurs. This method is called laminar dynamo [2.13, 6.7].

The other is called mean-field theory or turbulent dynamo [2.13–2.16]. As was seen in §6.1.2, Reynolds numbers encountered in astro/geophysical phenomena are very large, signifying that fluid motion is turbulent or highly asymmetric. In this method, our interest lies in the exploration of the mechanism under which symmetric fields are generated by highly asymmetric motion. This is the main theme of Part III.

6.3 Mean-Field Equations

In mean-field theory of dynamo, we are interested in large-scale properties of magnetic fields that are represented by the earth's dipole field and solar toroidal field as the origin of sunspots. In order to abstract such properties, we apply the ensemble averaging procedure. In a spherical body such as the earth and the sun, we may regard it as equivalent to the averaging around the axis of rotation. In the averaging, the mean field is inevitably axisymmetric around the axis, and asymmetric properties are beyond the scope of this method. We shall pay special attention to axisymmetric components of field and discuss the theoretical aspects of dynamo.

We use the ensemble average and divide a quantity f into the mean F and the fluctuation around it, f', the same as for electrically nonconducting fluids [see equation (3.22)]. Here f, F, and f' denote

$$f = (\boldsymbol{u}, p, \boldsymbol{\omega}, \boldsymbol{b}, \boldsymbol{j}, \boldsymbol{e}),$$
$$F = (\boldsymbol{U}, P, \boldsymbol{\Omega}, \boldsymbol{B}, \boldsymbol{J}, \boldsymbol{E}), \tag{6.62}$$
$$f' = (\boldsymbol{u}', p', \boldsymbol{\omega}', \boldsymbol{b}', \boldsymbol{j}', \boldsymbol{e}').$$

We take the ensemble average of equation (6.26) with the Coriolis and buoyancy terms dropped, and have

$$\frac{DU_i}{Dt} \equiv \left(\frac{\partial}{\partial t} + \boldsymbol{U} \cdot \boldsymbol{\nabla}\right) U_i$$
$$= -\frac{\partial}{\partial x_i}\left(P + \left\langle \frac{\boldsymbol{b}'^2}{2} \right\rangle\right) + (\boldsymbol{J} \times \boldsymbol{B})_i + \frac{\partial}{\partial x_j}(-R_{ij}) + \nu \boldsymbol{\nabla}^2 U_i, \tag{6.63}$$

with the solenoidal condition $\boldsymbol{\nabla} \cdot \boldsymbol{U} = 0$. Here the Reynolds stress R_{ij} is the MHD counterpart of equation (3.27) and is defined by

$$R_{ij} = \langle u'_i u'_j - b'_i b'_j \rangle. \tag{6.64}$$

We may insert the neglected Coriolis and buoyancy effects into equation (6.63) when necessary.

From equation (6.31), the mean magnetic field $\boldsymbol{B}$ obeys

$$\frac{\partial \boldsymbol{B}}{\partial t} = \boldsymbol{\nabla} \times (\boldsymbol{U} \times \boldsymbol{B} + \boldsymbol{E}_{\mathrm{M}}) + \lambda_{\mathrm{M}} \boldsymbol{\nabla}^2 \boldsymbol{B}, \tag{6.65}$$

where $\boldsymbol{E}_{\mathrm{M}}$ is defined by

$$\boldsymbol{E}_{\mathrm{M}} = \langle \boldsymbol{u}' \times \boldsymbol{b}' \rangle \tag{6.66}$$

and is called the turbulent electromotive force. In this context, the mean Ohm's law is written as

$$\boldsymbol{J} = \frac{1}{\lambda_{\mathrm{M}}}(\boldsymbol{E} + \boldsymbol{U} \times \boldsymbol{B} + \boldsymbol{E}_{\mathrm{M}}). \tag{6.67}$$

In the original Ohm's law (6.29), the electric current arising from the direct interaction between $\boldsymbol{u}$ and $\boldsymbol{b}$ is normal to $\boldsymbol{b}$, resulting in Cowling's anti-dynamo theorem. In the presence of the turbulence effect $\boldsymbol{E}_{\mathrm{M}}$, equation (6.60) is replaced with

$$\oint_C \boldsymbol{J}_{\mathrm{t}} \cdot \mathrm{d}\mathbf{s} = -\frac{1}{\lambda_{\mathrm{M}}} \int_S \frac{\partial \boldsymbol{B}}{\partial t} \cdot \boldsymbol{n}\, \mathrm{d}S + \frac{1}{\lambda_{\mathrm{M}}} \oint_C \boldsymbol{E}_{\mathrm{M}} \cdot \mathrm{d}\mathbf{s}. \tag{6.68}$$

The occurrence of the second term in equation (6.68) leads to the possibility that axisymmetric $\boldsymbol{J}_{\mathrm{t}}$ does not vanish under vanishing $\partial \boldsymbol{B}/\partial t$, escaping from Cowling's anti-dynamo theorem.

6.4 Turbulence Equations

In order to close equations (6.63) and (6.65), we need to express R_{ij} and $\boldsymbol{E}_{\mathrm{M}}$ in terms of the mean field and the quantities characterizing MHD turbulent state. This procedure is entirely the same as for turbulence modelling of electrically nonconducting fluids. Then mean-field theory of dynamo may be called MHD turbulence modelling. In chapters 4 and 5 we discussed turbulence modelling from both the heuristic and theoretical viewpoints. In those discussions, we stressed that the statistical quantities linked with conservation properties become very important in the construction of model equations with firm mathematical and physical bases. In §6.1.3, we showed that the MHD energy [equation (6.53)] and the cross helicity [equation (6.54)] are such quantities.

We attach special importance to the turbulent MHD energy and cross helicity, which are given by

$$K = \left\langle \frac{\boldsymbol{u}'^2 + \boldsymbol{b}'^2}{2} \right\rangle, \tag{6.69}$$

$$W = \langle \boldsymbol{u}' \cdot \boldsymbol{b}' \rangle. \tag{6.70}$$

The fluctuations $\boldsymbol{u}'$ and $\boldsymbol{b}'$ obey

$$\frac{\mathrm{D}u_i'}{\mathrm{D}t} + \frac{\partial}{\partial x_j}(u_i'u_j' - b_i'b_j' - R_{ij}) + \frac{\partial \varpi'}{\partial x_i} - \nu \nabla^2 u_i' = B_j \frac{\partial b_i'}{\partial x_j} - u_j' \frac{\partial U_i}{\partial x_j} + b_j' \frac{\partial B_i}{\partial x_j}, \tag{6.71}$$

$$\frac{\mathrm{D}b_i'}{\mathrm{D}t} + \frac{\partial}{\partial x_j}(u_j'b_i' - u_i'b_j' - \varepsilon_{ijl}E_{\mathrm{M}l}) - \lambda_{\mathrm{M}} \nabla^2 b_i' = B_j \frac{\partial u_i'}{\partial x_j} - u_j' \frac{\partial B_i}{\partial x_j} + b_j' \frac{\partial U_i}{\partial x_j}, \tag{6.72}$$

with the solenoidal conditions $\nabla \cdot \boldsymbol{u}' = \nabla \cdot \boldsymbol{b}' = 0$, where

$$\varpi' = p + \frac{\boldsymbol{b}^2}{2} - \left\langle p + \frac{\boldsymbol{b}^2}{2} \right\rangle. \tag{6.73}$$

From equations (6.71) and (6.72), we have the transport equations for K and W,

$$\frac{\mathrm{D}Z}{\mathrm{D}t} = P_Z - \varepsilon_Z + \nabla \cdot \boldsymbol{T}_Z \quad (Z = K \text{ or } W), \tag{6.74}$$

where

$$P_K = -R_{ij}\frac{\partial U_j}{\partial x_i} - \boldsymbol{E}_{\mathrm{M}} \cdot \boldsymbol{J} = -\frac{1}{2}R_{ij}S_{ij} - \boldsymbol{E}_{\mathrm{M}} \cdot \boldsymbol{J}, \tag{6.75}$$

$$\varepsilon \equiv \varepsilon_K = \nu\left\langle\left(\frac{\partial u_j'}{\partial x_i}\right)^2\right\rangle + \lambda_{\mathrm{M}}\left\langle\left(\frac{\partial b_j'}{\partial x_i}\right)^2\right\rangle, \tag{6.76}$$

$$\boldsymbol{T}_K = W\boldsymbol{B} + \left\langle -\left(\frac{\boldsymbol{u}'^2 + \boldsymbol{b}'^2}{2} + \varpi'\right)\boldsymbol{u}' + (\boldsymbol{u}' \cdot \boldsymbol{b}')\boldsymbol{b}'\right\rangle$$
$$+ \nu\boldsymbol{\nabla}\left\langle\frac{\boldsymbol{u}'^2}{2}\right\rangle + \lambda_{\mathrm{M}}\boldsymbol{\nabla}\left\langle\frac{\boldsymbol{b}'^2}{2}\right\rangle \tag{6.77}$$

$$P_W = -R_{ij}\frac{\partial B_j}{\partial x_i} - \boldsymbol{E}_{\mathrm{M}} \cdot \boldsymbol{\Omega}, \tag{6.78}$$

$$\varepsilon_W = (\nu + \lambda_{\mathrm{M}})\left\langle\frac{\partial u_j'}{\partial x_i}\frac{\partial b_j'}{\partial x_i}\right\rangle, \tag{6.79}$$

$$\boldsymbol{T}_W = K\boldsymbol{B} + \left\langle -(\boldsymbol{u}' \cdot \boldsymbol{b}')\boldsymbol{u}' + \left(\frac{\boldsymbol{u}'^2 + \boldsymbol{b}'^2}{2} - \varpi'\right)\boldsymbol{b}'\right\rangle$$
$$+ \nu\langle(\boldsymbol{b} \cdot \boldsymbol{\nabla}\boldsymbol{U})\rangle + \lambda_{\mathrm{M}}\langle(\boldsymbol{U} \cdot \boldsymbol{\nabla}\boldsymbol{b})\rangle. \tag{6.80}$$

In equation (6.75), we used equations (4.6)–(4.8). In equations (6.77) and (6.80) we shall hereafter drop the molecular diffusion terms that are usually much smaller than the terms expressing turbulent viscosity and diffusivity effects.

In equation (6.74), P_Z, ε_Z, and $\boldsymbol{T}_Z$ are the MHD counterparts of equations (3.40)–(3.42), respectively. Each role of the former is quite similar to that of the latter. Specifically, the MHD energy and the cross helicity are supplied from the mean to fluctuating field through P_K and P_W. These two terms will play a critical role in the discussion on the sustainment of the MHD turbulent state.

In the frame rotating with angular velocity $\boldsymbol{\omega}_{\mathrm{F}}$, the mean vorticity $\boldsymbol{\Omega}$ is subject to the transformation

$$\boldsymbol{\Omega} \to \boldsymbol{\Omega} + 2\boldsymbol{\omega}_{\mathrm{F}}, \tag{6.81}$$

as may be seen from equation (6.41). In this context, we have

$$P_W \to P_W = -R_{ij}\frac{\partial B_j}{\partial x_i} - \boldsymbol{E}_{\mathrm{M}} \cdot (\boldsymbol{\Omega} + 2\boldsymbol{\omega}_{\mathrm{F}}), \tag{6.82}$$

whereas the $\boldsymbol{\omega}_{\mathrm{F}}$ effect does not occur explicitly for P_K. Equation (6.82) indicates that W is sensitive to the frame rotation.

In the presence of the buoyancy force, P_K and P_W are supplemented with the thermal effects as

$$P_K: \quad -\alpha_{\rm T}\langle\theta' \boldsymbol{u}'\rangle \cdot \boldsymbol{g} = -\alpha_{\rm T}\mathbf{H}_\theta \cdot \boldsymbol{g}, \tag{6.83}$$

$$P_W: \quad -\alpha_{\rm T}\langle\theta' \boldsymbol{b}'\rangle \cdot \boldsymbol{g}, \tag{6.84}$$

with the definition (3.29). Specifically, the former may be expressed as

$$\alpha_{\rm T}\frac{\nu_{\rm T}}{\sigma_\theta}\nabla\Theta \cdot \boldsymbol{g}, \tag{6.85}$$

in terms of the turbulent-diffusivity model to $\boldsymbol{H}_\theta$, equation (4.70). This expression indicates that the buoyancy effect contributes to the generation of turbulent energy in a thermally unstable case, that is, in the presence of the temperature gradient parallel to the gravitational acceleration vector. This point will be further referred to later.

In correspondence to equation (6.55), we have

$$\frac{|W|}{K} \leq 1. \tag{6.86}$$

This relation will be later found to be very useful in the estimate of the strength of magnetic fields generated by turbulent motion.

References

[6.1] Roberts P H 1967 *An Introduction to Magnetohydrodynamics* (London: Longmans)

[6.2] Pedlosky J 1979 *Geophysical Fluid Dynamics* (New York: Springer)

[6.3] Phillips O M 1966 *The Dynamics of the Upper Ocean* (Cambridge: Cambridge University Press)

[6.4] Busse F 1970 *J. Fluid Mech.* **44** 441

[6.5] Cowling T G 1934 *Mon. Not. Astr. Roy. Soc.* **94** 39

[6.6] Hoyng P 1993 in *The Sun: A Laboratory for Astrophysics* ed J T Schmelz and J C Brown (Dordrecht: Kluwer)

[6.7] Braginsky S I 1991 *Geophys. Astrophys. Fluid Dyn.* **60** 89

Chapter 7

Theoretical Estimate of Turbulence Effects on Magnetic-Field Equations

For the estimate of the Reynolds stress R_{ij} and the turbulent electromotive force $\boldsymbol{E}_{\mathrm{M}}$, we start from a kinematic approach and then proceed to a magnetohydrodynamic (MHD) approach. The former consists of kinematic and counter-kinematic methods. In the kinematic method, the generation of magnetic fluctuation $\boldsymbol{b}'$ under given velocity fluctuation $\boldsymbol{u}'$ is examined through the magnetic induction equation, and R_{ij} and $\boldsymbol{E}_{\mathrm{M}}$ are evaluated. In the counter-kinematic method, this evaluation is made under given $\boldsymbol{b}'$ on the basis of the Navier–Stokes equation. In the MHD method, the generation process of R_{ij} and $\boldsymbol{E}_{\mathrm{M}}$ are examined through the simultaneous treatment of $\boldsymbol{u}'$ and $\boldsymbol{b}'$. These three approaches are summarized symbolically as follows:

$$\text{Kinematic method:} \quad \boldsymbol{u}' \Rightarrow \boldsymbol{b}' \Rightarrow R_{ij}, \boldsymbol{E}_{\mathrm{M}};$$

$$\text{Counter-kinematic method:} \quad \boldsymbol{b}' \Rightarrow \boldsymbol{u}' \Rightarrow R_{ij}, \boldsymbol{E}_{\mathrm{M}};$$

$$\text{MHD method:} \quad \boldsymbol{u}', \boldsymbol{b}' \Rightarrow R_{ij}, \boldsymbol{E}_{\mathrm{M}}.$$

In what follows, we shall drop effects of Coriolis and buoyancy forces for simplicity of discussion and refer to them only when necessary.

7.1 Kinematic Method

Once the statistics of $\boldsymbol{u}'$ are given, we may evaluate $\boldsymbol{E}_{\mathrm{M}}$ on the basis of equation (6.72) and examine its relationship with the generation process of $\boldsymbol{B}$. This approach is called kinematic dynamo. Here we follow the two-scale direct-interaction approximation (TSDIA) method [2.16, 4.8] in chapter 5 and perform kinematic dynamo modelling.

7.1.1 Introduction of Two Scales and Scale-Parameter Expansion

We use two spatial and temporal scales given by equation (5.48), and write f

[equation (6.62)] as equation (5.51). Then equation (6.72) may be rewritten as

$$\frac{\partial b'_i}{\partial \tau} + U_j \frac{\partial b'_i}{\partial \xi_j} + \frac{\partial}{\partial x_j}(u'_j b'_i - u'_i b'_j) - \lambda_{\mathrm{M}} \nabla_\xi^2 b'_i$$

$$= B_j \frac{\partial u'_i}{\partial \xi_j} + \delta_{\mathrm{S}} \left(-u'_j \frac{\partial B_i}{\partial X_j} + b'_j \frac{\partial U_i}{\partial X_j} - \frac{\mathrm{D} b'_i}{\mathrm{D}T} + B_j \frac{\partial b'_i}{\partial X_j} \right.$$

$$\left. - \frac{\partial}{\partial X_j}(u'_j b'_i - u'_i b'_j - \varepsilon_{ijl} E_{\mathrm{M}l}) \right), \tag{7.1}$$

where ∇_ξ and $\mathrm{D}/\mathrm{D}T$ were defined by equations (5.55) and (5.56), respectively, and we should note that $\boldsymbol{E}_{\mathrm{M}}$ is a function of $\boldsymbol{X}$ and T.

We apply the moving-frame Fourier representation, equation (5.57), to equation (7.1), and have

$$\frac{\partial b'_i(\boldsymbol{k};\tau)}{\partial \tau} + \lambda_{\mathrm{M}} k^2 b'_i(\boldsymbol{k};\tau) - \mathrm{i} N_{ijl}(\boldsymbol{k}) \iint u'_j(\boldsymbol{p};\tau) b'_l(\boldsymbol{q};\tau) \delta(\boldsymbol{k}-\boldsymbol{p}-\boldsymbol{q})\, \mathrm{d}\boldsymbol{p}\, \mathrm{d}\boldsymbol{q}$$

$$= -\mathrm{i}(\boldsymbol{k}\cdot\boldsymbol{B}) u'_i(\boldsymbol{k};\tau) + \delta_{\mathrm{S}} \left(-u'_j(\boldsymbol{k};\tau) \frac{\partial B_i}{\partial X_j} + b'_j(\boldsymbol{k};\tau) \frac{\partial U_i}{\partial X_j} \right.$$

$$\left. - \frac{\mathrm{D}^* b'_i(\boldsymbol{k};\tau)}{\mathrm{D}T^*} + B_j \frac{\partial^* u'_i(\boldsymbol{k};\tau)}{\partial X_j^*} + N_{Bi} \right), \tag{7.2}$$

where

$$N_{ijl}(\boldsymbol{k}) = k_j \delta_{il} - k_l \delta_{ij}, \tag{7.3}$$

$\mathrm{D}^*/\mathrm{D}T^*$ and ∇_X^* were defined by equation (5.60), and $\boldsymbol{N}_B$ expresses the terms that are nonlinear in $\boldsymbol{u}'$ and $\boldsymbol{b}'$ and are not dependent directly on the mean field $\boldsymbol{U}$ and $\boldsymbol{B}$. We focus attention on the interaction between the mean field and fluctuations, and neglect $\boldsymbol{N}_B$ in what follows. Moreover, the dependence of $\boldsymbol{u}'$ and $\boldsymbol{b}'$ on slow variables $\boldsymbol{X}$ and T are not written explicitly, except when necessary, as is similar to chapter 5.

We expand $\boldsymbol{b}'$ as equation (5.63). The first two parts are governed by

$$\frac{\partial b'_{0i}(\boldsymbol{k};\tau)}{\partial \tau} + \lambda_{\mathrm{M}} k^2 b'_{0i}(\boldsymbol{k};\tau) - \mathrm{i} N_{ijl}(\boldsymbol{k}) \iint u'_j(\boldsymbol{p};\tau) b'_{0l}(\boldsymbol{q};\tau) \delta(\boldsymbol{k}-\boldsymbol{p}-\boldsymbol{q})\, \mathrm{d}\boldsymbol{p}\, \mathrm{d}\boldsymbol{q}$$

$$= I_{B0}(\boldsymbol{k};\tau) \equiv -\mathrm{i}(\boldsymbol{k}\cdot\boldsymbol{B}) u'_i(\boldsymbol{k};\tau), \tag{7.4}$$

$$\frac{\partial b'_{1i}(\boldsymbol{k};\tau)}{\partial \tau} + \lambda_{\mathrm{M}} k^2 b'_{1i}(\boldsymbol{k};\tau) - \mathrm{i} N_{ijl}(\boldsymbol{k}) \iint u'_j(\boldsymbol{p};\tau) b'_{1l}(\boldsymbol{q};\tau) \delta(\boldsymbol{k}-\boldsymbol{p}-\boldsymbol{q})\, \mathrm{d}\boldsymbol{p}\, \mathrm{d}\boldsymbol{q}$$

$$= I_{B1i}(\boldsymbol{k};\tau)$$

$$\equiv -u'_j(\boldsymbol{k};\tau) \frac{\partial B_i}{\partial X_j} + b'_{0j}(\boldsymbol{k};\tau) \frac{\partial U_i}{\partial X_j} - \frac{\mathrm{D}^* b'_{0i}(\boldsymbol{k};\tau)}{\mathrm{D}T^*} + B_j \frac{\partial^* u'_i(\boldsymbol{k};\tau)}{\partial X_j^*}. \tag{7.5}$$

In order to solve equations (7.4) and (7.5), we introduce the Green's function $G'_{\mathrm{M}ij}$, which obeys

$$\frac{\partial G'_{\mathrm{M}ij}(\boldsymbol{k};\tau,\tau')}{\partial \tau} + \lambda_{\mathrm{M}} k^2 G'_{\mathrm{M}ij}(\boldsymbol{k};\tau,\tau') - \mathrm{i}N_{ilm}(\boldsymbol{k}) \iint u'_l(\boldsymbol{p};\tau) G'_{\mathrm{M}mj}(\boldsymbol{q};\tau,\tau')\delta(\boldsymbol{k}-\boldsymbol{p}-\boldsymbol{q})\,\mathrm{d}\boldsymbol{p}\,\mathrm{d}\boldsymbol{q} = \delta_{ij}\delta(\tau-\tau'), \tag{7.6}$$

as is quite similar to equation (5.81). With the aid of $G'_{\mathrm{M}ij}$, equation (7.4) may be integrated as

$$b'_{0i}(\boldsymbol{k};\tau) = b'_{00i}(\boldsymbol{k};\tau) + \int_{-\infty}^{\tau} G'_{\mathrm{M}ij}(\boldsymbol{k};\tau,\tau_1) I_{B0j}(\boldsymbol{k};\tau_1)\,\mathrm{d}\tau_1. \tag{7.7}$$

Here $\boldsymbol{b}'_{00}$ is governed by equation (7.4) with the $\boldsymbol{B}$-related part dropped, that is,

$$\frac{\partial b'_{00i}(\boldsymbol{k};\tau)}{\partial \tau} + \lambda_{\mathrm{M}} k^2 b'_{00i}(\boldsymbol{k};\tau) - \mathrm{i}N_{ijl}(\boldsymbol{k}) \iint u'_j(\boldsymbol{p};\tau) b'_{00l}(\boldsymbol{q};\tau)\delta(\boldsymbol{k}-\boldsymbol{p}-\boldsymbol{q})\,\mathrm{d}\boldsymbol{p}\,\mathrm{d}\boldsymbol{q} = 0. \tag{7.8}$$

This equation contains no factors generating and sustaining the statistical anisotropy of $\boldsymbol{b}'_{00}$. In what follows, the statistics related to $\boldsymbol{b}'_{00}$ will be assumed to be isotropic, but their nonmirror symmetry is taken into account when necessary.

Entirely similarly, equation (7.5) results in

$$b'_{1i}(\boldsymbol{k};\tau) = \int_{-\infty}^{\tau} G'_{\mathrm{M}ij}(\boldsymbol{k};\tau,\tau_1) I_{B1j}(\boldsymbol{k};\tau_1)\,\mathrm{d}\tau_1. \tag{7.9}$$

7.1.2 Evaluation of Turbulent Electromotive Force

Equations (7.7) and (7.9) indicate that $\boldsymbol{b}'$ is expressed in terms of $\boldsymbol{b}'_{00}$ under given $\boldsymbol{u}'$. Then $\boldsymbol{E}_{\mathrm{M}}$ may be evaluated once the statistics of $\boldsymbol{u}'$ are known. As the simplest turbulent state, we assume the isotropy of $\boldsymbol{u}'$. In this case, the covariance of $\boldsymbol{u}'$ is expressed as equation (3.108), that is,

$$\frac{\langle u'_i(\boldsymbol{k};\tau) u'_j(\boldsymbol{k}';\tau')\rangle}{\delta(\boldsymbol{k}+\boldsymbol{k}')} = D_{ij}(\boldsymbol{k}) Q_{\mathrm{K}}(k;\tau,\tau') + \frac{\mathrm{i}}{2}\frac{k_l}{k^2}\varepsilon_{ijl}\Gamma_{\mathrm{K}}(k;\tau,\tau'), \tag{7.10}$$

where subscript K denotes kinetic. Then the turbulent kinetic energy and the turbulent kinetic helicity are expressed by

$$K_{\rm K} = \left\langle \frac{\boldsymbol{u}'^2}{2} \right\rangle = \int Q_{\rm K}(k;\tau,\tau)\,{\rm d}\boldsymbol{k}, \tag{7.11}$$

$$H_{\rm K} = \langle \boldsymbol{u}' \cdot \boldsymbol{\omega}' \rangle = \int \Gamma_{\rm K}(k;\tau,\tau)\,{\rm d}\boldsymbol{k}, \tag{7.12}$$

respectively.

In relation to the second term of equation (7.10), we should recall convection columns referred to in §6.1.2. The columns are characterized by nonvanishing helicity. Equations (3.85) and (6.81) show that the generation of turbulent kinetic helicity $H_{\rm K}$ is associated with the angular velocity of a frame rotation. This fact indicates the necessity of keeping helicity effects in the statistical analysis of MHD flows in a rotating frame.

In the present discussion, we further introduce the correlation between $\boldsymbol{b}'_{00}$ and $\boldsymbol{u}'$, that is, the turbulent cross helicity. Its importance was discussed in §6.1.3 in the light of MHD conservation properties. We write

$$\frac{\langle u'_i(\boldsymbol{k};\tau) b'_{00j}(\boldsymbol{k}';\tau') \rangle}{\delta(\boldsymbol{k}+\boldsymbol{k}')} = D_{ij}(\boldsymbol{k})\Lambda(k;\tau,\tau'), \tag{7.13}$$

which gives

$$\langle \boldsymbol{u}' \cdot \boldsymbol{b}'_{00} \rangle = 2\int \Lambda(k;\tau,\tau)\,{\rm d}\boldsymbol{k}. \tag{7.14}$$

In purely kinematic dynamo, there is no necessity to retain Λ. In real dynamo, however, there is the close connection between magnetic field and velocity through equation (6.71). With this point in mind, we perform the analysis of $\boldsymbol{E}_{\rm M}$ on the basis of equations (7.10) and (7.13). The introduction of Λ will be shown to lead to a new dynamo effect that does not occur in the purely kinematic approach. This approach with equation (7.13) incorporated may be also called a quasi-kinematic dynamo, in contrast to the usual kinematic dynamo.

From equation (7.6), the Green's function $G'_{{\rm M}ij}$ is dependent on $\boldsymbol{u}'$ only that is statistically isotropic. Then we write

$$\langle G'_{{\rm M}ij}(\boldsymbol{k};\tau,\tau') \rangle = \delta_{ij} G_{\rm M}(k;\tau,\tau'). \tag{7.15}$$

We expand $\boldsymbol{E}_{\rm M}$ as

$$E_{{\rm M}i} = \varepsilon_{ijl}(\langle u'_j b'_{0l} \rangle + \delta_{\rm S} \langle u'_j b'_{1l} \rangle) + O(\delta_{\rm S}^2), \tag{7.16}$$

as is similar to equation (5.139). We substitute equations (7.7) and (7.9) into equation (7.16), and have

$$E_{Mi} = \varepsilon_{ijl}\left(\int \frac{\langle u'_j(\boldsymbol{k};\tau)b'_{00l}(\boldsymbol{k}';\tau)\rangle}{\delta(\boldsymbol{k}+\boldsymbol{k}')}\,\mathrm{d}\boldsymbol{k}\right)$$
$$+\varepsilon_{ijl}\left(-iB_n\int k'_n\,\mathrm{d}\boldsymbol{k}\int_{-\infty}^{\tau}\frac{\langle G'_{Mlm}(\boldsymbol{k}';\tau,\tau_1)u'_j(\boldsymbol{k};\tau)u'_m(\boldsymbol{k}';\tau_1)\rangle}{\delta(\boldsymbol{k}+\boldsymbol{k}')}\,\mathrm{d}\tau_1\right)$$
$$+\varepsilon_{ijl}\left(-\frac{\partial B_m}{\partial x_n}\int \mathrm{d}\boldsymbol{k}\int_{-\infty}^{\tau}\frac{\langle G'_{Mlm}(\boldsymbol{k}';\tau,\tau_1)u'_j(\boldsymbol{k};\tau)u'_n(\boldsymbol{k}';\tau_1)\rangle}{\delta(\boldsymbol{k}+\boldsymbol{k}')}\,\mathrm{d}\tau_1\right)$$
$$+\varepsilon_{ijl}\left(\frac{\partial U_m}{\partial x_n}\int \mathrm{d}\boldsymbol{k}\int_{-\infty}^{\tau}\frac{\langle G'_{Mlm}(\boldsymbol{k}';\tau,\tau_1)u'_j(\boldsymbol{k};\tau)b'_{00n}(\boldsymbol{k}';\tau_1)\rangle}{\delta(\boldsymbol{k}+\boldsymbol{k}')}\,\mathrm{d}\tau_1\right), \quad (7.17)$$

after the replacement (5.154) concerning $\boldsymbol{X}$ and T. In the context of the last two terms, we retained the first two terms in $\boldsymbol{I}_{B1}$ of equation (7.5) that are linearly dependent on the first spatial derivatives of $\boldsymbol{U}$ and $\boldsymbol{B}$.

In the first term of equation (7.17), we use equation (5.157a) and have

$$-\varepsilon_{ijl}\int D_{jl}(\boldsymbol{k})\Lambda(k;\tau,\tau)\,\mathrm{d}\boldsymbol{k} = \frac{2}{3}\varepsilon_{ijl}\delta_{jl}\int \Lambda(k;\tau,\tau)\,\mathrm{d}\boldsymbol{k} = 0. \quad (7.18)$$

To the second term, we apply the renormalization procedure leading to equation (5.145), and have

$$\varepsilon_{ijl}\left(B_n\int \mathrm{d}\boldsymbol{k}\int_{-\infty}^{\tau} G_{Mlm}(\boldsymbol{k}';\tau,\tau_1)\frac{(-\mathrm{i}k'_n)\langle u'_j(\boldsymbol{k};\tau)u'_m(\boldsymbol{k}';\tau_1)\rangle}{\delta(\boldsymbol{k}+\boldsymbol{k}')}\,\mathrm{d}\tau_1\right)$$
$$=\left(-\frac{1}{3}\int \mathrm{d}\boldsymbol{k}\int_{-\infty}^{\tau} G_M(k;\tau,\tau_1)\Gamma_K(k;\tau,\tau_1)\,\mathrm{d}\tau_1\right)B_i. \quad (7.19)$$

Here we should note that the second part of equation (7.10), which is related to the helicity effect, contributes to this result. The third and fourth terms in equation (7.17) may be evaluated in entirely the same way. Summarizing these results, we have

$$\boldsymbol{E}_M = \alpha_K\boldsymbol{B} - \beta_K\boldsymbol{J} + \gamma_K\boldsymbol{\Omega}. \quad (7.20)$$

Coefficients α_K, β_K, and γ_K are expressed in the form

$$\alpha_K = -\frac{1}{3}\int \mathrm{d}\boldsymbol{k}\int_{-\infty}^{\tau} G_M(k;\tau,\tau_1)\Gamma_K(k;\tau,\tau_1)\,\mathrm{d}\tau_1, \quad (7.21)$$

$$\beta_K = \frac{2}{3}\int \mathrm{d}\boldsymbol{k}\int_{-\infty}^{\tau} G_M(k;\tau,\tau_1)Q_K(k;\tau,\tau_1)\,\mathrm{d}\tau_1, \quad (7.22)$$

$$\gamma_K = \frac{2}{3}\int \mathrm{d}\boldsymbol{k}\int_{-\infty}^{\tau} G_M(k;\tau,\tau_1)\Lambda(k;\tau,\tau_1)\,\mathrm{d}\tau_1. \quad (7.23)$$

On dropping the third term on the left-hand side of equation (7.6), equations (7.21) and (7.22) are reduced to the expressions obtained by the first-order smoothing approximation [2.13]. We should note that the approximation is applicable to the case of low magnetic Reynolds number.

Equation (7.20) shows that turbulence effects generate the electromotive force aligned with the mean magnetic field, the mean electric current density, and the mean vorticity. Their physical meanings will be discussed later in detail.

7.1.3 Evaluation of Reynolds Stress

In the usual kinematic approach, our concern is focused on the evaluation of $\boldsymbol{E}_{\mathrm{M}}$, that is, the effect of velocity fluctuation on the equation for $\boldsymbol{B}$. In the present approach, we are also interested in the effect arising from the correlation between velocity and magnetic field or the turbulent cross helicity. By taking the effect into account, we may examine the feedback effect of the generated magnetic field on the equation for $\boldsymbol{U}$. The velocity-related part of R_{ij} is simply evaluated from equation (7.10) and is given in the isotropic form

$$\langle u'_i u'_j \rangle = \left(\frac{2}{3} \int Q_{\mathrm{K}}(k;\tau,\tau_1)\,\mathrm{d}\boldsymbol{k} \right) \delta_{ij}. \tag{7.24}$$

We use equations (7.7) and (7.9), and evaluate $\langle b'_i b'_j \rangle$. For the isotropic part of magnetic fluctuation, $\boldsymbol{b}'_{00}$, we write

$$\frac{\langle b'_{00i}(\boldsymbol{k};\tau) b'_{00j}(\boldsymbol{k}';\tau') \rangle}{\delta(\boldsymbol{k}+\boldsymbol{k}')} = D_{ij}(\boldsymbol{k}) Q_{\mathrm{M}}(k;\tau,\tau'). \tag{7.25}$$

Here we may add the nonmirror symmetric part, which is confirmed not to contribute to the following analysis. As is similar to equation (7.20), we pay special attention to the contributions linearly related to the first spatial derivatives of U and $\boldsymbol{B}$. After the combination with equation (7.24), R_{ij} is written as

$$R_{ij} = \tfrac{2}{3} K_{\mathrm{R}} \delta_{ij} - \nu_{\mathrm{TM}} S_{ij} + \nu_{\mathrm{MM}} M_{ij}, \tag{7.26}$$

where the mean velocity-strain tensor S_{ij} is given by equation (4.6), M_{ij} is its magnetic-field counterpart defined by

$$M_{ij} = \frac{\partial B_j}{\partial x_i} + \frac{\partial B_i}{\partial x_j}, \tag{7.27}$$

and the turbulent residual energy K_{R} is

$$K_{\mathrm{R}} = \left\langle \frac{\boldsymbol{u}'^2 - \boldsymbol{b}'^2}{2} \right\rangle = \int Q_{\mathrm{K}}(k;\tau,\tau)\,\mathrm{d}\boldsymbol{k} - \int Q_{\mathrm{M}}(k;\tau,\tau)\,\mathrm{d}\boldsymbol{k}. \tag{7.28}$$

Coefficients ν_{TM} and ν_{MM} are

$$\nu_{\mathrm{TM}} = \frac{2}{3} \int \mathrm{d}\boldsymbol{k} \int_{-\infty}^{\tau} G_{\mathrm{M}}(k;\tau,\tau_1) Q_{\mathrm{M}}(k;\tau,\tau_1)\,\mathrm{d}\tau_1, \tag{7.29}$$

$$\nu_{\mathrm{MM}} = \frac{2}{3} \int \mathrm{d}\boldsymbol{k} \int_{-\infty}^{\tau} G_{\mathrm{M}}(k;\tau,\tau_1) \Lambda(k;\tau,\tau_1)\,\mathrm{d}\tau_1. \tag{7.30}$$

From the comparison with equation (5.158), we may easily see that the ν_{TM}-related term in equation (7.26) expresses the turbulent-viscosity effect.

On the other hand, the ν_{MM}-related term denotes the feedback effect by the generated magnetic field on the mean flow. The importance of this effect will be clarified in later discussions.

7.2 Counter-Kinematic Method

In §7.1 we analyzed effects of velocity fluctuation on the $\boldsymbol{B}$ equation, which are represented by $\boldsymbol{E}_{\mathrm{M}}$. In what follows, we consider effects of magnetic-field fluctuation on the $\boldsymbol{B}$ and $\boldsymbol{U}$ equations. This approach, which may be called the counter-kinematic method, will be instrumental to understanding the interaction process between $\boldsymbol{U}$ and $\boldsymbol{B}$.

7.2.1 Scale-Parameter Expansion

In terms of two-scale variables, equation (6.71) is written as

$$\frac{\partial u_i'}{\partial \tau} + U_j \frac{\partial u_i'}{\partial \xi_j} + \frac{\partial}{\partial \xi_j}(u_i'u_j' - b_i'b_j') + \frac{\partial \varpi'}{\partial \xi_i} - \nu \nabla_\xi^2 u_i'$$
$$= B_j \frac{\partial b_i'}{\partial \xi_j} + \delta_{\mathrm{S}}\left(-u_j' \frac{\partial U_i}{\partial X_j} + b_j' \frac{\partial B_i}{\partial X_j} - \frac{\mathrm{D}u_i'}{\mathrm{D}T} + B_j \frac{\partial b_i'}{\partial X_j} \right.$$
$$\left. - \frac{\partial}{\partial X_j}(u_j'u_i' - b_j'b_i' - R_{ji}) - \frac{\partial \varpi'}{\partial X_i}\right), \tag{7.31}$$

with the solenoidal condition

$$\frac{\partial u_i'}{\partial \xi_i} = -\delta_{\mathrm{S}} \frac{\partial u_i'}{\partial X_i}. \tag{7.32}$$

We apply the moving-frame Fourier representation, equation (5.57), to equations (7.31) and (7.32), and have

$$\frac{\partial u_i'(\boldsymbol{k};\tau)}{\partial \tau} + \nu k^2 u_i'(\boldsymbol{k};\tau) - \mathrm{i}k_i \varpi'(\boldsymbol{k};\tau) - \mathrm{i}k_j \iint u_i'(\boldsymbol{p};\tau) u_j'(\boldsymbol{q};\tau) \delta(\boldsymbol{k}-\boldsymbol{p}-\boldsymbol{q})\, \mathrm{d}\boldsymbol{p}\, \mathrm{d}\boldsymbol{q}$$
$$= -\mathrm{i}(\boldsymbol{k}\cdot\boldsymbol{B}) b_i(\boldsymbol{k};\tau) + \mathrm{i}k_j \iint b_i'(\boldsymbol{p};\tau) b_j'(\boldsymbol{q};\tau) \delta(\boldsymbol{k}-\boldsymbol{p}-\boldsymbol{q})\, \mathrm{d}\boldsymbol{p}\, \mathrm{d}\boldsymbol{q}$$
$$+ \delta_{\mathrm{S}}\left(-u_j'(\boldsymbol{k};\tau) \frac{\partial U_i}{\partial X_j} + b_j'(\boldsymbol{k};\tau) \frac{\partial B_i}{\partial X_j} - \frac{\mathrm{D}^* u_i'(\boldsymbol{k};\tau)}{\mathrm{D}T^*} - \frac{\partial^* \varpi'(\boldsymbol{k};\tau)}{\partial X_i^*} \right.$$
$$+ B_j \frac{\partial^* b_i'(\boldsymbol{k};\tau)}{\partial X_j^*} - \iint \frac{\partial^*}{\partial X_j^*}[u_i'(\boldsymbol{p};\tau) u_j'(\boldsymbol{q};\tau)$$
$$\left. - b_i'(\boldsymbol{p};\tau) b_j'(\boldsymbol{q};\tau)] \delta(\boldsymbol{k}-\boldsymbol{p}-\boldsymbol{q})\, \mathrm{d}\boldsymbol{p}\, \mathrm{d}\boldsymbol{q} \right), \tag{7.33}$$

$$\boldsymbol{k}\cdot\boldsymbol{u}'(\boldsymbol{k};\tau) = \delta_{\mathrm{S}}\left(-\mathrm{i} \frac{\partial^* u_i'(\boldsymbol{k};\tau)}{\partial X_i^*} \right). \tag{7.34}$$

We expand $\boldsymbol{u}'$ and ϖ' as equation (5.63), and substitute them into equations (7.33) and (7.34). The leading parts $\boldsymbol{u}'_0$ and $\varpi'_0(\boldsymbol{k};\tau)$ obey

$$\frac{\partial u'_{0i}(\boldsymbol{k};\tau)}{\partial \tau} + \nu k^2 u'_{0i}(\boldsymbol{k};\tau) - \mathrm{i}k_i \varpi'_0(\boldsymbol{k};\tau) - \mathrm{i}k_j \iint u'_{0i}(\boldsymbol{p};\tau)u'_{0j}(\boldsymbol{q};\tau)\delta(\boldsymbol{k}-\boldsymbol{p}-\boldsymbol{q})\,\mathrm{d}\boldsymbol{p}\,\mathrm{d}\boldsymbol{q} = -\mathrm{i}k_j b_i(\boldsymbol{k};\tau)B_j + \mathrm{i}k_j \iint b'_i(\boldsymbol{p};\tau)b'_j(\boldsymbol{q};\tau)\delta(\boldsymbol{k}-\boldsymbol{p}-\boldsymbol{q})\,\mathrm{d}\boldsymbol{p}\,\mathrm{d}\boldsymbol{q}, \tag{7.35}$$

$$\boldsymbol{k}\cdot \boldsymbol{u}'_0(\boldsymbol{k};\tau) = 0. \tag{7.36}$$

We use equation (7.36) and eliminate ϖ'_0. The resulting equation is written as

$$\frac{\partial u'_{0i}(\boldsymbol{k};\tau)}{\partial \tau} + \nu k^2 u'_{0i}(\boldsymbol{k};\tau) - \mathrm{i}M_{ijl}(\boldsymbol{k})u'_{0j}(\boldsymbol{p};\tau)u'_{0l}(\boldsymbol{q};\tau)\delta(\boldsymbol{k}-\boldsymbol{p}-\boldsymbol{q})\,\mathrm{d}\boldsymbol{p}\,\mathrm{d}\boldsymbol{q} = -\mathrm{i}(\boldsymbol{k}\cdot\boldsymbol{B})b'_i(\boldsymbol{k};\tau) + \mathrm{i}M_{ijl}(\boldsymbol{k})\iint b'_j(\boldsymbol{p};\tau)b'_l(\boldsymbol{q};\tau)\delta(\boldsymbol{k}-\boldsymbol{p}-\boldsymbol{q})\,\mathrm{d}\boldsymbol{p}\,\mathrm{d}\boldsymbol{q}, \tag{7.37}$$

with $M_{ijl}(\boldsymbol{k})$ defined by equation (3.121).

In the present counter-kinematic approach, $\boldsymbol{b}'$ is assumed to be statistically known, and the right-hand side of equation (7.37) may be regarded as external forces imposed on the $\boldsymbol{u}'_0$ equation. It is difficult to exactly solve equation (7.37) owing to its nonlinearity. Then we treat those forces in the perturbational manner. We write $\boldsymbol{u}'_0$ as

$$\boldsymbol{u}'_0 = \boldsymbol{u}'_{00}(\boldsymbol{k};\tau) + \boldsymbol{u}'_{01}(\boldsymbol{k};\tau) + \cdots. \tag{7.38}$$

The first two terms are governed by

$$\frac{\partial u'_{00i}(\boldsymbol{k};\tau)}{\partial \tau} + \nu k^2 u'_{00i}(\boldsymbol{k};\tau) - \mathrm{i}M_{ijl}(\boldsymbol{k})\iint u'_{00j}(\boldsymbol{p};\tau)u'_{00l}(\boldsymbol{q};\tau)\delta(\boldsymbol{k}-\boldsymbol{p}-\boldsymbol{q})\,\mathrm{d}\boldsymbol{p}\,\mathrm{d}\boldsymbol{q} = 0, \tag{7.39}$$

$$\frac{\partial u'_{01i}(\boldsymbol{k};\tau)}{\partial \tau} + \nu k^2 u'_{01i}(\boldsymbol{k};\tau) - 2\mathrm{i}M_{ijl}(\boldsymbol{k})\iint u'_{00j}(\boldsymbol{p};\tau)u'_{01l}(\boldsymbol{q};\tau)\delta(\boldsymbol{k}-\boldsymbol{p}-\boldsymbol{q})\,\mathrm{d}\boldsymbol{p}\,\mathrm{d}\boldsymbol{q} = I_{K0i} \equiv -\mathrm{i}k_j b'_i(\boldsymbol{k};\tau)B_j + \mathrm{i}M_{ijl}(\boldsymbol{k})\iint b'_j(\boldsymbol{p};\tau)b'_l(\boldsymbol{q};\tau)\delta(\boldsymbol{k}-\boldsymbol{p}-\boldsymbol{q})\,\mathrm{d}\boldsymbol{p}\,\mathrm{d}\boldsymbol{q}. \tag{7.40}$$

Equation (7.40) is linear in $\boldsymbol{u}'_{01}$ and may be integrated in terms of the Green's function $G'_{\mathrm{K}ij}$ obeying

$$\frac{\partial G'_{\mathrm{K}ij}(\boldsymbol{k};\tau,\tau')}{\partial\tau}+\nu k^2 G'_{\mathrm{K}ij}(\boldsymbol{k};\tau,\tau')$$
$$-2\mathrm{i}M_{ilm}(\boldsymbol{k})\iint u'_{00l}(\boldsymbol{p};\tau)G'_{\mathrm{K}mj}(\boldsymbol{q};\tau,\tau')\delta(\boldsymbol{k}-\boldsymbol{p}-\boldsymbol{q})\,\mathrm{d}\boldsymbol{p}\,\mathrm{d}\boldsymbol{q}$$
$$=\delta_{ij}\delta(\tau-\tau'). \tag{7.41}$$

As a result, we have

$$u'_{01i}(\boldsymbol{k};\tau)=\int_{-\infty}^{\tau}G'_{\mathrm{K}il}(\boldsymbol{k};\tau,\tau_1)I_{K0l}(\boldsymbol{k};\tau_1)\,\mathrm{d}\tau_1. \tag{7.42}$$

For obtaining the $O(\delta_{\mathrm{S}})$ part of $\boldsymbol{u}'$, $\boldsymbol{u}'_1$, we write

$$\boldsymbol{u}'_1(\boldsymbol{k};\tau)=\boldsymbol{v}'_1(\boldsymbol{k};\tau)-\mathrm{i}\frac{\boldsymbol{k}}{k^2}\frac{\partial^* u'_{0i}(\boldsymbol{k};\tau)}{\partial X^*_i}, \tag{7.43}$$

as is the same as equation (5.71), leading to the solenoidal condition concerning $\boldsymbol{k}$,

$$\boldsymbol{k}\cdot\boldsymbol{v}'_1(\boldsymbol{k};\tau)=0. \tag{7.44}$$

The solenoidal part $\boldsymbol{v}'_1$ obeys

$$\frac{\partial v'_{1i}(\boldsymbol{k};\tau)}{\partial\tau}+\nu k^2 v'_{1i}(\boldsymbol{k};\tau)$$
$$-2\mathrm{i}M_{ilm}(\boldsymbol{k})\iint u'_{0l}(\boldsymbol{p};\tau)v'_{1m}(\boldsymbol{p};\tau)\delta(\boldsymbol{k}-\boldsymbol{p}-\boldsymbol{q})\,\mathrm{d}\boldsymbol{p}\,\mathrm{d}\boldsymbol{q}$$
$$=D_{il}(\boldsymbol{k})b'_j(\boldsymbol{k};\tau)\frac{\partial B_l}{\partial X_j}-D_{il}(\boldsymbol{k})u'_{0j}(\boldsymbol{k};\tau)\frac{\partial U_l}{\partial X_j}$$
$$-D_{ij}(\boldsymbol{k})\frac{\mathrm{D}^* u'_{0j}(\boldsymbol{k};\tau)}{\mathrm{D}T^*}+N_{Ki}. \tag{7.45}$$

Here $\boldsymbol{N}_K$ expresses the terms nonlinear in $\boldsymbol{u}'_0$ and $\boldsymbol{b}'$ related to the last term of the $O(\delta_{\mathrm{S}})$ part in equation (7.33). It will be dropped in the following discussion since our interest lies in the interaction between the mean field and fluctuation.

We substitute equation (7.38) into the right-hand side of equation (7.45) and retain the contributions from $\boldsymbol{u}'_{00}$ for minimizing the mathematical complexity. This approximation is helpful for abstracting the effects linear in $\boldsymbol{B}$ and $\boldsymbol{U}$ and neglecting the contribution from the nonlinear term $\boldsymbol{N}_K$. We may integrate the resulting equation by using equation

(7.41), and have

$$
\begin{aligned}
u'_{1i}(\boldsymbol{k};\tau) = & -\mathrm{i}\frac{k_i}{k^2}\frac{\partial^* v'_{00j}(\boldsymbol{k};\tau)}{\partial X_j^*} \\
& + D_{lm}(\boldsymbol{k})\frac{\partial B_l}{\partial X_j}\int_{-\infty}^{\tau} G'_{\mathrm{K}im}(\boldsymbol{k};\tau,\tau_1) b'_j(\boldsymbol{k};\tau_1)\,\mathrm{d}\tau_1 \\
& - D_{lm}(\boldsymbol{k})\frac{\partial U_l}{\partial X_j}\int_{-\infty}^{\tau} G'_{\mathrm{K}im}(\boldsymbol{k};\tau,\tau_1) u'_{00j}(\boldsymbol{k};\tau_1)\,\mathrm{d}\tau_1 \\
& - \int_{-\infty}^{\tau} G'_{\mathrm{K}ij}(\boldsymbol{k};\tau,\tau_1)\frac{\mathrm{D}^* u'_{00j}(\boldsymbol{k};\tau_1)}{\mathrm{D}T^*}\,\mathrm{d}\tau_1. \qquad (7.46)
\end{aligned}
$$

7.2.2 Evaluation of Turbulent Electromotive Force

For $\boldsymbol{b}'$, we also assume the same statistical property as for $\boldsymbol{u}'$ in the kinematic method. Namely, we write

$$
\frac{\langle b'_i(\boldsymbol{k};\tau) b'_j(\boldsymbol{k}';\tau')\rangle}{\delta(\boldsymbol{k}+\boldsymbol{k}')} = D_{ij}(\boldsymbol{k}) Q_{\mathrm{M}}(k;\tau,\tau') + \frac{\mathrm{i}}{2}\frac{k_l}{k^2}\varepsilon_{ijl}\Gamma_{\mathrm{M}}(k;\tau,\tau'). \qquad (7.47)
$$

Equation (7.47) gives

$$
K_{\mathrm{M}} = \left\langle \frac{\boldsymbol{b}'^2}{2} \right\rangle = \int Q_{\mathrm{M}}(k;\tau,\tau)\,\mathrm{d}\boldsymbol{k}, \qquad (7.48)
$$

$$
H_{\mathrm{M}} = \langle \boldsymbol{b}' \cdot \boldsymbol{j}' \rangle = \int \Gamma_{\mathrm{M}}(k;\tau,\tau)\,\mathrm{d}\boldsymbol{k}. \qquad (7.49)
$$

As is explained in §4.4.1, the helicity $\boldsymbol{u}\cdot\boldsymbol{\omega}$ is an indicator of helical or spiral structures of fluid motion. From the Elsasser's variables in §6.1.3, we may see the close correspondence between $\boldsymbol{u}$ and $\boldsymbol{b}$, resulting in the similar correspondence between $\boldsymbol{\omega}$ and $\boldsymbol{j}$. Then $\boldsymbol{b}\cdot\boldsymbol{j}$ expresses a helical property of magnetic-field lines. In what follows, the turbulent part of $\boldsymbol{b}\cdot\boldsymbol{j}$, H_{M}, will be shown to play an important role in mean-field theory of dynamo, through the combination with the turbulent kinetic helicity H_{K} defined by equation (7.12).

The leading part in equation (7.38), $\boldsymbol{u}'_{00}$, obeys equation (7.39), which is not explicitly dependent on the mean field generating the anisotropy of turbulent state. This situation is the same as for $\boldsymbol{u}'_0$ $(= \boldsymbol{v}'_0)$ obeying equation (5.68). Then we assume the statistical isotropy of $\boldsymbol{u}'_{00}$:

$$
\frac{\langle u'_{00i}(\boldsymbol{k};\tau) u'_{00j}(\boldsymbol{k}';\tau')\rangle}{\delta(\boldsymbol{k}+\boldsymbol{k}')} = D_{ij}(\boldsymbol{k}) Q_{\mathrm{K}}(k;\tau,\tau'), \qquad (7.50)
$$

as well as

$$
\langle G'_{\mathrm{K}ij}(\boldsymbol{k};\tau,\tau')\rangle = \delta_{ij} G_{\mathrm{K}}(k;\tau,\tau'). \qquad (7.51)
$$

In equation (7.50) we may add the nonmirror symmetric part, as in equation (7.10), which is confirmed to not contribute to the following analysis. Moreover we keep the correlation between $\boldsymbol{u}'_{00}$ and $\boldsymbol{b}'$ in the form

$$\frac{\langle u'_{00i}(\boldsymbol{k};\tau)b'_j(\boldsymbol{k}';\tau')\rangle}{\delta(\boldsymbol{k}+\boldsymbol{k}')} = D_{ij}(\boldsymbol{k})\Lambda(k;\tau,\tau'). \tag{7.52}$$

The physical meanings of equations (7.50) and (7.52) have already been mentioned in the kinematic method.

We expand $\boldsymbol{E}_{\mathrm{M}}$ as

$$\begin{aligned} E_{\mathrm{M}i} &= \varepsilon_{ijl}[\langle u'_{0j}b'_l\rangle + \delta_{\mathrm{S}}(\langle u'_{1j}b'_l\rangle)] + O(\delta_{\mathrm{S}}^2) \\ &= \varepsilon_{ijl}(\langle u'_{00j}b'_l\rangle + \langle u'_{01j}b'_l\rangle + \cdots) + \delta_{\mathrm{S}}(\varepsilon_{ijl}\langle u'_{1j}b'_l\rangle) + O(\delta_{\mathrm{S}}^2), \end{aligned} \tag{7.53}$$

by using equation (7.38). We substitute equations (7.42) and (7.46) into equation (7.53), and make use of the statistics designated by equations (7.47) and (7.50)–(7.52). The evaluation of $\boldsymbol{E}_{\mathrm{M}}$ is essentially the same as for equation (7.20), resulting in

$$\boldsymbol{E}_{\mathrm{M}} = \alpha_{\mathrm{M}}\boldsymbol{B} - \beta_{\mathrm{M}}\boldsymbol{J} + \gamma_{\mathrm{M}}\boldsymbol{\Omega}, \tag{7.54}$$

where

$$\alpha_{\mathrm{M}} = \frac{1}{3}\int \mathrm{d}\boldsymbol{k}\int_{-\infty}^{\tau} G_{\mathrm{K}}(k;\tau,\tau_1)\Gamma_{\mathrm{M}}(k;\tau,\tau_1)\,\mathrm{d}\tau_1, \tag{7.55}$$

$$\beta_{\mathrm{M}} = \frac{1}{3}\int \mathrm{d}\boldsymbol{k}\int_{-\infty}^{\tau} G_{\mathrm{K}}(k;\tau,\tau_1)Q_{\mathrm{M}}(k;\tau,\tau_1)\,\mathrm{d}\tau_1, \tag{7.56}$$

$$\gamma_{\mathrm{M}} = \frac{1}{3}\int \mathrm{d}\boldsymbol{k}\int_{-\infty}^{\tau} G_{\mathrm{K}}(k;\tau,\tau_1)\Lambda(k;\tau,\tau_1)\,\mathrm{d}\tau_1. \tag{7.57}$$

In equation (7.54) the explicit dependence of $\boldsymbol{E}_{\mathrm{M}}$ on the mean field is the same as for equation (7.20). A typical difference between the two expressions may be seen in equations (7.21) and (7.55). These coefficients are associated with helical statistical properties of velocity and magnetic-field fluctuations, respectively, but those properties occur with opposite signs attached.

7.2.3 Evaluation of Reynolds Stress

In the counter-kinematic method the magnetic part of R_{ij} is simply written as

$$\langle -b'_i b'_j\rangle = -\left(\frac{2}{3}\int Q_{\mathrm{M}}(k;\tau,\tau)\,\mathrm{d}\boldsymbol{k}\right)\delta_{ij}, \tag{7.58}$$

from equation (7.47).

The velocity counterpart of equation (7.53) is

$$\begin{aligned}\langle u'_i u'_j \rangle &= \langle u'_{0i} u'_{0j} \rangle + \delta_S (\langle u'_{0i} u'_{1j} \rangle + \langle u'_{1i} u'_{0j} \rangle) + O(\delta_S^2) \\ &= \langle u'_{00i} u'_{00j} \rangle + \langle u'_{00i} u'_{01j} \rangle + \langle u'_{01i} u'_{00j} \rangle + \cdots \\ &\quad + \delta_S (\langle u'_{00i} u'_{1j} \rangle + \langle u'_{1i} u'_{00j} \rangle + \cdots) + O(\delta_S^2), \end{aligned} \tag{7.59}$$

where use has been made of equation (7.38). We substitute equations (7.42) and (7.46) into equation (7.59), and use equations (7.47) and (7.50)–(7.52). After the combination with equation (7.50), we have

$$R_{ij} = \tfrac{2}{3} K_R \delta_{ij} - \nu_{TK} S_{ij} + \nu_{MK} M_{ij}, \tag{7.60}$$

where

$$\nu_{TK} = \frac{7}{15} \int d\boldsymbol{k} \int_{-\infty}^{\tau} G_K(k; \tau, \tau_1) Q_K(k; \tau, \tau_1) \, d\tau_1, \tag{7.61}$$

$$\nu_{MK} = \frac{7}{15} \int d\boldsymbol{k} \int_{-\infty}^{\tau} G_K(k; \tau, \tau_1) \Lambda(k; \tau, \tau_1) \, d\tau_1, \tag{7.62}$$

and the leading part of the turbulent residual energy K_R is written in the same form as equation (7.28).

7.3 Discussions on Dynamo Effects from Kinematic and Counter-Kinematic Methods

7.3.1 Mathematical Features of Obtained Expressions

From both the kinematic and counter-kinematic methods, we have the same types of expressions for the Reynolds stress R_{ij} and the turbulent electromotive force $\boldsymbol{E}_M$:

$$R_{ij} = \tfrac{2}{3} K_R \delta_{ij} - \nu_T S_{ij} + \nu_M M_{ij}, \tag{7.63}$$

$$\boldsymbol{E}_M = \alpha \boldsymbol{B} - \beta \boldsymbol{J} + \gamma \boldsymbol{\Omega}. \tag{7.64}$$

For R_{ij}, the kinematic method gives

$$\nu_T = \nu_{TM} \text{ [equation (7.29)]}, \qquad \nu_M = \nu_{MM} \text{ [equation (7.30)]}. \tag{7.65}$$

On the other hand, the counter-kinematic method leads to

$$\nu_T = \nu_{TK} \text{ [equation (7.61)]}, \qquad \nu_M = \nu_{MK} \text{ [equation (7.62)]}. \tag{7.66}$$

From the comparison among these expressions, we may see that ν_{TM} and ν_{TK} possess a common feature. They are related to the intensities of fluctuations, as is shown by the dependence on Q_M and Q_K. Such close relationship also holds between ν_{MM} and ν_{MK} since they are expressed in terms of Λ (the spectrum of the turbulent cross helicity).

For $\boldsymbol{E}_{\mathrm{M}}$, the kinematic and counter-kinematic methods give

$$\alpha = \alpha_{\mathrm{K}} \text{ [equation (7.21)]}, \qquad \beta = \beta_{\mathrm{K}} \text{ [equation (7.22)]}, \\ \gamma = \gamma_{\mathrm{K}} \text{ [equation (7.23)]}, \tag{7.67}$$

$$\alpha = \alpha_{\mathrm{M}} \text{ [equation (7.55)]}, \qquad \beta = \beta_{\mathrm{M}} \text{ [equation (7.56)]}, \\ \gamma = \gamma_{\mathrm{M}} \text{ [equation (7.57)]}, \tag{7.68}$$

respectively. Concerning β and γ, we may see the properties entirely similar to ν_{T} and ν_{M}. On the other hand, α is related to the helical properties of magnetic-field and velocity fluctuations in an opposite manner, as is seen from the difference of signs in equations (7.21) and (7.55).

In the investigation into real dynamos related to planetary magnetic fields, the interaction between velocity and magnetic field is very important. It is meaningful to infer such an interaction from the results by the separate treatment of velocity and magnetic-field fluctuations. In the simultaneous presence of those fluctuations, ν_{T} and β are inferred to be associated with their total spectrum. We symbolically write this situation as

$$Q_{\mathrm{K}} + Q_{\mathrm{M}} \Rightarrow \nu_{\mathrm{T}}, \beta. \tag{7.69}$$

Concerning ν_{M} and γ, their relationship with the cross-helicity effect is clear, which is

$$\Lambda \Rightarrow \nu_{\mathrm{M}}, \gamma. \tag{7.70}$$

On the other hand, we may infer

$$\Gamma_{\mathrm{M}} - \Gamma_{\mathrm{K}} \Rightarrow \alpha. \tag{7.71}$$

The difference $\Gamma_{\mathrm{M}} - \Gamma_{\mathrm{K}}$ is called the turbulent residual helicity. Its importance was first pointed out in the study of isotropic MHD turbulence by the EDQNMA referred to in §5.2.2.2 [7.1].

7.3.2 Physical Meanings of Obtained Expressions

We substitute equations (7.63) and (7.64) into equations (6.63), (6.65), and (6.67), and have

$$\frac{\mathrm{D}U_i}{\mathrm{D}t} = -\frac{\partial}{\partial x_i}\left(P + \frac{2}{3}K_{\mathrm{R}} + \left\langle\frac{\boldsymbol{b}'^2}{2}\right\rangle\right) + (\boldsymbol{J} \times \boldsymbol{B})_i \\ + \frac{\partial}{\partial x_j}(\nu_{\mathrm{T}} S_{ij} - \nu_{\mathrm{M}} M_{ij}) + \nu \nabla^2 U_i, \tag{7.72}$$

$$\frac{\partial \boldsymbol{B}}{\partial t} = \nabla \times (\boldsymbol{E} + \boldsymbol{U} \times \boldsymbol{B} + \alpha \boldsymbol{B} - \beta \boldsymbol{J} + \gamma \boldsymbol{\Omega}) + \lambda_{\mathrm{M}} \nabla^2 \boldsymbol{B}, \tag{7.73}$$

$$\boldsymbol{J} = \frac{1}{\beta + \lambda_{\mathrm{M}}}(\boldsymbol{E} + \boldsymbol{U} \times \boldsymbol{B} + \alpha \boldsymbol{B} + \gamma \boldsymbol{\Omega}). \tag{7.74}$$

In order to simply see the physical structures of these equations, we neglect the spatial dependence of ν_T, ν_M, α, β, and γ. Under this approximation, equations (7.72) and (7.73) are reduced to

$$\frac{DU}{Dt} = -\nabla\left(P + \tfrac{2}{3}K_R + \left\langle\frac{b'^2}{2}\right\rangle\right) + J \times B + (\nu_T + \nu)\nabla^2 U - \nu_M \nabla^2 B, \tag{7.75}$$

$$\frac{\partial B}{\partial t} = \nabla \times (E + U \times B) + \alpha J + (\beta + \lambda_M)\nabla^2 B - \gamma \nabla^2 U. \tag{7.76}$$

From equations (7.74) and (7.76), we may easily confirm that β expresses the enhancement of the magnetic diffusivity by fluctuations. It is called the turbulent magnetic diffusivity in the light of the turbulent viscosity ν_T that is familiar in turbulent flow of electrically nonconducting fluids. It is also named the anomalous or turbulent resistivity.

Next, we consider the physical meaning of the α-related effect with the aid of equation (7.74). There the α-related term, which was first proposed by Parker [7.2], expresses the occurrence of J parallel or anti-parallel to B, depending on the sign of α (see [2.13–2.15] for the historical development of the study of the effect). This point makes a sharp contrast with the original $U \times B$ term generating J normal to B. This α-related mechanism is usually called the alpha effect or dynamo. In equation (7.71), we saw that α is expressed in terms of two types of turbulent helical properties, namely, statistical helical properties of velocity and magnetic-field fluctuations.

Of the two helical properties, the relationship of α with the turbulent kinetic helicity $\langle u' \cdot \omega' \rangle$ is explained schematically by the use of figure 7.1. In the case of small λ_M or large R_{eM}, magnetic-field lines are frozen in fluid motion, and positive $\langle u' \cdot \omega' \rangle$ tends to generate J anti-parallel to B. This anti-parallelness is expressed by the negative sign in equation (7.21).

The alpha effect due to the helical magnetic property is expressed by equation (7.55) with equation (7.49). There positive $\langle b' \cdot j' \rangle$ contributes to positive α. Such α generates the occurrence of J parallel to B or positive $B \cdot J$. This process is not so easy to explain schematically, as is done for

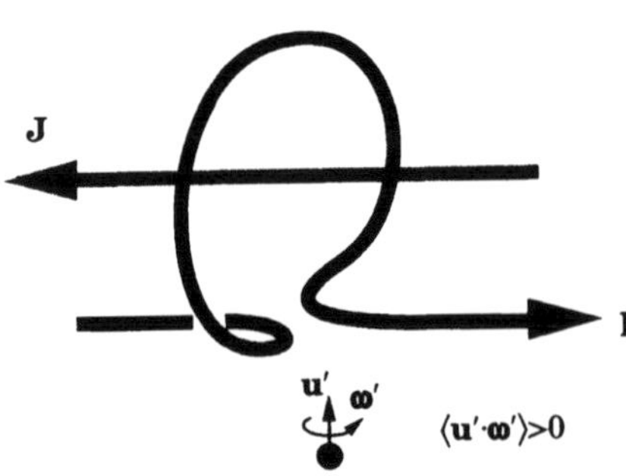

Figure 7.1. Alpha dynamo.

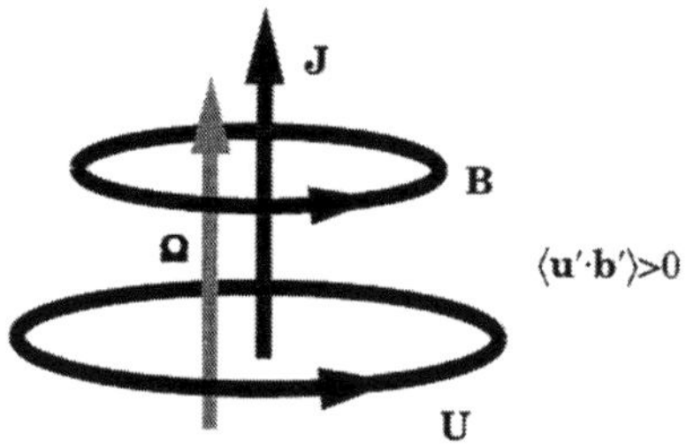

Figure 7.2. Cross-helicity dynamo.

$\langle \boldsymbol{u}' \cdot \boldsymbol{\omega}' \rangle$. The linkage between the signs of $\boldsymbol{B} \cdot \boldsymbol{J}$ and $\langle \boldsymbol{b}' \cdot \boldsymbol{j}' \rangle$, however, is understandable since the scale separation between $\boldsymbol{B}$ and $\boldsymbol{b}'$ is not so definite.

In equation (7.74) the γ-related term expresses the occurrence of $\boldsymbol{J}$ aligned with $\boldsymbol{\Omega}$, which is equivalent to the alignment between $\boldsymbol{B}$ and $\boldsymbol{U}$. The coefficient γ is related to the turbulent cross helicity $\langle \boldsymbol{u}' \cdot \boldsymbol{b}' \rangle$ through equations (7.23) and (7.57). This alignment may be schematically depicted in figure 7.2. The linkage between $\boldsymbol{U} \cdot \boldsymbol{B}$ and $\langle \boldsymbol{u}' \cdot \boldsymbol{b}' \rangle$ resembles the situation concerning $\boldsymbol{B} \cdot \boldsymbol{J}$ and $\langle \boldsymbol{b}' \cdot \boldsymbol{j}' \rangle$. The former linkage, however, is much closer from the conservation property referred to in §6.1.3. In the absence of molecular viscosity and diffusivity, the total amount of cross helicity, $\int_V \boldsymbol{u} \cdot \boldsymbol{b} \, \mathrm{d}V$, is conserved. In the presence of molecular viscosity and diffusivity, the cross helicity cascades from large- to small-scale components of MHD motion, as is the sum of kinetic and magnetic energy. Such a cascade process may be regarded as a cause of the linkage between $\boldsymbol{U} \cdot \boldsymbol{B}$ and $\langle \boldsymbol{u}' \cdot \boldsymbol{b}' \rangle$.

In the frame rotating with the angular velocity ω_{F}, the mean vorticity $\boldsymbol{\Omega}$ is subject to the transformation (6.81). As a result, we have

$$\boldsymbol{J} = \frac{1}{\beta + \lambda_{\mathrm{M}}} [\boldsymbol{E} + \boldsymbol{U} \times \boldsymbol{B} + \alpha \boldsymbol{B} + \gamma(\boldsymbol{\Omega} + 2\omega_{\mathrm{F}})]. \tag{7.77}$$

This indicates that the frame rotation may exert influence on the mean magnetic field in the presence of $\langle \boldsymbol{u}' \cdot \boldsymbol{b}' \rangle$. The importance of the fact will be discussed in the investigation into geodynamo. The other typical feature of the turbulent cross helicity is the explicit feedback effect on fluid motion through the Lorentz force

$$\boldsymbol{J} \times \boldsymbol{B} = \frac{1}{\beta + \lambda_{\mathrm{M}}} [\boldsymbol{E} + \boldsymbol{U} \times \boldsymbol{B} + \gamma(\boldsymbol{\Omega} + 2\omega_{\mathrm{F}})] \times \boldsymbol{B} \tag{7.78}$$

with no explicit alpha effect.

We have the other feedback effect on fluid motion through $\langle \boldsymbol{u}' \cdot \boldsymbol{b}' \rangle$, which occurs as the last term in equation (7.75). It arises from the effect of mean magnetic strain that is expressed by the last term in equation (7.26) or (7.60). In the situation that the magnetic-field line is deformed as in figure 7.3, we may write

$$\boldsymbol{B}_0 = (0, B_0, 0) \;\rightarrow\; \boldsymbol{B} = [B_x(y), B_y(y), 0], \tag{7.79}$$

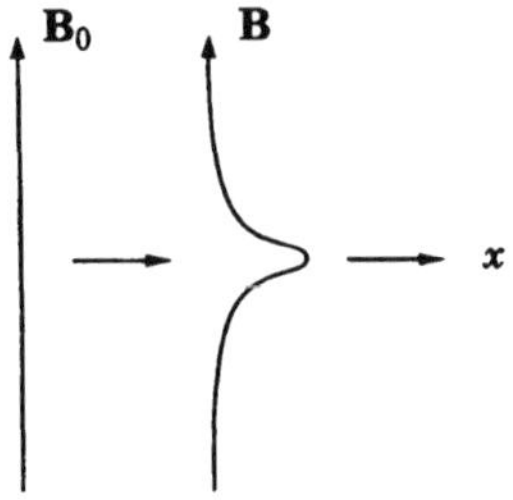

Figure 7.3. Feedback effect on fluid motion by tension of magnetic field.

with $d^2B_x/dy^2 < 0$. The last term in equation (7.75) gives

$$\nu_M > 0 \rightarrow -\nu_M \nabla^2 B_x = -\nu_M \frac{d^2 B_x}{dy^2} > 0, \qquad (7.80)$$

which signifies that the magnetic field deformed in the positive x direction drives the fluid in the direction.

7.4 Magnetohydrodynamic Method

In §7.1 and §7.2 we investigated the effects of magnetic-field and velocity fluctuations on the mean magnetic-field and velocity equations in a separative manner. Through the discussions, we reached expressions (7.69)–(7.71). Of these expressions, equations (7.69) and (7.71) remain as a conjecture in a fully MHD sense. In order to clarify this point, we need to treat the equations for $\boldsymbol{u}'$ and $\boldsymbol{b}'$ simultaneously. Such an analysis is very complicated. A method of alleviating the mathematical complexity is the use of Elsasser's variables introduced by equation (6.47). In what follows, we perform the TSDIA analysis based on these variables [2.16, 7.3, 7.4].

7.4.1 Elsasser's Variables and Two-Scale Description

We divide Elsasser's variables into mean and fluctuating parts, as in equation (6.62):

$$\phi = \Phi + \phi', \qquad \psi = \Psi + \psi'. \qquad (7.81)$$

We define

$$R_{ij}^{(E)} = \langle \phi_i' \psi_j' \rangle. \qquad (7.82)$$

In terms of $R_{ij}^{(E)}$, the turbulent electromotive force $\boldsymbol{E}_M$ and the Reynolds stress R_{ij} are rewritten as

$$E_{Mi} = -\tfrac{1}{2}\varepsilon_{ijl} R_{jl}^{(E)}, \qquad (7.83)$$

$$R_{ij} = \tfrac{1}{2}(R_{ij}^{(E)} + R_{ji}^{(E)}). \qquad (7.84)$$

We may call $R_{ij}^{(\mathrm{E})}$ Elsasser's Reynolds stress. In the equations for the mean field $(\boldsymbol{\Phi}, \boldsymbol{\Psi})$, it plays entirely the same role as R_{ij} in the equation for $\boldsymbol{U}$. From equations (7.83) and (7.84), however, $R_{ij}^{(\mathrm{E})}$ is found to be a more fundamental second-order correlation function in MHD turbulent flows.

In order to evaluate $R_{ij}^{(\mathrm{E})}$, we consider the equations for $\boldsymbol{\phi}'$ and $\boldsymbol{\psi}'$. There we take $\nu = \lambda_{\mathrm{M}}$ and simplify the mathematical manipulation. In the case of high kinetic and magnetic Reynolds numbers, this approximation gives rise to no critical inaccuracy. From equation (6.49), $\boldsymbol{\phi}'$ obeys

$$\frac{\partial \phi_i'}{\partial t} + \Psi_j \frac{\partial \phi_i'}{\partial x_j} + \frac{\partial}{\partial x_j}\left(\psi_j' \phi_i' - R_{ij}^{(\mathrm{E})}\right) + \frac{\partial \varpi'}{\partial x_i} - \nu \nabla^2 \phi_i' = -\psi_j' \frac{\partial \Phi_i}{\partial x_j}, \tag{7.85}$$

with the solenoidal condition

$$\nabla \cdot \boldsymbol{\phi}' = 0, \tag{7.86}$$

where ϖ' is defined by equation (6.73). The equation for $\boldsymbol{\psi}'$ may be obtained through the replacement

$$\boldsymbol{\phi}' \to \boldsymbol{\psi}', \quad \boldsymbol{\psi}' \to \boldsymbol{\phi}', \quad \boldsymbol{\Phi} \to \boldsymbol{\Psi}, \quad \boldsymbol{\Psi} \to \boldsymbol{\Phi}, \quad R_{ij}^{(\mathrm{E})} \to R_{ji}^{(\mathrm{E})}. \tag{7.87}$$

We apply the two-scale description to equations (7.85) and (7.86), and have

$$\frac{\partial \phi_i'}{\partial \tau} + U_j \frac{\partial \phi_i'}{\partial \xi_j} + \frac{\partial}{\partial \xi_j} \psi_j' \phi_i' + \frac{\partial \varpi'}{\partial \xi_i} - \nu \nabla_\xi^2 \phi_i'$$
$$= B_j \frac{\partial \phi_i'}{\partial \xi_j} + \delta_{\mathrm{S}} \left(-\psi_j' \frac{\partial \Phi_i}{\partial X_j} - \frac{\partial \phi_i'}{\partial T} - \Psi_j \frac{\partial \phi_i'}{\partial X_j} - \frac{\partial}{\partial X_j}\left(\psi_j' \phi_i' - R_{ij}^{(\mathrm{E})}\right) - \frac{\partial \varpi'}{\partial X_i} \right), \tag{7.88}$$

$$\frac{\partial \phi_i'}{\partial \xi_i} = -\delta_{\mathrm{S}} \frac{\partial \phi_i'}{\partial X_i}. \tag{7.89}$$

Following the approach in the kinematic and counter-kinematic methods, we introduce the moving-frame Fourier representation, equation (5.57). Then equations (7.88) and (7.89) are reduced to

$$\frac{\partial \phi_i'(\boldsymbol{k};\tau)}{\partial \tau} + \nu k^2 \phi_i'(\boldsymbol{k};\tau) - \mathrm{i} k_i \varpi'(\boldsymbol{k};\tau) - \mathrm{i} k_j \iint \psi_j'(\boldsymbol{p};\tau) \phi_i'(\boldsymbol{q};\tau) \delta(\boldsymbol{k}-\boldsymbol{p}-\boldsymbol{q})\, \mathrm{d}\boldsymbol{p}\, \mathrm{d}\boldsymbol{q}$$
$$= -\mathrm{i}(\boldsymbol{k} \cdot \boldsymbol{B}) \phi_i'(\boldsymbol{k};\tau) + \delta_{\mathrm{S}} \left(-\psi_j'(\boldsymbol{k};\tau) \frac{\partial \Phi_i}{\partial X_j} - \frac{\mathrm{D}^* \phi_i'(\boldsymbol{k};\tau)}{\mathrm{D} T^*} + B_j \frac{\partial^* \phi_i'(\boldsymbol{k};\tau)}{\partial X_j^*} \right.$$
$$\left. - \frac{\partial^* \varpi'(\boldsymbol{k};\tau)}{\partial X_i^*} - \frac{\partial^*}{\partial X_j^*} \iint \psi_j'(\boldsymbol{p};\tau) \phi_i'(\boldsymbol{q};\tau) \delta(\boldsymbol{k}-\boldsymbol{p}-\boldsymbol{q})\, \mathrm{d}\boldsymbol{p}\, \mathrm{d}\boldsymbol{q} \right), \tag{7.90}$$

$$\boldsymbol{k} \cdot \boldsymbol{\phi}'(\boldsymbol{k};\tau) = \delta \left(-\frac{\partial^* \phi_i'(\boldsymbol{k};\tau)}{\partial X_i^*} \right). \tag{7.91}$$

We write $\boldsymbol{\phi}'$ as

$$\boldsymbol{\phi}'(\boldsymbol{k};\tau) = \boldsymbol{\phi}^{+}(\boldsymbol{k};\tau) + \delta_{\mathrm{S}}\left(-\mathrm{i}\,\frac{\boldsymbol{k}}{k^{2}}\,\frac{\partial^{*}\phi'_{i}(\boldsymbol{k};\tau)}{\partial X^{*}_{i}}\right). \tag{7.92}$$

Then $\boldsymbol{\phi}^{+}$obeys the solenoidal condition concerning $\boldsymbol{k}$:

$$\boldsymbol{k}\cdot\boldsymbol{\phi}^{+}(\boldsymbol{k};\tau) = 0. \tag{7.93}$$

7.4.2 Perturbational Solution

We expand $\boldsymbol{\phi}'$ and ϖ' as

$$\boldsymbol{\phi}'(\boldsymbol{k};\tau) = \sum_{n=0}^{\infty}\delta_{\mathrm{S}}^{n}\boldsymbol{\phi}'_{n}(\boldsymbol{k};\tau), \qquad \boldsymbol{\phi}^{+}(\boldsymbol{k};\tau) = \sum_{n=0}^{\infty}\delta_{\mathrm{S}}^{n}\boldsymbol{\phi}^{+}_{n}(\boldsymbol{k};\tau), \tag{7.94}$$

$$\varpi'(\boldsymbol{k};\tau) = \sum_{n=0}^{\infty}\delta_{\mathrm{S}}^{n}\varpi'_{n}(\boldsymbol{k};\tau), \tag{7.95}$$

with similar expressions for $\boldsymbol{\psi}'$. In the $O(1)$ analysis, we have

$$\boldsymbol{\phi}'_{0}(\boldsymbol{k};\tau) = \boldsymbol{\phi}^{+}_{0}(\boldsymbol{k};\tau), \tag{7.96}$$

from equation (7.93). Then $\boldsymbol{\phi}'_{0}$ obeys

$$\begin{aligned}\frac{\partial\phi'_{0i}(\boldsymbol{k};\tau)}{\partial\tau} &+ \nu k^{2}\phi'_{0i}(\boldsymbol{k};\tau) - \mathrm{i}k_{i}\varpi'_{0}(\boldsymbol{k};\tau)\\ &- \mathrm{i}k_{j}\iint \psi'_{0j}(\boldsymbol{p};\tau)\phi'_{0i}(\boldsymbol{q};\tau)\delta(\boldsymbol{k}-\boldsymbol{p}-\boldsymbol{q})\,\mathrm{d}\boldsymbol{p}\,\mathrm{d}\boldsymbol{q}\\ &= -\mathrm{i}(\boldsymbol{k}\cdot\boldsymbol{B})\phi'_{0i}(\boldsymbol{k};\tau),\end{aligned} \tag{7.97}$$

$$\boldsymbol{k}\cdot\boldsymbol{\phi}'_{0}(\boldsymbol{k};\tau) = 0. \tag{7.98}$$

We apply equation (7.98) to equation (7.97), and have

$$\varpi'_{0}(\boldsymbol{k};\tau) = -\frac{k_{i}k_{j}}{k^{2}}\iint \psi'_{0i}(\boldsymbol{p};\tau)\phi'_{0j}(\boldsymbol{q};\tau)\delta(\boldsymbol{k}-\boldsymbol{p}-\boldsymbol{q})\,\mathrm{d}\boldsymbol{p}\,\mathrm{d}\boldsymbol{q}. \tag{7.99}$$

The substitution of equation (7.99) into equation (7.97) results in

$$\begin{aligned}\frac{\partial\phi'_{0i}(\boldsymbol{k};\tau)}{\partial\tau} &+ \nu k^{2}\phi'_{0i}(\boldsymbol{k};\tau) - \mathrm{i}Z_{ijl}(\boldsymbol{k})\iint \psi'_{0j}(\boldsymbol{p};\tau)\phi'_{0l}(\boldsymbol{q};\tau)\delta(\boldsymbol{k}-\boldsymbol{p}-\boldsymbol{q})\,\mathrm{d}\boldsymbol{p}\,\mathrm{d}\boldsymbol{q}\\ &= -\mathrm{i}(\boldsymbol{k}\cdot\boldsymbol{B})\phi'_{0i}(\boldsymbol{k};\tau),\end{aligned} \tag{7.100}$$

where

$$Z_{ijl}(\boldsymbol{k}) = k_{j}D_{il}(\boldsymbol{k}). \tag{7.101}$$

The ψ_0' counterpart of equation (7.100) is given by

$$\frac{\partial \psi_{0i}'(\boldsymbol{k};\tau)}{\partial \tau} + \nu k^2 \psi_{0i}'(\boldsymbol{k};\tau) - \mathrm{i}Z_{ijl}(\boldsymbol{k}) \iint \phi_{0j}'(\boldsymbol{p};\tau)\psi_{0l}'(\boldsymbol{q};\tau)\delta(\boldsymbol{k}-\boldsymbol{p}-\boldsymbol{q})\, \mathrm{d}\boldsymbol{p}\, \mathrm{d}\boldsymbol{q}$$
$$= -\mathrm{i}(\boldsymbol{k}\cdot\boldsymbol{B})\psi_{0i}'(\boldsymbol{k};\tau). \qquad (7.102)$$

Equations (7.100) and (7.102) are dependent linearly on $\boldsymbol{B}$, but they are coupled with each other, leading to the nonlinear dependence of ϕ_0' and ψ_0' on arbitrary $\boldsymbol{B}$ (for constant $\boldsymbol{B}$, see [7.5, 7.6]). It is difficult to solve exactly this coupled system of equations in an analytical manner. We expand

$$\phi_0'(\boldsymbol{k};\tau) = \sum_{m=0}^{\infty} \phi_{0m}'(\boldsymbol{k};\tau), \qquad (7.103)$$

with the similar expression for ψ_0', and solve the system by an iterative method with the right-hand sides regarded as perturbation terms. The leading term obeys

$$\frac{\partial \phi_{00i}'(\boldsymbol{k};\tau)}{\partial \tau} + \nu k^2 \phi_{00i}'(\boldsymbol{k};\tau)$$
$$- \mathrm{i}Z_{ijl}(\boldsymbol{k}) \iint \psi_{00j}'(\boldsymbol{p};\tau)\phi_{00l}'(\boldsymbol{q};\tau)\delta(\boldsymbol{k}-\boldsymbol{p}-\boldsymbol{q})\, \mathrm{d}\boldsymbol{p}\, \mathrm{d}\boldsymbol{q} = 0. \qquad (7.104)$$

We introduce the Green's function for equation (7.104) as

$$\frac{\partial G_{\phi ij}'(\boldsymbol{k};\tau,\tau')}{\partial \tau} + \nu k^2 G_{\phi ij}'(\boldsymbol{k};\tau,\tau')$$
$$- \mathrm{i}Z_{ilm}(\boldsymbol{k}) \iint \psi_{00l}'(\boldsymbol{p};\tau)G_{\phi mj}'(\boldsymbol{q};\tau,\tau')\delta(\boldsymbol{k}-\boldsymbol{p}-\boldsymbol{q})\, \mathrm{d}\boldsymbol{p}\, \mathrm{d}\boldsymbol{q}$$
$$= \delta_{ij}\delta(\tau-\tau'). \qquad (7.105)$$

With the aid of $G_{\phi ij}'$, we have

$$\phi_{01i}'(\boldsymbol{k};\tau) = -\mathrm{i}(\boldsymbol{k}\cdot\boldsymbol{B}) \int_{-\infty}^{\tau} G_{\phi ij}'(\boldsymbol{k};\tau,\tau_1)\phi_{00j}'(\boldsymbol{k};\tau_1)\, \mathrm{d}\tau_1. \qquad (7.106)$$

The terms ϕ_{0n}' $(n \geq 2)$ are expressed in terms of ϕ_{0m}' $(m \leq n-1)$ and ψ_{0m}' $(m \leq n-1)$, resulting in the nonlinear dependence on $\boldsymbol{B}$.

The foregoing, nonlinear dependence on $\boldsymbol{B}$ has been sought in relation to the $\boldsymbol{B}$ dependence of α in equation (7.64). The dependence, which arises from the interaction between the equations for $\boldsymbol{u}'$ and $\boldsymbol{b}'$, has been studied from the viewpoint of determining the saturation level of induced $\boldsymbol{B}$ [7.7–7.9]. On the other hand, the turbulent cross helicity is an important indicator of the correlation between $\boldsymbol{u}'$ and $\boldsymbol{b}'$. By taking this fact into account, we may examine the interaction effects in an entirely different manner, as is shown by the last terms of equations (7.63) and (7.64). In the present chapter, we prefer the latter methodology and do not consider ϕ'_{0n} $(n \geq 2)$. This point will be discussed in detail through the application to planetary dynamos.

For the $O(\delta_S)$ parts in equations (7.94) and (7.95), we have

$$\frac{\partial \phi'_{1i}(\boldsymbol{k};\tau)}{\partial \tau} + \nu k^2 \phi'_{1i}(\boldsymbol{k};\tau) - \mathrm{i}k_i \varpi'_1(\boldsymbol{k};\tau)$$
$$- \mathrm{i}k_j \iint \psi'_{0j}(\boldsymbol{p};\tau)\phi'_{1i}(\boldsymbol{q};\tau)\delta(\boldsymbol{k}-\boldsymbol{p}-\boldsymbol{q})\,\mathrm{d}\boldsymbol{p}\,\mathrm{d}\boldsymbol{q}$$
$$= -\mathrm{i}(\boldsymbol{k}\cdot\boldsymbol{B})\phi'_{1i}(\boldsymbol{k};\tau) - \psi'_{0j}(\boldsymbol{k};\tau)\frac{\partial \Phi_i}{\partial X_j} - \frac{\mathrm{D}^*\phi'_{0i}(\boldsymbol{k};\tau)}{\mathrm{D}T^*} + B_j \frac{\partial^* \phi'_{0i}(\boldsymbol{k};\tau)}{\partial X^*_j}$$
$$- \frac{\partial^* \varpi'_0(\boldsymbol{k};\tau)}{\partial X^*_i} - \mathrm{i}k_j \iint \psi'_{1j}(\boldsymbol{p};\tau)\phi'_{0i}(\boldsymbol{q};\tau)\delta(\boldsymbol{k}-\boldsymbol{p}-\boldsymbol{q})\,\mathrm{d}\boldsymbol{p}\,\mathrm{d}\boldsymbol{q}$$
$$- \frac{\partial^*}{\partial X^*_j} \iint \psi'_{0j}(\boldsymbol{p};\tau)\phi'_{0i}(\boldsymbol{q};\tau)\delta(\boldsymbol{k}-\boldsymbol{p}-\boldsymbol{q})\,\mathrm{d}\boldsymbol{p}\,\mathrm{d}\boldsymbol{q}, \tag{7.107}$$

$$\boldsymbol{\phi}'_1(\boldsymbol{k};\tau) = \boldsymbol{\phi}^+_1(\boldsymbol{k};\tau) - \mathrm{i}\frac{\boldsymbol{k}}{k^2}\frac{\partial^* \phi'_{0i}(\boldsymbol{k};\tau)}{\partial X^*_i}, \tag{7.108}$$

with similar expressions for $\boldsymbol{\psi}'_1$, where

$$\boldsymbol{k}\cdot\boldsymbol{\phi}^+_1(\boldsymbol{k};\tau) = 0. \tag{7.109}$$

As may be seen from equations (7.63) and (7.64), our main interest lies in the relationship of R_{ij} and $\boldsymbol{E}_M$ with the mean field $\boldsymbol{B}$ and $\boldsymbol{U}$. In solving (7.107), we focus attention on the first four terms on the right-hand side and drop the last two terms. We apply equations (7.108) and (7.109) to equation (7.107), and eliminate ϖ'_1. As a result, we have

$$\frac{\partial \phi^+_{1i}(\boldsymbol{k};\tau)}{\partial \tau} + \nu k^2 \phi^+_{1i}(\boldsymbol{k};\tau)$$
$$- \mathrm{i}Z_{ijl}(\boldsymbol{k}) \iint \psi'_{0j}(\boldsymbol{p};\tau)\phi^+_{1l}(\boldsymbol{q};\tau)\delta(\boldsymbol{k}-\boldsymbol{p}-\boldsymbol{q})\,\mathrm{d}\boldsymbol{p}\,\mathrm{d}\boldsymbol{q}$$
$$= -D_{il}(\boldsymbol{k})\psi'_{0j}(\boldsymbol{k};\tau)\frac{\partial \Phi_l}{\partial X_j} - \frac{\mathrm{D}^*\phi'_{0i}(\boldsymbol{k};\tau)}{\mathrm{D}T^*}$$
$$+ B_j \frac{\partial^* \phi'_{0i}(\boldsymbol{k};\tau)}{\partial X^*_j} - \mathrm{i}(\boldsymbol{k}\cdot\boldsymbol{B})\phi^+_{1i}(\boldsymbol{k};\tau). \tag{7.110}$$

In solving equation (7.110) we follow the method adopted for equation (7.102) with the right-hand side dealt with in a perturbational manner. We first substitute equation (7.103) and its counterpart for $\boldsymbol{\psi}'_0$ into the first three terms on the right-hand side of equation (7.110), and retain the contributions from their leading parts. We may integrate the resulting equation formally with the aid of equation (7.105), as

$$\phi_{1i}^{+}(\boldsymbol{k};\tau) = -\frac{\partial \Phi_l}{\partial X_j} D_{lm}(\boldsymbol{k}) \int_{-\infty}^{\tau} G'_{\phi im}(\boldsymbol{k};\tau,\tau_1)\psi'_{B\to 00}(\boldsymbol{k};\tau_1)\,\mathrm{d}\tau_1$$
$$- \int_{-\infty}^{\tau} G'_{\phi ij}(\boldsymbol{k};\tau,\tau_1)\frac{\mathrm{D}^*\phi'_{00j}(\boldsymbol{k};\tau_1)}{\mathrm{D}T^*}\,\mathrm{d}\tau_1$$
$$+ B_j \int_{-\infty}^{\tau} G'_{\phi il}(\boldsymbol{k};\tau,\tau_1)\frac{\partial^*\phi'_{00l}(\boldsymbol{k};\tau_1)}{\partial X_j^*}\,\mathrm{d}\tau_1. \quad (7.111)$$

The fourth term in equation (7.110) is estimated using equation (7.11). Its contributions to R_{ij} and $\boldsymbol{E}_{\mathrm{M}}$, however, are less important compared with those from equation (7.11) and are dropped here. Then the $O(\delta_{\mathrm{S}})$ solution is given by equation (7.108) combined with equation (7.111).

7.4.3 Evaluation of Elsasser's Reynolds Stress

From equations (7.106), (7.111), and their counterparts for ψ'_{01} and ψ'_1, the $O(1)$ and $O(\delta_{\mathrm{S}})$ solutions may be written in terms of ϕ'_{00}, ψ'_{00}, $G'_{\phi ij}$, and $G'_{\psi ij}$ in addition to $\boldsymbol{B}$ and $\boldsymbol{U}$. Equation (7.104) and its counterpart for ψ'_{00} are not dependent explicitly on the mean field that is a primary generator of the statistical anisotropy of ϕ' and ψ'. Then we assume their isotropic correlation functions, as is similar to equations (7.10) and (7.47):

$$\frac{\langle Y_i(\boldsymbol{k};\tau)Z_j(\boldsymbol{k}';\tau')\rangle}{\delta(\boldsymbol{k}+\boldsymbol{k}')} = D_{ij}(\boldsymbol{k})Q_{YZ}(k;\tau,\tau') + \frac{\mathrm{i}}{2}\frac{k_l}{k^2}\varepsilon_{ijl}H_{YZ}(k;\tau,\tau'), \quad (7.112)$$

$$\langle G'_{Yij}(\boldsymbol{k};\tau,\tau')\rangle = \delta_{ij}G_Y(k;\tau,\tau'). \quad (7.113)$$

Here $\boldsymbol{Y}$ and $\boldsymbol{Z}$ represent one of ϕ'_{00} and ψ'_{00}. For instance, we write

$$\frac{\langle \phi'_{00i}(\boldsymbol{k};\tau)\psi'_{00j}(\boldsymbol{k}';\tau')\rangle}{\delta(\boldsymbol{k}+\boldsymbol{k}')} = D_{ij}(\boldsymbol{k})Q_{\phi\psi}(k;\tau,\tau') + \frac{\mathrm{i}}{2}\frac{k_l}{k^2}\varepsilon_{ijl}H_{\phi\psi}(k;\tau,\tau'). \quad (7.114)$$

The velocity and the magnetic field corresponding to ϕ'_{00} and ψ'_{00} are given by

$$\boldsymbol{u}'_{00} = \frac{\phi'_{00}+\psi'_{00}}{2}, \qquad \boldsymbol{b}'_{00} = \frac{\phi'_{00}-\psi'_{00}}{2}. \quad (7.115)$$

They obey equations (7.8) and (7.39). We express their correlation functions by using equation (7.112) with one of $\boldsymbol{u}'_{00}$ and $\boldsymbol{b}'_{00}$ chosen as $\boldsymbol{Y}$ or $\boldsymbol{Z}$. For instance,

$$\frac{\langle u'_{00i}(\boldsymbol{k};\tau)u'_{00j}(\boldsymbol{k}';\tau')\rangle}{\delta(\boldsymbol{k}+\boldsymbol{k}')} = D_{ij}(\boldsymbol{k})Q_{uu}(k;\tau,\tau') + \frac{\mathrm{i}}{2}\frac{k_l}{k^2}\varepsilon_{ijl}H_{uu}(k;\tau,\tau') \quad (7.116)$$

corresponds to equation (7.10). The choice of $\boldsymbol{Y} = \boldsymbol{u}'_{00}$ and $\boldsymbol{Z} = \boldsymbol{b}'_{00}$ leads to

$$\frac{\langle u'_{00i}(\boldsymbol{k};\tau) b'_{00j}(\boldsymbol{k}';\tau')\rangle}{\delta(\boldsymbol{k}+\boldsymbol{k}')} = D_{ij}(\boldsymbol{k}) Q_{ub}(k;\tau,\tau') + \frac{\mathrm{i}}{2}\frac{k_l}{k^2}\varepsilon_{ijl} H_{ub}(k;\tau,\tau'), \tag{7.117}$$

which gives

$$\langle \boldsymbol{u}'_{00} \cdot \boldsymbol{b}'_{00}\rangle = 2\int Q_{ub}(k;\tau,\tau)\,\mathrm{d}\boldsymbol{k}, \tag{7.118}$$

$$\langle \boldsymbol{u}'_{00} \cdot \boldsymbol{j}'_{00}\rangle = \int H_{ub}(k;\tau,\tau)\,\mathrm{d}\boldsymbol{k}. \tag{7.119}$$

In the following discussions on dynamo effects, equation (7.119) will not play a significant role, compared with equation (7.118) related to the cross-helicity effect.

Among the correlation functions between (ϕ'_{00}, ψ'_{00}) and $(\boldsymbol{u}'_{00}, \boldsymbol{b}'_{00})$, we have the relations

$$Q_{\phi\phi}(k;\tau,\tau') + Q_{\psi\psi}(k;\tau,\tau') = 2[Q_{uu}(k;\tau,\tau') + Q_{bb}(k;\tau,\tau')], \tag{7.120a}$$

$$Q_{\phi\phi}(k;\tau,\tau') - Q_{\psi\psi}(k;\tau,\tau') = 2[Q_{ub}(k;\tau,\tau') + Q_{bu}(k;\tau,\tau')], \tag{7.120b}$$

$$Q_{\phi\psi}(k;\tau,\tau') + Q_{\psi\phi}(k;\tau,\tau') = 2[Q_{uu}(k;\tau,\tau') - Q_{bb}(k;\tau,\tau')], \tag{7.120c}$$

$$Q_{\phi\psi}(k;\tau,\tau') - Q_{\psi\phi}(k;\tau,\tau') = 2[Q_{bu}(k;\tau,\tau') - Q_{ub}(k;\tau,\tau')], \tag{7.120d}$$

$$H_{\phi\phi}(k;\tau,\tau') + H_{\psi\psi}(k;\tau,\tau') = 2[H_{uu}(k;\tau,\tau') + H_{bb}(k;\tau,\tau')], \tag{7.121a}$$

$$H_{\phi\phi}(k;\tau,\tau') - H_{\psi\psi}(k;\tau,\tau') = 2[H_{ub}(k;\tau,\tau') + H_{bu}(k;\tau,\tau')], \tag{7.121b}$$

$$H_{\phi\psi}(k;\tau,\tau') + H_{\psi\phi}(k;\tau,\tau') = 2[H_{uu}(k;\tau,\tau') - H_{bb}(k;\tau,\tau')], \tag{7.121c}$$

$$H_{\phi\psi}(k;\tau,\tau') - H_{\psi\phi}(k;\tau,\tau') = 2[H_{bu}(k;\tau,\tau') - H_{ub}(k;\tau,\tau')]. \tag{7.121d}$$

In the use of Elsasser's variables, Elsasser's Reynolds stress defined by equation (7.82) plays a central role and is expanded as

$$\begin{aligned} R^{(\mathrm{E})}_{ij} &= \langle \phi'_{0i}\psi'_{0j}\rangle + \delta_{\mathrm{S}}(\langle \phi'_{1i}\psi'_{0j}\rangle + \langle \phi'_{0i}\psi'_{1j}\rangle) + O(\delta_{\mathrm{S}}^2) \\ &= \langle \phi'_{00i}\psi'_{00j}\rangle + \langle \phi'_{01i}\psi'_{00j}\rangle + \langle \phi'_{00i}\psi'_{01j}\rangle + \cdots \\ &\quad + \delta_{\mathrm{S}}(\langle \phi'_{1i}\psi'_{00j}\rangle + \langle \phi'_{00i}\psi'_{1j}\rangle + \cdots) + O(\delta_{\mathrm{S}}^2). \end{aligned} \tag{7.122}$$

We substitute equations (7.103) and (7.108) into equation (7.122), and make use of equations (7.112) and (7.113). After a little lengthy mathematical manipulation, we have

$$
\begin{aligned}
R_{ij}^{(E)} &= \tfrac{2}{3}K^{(E)}\delta_{ij} + \left(\frac{1}{6}\int \mathrm{d}\boldsymbol{k}\int_{-\infty}^{\tau} G_\psi(k;\tau,\tau_1)H_{\phi\psi}(k;\tau,\tau_1)\,\mathrm{d}\tau_1\right)\varepsilon_{ijl}B_l \\
&+ \left(\frac{1}{6}\int \mathrm{d}\boldsymbol{k}\int_{-\infty}^{\tau} G_\phi(k;\tau,\tau_1)H_{\psi\phi}(k;\tau,\tau_1)\,\mathrm{d}\tau_1\right)\varepsilon_{ijl}B_l \\
&- \left(\frac{2}{3}\frac{\partial \Psi_j}{\partial x_i} + \frac{1}{15}\frac{\partial \Psi_i}{\partial x_j}\right)\int \mathrm{d}\boldsymbol{k}\int_{-\infty}^{\tau} G_\psi(k;\tau,\tau_1)Q_{\phi\phi}(k;\tau,\tau_1)\,\mathrm{d}\tau_1 \\
&- \left(\frac{2}{3}\frac{\partial \Phi_j}{\partial x_i} + \frac{1}{15}\frac{\partial \Phi_i}{\partial x_j}\right)\int \mathrm{d}\boldsymbol{k}\int_{-\infty}^{\tau} G_\phi(k;\tau,\tau_1)Q_{\psi\psi}(k;\tau,\tau_1)\,\mathrm{d}\tau_1,
\end{aligned}
\tag{7.123}
$$

with

$$
\begin{aligned}
K^{(E)} &= \tfrac{1}{2}\langle \boldsymbol{\phi}' \cdot \boldsymbol{\psi}' \rangle \\
&= \int Q_{\phi\psi}(k;\tau,\tau)\,\mathrm{d}\boldsymbol{k} - \frac{1}{2}\int \mathrm{d}\boldsymbol{k}\int_{-\infty}^{\tau} G_\psi(k;\tau,\tau_1)\frac{\mathrm{D}Q_{\phi\psi}(k;\tau,\tau_1)}{\mathrm{D}t}\,\mathrm{d}\tau_1 \\
&- \frac{1}{2}\int \mathrm{d}\boldsymbol{k}\int_{-\infty}^{\tau} G_\phi(k;\tau,\tau_1)\frac{\mathrm{D}Q_{\psi\phi}(k;\tau,\tau_1)}{\mathrm{D}t}\,\mathrm{d}\tau_1 \\
&- \frac{1}{4}\left(\int k^{-2}\,\mathrm{d}\boldsymbol{k}\int_{-\infty}^{\tau} G_\psi(k;\tau,\tau_1)\frac{\partial H_{\phi\psi}(k;\tau,\tau_1)}{\partial x_l}\,\mathrm{d}\tau_1\right)B_l \\
&- \frac{1}{4}\left(\int k^{-2}\,\mathrm{d}\boldsymbol{k}\int_{-\infty}^{\tau} G_\phi(k;\tau,\tau_1)\frac{\partial H_{\psi\phi}(k;\tau,\tau_1)}{\partial x_l}\,\mathrm{d}\tau_1\right)B_l.
\end{aligned}
\tag{7.124}
$$

7.4.4 Comparison with Kinematic and Counter-Kinematic Methods

The turbulent electromotive force $\boldsymbol{E}_\mathrm{M}$ and the Reynolds stress R_{ij} may be evaluated from Elsasser's Reynolds stress $R_{ij}^{(E)}$, equation (7.123), with the aid of equations (7.83) and (7.84). In the formalism based on Elsasser's variables, $R_{ij}^{(E)}$ is expressed in terms of the correlation functions concerning $\boldsymbol{\phi}'_{00}$ and $\boldsymbol{\psi}'_{00}$. In investigating astrophysical phenomena on the basis of these findings, it is more understandable to write $\boldsymbol{E}_\mathrm{M}$ and R_{ij} in terms of the original variables, that is, the velocity and magnetic field corresponding to $\boldsymbol{\phi}'_{00}$ and $\boldsymbol{\psi}'_{00}$, equation (7.115).

In correspondence to equation (7.115), we introduce

$$
G_+ = \frac{G_\phi + G_\psi}{2}, \qquad G_- = \frac{G_\phi - G_\psi}{2}. \tag{7.125}
$$

In order to see the physical meaning of these quantities, we consider the reflection of the coordinate system, $\boldsymbol{x} \to -\boldsymbol{x}$, under which we have

$$
\begin{gathered}
\boldsymbol{u} \to -\boldsymbol{u}, \qquad \boldsymbol{b} \to \boldsymbol{b}, \qquad \boldsymbol{\phi} \to -\boldsymbol{\psi}, \qquad \boldsymbol{\psi} \to -\boldsymbol{\phi}, \\
G'_{\phi ij} \to G'_{\psi ij}, \qquad G'_{\psi ij} \to G'_{\phi ij}.
\end{gathered}
\tag{7.126}
$$

As a result, G_+ and G_- are reflectionally symmetric and anti-symmetric, respectively.

We substitute equation (7.123) into equation (7.83), and make use of equations (7.120) and (7.121). Then we have

$$\boldsymbol{E}_{\mathrm{M}} = \alpha \boldsymbol{B} - \beta \boldsymbol{J} + \gamma \boldsymbol{\Omega}, \tag{7.127}$$

where

$$\begin{aligned}\alpha = &\frac{1}{3}\int \mathrm{d}\boldsymbol{k} \int_{-\infty}^{\tau} G_+(k,\boldsymbol{x};\tau,\tau_1,t)[-H_{uu}(k,\boldsymbol{x};\tau,\tau_1,t) + H_{bb}(k,\boldsymbol{x};\tau,\tau_1,t)]\,\mathrm{d}\tau_1 \\ &- \frac{1}{3}\int \mathrm{d}\boldsymbol{k} \int_{-\infty}^{\tau} G_-(k,\boldsymbol{x};\tau,\tau_1,t)[-H_{bu}(k,\boldsymbol{x};\tau,\tau_1,t) + H_{ub}(k,\boldsymbol{x};\tau,\tau_1,t)]\,\mathrm{d}\tau_1,\end{aligned} \tag{7.128}$$

$$\begin{aligned}\beta = &\frac{1}{3}\int \mathrm{d}\boldsymbol{k} \int_{-\infty}^{\tau} G_+(k,\boldsymbol{x};\tau,\tau_1,t)[Q_{uu}(k,\boldsymbol{x};\tau,\tau_1,t) + Q_{bb}(k,\boldsymbol{x};\tau,\tau_1,t)]\,\mathrm{d}\tau_1 \\ &- \frac{1}{3}\int \mathrm{d}\boldsymbol{k} \int_{-\infty}^{\tau} G_-(k,\boldsymbol{x};\tau,\tau_1,t)[Q_{ub}(k,\boldsymbol{x};\tau,\tau_1,t) + Q_{bu}(k,\boldsymbol{x};\tau,\tau_1,t)]\,\mathrm{d}\tau_1,\end{aligned} \tag{7.129}$$

$$\begin{aligned}\gamma = &\frac{1}{3}\int \mathrm{d}\boldsymbol{k} \int_{-\infty}^{\tau} G_+(k,\boldsymbol{x};\tau,\tau_1,t)[Q_{ub}(k,\boldsymbol{x};\tau,\tau_1,t) + Q_{bu}(k,\boldsymbol{x};\tau,\tau_1,t)]\,\mathrm{d}\tau_1 \\ &- \frac{1}{3}\int \mathrm{d}\boldsymbol{k} \int_{-\infty}^{\tau} G_-(k,\boldsymbol{x};\tau,\tau_1,t)[Q_{uu}(k,\boldsymbol{x};\tau,\tau_1,t) + Q_{bb}(k,\boldsymbol{x};\tau,\tau_1,t)]\,\mathrm{d}\tau_1,\end{aligned} \tag{7.130}$$

with the dependence on slow variables explicitly shown through $\boldsymbol{x}$ and t.

Entirely similarly, we may evaluate the Reynolds stress R_{ij} as

$$R_{ij} = \tfrac{2}{3} K_{\mathrm{R}} \delta_{ij} - \nu_{\mathrm{T}} S_{ij} + \nu_{\mathrm{M}} M_{ij}, \tag{7.131}$$

where K_{R} is defined by the first relation of equation (7.28), and

$$\nu_{\mathrm{T}} = \tfrac{7}{5}\beta, \tag{7.132}$$

$$\nu_{\mathrm{M}} = \tfrac{7}{5}\gamma. \tag{7.133}$$

The ratio ν_{T}/β is called the turbulent magnetic Prandtl number. In the TSDIA analysis, it is larger than one, and its physical meaning will become important in §10.2.2.

Both equations (7.127) and (7.131) possess the same dependence on the mean field as their counterparts by the kinematic and counter-kinematic methods that are summarized in §7.3.1. We are in a position to make the comparison between the coefficients by these three methods. In equation (7.128), the first part is related to the spectrum of the turbulent residual helicity that is defined by

$$-\boldsymbol{u}'_{00}\cdot\boldsymbol{\omega}'_{00}+\boldsymbol{b}'_{00}\cdot\boldsymbol{j}'_{00}. \tag{7.134}$$

This finding guarantees the conjecture by the kinematic and counter-kinematic methods, expression (7.71), and indicates the importance of the turbulent residual helicity for $\boldsymbol{E}_{\mathrm{M}}$, but not each of the kinetic and magnetic helicity. Such a conclusion was also confirmed by the method based on the low-Reynolds-number expansion [7.10]. In the case that the MHD turbulent state may be regarded as stationary in fast time τ, we have

$$H_{bu}(k;\tau,\tau')=H_{bu}(k;\tau',\tau)=H_{ub}(k;\tau,\tau'), \tag{7.135}$$

which leads to vanishing of the second part of equation (7.128).

In this context, we should refer to the concept of magnetic helicity. The representative quantity characterizing helical properties of magnetic field is the helicity based on the magnetic potential $\boldsymbol{a}$, which is defined by

$$\boldsymbol{a}\cdot\boldsymbol{b}\quad(\boldsymbol{b}=\nabla\times\boldsymbol{a}). \tag{7.136}$$

The total amount of $\boldsymbol{a}\cdot\boldsymbol{b}$ in a whole region, $\int_V \boldsymbol{a}\cdot\boldsymbol{b}\,\mathrm{d}V$, is conserved so long as there is neither net supply nor loss of magnetic helicity across the boundary. This point makes a sharp contrast with the other magnetic helicity $\boldsymbol{b}\cdot\boldsymbol{j}$ that is linked with the alpha effect. The absence of any conservation law concerning $\boldsymbol{b}\cdot\boldsymbol{j}$ is a stumbling block for constructing a self-consistent dynamo model applicable to real-world phenomena, as will be seen later.

For the turbulent resistivity β, the first part of equation (7.129) indicates that the conjecture given by equation (7.69) is plausible; namely, β is related to the spectrum of turbulent MHD energy (the sum of turbulent kinetic and magnetic energy). The second part expresses the contribution from the turbulent cross-helicity effect in the combination of G_-. In the state stationary in fast time τ, the situation entirely similar to equation (7.135) holds, and the second part in equation (7.129) vanishes. These facts concerning α and β suggest that the contributions from the G_+-related parts are more important. From the G_+-related part in equation (7.130) we may confirm expression (7.70).

As one of the prominent features of the formalism based on Elsasser's variables, we may mention equations (7.132) and (7.133). There we can see the clear relationship between the coefficients in $\boldsymbol{E}_{\mathrm{M}}$ and R_{ij}. In this context, we should note that no turbulent helicity effects corresponding to the alpha effect $\alpha\boldsymbol{B}$ enter R_{ij} in the analysis up to $O(\delta_{\mathrm{S}})$. There what connects the equations for $\boldsymbol{U}$ and $\boldsymbol{B}$ at the level of fluctuations is the turbulent cross-helicity effect (see [7.11] for the discussion about the cross-helicity effect on $\boldsymbol{E}_{\mathrm{M}}$ in the absence of mean field). In later discussions, the effect will be shown to play an important role of determining the saturation level of generated $\boldsymbol{B}$.

References

[7.1] Pouquet A, Frisch U and Léorat J 1976 *J. Fluid Mech.* **77** 321
[7.2] Parkar E N 1955 *Astrophys. J.* **122** 293
[7.3] Yoshizawa A 1985 *Phys. Fluids* **28** 3313
[7.4] Yoshizawa A 1990 *Phys. Fluids B* **2** 1589
[7.5] Nakayama K 1999 *Astrophys. J.* **523** 315
[7.6] Nakayama K 2000 *Astrophys. J.* **556** 1027
[7.7] Gruzinov A V and Diamond P H 1994 *Phys. Rev. Lett.* **72** 1651
[7.8] Gruzinov A V and Diamond P H 1996 *Phys. Plasmas* **3** 1853
[7.9] Field G B, Blackman E G and Chou H 1999 *Astrophys. J.* **513** 638
[7.10] Chen H and Montgomery D 1987 *Plasma Phys. Control. Fusion* **29** 205
[7.11] Blackman E G 2000 *Astrophys. J.* **529** 138

Chapter 8

One-Point Dynamo Modelling with Emphasis on Self-Consistency

8.1 Necessity and Significance of One-Point Modelling

In chapter 7 we examined the turbulence effects on the equations for the mean flow and magnetic field, that is, the Reynolds stress R_{ij} and the turbulent electromotive force $\boldsymbol{E}_{\mathrm{M}}$, from a few different viewpoints. The most orthodox method of closing equations (6.63) and (6.65) seems to construct the equations for the two-time spectral quantities such as $Q_{uu}(k;\tau,\tau')$ and $H_{uu}(k;\tau,\tau')$ in terms of which the coefficients in R_{ij} and $\boldsymbol{E}_{\mathrm{M}}$ are written. The method is not feasible in the study of highly inhomogeneous magnetohydrodynamic (MHD) turbulence at all.

The foregoing circumstances are well recognized in the study of electrically nonconducting turbulent flows. In the method based on the ensemble averaging procedure, the flow components except the mean velocity are eliminated, and their effects are taken into account through the modelling of R_{ij}. Its typical model is the turbulent-viscosity representation, equation (5.158). The concept of turbulent viscosity is physically useful in describing enhanced diffusion effects on fluid motion, but there are many flow properties beyond the scope of the concept alone (its typical instance is secondary flows in a square-duct flow of §4.2.5.3, and the work for adding other effects to the turbulent-viscosity representation is still in progress, as was stated in chapters 4 and 5.

In the study of real-world flows, various types of boundary conditions, which are the origin of inhomogeneity of turbulence, become a stumbling block to the use of spectral expressions such as equation (5.160). A great merit of the ensemble-mean method is the capability of examining complex flows at high Reynolds numbers that are far beyond the reach of the computer simulation of the primitive equations. The constraints on the method, such as the decrease in handled flow properties, should be compensated for by this merit. Then those spectral expressions need to be replaced with much more manageable one-point expressions in physical space.

The computer simulation of astrophysical and geophysical MHD flows is in progress with the advancement of a computer and a numerical scheme. The flows within the scope of the simulation, however, are still far from real-world flows in the magnitude of Reynolds and magnetic Reynolds numbers, Taylor number, etc. One of the major merits of the mean-field theory is to get a comprehensive understanding of magnetic-field generation and its feedback mechanisms, with no special constraints on the magnitude of those nondimensional parameters, although the subject of interest is limited to global MHD properties. For this purpose, we need to construct a manageable and self-consistent dynamo model mimicking real-world astrophysical phenomena, with the aid of the theoretical findings. In what follows we express the coefficients in R_{ij} and $\boldsymbol{E}_{\mathrm{M}}$ in terms of one-point quantities in physical space. Such a method may be called turbulent MHD or dynamo modelling after the terminology in the electrically nonconducting case.

8.2 Modelling Policy and Procedures

One-point modelling of electrically nonconducting turbulent flows in physical space is explained heuristically in chapter 4 and theoretically in chapter 5. The key task there is the choice of statistical quantities characterizing turbulent flows. It is necessary to make the number of those quantities as small as possible for reducing the mathematical burden in solving the resulting system of turbulence equations. The simplest model of the Reynolds stress R_{ij} in electrically nonconducting turbulent flows is the turbulent-viscosity (ν_{T}) representation, equation (4.63). Equation (7.131) is essentially the same level of approximation, apart from the newly-occurring magnetic-field effect. There we adopted the turbulent kinetic energy K [equation (3.38)] and its dissipation rate ε [equation (3.41)] as the characteristic turbulence quantities. This is the choice of the least number of turbulence quantities for constructing the characteristic time scale as K/ε.

The present MHD modelling consists of two key procedures. One is the choice of fundamental quantities characterizing MHD turbulence statistically, in terms of which the coefficients in R_{ij} and $\boldsymbol{E}_{\mathrm{M}}$ are expressed. The other is the construction of the transport equations for those quantities. In the traditional dynamo modelling with attention focused on the alpha effect, little attention has been paid to the latter point.

As the fundamental turbulence quantities, we choose the turbulent MHD energy K, the turbulent cross helicity W, the turbulent residual helicity H, and the MHD energy dissipation rate ε, which are given by

$$K = \left\langle \frac{\boldsymbol{u}'^2 + \boldsymbol{b}'^2}{2} \right\rangle, \tag{8.1}$$

$$W = \langle \boldsymbol{u}' \cdot \boldsymbol{b}' \rangle, \tag{8.2}$$

$$H = \langle -\boldsymbol{u}' \cdot \boldsymbol{\omega}' + \boldsymbol{b}' \cdot \mathbf{j}' \rangle, \tag{8.3}$$

$$\varepsilon = \nu \left\langle \left(\frac{\partial u_j'}{\partial x_i} \right)^2 \right\rangle + \lambda_{\mathrm{M}} \left\langle \left(\frac{\partial b_j'}{\partial x_i} \right)^2 \right\rangle, \tag{8.4}$$

respectively. In equation (7.128) for α, the first term is related to the spectrum of the turbulent residual helicity. Then the introduction of H is reasonable from the viewpoint of one-point modelling. The entirely similar situation holds for the choice of K and W in modelling β and γ.

Here we should note that the spectra of the turbulent residual helicity, MHD energy, and cross helicity occurring in equations (7.128)–(7.130) correspond to the lowest-order terms in the two-scale direct-interaction approximation (TSDIA) formalism, as may be seen from the use of equation (7.112). We denote those lowest-order contributions to K, W, H, and ε by K_0, W_0, H_0, and ε_0, respectively. In the present one-point modelling, we use their full counterparts, K, W, H, and ε through the replacement

$$K_0 \to K, \qquad W_0 \to W, \qquad H_0 \to H, \qquad \varepsilon_0 \to \varepsilon. \tag{8.5}$$

This procedure may be regarded as the physical-space renormalization in the comparison with the wavenumber-space renormalization such as equation (5.98). In addition to K, W, and H, ε is necessary for the construction of a characteristic time scale of MHD turbulent flow. Here W/H possesses the dimension of time, but it cannot be adopted as a characteristic time since W and H take both signs.

Under the choice of equations (8.1)–(8.4), the one-point modelling of equations (7.128)–(7.130) is straightforward. For instance, we write

$$H_0 = \int [-H_{uu}(k, \boldsymbol{x}; \tau, \tau, t) + H_{bb}(k, \boldsymbol{x}; \tau, \tau, t)] \, \mathrm{d}\boldsymbol{k}, \tag{8.6}$$

$$\tau_0(k) = \int_{-\infty}^{\tau} G_+(k, \boldsymbol{x}; \tau, \tau_1, t) \, \mathrm{d}\tau_1. \tag{8.7}$$

Equation (8.7) expresses the time scale associated with the spatial length k^{-1}. We take the energy-containing length

$$k_0^{-1} = \frac{K_0^{3/2}}{\varepsilon_0}, \tag{8.8}$$

and model

$$\tau_0(k_0) = \frac{k_0^{-1}}{\sqrt{K_0}} = \frac{K_0}{\varepsilon_0}. \tag{8.9}$$

As a result, equation (7.128) with only the first part retained is modelled as

$$\alpha = C_\alpha \frac{K}{\varepsilon} H, \tag{8.10}$$

after the renormalization, equation (8.5), where C_α is a positive model constant. Entirely similarly, we have

$$\beta = C_\beta \frac{K^2}{\varepsilon} \quad \left(= C_\beta \frac{K}{\varepsilon} K \right), \tag{8.11}$$

$$\gamma = C_\gamma \frac{K}{\varepsilon} W, \tag{8.12}$$

with positive constants C_β and C_γ.

In order to close the foregoing expressions, we need the transport equations governing K, W, H, and ε. The first two quantities are linked with the conservation laws, as was noted in §6.1.3. This property results in the equations for K and W, equation (6.74), whose mathematical structures are firm and clear. In equation (6.74) with $Z = K$, what is to be modelled is the second part of equation (6.77). We pay attention to the first term of the part and model it as

$$\left\langle \left(\frac{\mathbf{u}'^2 + \mathbf{b}'^2}{2} + \varpi \right) \mathbf{u}' \right\rangle = -\frac{\nu_\mathrm{T}}{\sigma_K} \nabla K, \tag{8.13}$$

where σ_K is a positive constant. This modelling is essentially the same as for equations (4.50) and (5.190).

In equation (6.74) with $Z = W$, we need to model equations (6.79) and (6.80). In the latter, the first term of the second part denotes the transport of W by turbulence, which is modelled as

$$\langle (\mathbf{u}' \cdot \mathbf{b}') \mathbf{u}' \rangle = -\frac{\nu_\mathrm{T}}{\sigma_W} \nabla W, \tag{8.14}$$

similar to equation (8.13), where σ_W is a positive constant.

Equation (6.79) is the destruction rate of W due to molecular viscous and resistive effects. As the characteristic time scale of MHD turbulence, we may consider two time scales. One is the foregoing time scale based on K and ε and is given by K/ε. The other is its counterpart of W, that is, W/ε_W. We assume that they are close to each other. Then we have

$$\varepsilon_W = C_W \frac{K}{\varepsilon} W, \tag{8.15}$$

where C_W is a positive constant close to one.

In hydrodynamic one-point modelling, we have already seen that ε obeys equation (4.52) with quite a complicated mathematical structure since it is related to no conservation constraint, unlike K and W. There we adopted a phenomenological model equation such as equation (4.60). This type of equation may also be derived by a statistical method, independent of equation (4.52) [2.16]. In MHD one-point modelling we follow the same line of modelling and adopt

$$\frac{\mathrm{D}\varepsilon}{\mathrm{D}t} = C_{\varepsilon 1}\frac{\varepsilon}{K}P_K - C_{\varepsilon 1}\frac{\varepsilon^2}{K} + \nabla \cdot \left(\frac{\nu_\mathrm{T}}{\sigma_\varepsilon}\nabla\varepsilon\right), \tag{8.16}$$

where $C_{\varepsilon 1}$, $C_{\varepsilon 2}$, and σ_ε are positive model constants. This model is the simplest extension of the hydrodynamic model (4.60) through the replacement of P_K [equation (3.40)] with MHD P_K [equation (6.75)].

Finally, we refer to the equation for the turbulent residual helicity H. Its equation may be written as [2.16]

$$\begin{aligned}\frac{\mathrm{D}H}{\mathrm{D}t} &= \langle u'_j u'_i + b'_j b'_i \rangle \frac{\partial \Omega_i}{\partial x_j} + \langle b'_j j'_i \rangle \frac{\partial U_i}{\partial x_j} + \langle \boldsymbol{u}' \times \boldsymbol{\omega}' \rangle \cdot \boldsymbol{\Omega} - \nabla \cdot (\langle \boldsymbol{u}'^2/2 \rangle \boldsymbol{\Omega}) \\ &\quad + \left\langle \frac{\partial u'_i}{\partial x_j} j'_i + \frac{\partial \omega'_i}{\partial x_j} b'_i - \omega'_i \frac{\partial b'_i}{\partial x_j} - u'_i \frac{\partial j'_i}{\partial x_j} \right\rangle B_j - \langle \omega'_i b'_j \rangle \frac{\partial B_i}{\partial x_j} \\ &\quad + \langle u'_j b'_i + u'_i b'_j \rangle \frac{\partial J_i}{\partial x_j} - \langle \boldsymbol{u}' \times \boldsymbol{j}' \rangle \cdot \boldsymbol{J} + \langle u'_j b'_i + u'_i b'_j \rangle \frac{\partial J_i}{\partial x_j} + R_H, \end{aligned} \tag{8.17}$$

where R_H denotes the remaining part that is not dependent explicitly on the mean field and is composed of the correlation functions of the third order in $\boldsymbol{u}'$, $\boldsymbol{b}'$, $\boldsymbol{\omega}'$, and $\boldsymbol{j}'$. This complicated form makes a sharp contrast with the case for the turbulent kinetic helicity in electrically nonconducting turbulent flows [see equation (3.84)].

The foregoing fact is a big stumbling block for the construction of a self-consistent dynamo model that is applicable to various types of astrophysical phenomena. In the current study of mean-field theory with the alpha effect as a cornerstone, it is rather curious that little attention has been paid to this difficulty. We may mention two reasons for this situation. One is that the alpha coefficient α is often assumed to be a given parameter and that its self-consistent determination is of no interest. The other is that attention is paid to the kinetic part of H from the kinematic viewpoint. In this case, we have equation (3.84) with the firm mathematical basis.

8.3 Summary of Dynamo Model

We summarize the one-point dynamo model that has been obtained with the aid of statistical methods [2.16, 7.3, 7.4]. We give the model by explicitly including effects of frame rotation and the buoyancy force based on the Boussinesq approximation.

8.3.1 System of Model Equations

The mean velocity $\boldsymbol{U}$, the mean magnetic field $\boldsymbol{B}$, and the mean temperature Θ obey

$$\frac{\mathrm{D}U_i}{\mathrm{D}t} = -\frac{\partial}{\partial x_i}\left(P + \left\langle \frac{\boldsymbol{b}'^2}{2} \right\rangle\right) + (\boldsymbol{J} \times \boldsymbol{B})_i + \frac{\partial}{\partial x_j}(-R_{ij}) + \nu \nabla^2 U_i$$
$$+ 2(\boldsymbol{U} \times \boldsymbol{\omega}_{\mathrm{F}})_i - \alpha_{\mathrm{T}}(\Theta - \Theta_{\mathrm{R}}) g_i, \tag{8.18}$$

$$\frac{\partial \boldsymbol{B}}{\partial t} = \nabla \times (\boldsymbol{U} \times \boldsymbol{B} + \boldsymbol{E}_{\mathrm{M}}) + \lambda_{\mathrm{M}} \nabla^2 \boldsymbol{B}, \tag{8.19}$$

$$\frac{\mathrm{D}\Theta}{\mathrm{D}t} = \nabla \cdot (-\boldsymbol{H}_\theta) + \lambda_\theta \nabla^2 \Theta, \tag{8.20}$$

with the solenoidal condition $\nabla \cdot \boldsymbol{U} = \nabla \cdot \boldsymbol{B} = 0$. In equation (8.18), $\boldsymbol{\omega}_{\mathrm{F}}$ is the angular velocity of frame rotation, Θ_{R} is the reference temperature, α_{T} is the coefficient of thermal expansion, and $\boldsymbol{g}$ is the vector of gravitational acceleration.

The Reynolds stress R_{ij}, the turbulent electromotive force $\boldsymbol{E}_{\mathrm{M}}$, and the turbulent heat flux $\boldsymbol{H}_\theta$ are expressed as

$$R_{ij} \equiv \langle u'_i u'_j - b'_i b'_j \rangle = \tfrac{2}{3} K_{\mathrm{R}} \delta_{ij} - \nu_{\mathrm{T}} S_{ij} + \nu_{\mathrm{M}} M_{ij}, \tag{8.21}$$

$$\boldsymbol{E}_{\mathrm{M}} \equiv \langle \boldsymbol{u}' \times \boldsymbol{b}' \rangle = \alpha \boldsymbol{B} - \beta \boldsymbol{J} + \gamma(\boldsymbol{\Omega} + 2\boldsymbol{\omega}_{\mathrm{F}}), \tag{8.22}$$

$$\boldsymbol{H}_\theta \equiv \langle \boldsymbol{u}' \theta' \rangle = -\frac{\nu_{\mathrm{T}}}{\sigma_\theta} \nabla \Theta, \tag{8.23}$$

where

$$K_{\mathrm{R}} = \left\langle \frac{\boldsymbol{u}'^2 - \boldsymbol{b}'^2}{2} \right\rangle, \tag{8.24}$$

$$S_{ij} = \frac{\partial U_j}{\partial x_i} + \frac{\partial U_i}{\partial x_j}, \tag{8.25}$$

$$M_{ij} = \frac{\partial B_j}{\partial x_i} + \frac{\partial B_i}{\partial x_j}. \tag{8.26}$$

In relation to equation (8.22), we should note the vorticity transformation rule in the rotating frame,

$$\boldsymbol{\Omega} \rightarrow \boldsymbol{\Omega} + 2\boldsymbol{\omega}_{\mathrm{F}}. \tag{8.27}$$

In the presence of buoyancy force, its effect may be incorporated into R_{ij} in the form proportional to $H_{\theta i} g_j + H_{\theta j} g_i$. Here this buoyancy effect is neglected with more emphasis on the magnetic effect.

The coefficients in equations (8.21)–(8.23) are given by

$$\nu_{\mathrm{T}} = \tfrac{7}{5} \beta, \tag{8.28}$$

$$\nu_{\mathrm{M}} = \tfrac{7}{5}\gamma, \tag{8.29}$$

$$\alpha = C_\alpha \frac{K}{\varepsilon} H, \tag{8.30}$$

$$\beta = C_\beta \frac{K^2}{\varepsilon}, \tag{8.31}$$

$$\gamma = C_\gamma \frac{K}{\varepsilon} W, \tag{8.32}$$

where

$$K = \left\langle \frac{\boldsymbol{u}'^2 + \boldsymbol{b}'^2}{2} \right\rangle, \tag{8.33}$$

$$W = \langle \boldsymbol{u}' \cdot \boldsymbol{b}' \rangle, \tag{8.34}$$

$$H = \langle -\boldsymbol{u}' \cdot \boldsymbol{\omega}' + \boldsymbol{b}' \cdot \boldsymbol{j}' \rangle, \tag{8.35}$$

$$\varepsilon = \nu \left\langle \left(\frac{\partial u_j'}{\partial x_i} \right)^2 \right\rangle + \lambda_{\mathrm{M}} \left\langle \left(\frac{\partial b_j'}{\partial x_i} \right)^2 \right\rangle. \tag{8.36}$$

Between K and W we have a very important relationship

$$\frac{|W|}{K} = \frac{2|\langle \boldsymbol{u}' \cdot \boldsymbol{b}' \rangle|}{\langle \boldsymbol{u}'^2 + \boldsymbol{b}'^2 \rangle} \le 1. \tag{8.37}$$

It will be later confirmed to play a critical role in the discussions on the generation processes of astro/geophysical magnetic fields.

Of equations (8.33)–(8.36), the turbulent MHD energy K and the turbulent cross helicity W are governed by

$$\frac{\mathrm{D}Z}{\mathrm{D}t} = P_Z - \varepsilon_Z + \nabla \cdot \boldsymbol{T}_Z \quad (Z = K \text{ or } W), \tag{8.38}$$

where $\varepsilon_z \equiv \varepsilon$, and

$$P_K = -R_{ij} \frac{\partial U_j}{\partial x_i} - \boldsymbol{E}_{\mathrm{M}} \cdot \boldsymbol{J} - \alpha_{\mathrm{T}} \boldsymbol{H}_\theta \cdot \boldsymbol{g}, \tag{8.39}$$

$$\boldsymbol{T}_K = W\boldsymbol{B} + \frac{\nu_{\mathrm{T}}}{\sigma_K} \nabla K, \tag{8.40}$$

$$P_W = -R_{ij} \frac{\partial B_j}{\partial x_i} - \boldsymbol{E}_{\mathrm{M}} \cdot (\boldsymbol{\Omega} + 2\boldsymbol{\omega}_{\mathrm{F}}) - \alpha_{\mathrm{T}} \frac{W}{K} \boldsymbol{H}_\theta \cdot \boldsymbol{g}, \tag{8.41}$$

$$\varepsilon_W = C_W \frac{K}{\varepsilon} W, \tag{8.42}$$

$$\boldsymbol{T}_W = K\boldsymbol{B} + \frac{\nu_{\mathrm{T}}}{\sigma_W} \nabla W. \tag{8.43}$$

In equation (8.39) we should note that the buoyancy-force effect is included explicitly through the last term [see equation (6.83)], but not through the buoyancy effect on R_{ij}. In equation (8.41), the third term arises from equation (6.84) in the combination with a simple model

$$\boldsymbol{b}' = \frac{W}{K}\boldsymbol{u}'. \tag{8.44}$$

This model is plausible in the sense that its product with $\boldsymbol{u}'$ leads to the identity relation. It is used only for the estimate of expression (6.84).

As the model equation for ε, we adopt

$$\frac{\mathrm{D}\varepsilon}{\mathrm{D}t} = C_{\varepsilon 1}\frac{\varepsilon}{K}P_K - C_{\varepsilon 1}\frac{\varepsilon^2}{K} + \nabla\cdot\left(\frac{\nu_\mathrm{T}}{\sigma_\varepsilon}\nabla\varepsilon\right). \tag{8.45}$$

This is a phenomenological equation, unlike equation (8.38), but its hydrodynamic version has been tested in various types of flow and shown to be an acceptable model, although there is still room for improvement.

Concerning H, we have no model equation whose reliability is comparable to equations (8.38) and (8.45). The sole model equation was constructed for the case of no mean velocity and was studied in the investigation into reversed-field pinches of plasmas [8.1, 8.2] (see §10.4). Modelling the equation for H in the presence of both the mean velocity and magnetic field is a primary unresolved subject in mean-field theory of dynamo. In the following applications of the present dynamo model to the study of astro/geophysical magnetic fields, we shall make discussions without resort to the details of H-transport processes.

8.3.2 Model Constants

The model constants in the present dynamo model are adopted as follows:

$$R_{ij},\ \boldsymbol{E}_\mathrm{M}:\ C_\alpha \cong 0.02,\quad C_\beta \cong 0.05,\quad C_\gamma \cong 0.04, \tag{8.46a}$$

$$\boldsymbol{H}_\theta:\ \sigma_\theta \cong 1, \tag{8.46b}$$

$$\text{Equation for } K:\ \sigma_K \cong 1, \tag{8.46c}$$

$$\text{Equation for } W:\ C_W \cong 1.1,\quad \sigma_W \cong 1, \tag{8.46d}$$

$$\text{Equation for } \varepsilon:\ C_{\varepsilon 1} \cong 1.4,\quad C_{\varepsilon 2} \cong 1.9,\quad \sigma_\varepsilon \cong 1.3. \tag{8.46e}$$

Of these ten model constants, σ_θ, σ_K, $C_{\varepsilon 1}$, $C_{\varepsilon 2}$, and σ_ε survive in the hydrodynamic case. Then we adopt the same values as for the case.

The model constants given by equation (8.46a) were estimated with the aid of the computer experiment of MHD turbulent flow in a cubic region [8.3]. There the turbulent state is inhomogeneous in one direction and homogeneous in the other two directions. The inhomogeneity of state is sustained through the imposition of an external force.

The modelling of ε_W in the form of equation (8.15) or (8.42) arises from the assumption that C_W is close to one. We consider the homogeneous MHD turbulent state with vanishing $\boldsymbol{U}$ and $\boldsymbol{B}$. In this situation, equation (8.38) results in

$$\frac{\partial K}{\partial t} = -\varepsilon, \tag{8.47}$$

$$\frac{\partial W}{\partial t} = -C_W \frac{\varepsilon}{K} W, \tag{8.48}$$

which gives

$$\frac{\partial}{\partial t}\frac{W}{K} = -(C_W - 1)\frac{\varepsilon}{K^2} W. \tag{8.49}$$

For $C_W < 1$, initially nonvanishing $|W|/K$ continues to grow and eventually exceeds one. This result violates the constraint on $|W|/K$, equation (8.37). Then we need $C_W > 1$ and adopt the first of equation (8.46d).

8.3.3 Remarks on Characteristic Time Scales

Equations (8.10)–(8.12) show that the time scale characterizing energy-containing eddies of MHD fluctuations

$$\tau_{\mathrm{M}} = \frac{K}{\varepsilon} \tag{8.50}$$

plays a critical role in modelling α, β, and γ. This time scale was obtained from equation (8.9) or the lowest-order time scale in the TSDIA formalism, with the aid of the renormalization (8.5). We should note that mean-flow and external-force effects do not enter τ_{M} in an explicit manner. Then those effects need to be treated properly in the equations governing K and ε.

Let us seek a method of explicitly incorporating the foregoing effects into a characteristic time scale. From the procedure of derivation in chapter 7 we may see that equation (8.21) corresponds to the turbulent-viscosity representation in electrically nonconducting flows, that is, the first part in equation (4.17) with C_ν taken as constant. Such a representation is the lowest-order approximation to R_{ij}, as was discussed in §4.2.1 and §5.2.3.2. Higher-order contributions to R_{ij} may be classified into two categories. One is mean-flow and nonequilibrium effects on nondimensional coefficients C_ν, etc. [recall equation (4.25) and the discussion below it]. The other is the contribution represented by nonlinear effects concerning the mean velocity strain tensor S_{ij} [equation (4.6)] and the mean vorticity tensor Ω_{ij} [equation (4.7)], as is given by the terms except the first in equation (4.20). In what follows we pay special attention to the former category.

We introduce the time-scales intrinsic to S_{ij} and Ω_{ij} as

$$\tau_{\mathrm{S}} = \|S\|^{-1}, \tag{8.51}$$

$$\tau_{\mathrm{V}} = \|\Omega\|^{-1}, \tag{8.52}$$

with the definition

$$\|A\| = \sqrt{A_{ij}^2}. \tag{8.53}$$

In the frame rotating with angular velocity ω_F, Ω_{ij} is subject to the transformation

$$\Omega_{ij} \rightarrow \Omega_{ij} + 2\varepsilon_{ijl}\omega_{Fl}, \tag{8.54}$$

in correspondence to equation (8.27). This fact indicates that τ_V is the time scale closely related to frame rotation. In the context of planetary dynamo that is a primary concern of Part III, frame rotation is one of the key ingredients in the process of magnetic-field generation. Then we shall focus on τ_V and consider the combined effect of τ_M and τ_V.

We define the relative magnitude of τ_M to τ_V by

$$r = \frac{\tau_M}{\tau_V} = \frac{K}{\varepsilon}\|\Omega\|. \tag{8.55}$$

In the case of $r \gg 1$, the turn-over time of mean flow is much shorter than that of energy-containing components of fluctuations or eddies, τ_M. In this situation, τ_V is the time scale more appropriate for describing the evolution of turbulence properties. In order to take this fact into account, we define the combined time scale τ_{MV} by

$$\frac{1}{\tau_{MV}} = \frac{1}{\tau_M} + \frac{C_{MV}}{\tau_V} \tag{8.56}$$

or

$$\tau_{MV} = \frac{1}{1 + C_{MV}r}\frac{K}{\varepsilon}, \tag{8.57}$$

with C_{MV} as a positive constant [8.4]. On the basis of equation (8.57), we model α, β, and γ as

$$\alpha = C_\alpha \tau_{MV} H, \tag{8.58}$$

$$\beta = C_\beta \tau_{MV} K, \tag{8.59}$$

$$\gamma = C_\gamma \tau_{MV} W, \tag{8.60}$$

in correspondence to equations (8.10)–(8.12).

The discussions in chapters 9 and 10 will be made on the basis of equations (8.10)–(8.12). There we shall be able to see that the relative magnitude among α, β, and γ, such as γ/β, plays an important role in the light of the magnetic-field generation and its feedback effect on fluid motion. As a result, such discussions are not dependent on the choice of τ_M or τ_{MV} as long as the spatial variation of α, β, and γ are dealt with implicitly.

References

[8.1] Yoshizawa A and Hamba F 1988 *Phys. Fluids* **31** 2276
[8.2] Hamba F 1990 *Phys. Fluids B* **2** 3064
[8.3] Hamba F 1992 *Phys. Fluids A* **4** 441
[8.4] Zhou Y 1995 *Phys. Fluids* **7** 2092

Chapter 9

Typical Magnetic-Field Generation Processes

In the dynamo model summarized in §8.3 we have two typical generation mechanisms of magnetic fields; one is the alpha or turbulent residual-helicity effect, and the other is the turbulent cross-helicity effect. In what follows we shall show that entirely different magnetic-field generation processes occur according to the relative importance of these two effects.

9.1 Dominant-Helicity Dynamo

9.1.1 Convection Columns and Helicity

In §2.2 the characteristics of the geomagnetic and solar magnetic fields were explained in the light of the geometrical features of the earth's outer core (figure 2.6) and the solar convective zone (figure 2.7). The magnitude of nondimensional parameters of flows in the outer core and the convective zone was discussed in §6.1.2. There the largeness of the Taylor number of the outer core was emphasized, compared with its solar counterparts, indicating the greater importance of frame-rotation effects on the geomagnetic-field generation process.

Effects of frame rotation occur typically through convection columns along the axis of rotation, as was mentioned in §6.1.2. Each column is composed of the fluid motion coming up or down while rotating, leading to the occurrence of helicity. The spherical-shell region of the earth's outer core is relatively much wider than the solar convective zone. Then the convection columns may occur more clearly in the earth's outer core.

A computer simulation based on a system of primitive equations is a method appropriate for the study of highly three-dimensional, time-dependent magnetodydrodynamic (MHD) flows, unlike the mean-field theory focusing on a stationary or quasi-stationary global MHD state. In a number of computer simulations mimicking geodynamo [2.10–2.12, 9.1–9.5], the column-like or elongated flow structure along the rotation axis has been detected clearly. The structure consists of a few pairs of distinct convection columns, as is illustrated in figure 6.1. In each pair, the fluid in one column rotating in the same direction as the rotating shell sinks from the column ends towards the equatorial plane, whereas the fluid in the other rotating in the opposite direction rises from the equatorial plane towards the ends (recall figure 6.1). As a result, the kinetic helicity tends to be negative and positive in the northern and southern hemispheres, respectively. These findings by the computer simulations signify an important role of helicity effects on MHD flows in a spherical-shell region such as the earth's outer core (readers may consult [9.6, 9.7] for recent reviews of geodynamo). In the solar convective zone, the shell region is much thinner, and the column-like structure is restricted to the low-latitude region, compared with the earth's outer core. Then the role of column-like structures is considered to be much different in geodynamo and solar dynamo [9.8].

In what follows, we bear the earth's outer core in mind and examine the situation in which effects of helicity play a dominant role in the magnetic-field generation process. The thin-shell counterpart with the solar convective zone in mind will be discussed in §9.2. In this context, we should note that each convection column observed in the computer simulations is beyond the scope of the mean-field theory of dynamo. What may be dealt with explicitly by the theory is the helicity effect arising from the ensemble mean of such convection-column flows or their average around the axis of frame rotation. We focus attention on the helicity effect through the turbulent electromotive force $\boldsymbol{E}_{\mathrm{M}}$ and the Reynolds stress R_{ij}.

9.1.2 Mean-Field Equations

We consider the mean magnetic induction equation (8.19) with equation (8.22) or

$$\frac{\partial \boldsymbol{B}}{\partial t} = \nabla \times [\boldsymbol{U} \times \boldsymbol{B} + \alpha \boldsymbol{B} - \beta \boldsymbol{J} + \gamma(\boldsymbol{\Omega} + 2\boldsymbol{\omega}_{\mathrm{F}})], \tag{9.1}$$

where the molecular magnetic diffusivity λ_{M} was neglected, compared with its turbulent counterpart β. We seek the stationary state of the magnetic field $\boldsymbol{B}$ that is subject to

$$\nabla \times [\boldsymbol{U} \times \boldsymbol{B} + \alpha \boldsymbol{B} - \beta \boldsymbol{J} + \gamma(\boldsymbol{\Omega} + 2\boldsymbol{\omega}_{\mathrm{F}})] = 0. \tag{9.2}$$

Here the turbulent-resistivity part $-\beta \boldsymbol{J}$ contributes to the diffusion of magnetic-field structure due to turbulent motion. In order that a distinctive global structure of $\boldsymbol{B}$ may continue to persist, there needs to be the effect that balances with $-\beta \boldsymbol{J}$ and cancels the diffusion effect arising from it. We assume the dominant-helicity state in which the alpha term $\alpha \boldsymbol{B}$ due to the helicity effect balances with $-\beta \boldsymbol{J}$; namely, we put

$$\boldsymbol{J} = \frac{\alpha}{\beta} \boldsymbol{B}, \tag{9.3}$$

where

$$\frac{\alpha}{\beta} = \frac{C_\alpha}{C_\beta} \frac{H}{K}, \tag{9.4}$$

from equations (8.30) and (8.31).

Equation (9.3) represents the typical manifestation of the turbulent-helicity or alpha effect. There $\boldsymbol{J}$ is aligned with $\boldsymbol{B}$, which results in vanishing of the Lorentz force $\boldsymbol{J} \times \boldsymbol{B}$ in the mean-flow equation (8.18). From this property, the magnetic field obeying equation (9.3) is called the force-free field.

Of the remaining two parts in equation (9.2), we consider the γ-related term. A primary cause generating helicity effects is the frame rotation. Then we retain the ω_F-related part and put aside the first term or $\boldsymbol{U} \times \boldsymbol{B}$. We shall later refer to it. Equation (9.2) is satisfied by

$$\boldsymbol{J} = \frac{\alpha}{\beta} \boldsymbol{B} + 2 \frac{\gamma}{\beta} \boldsymbol{\omega}_\mathrm{F}, \tag{9.5}$$

with

$$\frac{\gamma}{\beta} = \frac{C_\gamma}{C_\beta} \frac{W}{K}, \tag{9.6}$$

from equations (8.31) and (8.32).

Let us see equation (9.5) in light of the magnetic-field growth. From equation (9.1) with the $\boldsymbol{\Omega}$ effect dropped, we have

$$\frac{\partial}{\partial t} \int_V \frac{\boldsymbol{B}^2}{2} \mathrm{d}V = - \int_S \frac{\boldsymbol{B}^2}{2} \boldsymbol{U} \cdot \boldsymbol{n} \, \mathrm{d}S + \int_V \boldsymbol{B} \cdot [(\boldsymbol{B} \cdot \nabla) \boldsymbol{U}] \, \mathrm{d}V$$
$$+ \int_V \boldsymbol{B} \cdot [\nabla \times (\alpha \boldsymbol{B} - \beta \boldsymbol{J} + 2\gamma \boldsymbol{\omega}_\mathrm{F})] \, \mathrm{d}V, \tag{9.7}$$

where V and S denote the volume and surface of a spherical-shell region, respectively, and $\boldsymbol{n}$ is the outward unit vector normal to the surface. On the right-hand side, the first and second terms arise from the $\boldsymbol{U} \times \boldsymbol{B}$-related part in equation (9.1). The first term, which was rewritten using the Stokes' integral theorem, does not contribute to the net energy increase so long as there is no net energy inflow across the boundary. The second term represents the energy increase due to the stretching of magnetic field lines by fluid motion. Under equation (9.5), the alignment between $\boldsymbol{B}$ and $\boldsymbol{U}$, which means vanishing of $\boldsymbol{U} \times \boldsymbol{B}$, leads to the saturation of the magnetic-energy growth and the stationary state of $\boldsymbol{B}$.

Next, we consider the mean-flow equation (8.18) with equation (8.21). For simplicity of discussion, we assume the quasi-homogeneity of turbulent state; namely, the spatial derivatives of α, β, and γ are neglected, and their spatial variation is taken into account through the implicit dependence on location. Under equation (9.5), it may be rewritten as

$$\frac{\mathrm{D}\boldsymbol{U}}{\mathrm{D}t} = -\nabla\left(P + \left\langle\frac{\boldsymbol{b}'^2}{2}\right\rangle + \frac{2}{3}K_\mathrm{R}\right) - \alpha_\mathrm{T}(\Theta - \Theta_\mathrm{R})\boldsymbol{g}$$

$$+ 2\left(\boldsymbol{U} - \frac{\gamma}{\beta}\boldsymbol{B}\right) \times \boldsymbol{\omega}_\mathrm{F} + \nu_\mathrm{T}\nabla^2\left(\boldsymbol{U} - \frac{\gamma}{\beta}\boldsymbol{B}\right), \tag{9.8}$$

where equations (8.28) and (8.29) were used. In equation (9.8) the third term on the right-hand side comes from the combination of the Lorentz and Coriolis forces. We mentioned above that the alignment between $\boldsymbol{B}$ and $\boldsymbol{U}$ is important for the saturation of $\boldsymbol{B}$. In equation (9.8), the alignment

$$\boldsymbol{U} = \frac{\gamma}{\beta}\boldsymbol{B} \quad \text{or} \quad \boldsymbol{B} = \frac{\beta}{\gamma}\boldsymbol{U}. \tag{9.9}$$

signifies the cancellation of the momentum-diffusion effect due to turbulent motion, $\nu_\mathrm{T}\nabla^2\boldsymbol{U}$, by the magnetic feedback effect $-\nu_\mathrm{M}\nabla^2\boldsymbol{B}$. The cancellation leads to the persistence of the mean-flow structure that is coupled with a distinct global structure of magnetic field. We should note that vanishing of $\boldsymbol{U} \times \boldsymbol{B}$ on which equation (9.5) is founded is assured by equation (9.9).

Equation (9.9) expresses the balance between the Lorentz and Coriolis forces. The ratio of Lorentz to Coriolis forces is called the Elsasser number. From the computer simulation of the magnetoconvection [2.12], the persistence of column structures is tightly linked with the state with the Elsasser number close to one. This state is realized by equation (9.9).

Equation (9.9) also stipulates the saturation level of generated $\boldsymbol{B}$. The alpha effect $\alpha\boldsymbol{B}$ is linear in $\boldsymbol{B}$ so long as α is not dependent explicitly on $\boldsymbol{B}$. As a result, equation (8.19) with the alpha and turbulent-resistivity effects embedded cannot determine the saturation level. A method for overcoming this difficulty is the incorporation of effects nonlinear in $\boldsymbol{B}$ into α, as was noted in §7.4.2 [7.7–7.9]. Equation (9.9) indicates that the inclusion of the turbulent cross-helicity effect results in the automatic determination of the level through the interaction with the mean-flow equation.

9.1.3 Turbulence Equations

In §9.1.2 we made discussions on $\boldsymbol{B}$ and $\boldsymbol{U}$ on the basis of nonvanishing α, β, and γ. These quantities are expressed in terms of the turbulent energy K, the turbulent cross helicity W, the turbulent residual helicity H, and the energy dissipation rate ε, as in equations (8.30)–(8.32). To show how these turbulence quantities are sustained consistently with $\boldsymbol{B}$ and $\boldsymbol{U}$ is indispensable for the self-consistent understanding of dynamo processes. In the past study of mean-field theory of dynamo, the sustainment of nonvanishing α and β was assumed, and little attention was paid to this aspect.

The turbulent energy K is generated by the production term in equation (8.38), P_K [equation (8.39)]. Under equations (9.5) and (9.9), we have

$$R_{ij} = \tfrac{2}{3} K \delta_{ij}, \qquad \boldsymbol{E}_{\mathrm{M}} = 0, \tag{9.10}$$

which results in

$$P_K = -\alpha_{\mathrm{T}} \boldsymbol{H}_\theta \cdot \boldsymbol{g}. \tag{9.11}$$

Equation (9.11) indicates that the turbulent MHD state subject to dominant helicity effects is sustained through the turbulent-energy supply by an external force such as the buoyancy force. In the case of a spherical-shell region like the earth's outer core, we have

$$\boldsymbol{H}_\theta = -\frac{\nu_{\mathrm{T}}}{\sigma_\theta}\left(\frac{\partial \Theta}{\partial r}, 0, 0\right), \qquad \boldsymbol{g} = (-g, 0, 0), \tag{9.12}$$

in the spherical coordinates (r, θ, ϕ) (figure 6.2), where use has been made of equation (8.23). As a result, P_K is written as

$$P_K = -\alpha_{\mathrm{T}} g \frac{\nu_{\mathrm{T}}}{\sigma_\theta} \frac{\partial \Theta}{\partial r} > 0, \tag{9.13}$$

since $\partial \Theta / \partial r < 0$. Namely, K continues to be generated by P_K. The energy dissipation rate ε obeys equation (8.45). Nonvanishing ε is sustained by the first term proportional to P_K under equation (9.13).

Entirely similarly, the production term for W, equation (8.41), is given by

$$P_W = \frac{W}{K} P_K, \tag{9.14}$$

which leads to

$$W > 0 \rightarrow P_W > 0 \qquad \text{and} \qquad W < 0 \rightarrow P_W < 0. \tag{9.15}$$

This finding signifies that positive and negative W are generated in the region with positive and negative W, respectively, resulting in the sustainment of nonvanishing W.

In the above discussions, the buoyancy force plays a key role of sustaining the turbulent state represented by K and W. Is the dominant-helicity state possible in the absence of an external force such as the buoyancy force? In this case, we have no production mechanisms of K and W. This point will be later discussed in the context of the collimation of accretion jets that was referred to in §2.3.

In the present stage of mean-field theory, the discussion comparable to those on K and W cannot be made on the turbulent residual helicity H, owing to the lack of the reliable model equation for it. The construction of the model equation is a major unresolved subject of mean-field theory.

9.2 Dominant/Cross-Helicity State

In the solar convective zone (figure 2.7), the spherical-shell region is relatively much thinner than that of the earth's outer core. This fact suggests that convection column structures are difficult to be clearly formed there, compared with the outer core, as was referred to in §9.1.1. With this point in mind, we shall consider the situation opposite to §9.1, that is, the dominant/cross-helicity case.

9.2.1 Mean-Field Equations

As one of the prominent differences between the fluid motions in the solar convective zone and the earth's outer core, we may mention the strong differential rotation in the former [9.9]. In the zone, the angular velocity of the rotational motion is highly dependent on location. The angular velocity decreases by about 10% at the bottom near the equatorial plane; namely, there is a steep radial velocity gradient there. In order to properly treat this situation in an analytical manner, we adopt the inertial frame (the frame not rotating) with no translational velocity relative to the astronomical object concerned (for instance, the sun).

The mean magnetic field $\boldsymbol{B}$ obeys

$$\frac{\partial \boldsymbol{B}}{\partial t} = \nabla \times (\boldsymbol{U} \times \boldsymbol{B} + \alpha \boldsymbol{B} - \beta \boldsymbol{J} + \gamma \boldsymbol{\Omega}). \tag{9.16}$$

We seek its stationary state in the sense of a special solution, similar to the discussions on equation (9.1), as

$$\nabla \times (\boldsymbol{U} \times \boldsymbol{B} + \alpha \boldsymbol{B} - \beta \boldsymbol{J} + \gamma \boldsymbol{\Omega}) = 0. \tag{9.17}$$

We consider the dominant/cross-helicity state and deal with the alpha effect in a perturbational manner. We write

$$\boldsymbol{B} = \sum_{n=0}^{\infty} \boldsymbol{B}_n, \quad \boldsymbol{J} = \sum_{n=0}^{\infty} \boldsymbol{J}_n, \tag{9.18}$$

where $\boldsymbol{B}_n$ and $\boldsymbol{J}_n$ are of the nth order in α, which are symbolically written as

$$\boldsymbol{B}_n, \ \boldsymbol{J}_n = O(\alpha^n). \tag{9.19}$$

The first two parts obey

$$\nabla \times (\boldsymbol{U} \times \boldsymbol{B}_0 - \beta \boldsymbol{J}_0 + \gamma \boldsymbol{\Omega}) = 0, \tag{9.20}$$

$$\nabla \times (\boldsymbol{U} \times \boldsymbol{B}_1 + \alpha \boldsymbol{B}_0 - \beta \boldsymbol{J}_1) = 0. \tag{9.21}$$

As was done in §9.1.2, we assume the quasi-homogeneity of turbulent state and neglect the spatial derivatives of α, β, and γ. Under this approximation, a special solution of equation (9.20) is given by

$$\boldsymbol{B}_0 = \frac{\gamma}{\beta}\boldsymbol{U}, \tag{9.22}$$

$$\boldsymbol{J}_0 = \frac{\gamma}{\beta}\boldsymbol{\Omega}, \tag{9.23}$$

with equation (9.6) for γ/β. Here we should note $\boldsymbol{U} \times \boldsymbol{B}_0 = 0$. Equation (9.22) shows that the toroidal field is generated from the toroidal velocity. This point will be discussed in detail in the context of solar magnetic fields.

Let us consider equation (9.21) for $\boldsymbol{B}_1$. We first drop the $\boldsymbol{U} \times \boldsymbol{B}_1$-related part and examine this approximation below. Then we take

$$\boldsymbol{J}_1 = \frac{\alpha}{\beta}\boldsymbol{B}_0 = \frac{\alpha\gamma}{\beta^2}\boldsymbol{U}. \tag{9.24}$$

Since the primary global motion in the convection zone is the toroidal or rotational motion, we write

$$\boldsymbol{U} = [0, 0, U_\phi(r, \theta)], \tag{9.25}$$

in the spherical coordinates (r, θ, ϕ) (figure 6.2). In this situation, $\boldsymbol{J}_1$ is toroidal. From Ampère's law $\boldsymbol{J}_1 = \nabla \times \boldsymbol{B}_1$, $\boldsymbol{B}_1$ is poloidal and is expressed as

$$\boldsymbol{B}_1 = B_{1r}\boldsymbol{e}_r + B_{1\theta}\boldsymbol{e}_\theta, \tag{9.26}$$

with $\boldsymbol{e}_r$ and $\boldsymbol{e}_\theta$ as the unit vectors in the r and θ directions, respectively. Here we should recall that $\boldsymbol{B}_1$ is of $O(\alpha)$ in the α expansion, equation (9.18).

We examine the foregoing approximation of dropping the $\boldsymbol{U} \times \boldsymbol{B}_1$-related part. From equations (9.25) and (9.26), we have

$$\nabla \times (\boldsymbol{U} \times \boldsymbol{B}_1) = \left(0, 0, B_{1r}\frac{\partial U_\phi}{\partial r} + B_{1\theta}\frac{1}{r}\frac{\partial U_\phi}{\partial \theta}\right), \tag{9.27}$$

where use has been made of $\nabla \cdot \boldsymbol{B}_1 = \nabla \cdot \boldsymbol{U} = 0$ and $\boldsymbol{U} \cdot \nabla = 0$. In the case that the primary part of $\boldsymbol{B}_1$ is of dipole type (see figure 10.1), it is nearly along the rotation axis in the low-latitude region or near the equatorial plane (the occurrence of the dipole component of $\boldsymbol{B}_1$ will be shown when discussing the solar field in §10.1). In this situation, the radial component B_{1r} is small in the low-latitude region, indicating the smallness of $B_{1r}(\partial U_\phi/\partial r)$ there. The rotational velocity $\boldsymbol{U}$ is symmetric across the equatorial plane, and $B_{1\theta}(\partial U_\phi/(r\partial\theta))$ also becomes small in the low-latitude region. As a result, equation (9.24) is an approximate solution of equation (9.21) in the region. With this point in mind, we use equation (9.24) in later discussions.

Let us see equation (9.24) in the context of convection-column structures noted in §9.1.1. In a thin spherical-shell region, these structures are limited to the low-latitude region. Under the alpha effect arising from them, the toroidal current $\boldsymbol{J}_1$ occurs in the region from the toroidal field $\boldsymbol{B}_0$, as in figure 7.1.

We consider the mean-flow equation (8.18) with equation (8.21). We drop the Coriolis term and have

$$\frac{\partial \boldsymbol{U}}{\partial t} = -\nabla\left(P + \frac{1}{2}U^2 + \frac{2}{3}K_{\rm R} + \left\langle \frac{\boldsymbol{b}'^2}{2} \right\rangle\right) - \alpha_{\rm T}(\Theta - \Theta_{\rm R})\boldsymbol{g}$$
$$+ \boldsymbol{U} \times \boldsymbol{\Omega} + \boldsymbol{J} \times \boldsymbol{B} + \nu_{\rm T}\nabla^2\boldsymbol{U} - \nu_{\rm M}\nabla^2\boldsymbol{B}. \tag{9.28}$$

In equation (9.18), we retain the leading terms, which are given by equations (9.22) and (9.23). Then equation (9.28) may be rewritten as

$$\frac{\partial \boldsymbol{U}}{\partial t} = -\nabla\left(P + \frac{1}{2}U^2 + \frac{2}{3}K_{\rm R} + \left\langle \frac{\boldsymbol{b}'^2}{2} \right\rangle\right) - \alpha_{\rm T}(\Theta - \Theta_{\rm R})\boldsymbol{g}$$
$$+ \left[1 - \left(\frac{\gamma}{\beta}\right)^2\right]\boldsymbol{U} \times \boldsymbol{\Omega} + \nu_{\rm T}\left[1 - \left(\frac{\gamma}{\beta}\right)^2\right]\nabla^2\boldsymbol{U}. \tag{9.29}$$

As will be shown later, we have $\gamma/\beta < 1$. Then the feedback influence on the fluid motion due to generated $\boldsymbol{B}$ is weak in this case. Such a situation makes a sharp contrast with the dominant-helicity state in which the saturation level of $\boldsymbol{B}$ is determined through the interaction between the magnetic field and the fluid motion.

9.2.2 Turbulence Equations

In §9.1.3 we investigated the production term P_K [equation (8.39)] to understand how nonvanishing K is sustained in the dominant-helicity case. We now retain the leading parts in equation (9.18), that is, equations (9.22) and (9.23), while neglecting the alpha effect. Under this approximation, we have

$$\boldsymbol{E}_{\rm M} = 0. \tag{9.30}$$

The combination of equation (8.21) with equation (9.22) gives

$$P_K = \frac{1}{2}\nu_{\rm T}\left[1 - \left(\frac{\gamma}{\beta}\right)^2\right]S_{ij}^2 - \alpha_{\rm T}\boldsymbol{H}_\theta \cdot \boldsymbol{g}, \tag{9.31}$$

under equation (9.30). From the remark on γ/β below equation (9.29), we drop the γ/β-related part and have

$$P_K = \tfrac{1}{2}\nu_{\rm T}S_{ij}^2 - \alpha_{\rm T}\boldsymbol{H}_\theta \cdot \boldsymbol{g}. \tag{9.32}$$

Both of these two terms are positive [see equation (9.13)] and contribute to the sustainment of the turbulent state.

A big difference between the P_K in the dominant-helicity and dominant/cross-helicity dynamos lies in the first term in equation (9.32). In the dominant/cross-helicity case, the supply of energy is made through the mean flow and the buoyancy force, unlike the dominant-helicity case with the buoyancy force as the sole turbulent-energy source. This mechanism is essentially the same as for electrically nonconducting flows. A similar situation holds for ε.

In relation to thc generation of turbulent energy due to P_K, we should emphasize the differential rotation, which is specifically prominent near the bottom of the solar convective zone. The toroidal velocity given by equation (9.25) is written as

$$\boldsymbol{U} = [0, U_\vartheta(\sigma, z), 0] \tag{9.33}$$

in the cylindrical coordinates (σ, ϑ, z) (figure 6.2). Under equation (9.33), equation (9.32) is expressed as

$$P_K = \nu_\mathrm{T}\left[\sigma\left(\frac{\partial}{\partial\sigma}\frac{U_\vartheta}{\sigma}\right)\right]^2 - \alpha_\mathrm{T}\boldsymbol{H}_\theta\cdot\boldsymbol{g}. \tag{9.34}$$

The first term clearly indicates that the differential part of rotation plays the role of energy supply from mean to fluctuating flows. This point makes a sharp contrast with the relationship of the mean magnetic field with the mean flow. In the latter, the whole rotational motion becomes important, as is seen from equation (9.22).

The production term for the turbulent cross helicity W, P_W [equation (8.41)], is written as

$$P_W = \frac{W}{K}\left(\frac{1}{2}\frac{C_\gamma}{C_\beta}\nu_\mathrm{T}S_{ij}^2 - \alpha_\mathrm{T}\boldsymbol{H}_\theta\cdot\boldsymbol{g}\right), \tag{9.35}$$

from equations (8.31) and (8.32). Inside the parentheses of equation (9.35), both parts are positive from the discussion on P_K, leading to equation (9.15). Namely, W is sustained through the velocity-strain and buoyancy effects. As is the case for P_K, the occurrence of the strain effect is the feature not shared by the dominant-helicity case.

9.3 Traditional Kinematic Dynamos

In §9.1 and §9.2 we discussed two different magnetic-field generation mechanisms. One is the alpha dynamo arising from the turbulent helicity effect, and the other is the cross-helicity dynamo linked with the mean rotational motion. In the former, the cross-helicity effect becomes important in the stage of determining the saturation level of generated magnetic field. The cross-helicity effects are not treated explicitly in the traditional kinematic dynamo. In what follows, we shall scrutinize the primary differences between the present and traditional dynamos.

9.3.1 Alpha–Alpha Dynamo

For the comparison with the traditional kinematic dynamo, we consider equation (9.1) with the cross-helicity effect dropped; namely, we have

$$\frac{\partial \boldsymbol{B}}{\partial t} = \nabla \times (\boldsymbol{U} \times \boldsymbol{B} + \alpha \boldsymbol{B} - \beta \boldsymbol{J}). \tag{9.36}$$

For $\boldsymbol{U}$ and $\boldsymbol{B}$ axisymmetric around the axis of frame rotation, we may write

$$\boldsymbol{U} = U_\phi \boldsymbol{e}_\phi + \boldsymbol{U}_{\rm p}, \tag{9.37}$$

$$\boldsymbol{B} = B_\phi \boldsymbol{e}_\phi + \boldsymbol{B}_{\rm p}, \tag{9.38}$$

in the spherical coordinate system (r, θ, ϕ) (figure 6.2), where the poloidal components $\boldsymbol{U}_{\rm p}$ and $\boldsymbol{B}_{\rm p}$ are given by

$$\boldsymbol{U}_{\rm p} = U_r \boldsymbol{e}_r + U_\theta \boldsymbol{e}_\theta, \tag{9.39}$$

$$\boldsymbol{B}_{\rm p} = B_r \boldsymbol{e}_r + B_\phi \boldsymbol{e}_\phi = \nabla \times (A_\phi \boldsymbol{e}_\phi) \tag{9.40}$$

(A_ϕ is the toroidal component of the vector potential $\boldsymbol{A}$). From equations (9.37)–(9.40), we may rewrite equation (9.36) as [2.7, 2.13]

$$\frac{\partial B_\phi}{\partial t} + \sigma(\boldsymbol{U}_P \cdot \nabla)\frac{B_\phi}{\sigma} = \alpha(\nabla \times \boldsymbol{B}_{\rm p})_\phi + \sigma(\boldsymbol{B}_{\rm p} \cdot \nabla)\frac{U_\phi}{\sigma} + \beta\left(\nabla^2 - \frac{1}{\sigma^2}\right)B_\phi, \tag{9.41}$$

$$\frac{\partial A_\phi}{\partial t} + \frac{1}{\sigma}(\boldsymbol{U}_{\rm p} \cdot \nabla)(\sigma A_\phi) = \alpha B_\phi + \beta\left(\nabla^2 - \frac{1}{\sigma^2}\right)A, \tag{9.42}$$

where the spatial variation of α and β has been dropped, as is similar to §9.1 and §9.2, and use has been made of the relation $\sigma = r \sin\theta$.

Let us consider the role of each term in equations (9.41) and (9.42). The second terms on the left-hand sides represent the advection effect. The generation of magnetic fields arises from the first two terms and the first term on the right-hand sides, respectively. We focus attention on the two α-related terms. They represent the following generation cycle of magnetic field:

Poloidal field $\boldsymbol{B}_{\rm p}$

→ Toroidal current $\nabla \times \boldsymbol{B}_{\rm p}$ by the Ampère law (9.43a)

→ Toroidal field B_ϕ by the alpha effect $\alpha(\nabla \times \boldsymbol{B}_{\rm p})_\phi$ (9.43b)

→ Toroidal current J_ϕ by the alpha effect αB_ϕ (9.43c)

→ Poloidal field $\boldsymbol{B}_{\rm p}$ by the Ampère law. (9.43d)

In the above processes, the helicity dynamo has been used twice, completing the magnetic generation cycle [2.13–2.15]. This process is called the alpha–alpha dynamo.

In the stationary state, the processes (9.43a–d) correspond to equation (9.3). Therefore the saturation level of $\boldsymbol{B}$ cannot be determined within the framework of equation (9.41) and (9.42), as was stressed in §9.1.2. To overcome this difficulty, the inclusion of nonlinear $\boldsymbol{B}$ effects on α has been studied in the kinematic approach [7.7–7.9].

9.3.2 Alpha–Omega Dynamo

In equation (9.41) we have one more term that leads to the generation of the toroidal magnetic field B_ϕ. It is the second term on the right-hand side, which is linked with the spatial nonuniformity of the angular velocity U_ϕ/σ or the differential rotation. In §9.2.2 its importance was stressed in the light of the sustainment mechanism of turbulent MHD state [recall equations (9.34) and (9.35)]. In the context of equation (9.41), the differential rotation signifies

$$\begin{array}{l}\text{Generation of the toroidal field } B_\phi \\ \text{by the distortion of the poloidal field } \boldsymbol{B}_\mathrm{p}.\end{array} \tag{9.44}$$

The physical meaning of the process (9.44) may be explained as follows. As a typical case, we consider a magnetic-field tube parallel to the rotation axis. The term $(\boldsymbol{B}_\mathrm{p}\cdot\nabla)U_\phi/\sigma$ signifies the change of the angular velocity U_ϕ/σ along the poloidal component of the tube. At high magnetic Reynolds numbers, magnetic fields are nearly frozen in a fluid and move with it. For positive $(\boldsymbol{B}_\mathrm{p}\cdot\nabla)U_\phi/\sigma$, the upper part of the poloidal magnetic tube moves faster in the toroidal direction than its lower counterpart. The resulting deformation of the tube leads to the occurrence of the toroidal component of magnetic field. This process is called the omega dynamo. In the case that the tube is stretched, its cross-section decreases and the strength of magnetic tension increases. The overstretched tube overcomes the stretching by the fluid motion and, in turn, shrinks (recall the motion of a spring). As a result, the omega dynamo indicates an oscillatory behavior in time. The cycle consisting of the processes (9.43c,d) and (9.44) is named the alpha–omega dynamo.

In the foregoing, two different types of generation circle, which works preferentially is dependent on the relative magnitude of the first to second terms on the right-hand side of equation (9.41). We denote the reference values of α, the length, and the angular velocity of the mean fluid motion by α_R, L_R, and Ω_R, respectively. The relative magnitude is given by

$$D_{\alpha\omega} = \frac{\alpha_\mathrm{R}}{L_\mathrm{R}\Omega_\mathrm{R}}. \tag{9.45}$$

In solar magnetic fields, the differential rotation is prominent specifically near the bottom of the convective zone, as was noted in §9.2.2. In the past study of solar fields, much attention was paid to the alpha–omega dynamo in close relation to the solar polarity reversal [2.9, 9.10].

References

[9.1] Glatzmaier G A and Roberts P H 1995 *Nature* **377** 203
[9.2] Kageyama A and Sato T 1993 *Phys. Fluids B* **5** 2793
[9.3] Kitauchi H and Kida S 1998 *Phys. Fluids* **10** 457
[9.4] Sarson G R, Jones C A and Longbottom A W 1998 *Geophys. Astrophys. Fluid Dyn.* **88** 225
[9.5] Katayama J S, Matsushima M and Honkura Y 1999 *Phys. Earth Planet. Inter.* **111** 141
[9.6] Busse F H G 2000 *Ann. Rev. Fluid Mech.* **32** 383
[9.7] Zhang K and Schubert G 2000 *Ann. Rev. Fluid Mech.* **32** 409
[9.8] Yoshizawa A and Yokoi N 1996 *Phys. Plasmas* **3** 3604
[9.9] Schou J *et al* 1998 *Astrophys. J.* **505** 390
[9.10] Yoshimura H 1983 *Astrophys. J. Suppl.* **52** 36

Chapter 10

Application to Astro/Geophysical and Fusion Dynamos

10.1 Solar Magnetic Fields

In §2.2 we summarized some representative observational properties associated with sunspots [2.7, 2.8]. In this section, we shall consider how those properties may be interpreted with the aid of the findings by the mean-field dynamo model given in chapter 9, specifically, the model based on the dominant/cross-helicity concept [10.1, 10.2].

10.1.1 Sunspot's Magnetic Field

Sunspots represent the cross-sections of an intense toroidal magnetic-field tube when it rises up owing to buoyancy forces and breaks through the photosphere, the thin layer adjacent to the outer boundary of the convective zone. In the light of equation (9.22), the mean toroidal magnetic field B_ϕ is related to the mean toroidal velocity U_ϕ as

$$B_\phi = \frac{C_\gamma}{C_\beta} \frac{W}{K} U_\phi \tag{10.1}$$

in the spherical coordinates (r, θ, ϕ) (figure 6.2), where use has been made of equation (9.6).

One of the prominent solar polarity properties is that the polarity of a pair of sunspots is opposite in the northern and southern hemispheres (see figure 2.8). This property signifies that the direction of the toroidal magnetic field in the convective zone is opposite in the two hemispheres. We may seek its cause in the turbulent cross helicity W in equation (10.1). The quantity changes its sign under the reflection of a coordinate system, namely, it is a pseudo-scalar.

In rotating spherical objects such as the sun and the earth, the axis of rotation is the sole factor distinguishing between the northern and southern hemispheres in a dynamical sense. Scalars are statistically symmetric with respect to the equatorial plane, but pseudo-scalars become statistically anti-symmetric. Then we have

$$W(r, \pi - \theta) = -W(r, \theta), \tag{10.2}$$

where the dependence on ϕ was dropped since the magnetohydrodynamic (MHD) state is assumed to be axisymmetric. The same situation holds for the turbulent residual helicity H as

$$H(r, \pi - \theta) = -H(r, \theta). \tag{10.3}$$

From equations (10.1) and (10.2), the toroidal field B_ϕ is anti-symmetric with respect to the equatorial plane, resulting in the sunspot's polarity depicted in figure 2.8.

Let us examine equation (10.1) from a quantitative viewpoint. A typical velocity associated with the solar rotational motion is the equatorial speed, which is about 2000 m s^{-1}. Then we adopt

$$U_\phi \cong 1000\,\mathrm{m\,s^{-1}}, \qquad \frac{C_\gamma}{C_\beta} \cong 1 \tag{10.4}$$

[see equation (8.46a) for the latter]. Equations (10.1) and (10.4) give

$$B_\phi = 10^3 \frac{W}{K} \quad (\mathrm{m\,s^{-1}}). \tag{10.5}$$

The magnetic field in original Gauss units, B_ϕ^*, is related to B_ϕ as

$$B_\phi^* = 0.4 \times 10^{-12} \sqrt{n} B_\phi \quad (\mathrm{G}), \tag{10.6}$$

from expression (6.25), where n (m^{-3}) is the number density of hydrogen [2.7]. We combine equation (10.5) with equation (10.6), and have

$$B_\phi^* = 0.4 \times 10^{-9} \frac{W}{K} \sqrt{n} \quad (\mathrm{G}). \tag{10.7}$$

Concerning the magnitude of W/K, we have the strong constraint, equation (8.37). In the convective zone, the fluid is highly electrically conducting owing to the high temperature, resulting in large magnetic Reynolds number in addition to large Reynolds number. This situation suggests that the correlation between velocity and magnetic field is not low. Then we assume

$$\frac{|W|}{K} = O(10^{-1}), \tag{10.8}$$

which results in

$$B_\phi^* = O(10^{-11}) \sqrt{n} \quad (\mathrm{G}). \tag{10.9}$$

As a typical magnetic field of large sunspots, we choose $B_\phi^* = O(10^3)$ G, which gives

$$n = O(10^{28})\,\mathrm{m}^{-3}. \tag{10.10}$$

From observations [2.7], n is estimated as

$$n = O(10^{32})\,\mathrm{m}^{-3} \text{ in the core,} \tag{10.11a}$$

$$n = O(10^{23})\,\mathrm{m}^{-3} \text{ in the photosphere.} \tag{10.11b}$$

Equation (10.10) falls between these two values. Then equations (10.1) and (10.8) are consistent with the occurrence of the toroidal field of $O(10^3)$ in sunspots.

10.1.2 Relationship of Sunspot's Polarity with Polar Field

As is shown in figure 2.8, the polarity of the leading sunspot is coincident with the polarity of the polar field (the magnetic field in the pole region). In order to see this relationship, we consider equation (9.17) in the pole region. There we divide $\boldsymbol{\Omega}$ into the uniform rotation part $\boldsymbol{\omega}_{\mathrm{F}}$ and the deviation from it, $\boldsymbol{\Omega}_{\mathrm{D}}$, as

$$\boldsymbol{\Omega} = 2\boldsymbol{\omega}_{\mathrm{F}} + \boldsymbol{\Omega}_{\mathrm{D}} \tag{10.12}$$

(subscript D denotes the differential-rotation part). In the pole region, $\boldsymbol{\Omega}_{\mathrm{D}}$ is small, compared with its counterpart near the equatorial plane, and the rotational velocity $\boldsymbol{U}$ itself is low. Then equation (9.17) is approximated by

$$\nabla \times (\alpha\boldsymbol{B} - \beta\boldsymbol{J} + 2\gamma\boldsymbol{\omega}_{\mathrm{F}}) = 0. \tag{10.13}$$

It is satisfied by

$$\boldsymbol{B} = -\frac{2\gamma}{\alpha}\boldsymbol{\omega}_{\mathrm{F}} \tag{10.14}$$

since this $\boldsymbol{B}$ leads to vanishing of $\boldsymbol{J}$ (the spatial derivatives of α and γ are neglected). Equation (10.14) shows that the polar field is aligned with the rotation axis. Such alignment was really observed in the computer simulation [10.3].

We examine the polarity rule with the aid of equation (10.14). With the left side of figure 2.8 in mind, we consider

$$\text{Northern hemisphere:} \quad W > 0, \quad H > 0, \tag{10.15a}$$

$$\text{Southern hemisphere:} \quad W < 0, \quad H < 0, \tag{10.15b}$$

which is equivalent to

$$\text{Northern hemisphere:} \quad \gamma > 0, \quad \alpha > 0, \tag{10.16a}$$

$$\text{Southern hemisphere:} \quad \gamma < 0, \quad \alpha < 0, \tag{10.16b}$$

from equations (8.30) and (8.32). The assumption about the sign of H is based on the computer experiment mimicking the solar convection zone. It suggests that $-\boldsymbol{u}\cdot\boldsymbol{\omega}$ and $\boldsymbol{b}\cdot\boldsymbol{j}$ tend to be positive in the northern hemisphere [10.4]; namely, the residual helicity $-\boldsymbol{u}\cdot\boldsymbol{\omega} + \boldsymbol{b}\cdot\boldsymbol{j}$ tends to be positive there.

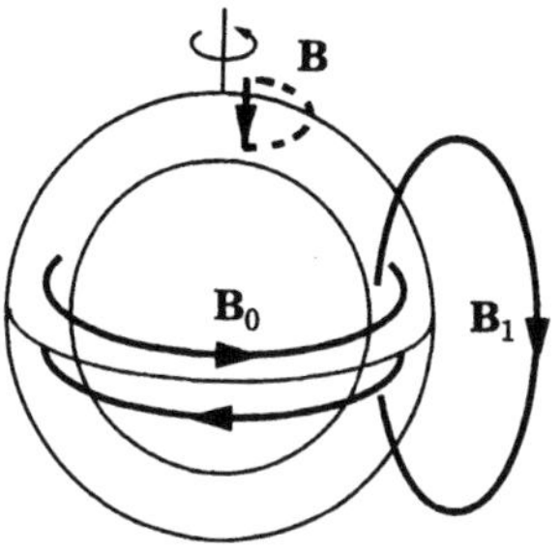

Figure 10.1. Configuration of generated magnetic fields.

From equation (10.1) the positive toroidal magnetic field B_ϕ is induced in the northern hemisphere under condition (10.15). Rising-up of the loops of the field under the buoyancy effect leads to the sunspot's polarity at the left side of figure 2.8. In the pole region, equation (10.14) shows that the negative poloidal field or the field anti-parallel to the rotation axis is induced. This polarity is coincident with the leading-sunspot polarity.

From equation (9.24) $\boldsymbol{J}_1$ is symmetric across the equatorial plane since $\alpha\gamma > 0$. The field $\boldsymbol{B}_1$ occurring from $\boldsymbol{J}_1$ under the Ampère law is anti-symmetric; namely it is of dipole type with the axis parallel to the axis of rotation [recall the reference to the dipole field below equation (9.27)]. The polar field is anti-parallel to the rotation axis. The configuration of these fields is schematically summarized in figure 10.1.

The right side of figure 2.8 corresponds to

$$\text{Northern hemisphere:} \quad W < 0, \quad H > 0, \tag{10.17a}$$

$$\text{Southern hemisphere:} \quad W > 0, \quad H < 0. \tag{10.17b}$$

Namely, the transition of the left-side to right-side polarity occurs through the change of the sign of W, but not H. This conclusion is reasonable from the fact that the sign of the residual helicity $-\boldsymbol{u}\cdot\boldsymbol{\omega} + \boldsymbol{b}\cdot\boldsymbol{j}$ is invariant under the reversal of the sign of $\boldsymbol{b}$, that is, $\boldsymbol{b} \rightarrow -\boldsymbol{b}$.

10.1.3 Lorentz Force and Meridional Flow

One of the representative flows observed at the solar surface is the meridional flow or circulation [2.7, 10.5–10.7]. The flow, which is of 20–30 m s^{-1}, is nearly stationary and is towards the poles. It is considered to be a part of large convection cells.

In §10.1.1 and §10.1.2, attention was focused on effects of turbulence on the mean magnetic field. One of the merits of introducing cross-helicity effects is that feedback effects of induced magnetic fields on fluid motion may be treated efficiently. They occur twofold; one is the effect through the Reynolds stress R_{ij}, and another is the Lorentz force $\boldsymbol{J} \times \boldsymbol{B}$.

We examine the poleward meridional flow from the viewpoint of the Lorentz force exerted to the fluid motion. We approximate $\boldsymbol{B}$ and $\boldsymbol{J}$ by $\boldsymbol{B}_0$ [equation (9.22)] and $\boldsymbol{J}_0$ [equation (9.23)]. We retain the solid-rotation part in equation (10.12), and have

$$\boldsymbol{J}_0 \times \boldsymbol{B}_0 = \left(\frac{\gamma}{\beta}\right)^2 r\omega_{\mathrm{F}}^2(-2\sin^2\theta, -\sin 2\theta, 0), \tag{10.18}$$

in the spherical coordinates (r, θ, ϕ), which consists of the r and θ components. Of these components, what is interesting in relation to the meridional flow is the latter, that is,

$$(\boldsymbol{J}_0 \times \boldsymbol{B}_0)_\theta = -\left(\frac{\gamma}{\beta}\right)^2 r\omega_{\mathrm{F}}^2 \sin 2\theta. \tag{10.19}$$

Irrespective of the sign of W in γ, equation (10.19) is negative (positive) in the northern (southern) hemisphere. This force contributes to the driving of fluid towards the poles in both the hemispheres.

The velocity of observed meridional flow is approximately proportional to [10.5]

$$C \sin\theta - \sin 2\theta, \tag{10.20}$$

with

$$C \cong 0.2. \tag{10.21}$$

Namely, the velocity is nearly proportional to $\sin 2\theta$. It is noteworthy that such θ dependence is coincident with its Lorentz-force counterpart, equation (10.19). This fact indicates that the force is a candidate for the cause of the meridional flow.

10.1.4 Mean-Field-Theory Interpretation of Polarity Reversal

We use equations (9.22)–(9.24) and seek a cause of the polarity reversal from the viewpoint of mean-field theory. We examine an initial stage in which the magnetic field $\boldsymbol{B}$ is weak, with positive (negative) turbulent cross helicity W in the northern (southern) hemisphere [condition (10.15)]. In this stage, the alpha effect $\alpha\boldsymbol{B}$ is considered smaller than the cross-helicity effect $\gamma\boldsymbol{\Omega}$, owing to weak $\boldsymbol{B}$. Such a situation is described by equations (9.22) and (9.23) in a quasi-stationary sense. The alpha effect, in turn, induces the toroidal current $\boldsymbol{J}_1$ through equation (9.24), resulting in the occurrence of the poloidal field $\boldsymbol{B}_1$ from the Ampère law (see figure 10.1).

In the foregoing generation process of the mean magnetic field, we pay attention to the change of the cross helicity. The cross helicity of the mean field given by equations (9.22) and (9.23) is $\boldsymbol{U} \cdot \boldsymbol{B}_0$. In figure 10.1 the poloidal field $\boldsymbol{B}_1$ generates the poloidal current $\boldsymbol{J}_2$ through the alpha effect. This current is aligned with $\boldsymbol{B}_1$ from the condition (10.15). The occurrence of such $\boldsymbol{J}_2$ leads to the toroidal field $\boldsymbol{B}_2$ under the Ampère law, and the direction is the same as $\boldsymbol{B}_0$. As a result, the cross helicity of the mean field increases as

$$\boldsymbol{U} \cdot \boldsymbol{B}_0 \rightarrow \boldsymbol{U} \cdot \boldsymbol{B}_0 + \boldsymbol{U} \cdot \boldsymbol{B}_2, \tag{10.22}$$

in the northern hemisphere.

It was noted in §6.1.3 that the total amount of cross helicity is conserved in the absence of molecular viscosity and magnetic diffusivity so long as there is no supply of cross helicity by external effects. Nonvanishing molecular viscosity and magnetic diffusivity give rise to the decrease in the total amount. In the solar convective zone, the buoyancy term in equation (8.18) contributes to the supply of cross helicity, enabling the existence of the quasi-stationary state of the total amount of cross helicity. In the northern hemisphere, the increase in the cross helicity of the mean field given by equation (10.22) needs to be balanced with the decrease in the cross helicity of the turbulent part (this situation is similar to the decrease in the turbulent energy under the rapid growth of the mean velocity in the electrically nonconducting case). In other words, the occurrence of $\boldsymbol{B}_2$ gives rise to the destruction of W, which also means the destruction of $\boldsymbol{B}_0$ since the latter is generated from $\boldsymbol{U}$ through the intermediary of W. The weakening of $\boldsymbol{B}_0$ leads to the decrease in the cross helicity of the mean field, which, in turn, brings the increase in W connecting $\boldsymbol{B}_0$ and $\boldsymbol{U}$. These features of the cross helicity suggest a periodic variation of the solar mean field. Its more detailed investigation into the polarity reversal may be done with the aid of the production term of W, equation (8.41) [10.2].

10.2 Geomagnetic Fields

10.2.1 Computer Simulation of Geodynamo

In §9.1.1 we mentioned the accomplishments by computer simulations mimicking the geodynamo. The most prominent feature there is the occurrence of the elongated flow structure along the rotation axis. Its details, however, differ from one computer simulation to another. Such a difference is considered to arise from the relative strength of the buoyancy to Coriolis effects, that is, the relative magnitude of R_a (Rayleigh number) to T_a (Taylor number). In the earth's outer core, R_a and T_a are $O(10^{15})$ and $O(10^{26})$, respectively [see equation (6.46)].

The elongated flow structure in the simulation by Glatzmaier and Roberts [2.10] with $R_a = O(10^7)$ and $T_a = O(10^{11})$ is more irregular, compared with that by Kageyama and Sato [2.11] with $R_a = O(10^4)$ and $T_a = O(10^6)$. This finding suggests that larger T_a results in more irregular elongated structures. In these two simulations, it is concluded that the generation process of magnetic fields is closer to the alpha–omega dynamo of §9.3.2. The simulation by Olson *et al* [2.12] with $R_a = O(10^2)$ and $T_a = O(10^8)$ is closer to the earth-like condition in the relative magnitude of T_a to R_a. There the magnetic-field generation process resembles the alpha–alpha dynamo of §9.3.1.

The molecular magnetic Prandtl number P_{rM} is an indicator giving the relative strength of the molecular momentum diffusion to the magnetic counterpart. From the viewpoint of molecular diffusion, $P_{rM} > 1$ is necessary for the spontaneous growth of small magnetic perturbations [10.8]. In the earth's outer core, P_{rM} is $O(10^{-6})$, signifying that the magnetic field is much more dissipative, compared with the momentum. This situation seems to lead to the conclusion that the geomagnetic energy is much smaller than the kinetic energy. The geomagnetic energy, on the contrary, is inferred to be $O(10^4)$–$O(10^6)$ times the kinetic energy, as was noted in §2.2. In the current simulations, the energy of the generated magnetic field is $O(1)$–$O(10^2)$ times larger than the energy of fluid motion in the rotating frame. This result is consistent with the above inference. In the simulations, however, P_{rM} is usually chosen to be larger than one; namely, the fluid motion is more dissipative. The simulation by Olson *et al* adopting $P_{rM} = 1$ is closer to the outer-core situation, but it is still much larger than its real value.

With this reservation concerning the nondimensional parameters, the computer-simulation findings about the geodynamo may be summarized as follows.

(i) A prominent feature of MHD flows in a spherical-shell region is the occurrence of the column-like or elongated structure owing to the frame-rotation or Coriolis-force effect. Increasing T_a with fixed R_a, however, leads to larger deformation or higher irregularity of the structure.
(ii) The flow inside the column is linked with the generation of the poloidal component of magnetic field, specifically, the dipole one.
(iii) The energy of induced magnetic fields is much larger than the kinetic energy of the flow driven by the buoyancy force.
(iv) The toroidal magnetic field is stronger than the poloidal magnetic field.

The mean-field theory is founded on the ensemble averaging or the averaging around the rotation axis. Then it cannot detect the elongated flow structure consisting of a few pairs of convection columns. It should be stressed that the concern of the theory as to the geodynamo is the resultant fluid motion, the resultant turbulent helicity, etc., after the averaging of flow [10.9].

10.2.2 Saturation of Generated Magnetic Field

In §9.1 we discussed the MHD state subject to dominant effects of helicity. In this state, the mean magnetic field $\boldsymbol{B}$ obeys equation (9.3) that is linear in $\boldsymbol{B}$.

The saturation level of $\boldsymbol{B}$ is determined by equation (9.9) through the interaction with the mean flow $\boldsymbol{U}$. In the mean-field theory based on the ensemble averaging, the fluid motion consists of the axisymmetric poloidal and toroidal components, $\boldsymbol{U}_{\rm p}$ and $\boldsymbol{U}_{\rm t}$, which come from the ensemble average of the flow in the column structure. They are written as

$$\boldsymbol{U}_{\rm p} = U_{\rm r}(r,\theta)\boldsymbol{e}_{\rm r} + U_\theta(r,\theta)\boldsymbol{e}_\theta, \tag{10.23}$$

$$\boldsymbol{U}_{\rm t} = U_\phi(r,\theta)\boldsymbol{e}_\phi, \tag{10.24}$$

respectively, in the spherical coordinates (r,θ,ϕ). Equation (9.9) indicates that $\boldsymbol{B}_{\rm p}$ is determined by $\boldsymbol{U}_{\rm p}$, as is consistent with item (ii) in §10.2.1. Specifically, $\boldsymbol{U}_{\rm p}$ is reflectionally symmetric with respect to the equatorial plane. Then $\boldsymbol{B}_{\rm p}$ is reflectionally anti-symmetric since W in γ is a pseudo-scalar, signifying that $\boldsymbol{B}_{\rm p}$ contains the dipole component.

From equation (9.9) the ratio of the magnetic energy $T_{\rm M}$ to the flow energy $T_{\rm K\to F}$ is

$$\frac{T_{\rm M}}{T_{\rm K\to F}} = \frac{\boldsymbol{B}^2/2}{\boldsymbol{U}^2/2} = \left(\frac{\beta}{\gamma}\right)^2 = \left(\frac{C_\beta}{C_\gamma}\right)^2\left(\frac{K}{W}\right)^2. \tag{10.25}$$

It is written as

$$\frac{T_{\rm M}}{T_{\rm K\to F}} \cong \left(\frac{K}{W}\right)^2 \geq 1, \tag{10.26}$$

with the aid of equations (8.37) and (8.46a). The energy of the induced magnetic field is larger than the energy of the fluid motion that is the generator of the former.

In order to estimate the magnitude of $T_{\rm M}/T_{\rm K\to F}$, we consider the toroidal and poloidal components of the magnetic field, $\boldsymbol{B}_{\rm t}$ and $\boldsymbol{B}_{\rm p}$. The flow velocity in the outer core is inferred to be $O(10^{-4})\,{\rm m\,s^{-1}}$ through indirect observations, for instance, the westward immigration of magnetic fields. It corresponds to $\boldsymbol{U}_{\rm t}$, and $\boldsymbol{U}_{\rm p}$ is inferred to be by one order smaller than $\boldsymbol{U}_{\rm t}$. Then we assume

$$|\boldsymbol{U}_{\rm t}| = 10^{-4}\,{\rm m\,s^{-1}}, \qquad |\boldsymbol{U}_{\rm p}| = 10^{-5}\,{\rm m\,s^{-1}}. \tag{10.27}$$

The primary part of the poloidal field is the dipole component, whose magnitude at the surface is $O(1)\,{\rm G}$. We may write the poloidal field in original Gauss units, $\boldsymbol{B}_{\rm p}^*$, as

$$|\boldsymbol{B}_{\rm p}^*| = |\sqrt{\rho\mu_0}||\boldsymbol{B}_{\rm p}| = O(1)\,{\rm G}, \tag{10.28}$$

from expression (6.25). We use the physical parameters [2.5]

$$\rho = 0.8\times 10^4\,{\rm kg\,m^{-3}}\ \text{(iron)}, \tag{10.29a}$$

$$\mu_0 = 1.3\times 10^{-6}\,{\rm henry\,m^{-1}}\ \text{(vacuum)}. \tag{10.29b}$$

Under equation (10.29), equation (10.28) gives

$$|\boldsymbol{B}_{\mathrm{p}}| = 10^{-3}\,\mathrm{m\,s^{-1}}. \tag{10.30}$$

From the latter of equation (10.27), this suggests

$$\frac{|W|}{K} = O(10^{-2}) \tag{10.31}$$

in the outer core. On the other hand, the former gives

$$|\boldsymbol{B}_{\mathrm{t}}| = \frac{C_\gamma}{C_\beta}\frac{W}{K}|\boldsymbol{U}_{\mathrm{t}}| = O(10)\,\mathrm{G}, \tag{10.32}$$

which is consistent with the observational inference about the toroidal field [recall item (iv) in §10.2.1].

In relation to equation (10.31) we should recall equation (10.8) for the solar convective zone. The zone mainly consists of hydrogen gases that are highly ionized owing to high temperature. On other hand, the outer core consists of melted iron, and the magnetic Prandtl number $P_{r\mathrm{M}}$ is much lower than one. This fact indicates that the correlation between the magnetic field and the velocity, which may be characterized by W, is lower than its solar counterpart. Considering the difference between these two regions, equation (10.31) is reasonable, and equation (10.26) results in

$$\frac{T_{\mathrm{M}}}{T_{\mathrm{F}}} = O(10^4), \tag{10.33}$$

in correspondence to item (iii) in §10.2.1.

From the viewpoint of the molecular magnetic Prandtl number $P_{r\mathrm{M}}$, a conclusion such as equation (10.33) is rather curious since $P_{r\mathrm{M}} \ll 1$, as was stated in §10.2.1. In a turbulent MHD state, however, the turbulent magnetic Prandtl number

$$P_{r\mathrm{MT}} = \frac{\nu_{\mathrm{T}}}{\beta} \tag{10.34}$$

is more important. From equation (8.28), we have

$$P_{r\mathrm{MT}} = \tfrac{7}{5}, \tag{10.35}$$

in sharp contrast to $P_{r\mathrm{M}} \ll 1$. This reversal of magnitude between $P_{r\mathrm{M}}$ and $P_{r\mathrm{MT}}$ may be considered to be a cause of $T_{\mathrm{M}} \gg T_{\mathrm{F}}$.

10.2.3 Frame-Rotation Effect on Magnetic Field

In item (i) of §10.2.1 it was stated that effects of frame rotation occur twofold; one leads to the formation of convection columns, and the other gives rise to their deformation or irregularity. In the present mean-field theory, the former signifies the importance of the resultant helicity, whereas the latter may be interpreted as the disturbance due to the Coriolis force. The explicit effect of frame rotation on the magnetic field in the dominant-helicity state occurs through the second term of the right-hand side of equation (9.5). It gives rise to the nonvanishing Lorentz force

$$\boldsymbol{J} \times \boldsymbol{B} = -2\frac{\gamma}{\beta}\boldsymbol{B} \times \boldsymbol{\omega}_{\mathrm{F}}, \tag{10.36}$$

resulting in parts of the third term on the right-hand side of equation (9.8).

As the magnetic field $\boldsymbol{B}$ approaches the saturation level given by equation (9.9), both the original frame-rotation effect and the turbulent diffusion effect weaken in equation (9.8), owing to the feedback effect by the generated magnetic field. The weakening of the frame-rotation effect at the level of mean motion is considered to be linked with that of a clear column-like structure before averaging. In short, the Coriolis force combined with the Lorentz force may play a role different from the Coriolis effect in the absence of magnetic field. This may be considered to be a cause of the T_a effect that was noted in item (i) of §10.2.1. The picture presented here for the velocity and magnetic-field interaction processes is summarized in figure 10.2.

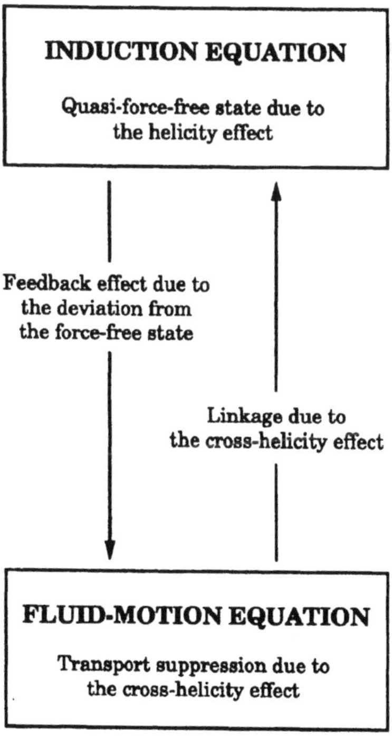

Figure 10.2. Magnetic-field generation and feedback effect on fluid motion.

10.3 Collimation of Accretion-Disk Jets

10.3.1 Computer Simulation and Mean-Field Theory

In §2.3 we remarked that high-speed jets are ubiquitously observed around high-mass astronomical objects such as galactic nuclei, protostars, binary X-ray sources, etc. [2.17, 2.18, 10.10]. These jets are often flows of electrically conducting gases and are highly collimated. As the representative interests of accretion-disk jets, we may mention their generation and collimation mechanisms.

A computer simulation based on a system of primitive MHD equations is a powerful tool for the investigation into flows around an accretion disk. The generation mechanism of jets has been explored by computer simulations [10.11–10.13]. There magnetic field lines threading through the disk are twisted by accreting, rotating gases, and the reconnection of these lines is the cause of the force driving the jets in the two directions normal to the disk. At present, the simulation of the occurrence of jets is limited to their initial stage since the simulation of their stationary behavior needs a big computational domain and a large amount of computational time. A similar difficulty is encountered in the simulation of jet collimation.

With these situations of a computer simulation in mind, it is meaningful to examine accretion-disk jets, specifically, the collimation mechanism of jets from the viewpoint of the mean-field theory [10.14]. In the mean-field model developed above, no effect of fluid compressibility is taken into account. In §2.3 and §5.3.2 we noted that fluid compressibility has influence on the suppression of turbulence. This fact signifies that the growth of jet width is suppressed by the compressibility effect. Therefore it is necessary to seek the relationship of jet collimation with both compressibility and magnetic-field effects. Moreover, relativistic effects become critical for the jets related to active galactic nuclei whose speed approaches light speed. In what follows, we shall confine ourselves to the magnetic effect and seek the collimation mechanism.

10.3.2 Driving Force of Bipolar Jets

To understand the later analysis of jet collimation, we simply refer to the toroidal-field generation process due to the cross-helicity effect in disk geometry. We adopt the cylindrical coordinates (σ, ϑ, z) (figure 10.3). Accretion disks obey the so-called Keplerian motion, whose angular velocity is proportional to $\sigma^{-1/2}$ and is highly differential. Then we adopt the inertial frame, as in §9.2.1. The mean magnetic field $\boldsymbol{B}$ is governed by equation (9.17), that is,

$$\nabla \times (\boldsymbol{U} \times \boldsymbol{B} + \alpha \boldsymbol{B} - \beta \boldsymbol{J} + \gamma \boldsymbol{\Omega}) = 0. \tag{10.37}$$

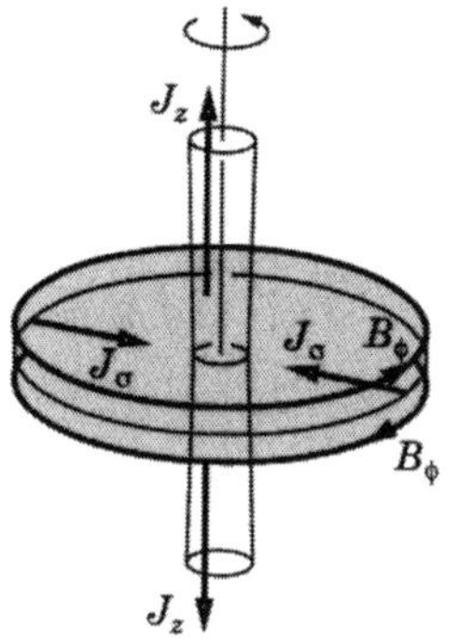

Figure 10.3. Magnetic-field configuration in an accretion disk.

An accretion disk is thin in the vertical or z direction, compared with its horizontal extent. Owing to the absence of the vertical flow comparable to the horizontal motion, there is little room for the occurrence of strong helicity effects. Then we neglect the helicity effect in equation (10.37) and have

$$\mathbf{B} = \frac{\gamma}{\beta}\mathbf{U}, \tag{10.38}$$

which is the same as equation (9.22). Equation (10.38) shows that the rotational motion generates the toroidal field B_ϑ in the presence of non-vanishing β and γ. The sign of γ proportional to W is opposite in the upper and lower halves of the disk since W is a pseudoscalar. In the case of positive W in the upper half, the toroidal field is shown in figure 10.3.

The other important feature in relation to equation (10.38) is the occurrence of the radial electric current towards the center of the disk, J_σ. Parts of J_σ become the current along the rotation axis, J_z. This J_z generates the toroidal field under the Ampère law, which gives rise to a strong magnetic pressure near the central part in the disk and contributes to the driving of jets [10.15, 10.16].

The foregoing discussions are based on nonvanishing K and W. We need to consider their sustainment mechanism. In equation (8.38) they are generated by P_K [equation (8.39)] and P_W [equation (8.41)], in both of which the buoyancy effects are dropped. In the stationary state, we have equation (9.30), that is, vanishing $\mathbf{E}_\mathrm{M}$. Then we have

$$P_K = \frac{1}{2}\nu_\mathrm{T}\left[1 - \left(\frac{\gamma}{\beta}\right)^2\right] s_{ij}^2 \cong \nu_\mathrm{T}\left[\sigma\left(\frac{\partial}{\partial\sigma}\frac{U_\vartheta}{\sigma}\right)\right]^2, \tag{10.39}$$

$$P_W = \frac{W}{K}P_K, \tag{10.40}$$

from equations (9.31), (9.34), and (9.35). In the second relation of equation (10.39), the γ/β-related part was neglected owing to the reason explained below. The Keplerian rotation leads to nonvanishing P_K and sustains K. As a result, W is also sustained under equation (10.40).

In this context, we refer to $|W|/K$ in galactic magnetic fields. Galaxies are rotating with constant velocity in their great portion, except the central part. The magnitude of magnetic fields observed in each galaxy is nearly proportional to its rotational speed. The application of equation (10.38) to this relation gives the estimate [10.17]

$$\frac{|W|}{K} = O(10^{-1}), \tag{10.41}$$

which is similar to its solar counterpart, equation (10.8). The second relation in equation (10.39) is guaranteed under equation (10.41).

The motion of gases is turbulent inside the disk. Then jets carry away nonvanishing K and W from the disk region, in addition to the angular momentum associated with the Keplerian rotation. As a result, the jet region continues to be in an MHD turbulent state even if K and W are not produced through P_K and P_W inside the jet. This point will become important in the later discussion on the jet collimation.

10.3.3 Collimation Mechanism Due to Magnetic Effect

We focus attention on the intermediate region far from the top of the jet and the root adjacent to the disk, and consider its stationary MHD state. Vanishing of $\partial \boldsymbol{B}/\partial t$ is guaranteed by

$$\boldsymbol{J} = \frac{\alpha}{\beta}\boldsymbol{B} + \frac{\gamma}{\beta}\boldsymbol{\Omega} + \frac{1}{\beta}\boldsymbol{U} \times \boldsymbol{B}, \tag{10.42}$$

as is confirmed from equation (10.37). The occurrence of bipolar jets is considered to be related to the release of the angular momentum possessed by the gases accreting onto a central high-mass body. Then jets are composed of the longitudinal and rotational motion, resulting in the helicity at the mean-flow level. This situation resembles the motion inside each convection column discussed in geodynamo, and indicates the importance of helicity effects in the study of jet collimation.

From equation (10.42) the mean Lorentz force is

$$\boldsymbol{J} \times \boldsymbol{B} = -\frac{\gamma}{\beta}\boldsymbol{B} \times \boldsymbol{\Omega} + \frac{1}{\beta}(\boldsymbol{U} \times \boldsymbol{B}) \times \boldsymbol{B}. \tag{10.43}$$

We now assume

$$\boldsymbol{U} \times \boldsymbol{B} = 0, \tag{10.44}$$

which reduces equation (10.43) to

$$\boldsymbol{J} \times \boldsymbol{B} = -\frac{\gamma}{\beta}\boldsymbol{B} \times \boldsymbol{\Omega}. \tag{10.45}$$

Equation (10.44) will be found to be consistent with the later discussion.

The mean rotational motion of jets is characterized by the mean vorticity $\boldsymbol{\Omega}$. In order to examine its behavior, we consider the mean-flow equation (8.18) with both the Coriolis and buoyancy effects dropped, which is given by

$$\frac{\partial \boldsymbol{U}}{\partial t} = -\boldsymbol{\nabla}\left(P + \frac{1}{2}U^2 + \frac{2}{3}K_{\mathrm{R}} + \left\langle \frac{\boldsymbol{b}'^2}{2} \right\rangle\right)$$
$$+ \boldsymbol{U} \times \boldsymbol{\Omega} + \boldsymbol{J} \times \boldsymbol{B} + \nu_{\mathrm{T}} \boldsymbol{\nabla}^2 \boldsymbol{U} - \nu_{\mathrm{M}} \boldsymbol{\nabla}^2 \boldsymbol{B} \qquad (10.46)$$

(the spatial variation of ν_{T} and ν_{M} is neglected from the assumption of local homogeneity). We substitute equation (10.45) into equation (10.46), and take the curl of the resulting equation. Then we have

$$\frac{\partial \boldsymbol{\Omega}}{\partial t} = \boldsymbol{\nabla} \times \left[\left(\boldsymbol{U} - \frac{\gamma}{\beta}\boldsymbol{B}\right) \times \boldsymbol{\Omega} + \nu_{\mathrm{T}} \boldsymbol{\nabla}^2 \left(\boldsymbol{U} - \frac{\gamma}{\beta}\boldsymbol{B}\right)\right], \qquad (10.47)$$

where use has been made of equations (8.28) and (8.29).

The stationary state of equation (10.47), that is,

$$\frac{\partial \boldsymbol{\Omega}}{\partial t} = 0 \qquad (10.48)$$

is guaranteed by

$$\boldsymbol{B} = \frac{\beta}{\gamma}\boldsymbol{U}. \qquad (10.49)$$

The growth of jet width arises from the diffusion of $\boldsymbol{\Omega}$, which is expressed by the term

$$\boldsymbol{\nabla} \times (\nu_{\mathrm{T}} \boldsymbol{\nabla}^2 \boldsymbol{U}) = \nu_{\mathrm{T}} \boldsymbol{\nabla}^2 \boldsymbol{\Omega} \qquad (10.50)$$

in equation (10.47). Under equation (10.49), this $\boldsymbol{\Omega}$ diffusion effect is canceled by the feedback effect of the generated magnetic field. Equation (10.49) also guarantees equation (10.44) previously assumed, reducing equation (10.42) to

$$\boldsymbol{J} = \frac{1}{1 - (\gamma/\beta)^2} \frac{\alpha}{\beta} \boldsymbol{B}, \qquad (10.51)$$

since $\boldsymbol{\Omega} = (\gamma/\beta)\boldsymbol{J}$ from equation (10.49).

Under equation (10.41), equation (10.51) is written as

$$\boldsymbol{J} = \frac{\alpha}{\beta}\boldsymbol{B}. \qquad (10.52)$$

This relation is a typical manifestation of the helicity or alpha effect, as is entirely similar to the geodynamo. Equation (10.51) is linear in $\boldsymbol{B}$ and cannot determine the magnitude of $\boldsymbol{B}$. The coupling with the fluid motion through the cross-helicity effect relates $\boldsymbol{B}$ to $\boldsymbol{U}$, as in equation (10.49), leading to the determination of its saturation level. This situation is also the same as for the geodynamo.

Under equation (10.49) guaranteeing the stationary state of $\boldsymbol{\Omega}$, equation (10.46) results in

$$\nabla\left(P+\frac{1}{2}U^2+\frac{2}{3}K_{\mathrm{R}}+\left\langle\frac{\boldsymbol{b}'^2}{2}\right\rangle\right)=0, \tag{10.53}$$

and the mean pressure P is determined by the Bernoulli theorem

$$P+\tfrac{1}{2}U^2+\tfrac{1}{2}K+\tfrac{1}{6}K_{\mathrm{R}}=\mathrm{const}, \tag{10.54}$$

where use has been made of $\langle \boldsymbol{b}'^2\rangle = K - K_{\mathrm{R}}$.

10.3.4 Sustainment of Turbulent State

In §10.3.3 we investigated the MHD state that is free from the diffusion effect, although the state is turbulent. The MHD turbulent state characterized by nonvanishing K and W is the most important ingredient in the present jet-collimation mechanism, as is seen from the dependence of β and γ on these two quantities. The elucidation of the mechanism of sustaining K and W is critical for understanding the jet collimation from the viewpoint of mean-field theory. As was emphasized in §8.2, we still do not have the model equation for the turbulent residual helicity H whose mathematical and physical bases are as firm as the equations for K and W. In order to avoid the uncertainty arising from this situation, we shall make the following discussion with the full use of the equations for K and W.

From equations (10.49) and (10.51), we have

$$R_{ij}=\tfrac{2}{3}K_{\mathrm{R}}\delta_{ij}, \tag{10.55}$$

$$\boldsymbol{E}_{\mathrm{M}}=0. \tag{10.56}$$

In the absence of buoyancy effects, equations (8.39) and (8.41) result in

$$P_K=P_W=0, \tag{10.57}$$

and the most typical production mechanisms of K and W are lost in the jet-collimation process sought in §10.3.3. This situation seems to contradict the foregoing discussions and should be addressed consistently.

Gases in a disk accrete onto a central high-mass body, while differentially rotating, and they are in a turbulent state, as was noted in §10.3.2. Some of those gases are ejected as bipolar jets. Then the jets are also turbulent; namely, the turbulent energy K is transported from the disk to the jet region. This transfer is the source of K in the jet.

In order to see this situation in mathematical terms, we use equations (8.42), (10.55), and (10.56), and rewrite equation (8.38) as

$$\nabla(K\boldsymbol{U}-W\boldsymbol{B})=-\varepsilon, \tag{10.58}$$

$$\nabla(W\boldsymbol{U}-K\boldsymbol{B})=-C_W\frac{K}{\varepsilon}W, \tag{10.59}$$

where the transport effects due only to the fluctuations in $\boldsymbol{T}_K$ and $\boldsymbol{T}_W$ were discarded, compared with the mean-field/dependent parts $W\boldsymbol{B}$ and $K\boldsymbol{B}$ ($|\boldsymbol{B}|$ and $|\boldsymbol{U}|$ are usually much larger than $|\boldsymbol{b}'|$ and $|\boldsymbol{u}'|$). We substitute equation (10.49) into equations (10.58) and (10.59), and have

$$\left(\frac{C_\beta}{C_\gamma} - 1\right)(\boldsymbol{U}\cdot\nabla)K = \varepsilon, \tag{10.60}$$

$$(\boldsymbol{U}\cdot\nabla)\left[\left(\frac{C_\beta}{C_\gamma}\left(\frac{K}{W}\right)^2 - 1\right)W\right] = C_W\frac{K}{\varepsilon}W, \tag{10.61}$$

where use has been made of equations (8.31) and (8.32). In this context, we should note

$$\frac{C_\beta}{C_\gamma} > 1, \tag{10.62}$$

$$\frac{C_\beta}{C_\gamma}\left(\frac{K}{W}\right)^2 > 1, \tag{10.63}$$

from equations (8.37) and (8.46a).

In accretion-disk jets, mean MHD turbulent flows are statistically axisymmetric around the rotation or z axis (figure 10.3). Considering that U_z is the main component of mean flow, we rewrite (10.60) as

$$\left(\frac{C_\beta}{C_\gamma} - 1\right)U_z\frac{\partial K}{\partial z} = \varepsilon. \tag{10.64}$$

The disk is in a turbulent state with nonvanishing K. Equation (10.64) indicates that this K is transported to the jet region, balancing with the energy dissipation ε.

In equation (10.61) we discard the spatial variation of the nondimensional quantity $(C_\beta/C_\gamma)(K/W)^2 - 1$. Then the mathematical structure of the resulting equation is similar to equation (10.64), and W is transported from the disk to the jet region.

Finally, we refer to the sustainment of the turbulent residual helicity H that is related to α as equation (8.30). In the present stage of the progress in mean-field theory, we have no reliable model equation for it, as has already been noted. This situation arises from the fact that no conservation property holds for the residual helicity $-\boldsymbol{u}\cdot\boldsymbol{\omega} + \boldsymbol{b}\cdot\boldsymbol{j}$ in ideal MHD flow. We seek a cause of nonvanishing H from the residual helicity of mean field

$$H_m = -\boldsymbol{U}\cdot\boldsymbol{\Omega} + \boldsymbol{B}\cdot\boldsymbol{J}. \tag{10.65}$$

It may be rewritten as

$$H_m = \left[\frac{C_\beta}{C_\gamma}\left(\frac{K}{W}\right)^2 - 1\right]\boldsymbol{U}\cdot\boldsymbol{\Omega} \cong \frac{C_\beta}{C_\gamma}\left(\frac{K}{W}\right)^2\boldsymbol{U}\cdot\boldsymbol{\Omega}, \tag{10.66}$$

from equations (8.31), (8.32), (10.49), and (10.52). One of the prominent characteristics of accretion-disk jets is that gases flow while rotating. This fact signifies nonvanishing of $\boldsymbol{U} \cdot \boldsymbol{\Omega}$ or H_m. As a result, it is highly probable that nonvanishing H is also sustained in the presence of nonvanishing H_m.

10.3.5 Physical Interpretation of Jet Collimation

In §10.3.3 we showed that the turbulent diffusion of the momentum and magnetic field may be suppressed under the combined cross-helicity and helicity effects. The suppression of turbulent diffusion is a cause of jet collimation. The saturation level of $\boldsymbol{B}$ is given by equation (10.49). Here we should note that the primary component of $\boldsymbol{B}$ is of the dipole type since W or γ is antisymmetric with respect to the midplane of the disk. From equations (8.37) and (10.49), we have

$$\frac{\boldsymbol{B}^2}{2} > \frac{U^2}{2}. \tag{10.67}$$

Namely, the magnetic energy is larger than the kinetic energy. Specifically, the magnetic field may become very large under equation (10.41). One of the prominent properties of a magnetic field is the force of tension, under which magnetic field lines resist bending due to fluid motion. It is highly probable that the jet possessing such an intense magnetic field inside may strongly resist bending.

10.4 Reversed-Field Pinches of Plasmas

10.4.1 Magnetic Plasma Confinement in a Torus

The representative approach to plasma confinement by magnetic fields is tokamaks in torus geometry (figure 10.4). In this approach, the toroidal magnetic field $\boldsymbol{B}_t$ generated by external coils wrapped around a torus is much stronger than the poloidal one, $\boldsymbol{B}_p$, coming from a toroidal plasma

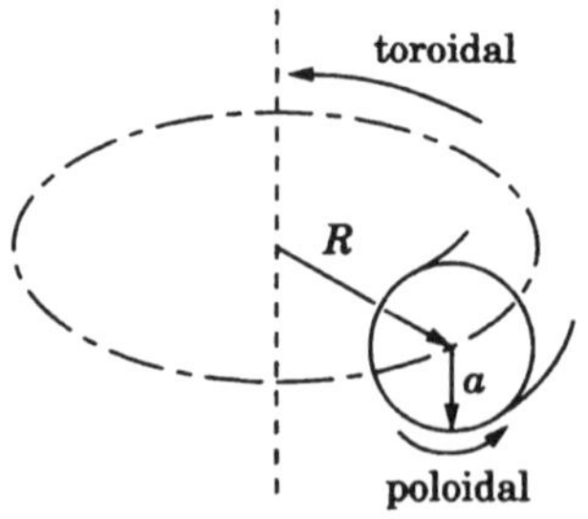

Figure 10.4. Toroidal geometry in controlled fusion.

current. Compared with the motion of plasmas along this strong toroidal field, the motion normal to it is highly suppressed owing to the tension of the toroidal field. The relationship between $\boldsymbol{B}_t$ and $\boldsymbol{B}_p$ in tokamaks is characterized by

$$q = \frac{a}{R}\frac{|\boldsymbol{B}_t|}{|\boldsymbol{B}_p|} > 1, \tag{10.68}$$

where R and a are the major and minor radii, respectively. The quantity q is called the safety factor [6.1]. From this constraint, the minor radius a cannot be made much smaller than R.

In the context of q, the confinement approach that is in a directly opposite position to tokamaks is reversed-field pinches (RFPs) of plasmas [10.18–10.20]. In this approach, the poloidal magnetic field is nearly comparable to the toroidal field. The RFP state is characterized by

$$q \ll 1, \tag{10.69}$$

and a can be chosen to be much smaller than R. In the theoretical investigation into RFPs, the torus is often approximated by a circular cylinder.

The most prominent feature of RFPs is the reversal of the toroidal magnetic field at the outer edge of the plasma. This phenomenon may be explained intuitively as follows. In the initial setting-up phase, the poloidal field generated by a strong toroidal plasma current interacts with the toroidal magnetic field, and a large deformation of the latter is induced by kink and sausage instability. Such deformation leads to the formation of the loop of a magnetic-field line because of its tension, as in figure 10.5, and the reconnection of the field line results in the reversal of the toroidal magnetic field at the outer edges of plasmas. In this situation, the continuation of plasma currents is equivalent to that of the reversed toroidal magnetic field. As a result, the alignment between the imposed current and the generated reversed magnetic field is a key ingredient of RFPs. Mathematically, this property is closely associated with equation (9.3). The process of magnetic-field reversal was examined in detail by the computer simulation of the MHD equations [10.21, 10.22].

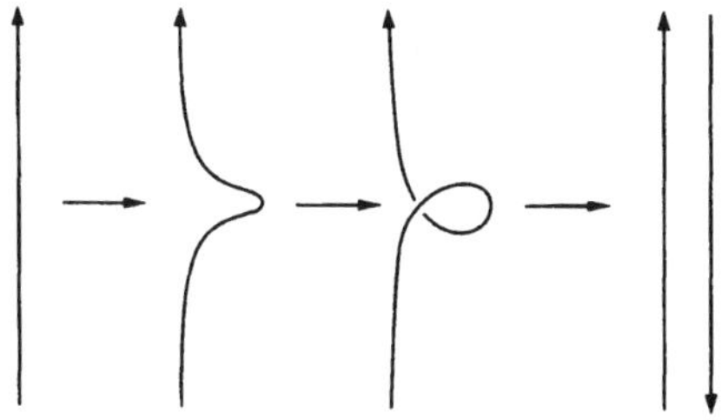

Figure 10.5. Reversal of toroidal magnetic field.

10.4.2 Derivation of Force-Free Field by Mean-Field Theory

In RFPs there is no macroscopic plasma flow. Then we have no mechanism of generating and sustaining the turbulent cross helicity, as is seen from equation (8.41) with the frame-rotation term dropped. We neglect the effect and write

$$\boldsymbol{E}_{\mathrm{M}} = \alpha \boldsymbol{B} - \beta \boldsymbol{J}. \tag{10.70}$$

From equation (10.70), the equation for the mean magnetic field $\boldsymbol{B}$ obeys

$$\frac{\partial \boldsymbol{B}}{\partial t} = \nabla \times (\alpha \boldsymbol{B} - \beta \boldsymbol{J}). \tag{10.71}$$

The solution

$$\boldsymbol{J} = \kappa_m \boldsymbol{B} \tag{10.72}$$

guarantees the stationary state of $\boldsymbol{B}$, where

$$\kappa_m = \frac{\alpha}{\beta}. \tag{10.73}$$

The close relationship between RFPs and the alpha effect in the mean-field theory was first pointed out by Gimblett and Watkins [10.23].

We seek the solution of equation (10.72) in cylindrical geometry and examine its relationship with RFPs. To this end, we employ the cylindrical coordinates (σ, ϑ, z), where ϑ and z represent the poloidal and toroidal directions, respectively. We assume the constancy of α and take the curl of equation (10.72). Then we have

$$\nabla^2 \boldsymbol{B} + \kappa_m^2 \boldsymbol{B} = 0. \tag{10.74}$$

Under the condition of the axisymmetry and z independence of $\boldsymbol{B}$, we are led to

$$\frac{\mathrm{d}^2 B_z}{\mathrm{d}\sigma^2} + \frac{1}{\sigma}\frac{\mathrm{d}B_z}{\mathrm{d}\sigma} + \kappa_m^2 B_z = 0, \tag{10.75}$$

$$B_\vartheta = -\frac{1}{\kappa_m}\frac{\mathrm{d}B_z}{\mathrm{d}\sigma}, \tag{10.76}$$

from equations (10.72) and (10.74). The solution of equation (10.75) and (10.76) is given by

$$B_z = B_0 J_0(\kappa_m \sigma), \tag{10.77a}$$

$$B_\vartheta = B_0 J_1(\kappa_m \sigma), \tag{10,77b}$$

where $B_0 = J_0(0)$, and J_n is the first-kind Bessel function of the nth order.

The Bessel function $J_n(s)$ possesses an infinite number of zero points, the s_m or the points obeying $J_n(s_m) = 0$. The first zero point s_1 is about 2.4. This fact indicates that the toroidal magnetic field reverses its sign at the edge of

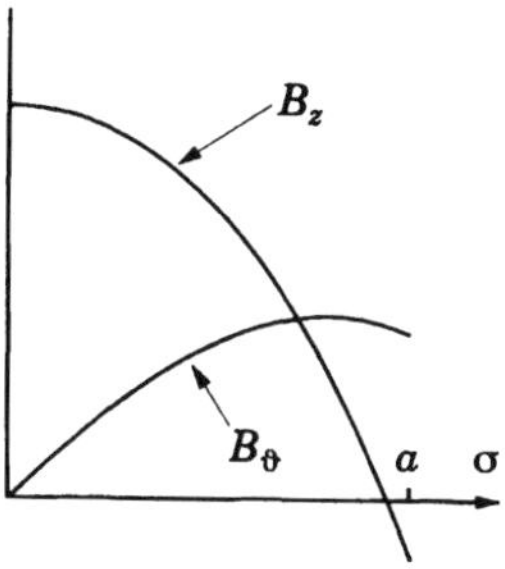

Figure 10.6. Global magnetic fields in RFPs.

plasma under the condition

$$|\kappa_m a| > 2.4. \qquad (10.78)$$

This situation is depicted schematically in figure 10.6. The solution of the Bessel-function type, equation (10.77), may capture well the feature of the global magnetic fields in RFPs and has been a guiding principle in studying RFPs experimentally and theoretically.

Equation (10.77) with constant κ_m suffers from some shortfalls near the edge of plasma. For instance, it cannot satisfy the condition of vanishing $\boldsymbol{J}$ there. Within the framework of mean-field theory, α and β are not constant, but they are determined by the equations for K, H, and ε. A model equation for H was proposed as [8.1, 8.2]

$$\frac{\partial H}{\partial t} = -C_{H1}\frac{\varepsilon^2}{K^3}\boldsymbol{B}\cdot\boldsymbol{E}_{\mathrm{M}} - C_{H2}\frac{\varepsilon}{K}H + \nabla\cdot\left(\frac{\beta}{\sigma_H}\nabla H\right), \qquad (10.79)$$

where C_{H1}, C_{H2}, and σ_H are positive constants. The mean-field theory with equation (10.79) added was confirmed to reproduce the primary features of RFPs under spatially varying κ_m [8.2].

10.4.3 Derivation of Force-Free Field by Variational Method

The force-free field, equation (10.72), was originally derived by Taylor [10.24] from an entirely different viewpoint. There RFP's were regarded as a state relaxing from an initial unstable state subject to a proper constraint. From the fact that the total amount of magnetic helicity, $\int_V \boldsymbol{a}\cdot\boldsymbol{b}\,\mathrm{d}V$, is conserved in the absence of molecular effects, Taylor considered that the final state in the relaxation process may be described by the condition

$$\text{minimum} \int_V \tfrac{1}{2}\boldsymbol{b}^2\,\mathrm{d}V \text{ under constant} \int_V \boldsymbol{a}\cdot\boldsymbol{b}\,\mathrm{d}V. \qquad (10.80)$$

Here V denotes the region occupied by plasma, and $\boldsymbol{a}$ is the vector potential related to $\boldsymbol{b}$ as $\boldsymbol{b} = \nabla\times\boldsymbol{a}$. In the comparison between these two integrals, the former contains higher-wavenumber components, owing to the relation $\boldsymbol{b} = \nabla \times \boldsymbol{a}$. As a result, the former decays faster under stronger small-scale destruction effects.

We write the condition (10.80) in the variational form

$$\delta\left[\int_V \left(\frac{\boldsymbol{b}^2}{2} - \frac{\kappa_m}{2}\boldsymbol{a}\cdot\boldsymbol{b}\right)\mathrm{d}V\right] = 0, \tag{10.81}$$

with $\boldsymbol{a}$ fixed at the surface of V, where $\kappa_m/2$ is a constant Lagrange multiplier. By partial integration, equation (10.81) is reduced to

$$\int_V (\nabla \times \boldsymbol{b} - \kappa_m \boldsymbol{b})\cdot\delta\boldsymbol{a}\,\mathrm{d}V = 0, \tag{10.82}$$

resulting in the same type of expression as equation (10.72). We should note that the constancy of κ_m is a key factor in this derivation.

10.5 Plasma Rotation in Tokamaks

A big breakthrough of tokamaks was attained through the so-called high-confinement (H) modes [2.35]. The transition of plasma state from low-confinement (L) to H modes may be characterized by a steep radial electric field and a narrow poloidal plasma rotation, just inside the separatrix [2.32, 2.33, 2.44]. Their occurrence resulting in the formation of transport barriers of thermal energy may be regarded as a kind of structure formation in plasmas [2.28]. Understanding of this mechanism is expected to pave the way for the further improvement of plasma confinement and is a central subject in the study of magnetically confined plasmas.

In sharp contrast to the foregoing edge transport barriers, the discharges with transport barriers in a core region, which are called the internal or core transport barriers, have recently attracted much attention [10.25–10.28]. Such discharges are characterized by the negative magnetic shear $\mathrm{s_M}$, that is,

$$\mathrm{s} \to \mathrm{s_M} = \frac{\sigma}{q}\frac{\mathrm{d}q}{\mathrm{d}\sigma} < 0 \tag{10.83}$$

where q is the safety factor defined by equation (10.68), and σ is the minor radial coordinate. The transport barrier in a core region is accompanied by a steep gradient of poloidal flow, as in an edge region of H modes. Here the discharge characterized by negative s is simply called reversed-shear (RS) modes. The inhomogeneous plasma rotation, which is observed in the region with minimum q, is considered to play a critical role for those modes. The internal transport barriers are very similar to the H-mode counterparts in the sense that both are accompanied by the steep spatial variation of a radial electric field and a poloidal flow.

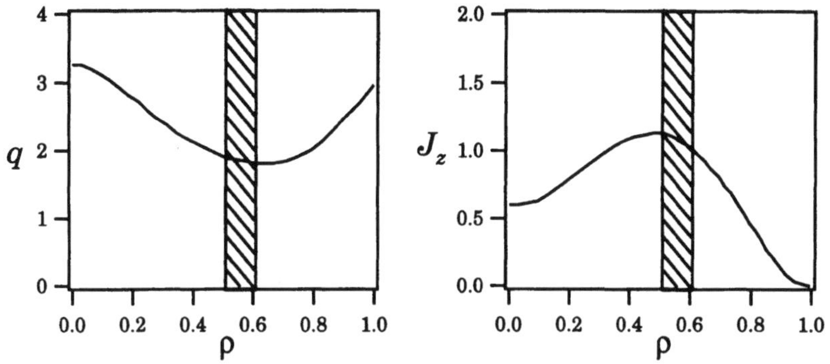

Figure 10.7. Safety factor q and current density J_z in the RS mode.

One of the prominent characteristics in the RS mode is the existence of minimum q, as in figure 10.7, where $\rho = \sigma/a$ (a is the minor radius). This q profile comes from the concave profile of the electric current density J_z that is also shown in figure 10.7 [10.29]. The transport barrier accompanied by the poloidal rotation of plasma is formed near the minimum-q point. We examine the relationship of concave J_z with the occurrence of plasma rotation, with the aid of the mean-field theory [10.30]. It should be recalled, however, that many interesting phenomena associated with electric-field effects on tokamaks are beyond the scope of the one-fluid MHD system on which the mean-field theory is founded.

We assume the axisymmetry of all statistical quantities and neglect the dependence on the toroidal (z) direction. In the mean-flow equation (8.18), we drop the Coriolis and buoyancy terms. The resulting equation and its counterpart for the mean vorticity $\mathbf{\Omega}$ are given by equations (10.46) and (10.47), respectively. In the latter equation, $\mathbf{\Omega}$ is small at the onset stage of the plasma rotation. We drop the first term on the right-hand side. The poloidal rotation is characterized by the z component of the mean vorticity, which obeys

$$\frac{\partial \Omega_z}{\partial t} = \nu_{\mathrm{T}} \nabla^2 \Omega_z - \nu_{\mathrm{M}} \nabla^2 J_z, \tag{10.84}$$

where the spatial dependence of ν_{T} and ν_{M} was neglected for simplicity of discussion.

In the absence of the second ν_{M}-related term in equation (10.84), Ω_z is subject to the resistive effect only, and there is no room for its autonomous generation. We now seek the Ω_z generation process due to the ν_{M}-related effect. For this purpose, we pick up its contribution and write

$$\frac{\partial \Omega_z}{\partial t} = -\frac{5C_\gamma}{7}\frac{K}{\varepsilon} W \nabla^2 J_z + R_{\Omega 1}, \tag{10.85}$$

where use has been made of equations (8.29) and (8.32), and $R_{\Omega 1}$ denotes the remaining contribution. Of three turbulence quantities K, ε, and W, the last quantity is a pseudoscalar that changes its sign under the reflection of the coordinate system, $\boldsymbol{x} \rightarrow -\boldsymbol{x}$. This fact indicates that W plays a critical role in the Ω_z generation since the sign of rotational motion is dependent on the choice of a coordinate system.

The quantity W is generated by P_W [equation (8.41)]. At the onset of Ω_z, the second term in equation (8.22) is primary, resulting in

$$\boldsymbol{E}_{\mathrm{M}} \cong -\beta \boldsymbol{J}. \tag{10.86}$$

We combine equation (10.86) with P_W and have

$$P_W \cong \beta J_z \Omega_z. \tag{10.87}$$

In obtaining equation (10.87), we retained only the $\boldsymbol{\Omega}$-dependent part of equation (8.41) since our interest lies in the generation process of $\boldsymbol{\Omega}$. The contribution of equation (10.87) to $\partial W / \partial t$ [equation (8.38)] is

$$\frac{\partial W}{\partial t} = \beta J_z \Omega_z + R_W = C_\beta \frac{K^2}{\varepsilon} J_z \Omega_z + R_W, \tag{10.88}$$

in correspondence to equation (10.85), where equation (8.31) was used.

We eliminate W from equations (10.85) and (10.88), and connect Ω_z directly with J_z. Here we focus attention on the temporal growth of W, and neglect the temporal change of J_z, K, and ε. Then we have

$$\frac{\partial^2 \Omega_z}{\partial t^2} - \left(-\frac{5 C_\beta C_\gamma}{7} \frac{K^3}{\varepsilon^2} J_z \nabla^2 J_z \right) \Omega_z = R_{\Omega 2}, \tag{10.89}$$

where $R_{\Omega 2}$ expresses all the remaining contributions and is not discussed here. Equation (10.89) indicates that Ω_z may grow under the condition

$$\chi_\Omega^2 \cong -\frac{5 C_\beta C_\gamma}{7} \frac{K^3}{\varepsilon^2} J_z \nabla^2 J_z > 0, \tag{10.90}$$

where χ_Ω represents its temporal growth rate.

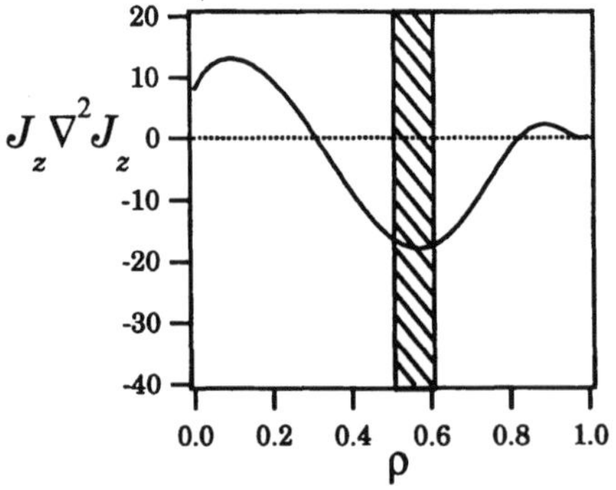

Figure 10.8. Magnitude of $J_z \nabla^2 J_z$ corresponding to figure 10.7.

We examine equation (10.90) in the light of the concave profile of J_z. The latter profile is shown in figure 10.7 [10.29], and corresponding $J_z \nabla^2 J_z$ is given in figure 10.8. There large negative $J_z \nabla^2 J_z$ occurs near the location of minimum q or at $\rho \cong 0.6$. Equations (10.89) and (10.90) suggest that the poloidal rotation starts to be driven there. This finding is consistent with the fact that a poloidal flow in RS modes is observed near the minimum-q point or the transport barrier.

10.6 Transport Suppression Due to Electric-Field Effects

In all the foregoing discussions of chapters 7–10 we assumed the quasi-neutrality of charge density, equation (6.13). Then the electric field has no direct linkage with the velocity and magnetic-field equations (6.16) and (6.31). As was noted in §10.5, however, the electric field is confirmed to be closely related to transport barriers in tokamaks, where the transports of heat and particles are highly suppressed. In what follows we shall pay attention to electric-field effects on turbulent motion and partially see the relationship between the transport suppression and the electric field [10.31]. Detailed discussions about the role of electric field will be given in Part IV (Plasma Turbulence), specifically, chapters 19 and 20.

10.6.1 Equations with Electric-Field Effects Supplemented

We abandon the quasi-neutrality assumption, equation (6.13). Then equation (6.16) is replaced with

$$\frac{\partial}{\partial t}\rho u_i + \frac{\partial}{\partial x_j}\rho u_j u_i = -\frac{\partial p}{\partial x_i} + \rho_C e_i + (\boldsymbol{j} \times \boldsymbol{b})_i + \frac{\partial}{\partial x_j}\left[\mu\left(\frac{\partial u_j}{\partial x_i} + \frac{\partial u_i}{\partial x_j} - \tfrac{2}{3}\nabla \cdot \boldsymbol{u}\delta_{ij}\right)\right]. \quad (10.91)$$

In order to see only electric-field effects on turbulence, we drop magnetic-field and molecular-diffusion effects in equation (10.91) and write

$$\frac{\partial}{\partial t}\rho u_i + \frac{\partial}{\partial x_j}\rho u_i u_j = -\frac{\partial p}{\partial x_i} + \rho_C e_i. \quad (10.92)$$

The mass density ρ and the internal energy ζ obey

$$\frac{\partial \rho}{\partial t} + \nabla \cdot (\rho \boldsymbol{u}) = 0, \quad (10.93)$$

$$\frac{\partial}{\partial t}\rho\zeta + \nabla \cdot (\rho\zeta\boldsymbol{u}) = -p\nabla \cdot \boldsymbol{u}. \quad (10.94)$$

In equation (10.94) the molecular-diffusion effect on ζ and the conversion of kinetic and magnetic energy to heat were neglected. From the second relation in equation (3.14), equation (10.94) is reduced to the equation for the pressure, that is,

$$\frac{\partial p}{\partial t} + \nabla \cdot (p\boldsymbol{u}) = -(\gamma - 1)p\nabla \cdot \boldsymbol{u}. \tag{10.95}$$

For closing equations (10.92), (10.93), and (10.95), we need the equation for the charge density ρ_{C}. From the first relation in equation (6.13) and the number-density equation

$$\frac{\partial n_{\mathrm{S}}}{\partial t} + \nabla \cdot (n_{\mathrm{S}}\boldsymbol{v}_{\mathrm{S}}) = 0, \tag{10.96}$$

we have

$$\frac{\partial \rho_{\mathrm{C}}}{\partial t} + \nabla \cdot (\rho_{\mathrm{C}}\boldsymbol{u}) = s_{\mathrm{C}}, \tag{10.97}$$

under equation (6.12), where

$$s_{\mathrm{C}} = \nabla \cdot (en_{\mathrm{E}}(\boldsymbol{v}_{\mathrm{E}} - \boldsymbol{u})). \tag{10.98}$$

In what follows, we pay attention to the fluctuation of ρ_{C}, ρ'_{C}, by turbulence motion, and neglect effects of s_{C}.

10.6.2 Analysis of Turbulent Transport Rate of Thermal Energy

10.6.2.1 Thermal-energy transport rate due to turbulence

We divide a quantity f into the ensemble-mean part F and the fluctuation around it, f'. Here f, F, and f' denote

$$f = (\rho, \boldsymbol{u}, p, \boldsymbol{e}, \rho_{\mathrm{C}}), \qquad F = (\bar{\rho}, \boldsymbol{U}, P, \boldsymbol{E}, \bar{\rho}_{\mathrm{C}}), \qquad f' = (0, \boldsymbol{u}', p', 0, \rho'_{\mathrm{C}}). \tag{10.99}$$

Here, in addition to the mass-density fluctuation ρ', we neglected the electric-field fluctuation $\boldsymbol{e}'$ since the mean electric field $\boldsymbol{E}$ is our primary concern in the light of transport barriers in tokamaks. The mean pressure P, which is equivalent to the mean internal energy through equation (3.14), obeys

$$\frac{\partial P}{\partial t} + \nabla \cdot (P\boldsymbol{U}) = \nabla \cdot (-\boldsymbol{H}_{\mathrm{P}}) - (\gamma - 1)P\nabla \cdot \boldsymbol{U}, \tag{10.100}$$

where $\boldsymbol{H}_{\mathrm{P}}$ is defined by

$$\boldsymbol{H}_{\mathrm{P}} \equiv \langle p'\boldsymbol{u}' \rangle = (\gamma - 1)\bar{\rho}\langle \zeta'\boldsymbol{u}' \rangle \tag{10.101}$$

and expresses the transport rate of thermal energy due to turbulence. The primary interest of the following discussion is to investigate the electric-field effect on $\boldsymbol{H}_{\mathrm{P}}$.

10.6.2.2 Statistical analysis of transport rate

The fluctuation f' is governed by

$$\frac{\mathrm{D}u_i'}{\mathrm{D}t} + (\boldsymbol{u}' \cdot \nabla)u_i' + (\boldsymbol{u}' \cdot \nabla)U_i = -\frac{1}{\bar{\rho}}\frac{\partial p'}{\partial x_i} + \rho_\mathrm{C}' \frac{E_i}{\bar{\rho}}, \tag{10.102}$$

$$\frac{\mathrm{D}p'}{\mathrm{D}t} + (\boldsymbol{u}' \cdot \nabla)p' + (\boldsymbol{u}' \cdot \nabla)P = -\gamma p' \nabla \cdot \boldsymbol{U}, \tag{10.103}$$

$$\frac{\mathrm{D}\rho_\mathrm{C}'}{\mathrm{D}t} + (\boldsymbol{u}' \cdot \nabla)\rho_\mathrm{C}' + (\boldsymbol{u}' \cdot \nabla)\bar{\rho}_\mathrm{C} + \rho_\mathrm{C}' \nabla \cdot \boldsymbol{U} = 0, \tag{10.104}$$

from equations (10.92), (10.95), and (10.97). Here we dropped $\nabla \cdot \boldsymbol{u}'$ in correspondence to the neglect of the density fluctuation ρ'.

We apply simplified version of the TSDIA detailed in chapter 7 to equations (10.102)–(10.104), and evaluate $\boldsymbol{H}_\mathrm{P}$. The resulting expression is reduced to the one-point form in physical space through the procedures given by equations (8.6)–(8.9). After performing the renormalization such as equation (8.5), we have an algebraic equation for $\boldsymbol{H}_\mathrm{P}$ [10.31]:

$$H_{\mathrm{P}i} + \tau_u \tau_c H_{\mathrm{P}j} \frac{\partial \bar{\rho}_\mathrm{C}}{\partial x_j} \frac{E_i}{\bar{\rho}} + \tau_u H_{\mathrm{P}j} \frac{\partial U_i}{\partial x_j} + \gamma \tau_p H_{\mathrm{P}i} \nabla \cdot \boldsymbol{U} = -\tfrac{1}{3}\tau_p \langle \boldsymbol{u}'^2 \rangle \frac{\partial P}{\partial x_i}. \tag{10.105}$$

Here the fluctuation $\boldsymbol{u}'$ was assumed to be statistically isotropic, and τ_u, τ_p, and τ_c are the characteristic time scales of $\boldsymbol{u}'$, p', and ρ_C', respectively.

10.6.3 Effect of Radial Electric Field on Thermal-Energy Transport

With tokamak plasmas in mind, we consider equation (10.105) in the cylindrical coordinates (σ, ϑ, z). We assume the axisymmetry and neglect the dependence on z. Under this approximation, the radial mean velocity U_σ may be neglected, and the radial component of equation (10.105) is written as

$$H_{\mathrm{P}\sigma} = -\lambda_\mathrm{T} \frac{\mathrm{d}P}{\mathrm{d}\sigma}. \tag{10.106}$$

The turbulent diffusivity of thermal energy, λ_T, is given by

$$\lambda_\mathrm{T} = \frac{\lambda_\mathrm{T0}}{1 + \Lambda}, \tag{10.107}$$

where

$$\lambda_\mathrm{T0} = \tfrac{1}{3}\tau_p \langle \boldsymbol{u}'^2 \rangle, \tag{10.108}$$

$$\Lambda = \frac{\tau_u \tau_c}{\bar{\rho}} E_\sigma \frac{\mathrm{d}\bar{\rho}_\mathrm{C}}{\mathrm{d}\sigma}. \tag{10.109}$$

The mean charge density $\bar{\rho}_C$ is related to the mean electric field $\boldsymbol{E}$ as

$$\bar{\rho}_C = \varepsilon_0 \nabla \cdot \boldsymbol{E} = \varepsilon_0 \left(E_\sigma E''_\sigma + \frac{1}{\sigma} E_\sigma E'_\sigma - \frac{2}{\sigma^2} E_\sigma^2 \right), \tag{10.110}$$

where $E'_\sigma = \mathrm{d}E_\sigma/\mathrm{d}\sigma$ and $E''_\sigma = \mathrm{d}^2 E_\sigma/\mathrm{d}\sigma^2$. Then equation (10.109) is expressed in terms of E_σ as

$$\Lambda = \frac{\varepsilon_0 \tau_u \tau_c}{\bar{\rho}} \left(E_\sigma E''_\sigma + \frac{1}{\sigma} E_\sigma E'_\sigma - \frac{2}{\sigma^2} E_\sigma^2 \right). \tag{10.111}$$

Equation (10.108), that is, λ_{T0}, corresponds to equation (4.32) and represents the enhancement of thermal-energy transport due to turbulence. Then positive Λ is a promising candidate for the indicator of transport suppression due to E_σ effects.

We are in a position to examine Λ in the light of transport suppressions in H modes. Figure 10.9 shows the radial electric field near the separatrix in the JFT-2M tokamak with 130 cm and 33 cm as the major and minor radii, respectively (s is measured with the separatrix as the origin) [10.32]. We use figure 10.9 and calculate

$$\Gamma_1 = E_\sigma E''_\sigma, \qquad \Gamma_2 = \frac{1}{\sigma} E_\sigma E'_\sigma, \qquad \Gamma_3 = -\frac{2}{\sigma^2} E_\sigma^2. \tag{10.112}$$

Figure 10.10 shows the Γ_n near the separatrix. From it, we may see that Γ_1, which is the combination of E_σ and E''_σ, is positive for $s < -1.5$ and dominant, compared with the other two; namely, we may write

$$\Lambda \cong \frac{\varepsilon_0 \tau_u \tau_c}{\bar{\rho}} E_\sigma E''_\sigma > 0 \quad (s < -1.5) \tag{10.113}$$

(note that $\varepsilon_0 \tau_u \tau_c / \bar{\rho}$ is positive). This finding suggests that the curvature of E_σ is linked with the suppression of the turbulent transport of thermal energy. In the L mode, all the Γ_n are much smaller than the H-mode counterparts and

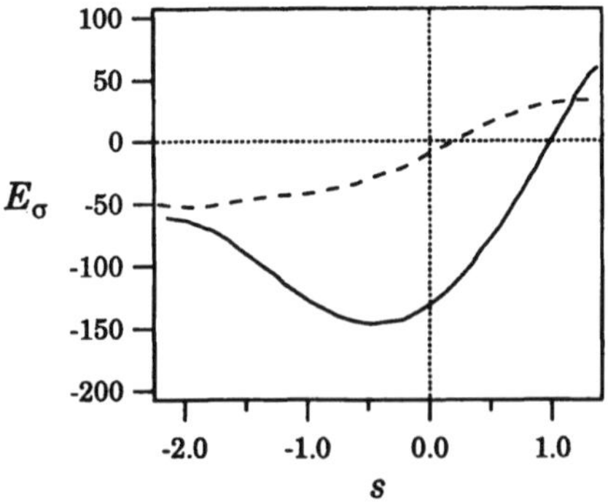

Figure 10.9. Radial electric fields near the separatrix in JFT-2M [10.32]: ——, H mode; – – –, L mode (units: E_σ, V cm^{-1}; s, cm).

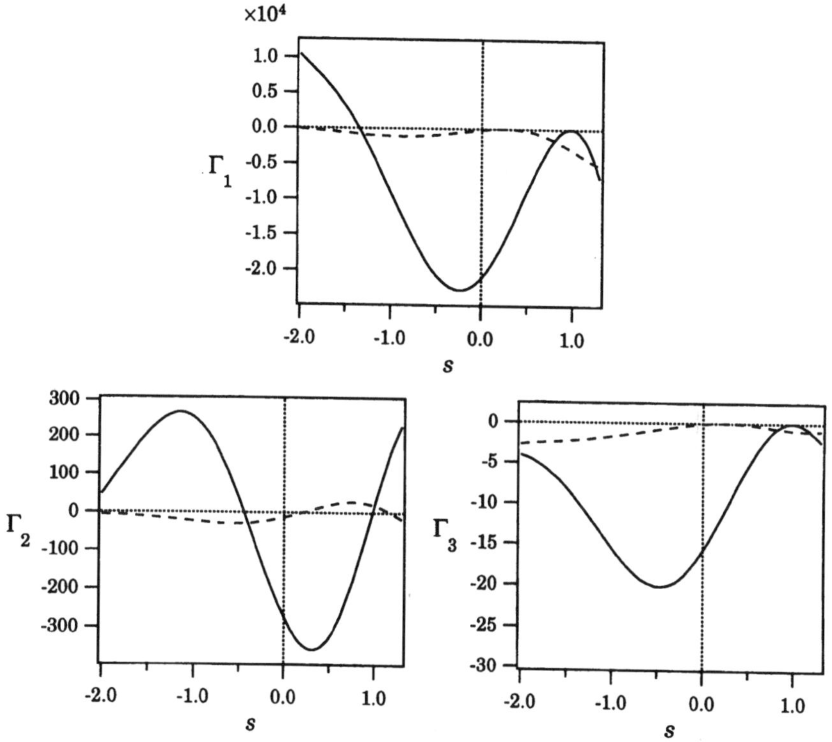

Figure 10.10. Magnitude of Γ_n near the separatrix in JFT-2M [10.32]: ——, H mode; - - -, L mode (units: Γ_n, $\mathrm{V\,cm^{-3}}$; s, cm).

are negative inside the separatrix. Then Λ becomes negative there, which tends to intensify the thermal-energy transport.

References

[10.1] Yoshizawa A 1993 *Publ. Astron. Soc. Jpn.* **45** 129

[10.2] Yoshizawa A, Kato H and Yokoi N 2000 *Astrophys. J.* **537** 1039

[10.3] Gilman P A 1983 *Astrophys. J. Suppl.* **53** 243

[10.4] Glatzmaier G A 1985 *Astrophys. J.* **291** 300

[10.5] Hathaway D H *et al* 1996 *Science* **272** 130

[10.6] Giles P M, Duvall Jr T L, Scherrer P H and Bogart R S 1997 *Nature* **390** 52

[10.7] Basu S, Antia H M and Tripathy S C 1999 *Astrophys. J.* **512** 458

[10.8] Landau L D and Lifshitz E M 1960 *Electrodynamics of Continuous Media* (Oxford: Pergamon) p 234

[10.9] Yoshizawa A, Yokoi N and Kato H 1999 *Phys. Plasmas* **6** 4586

[10.10] Kato S, Fukue J and Mineshige S 1998 *Black-Hole Accretion Disks* (Kyoto: Kyoto University Press)

[10.11] Uchida Y and Shibata K 1985 *Publ. Astron. Soc. Jpn.* **37** 515
[10.12] Stone J M and Norman M L 1994 *Astrophys. J.* **433** 746
[10.13] Matsumoto R *et al* 1996 *Astrophys. J.* **461** 115
[10.14] Yoshizawa A, Yokoi N and Kato H 2000 *Phys. Plasmas* **7** 2646
[10.15] Yoshizawa A and Yokoi N 1993 *Astrophys. J.* **407** 540
[10.16] Nishino S and Yokoi N 1998 *Publ. Astron. Soc. Jpn.* **50** 653
[10.17] Yokoi N 1996 *Astron. Astrophys.* **311** 731
[10.18] Bodin H A B and Newton A A 1980 *Nucl. Fusion* **20** 1255
[10.19] Bodin H A B 1987 *Plasma Phys. Control. Fusion* **29** 1297
[10.20] Miyamoto K 1988 *Plasma Phys. Control. Fusion* **30** 1493
[10.21] Schnack D D, Caramana C J and Nebel R A 1985 *Phys. Fluids* **28** 321
[10.22] Kusano K and Sato T 1986 *Nucl. Fusion* **26** 1051
[10.23] Gimblett C G and Watkins M K 1975 *Proc. Seventh European Conf. on Controlled Fusion and Plasma Physics* Vol I (Lausanne: École de Polytechnique Fédéral de Lausanne) p 63
[10.24] Taylor J B 1974 *Phys. Rev. Lett.* **33** 1139
[10.25] Koide Y *et al* 1994 *Phys. Rev. Lett.* **72** 3662
[10.26] Levinton F M *et al* 1995 *Phys. Rev. Lett.* **75** 4417
[10.27] Strait E J 1995 *Phys. Rev. Lett.* **75** 4421
[10.28] Synakowski E J 1998 *Phys. Control. Fusion* **40** 581
[10.29] Fujita F 1997 *J. Plasma Fusion Res.* **73** 549
[10.30] Yoshizawa A, Yokoi N, Itoh S-I and Itoh K 1999 *Phys. Plasmas* **6** 3194
[10.31] Yoshizawa A and Yokoi N 1998 *Phys. Plasmas* **5** 1998
[10.32] Ida K *et al* 1992 *Phys. Fluids B* **4** 2552

Part IV

Plasma Turbulence

Turbulence in plasmas has several characteristic features. One is that the fluctuation level becomes high through the instabilities driven by strong inhomogeneity. The turbulent level and spectrum are strongly influenced by the spatial inhomogeneity and plasma configuration. Inhomogeneities exist for plasma parameters (e.g., density and temperature) as well as for the fields (e.g., magnetic field and radial electric field). These inhomogeneities couple so as to drive and/or suppress instabilities and turbulent fluctuations. In particular, the anisotropy along and perpendicular to the strong magnetic field induces the varieties of nature of possible fluctuations: fluctuations often have a very long correlation length along the magnetic field line and are quasi-two-dimensional. In addition, mobilities of electrons and ions differ prominently. The inhomogeneities, anisotropy due to the strong magnetic field, and the difference of ion and electron mobilities, have strong influences on plasma turbulence as well as on the linear properties of plasma waves. In many cases, instabilities develop into strong turbulence, so that the decorrelation rate caused by the nonlinear interactions is usually of the same order of or much larger than the damping rate (growth rate) of the linear eigenmode: a theoretical method like fluid turbulence is required. In some cases, only a few modes are excited, and an analysis based on the weak turbulence suffices. (General discussion is given in, e.g. [2.22, 2.26, 2.39, 2.40].)

In order to illustrate the common features and different characteristics between the neutral fluid turbulence and plasma turbulence, various theoretical approaches are illustrated with examples of applications in Part IV. Emphasis is made of how the theory of turbulence is applied to the system which is composed of components of different mobility with strong inhomogeneity and anisotropy.

One important difference of plasma turbulence from neutral fluid turbulence is that plasma particles sometimes respond as 'collisionless' particles. In neutral fluids, like liquids or gases, molecules and atoms collide with each other in a short distance. (These collisions are the origin of molecular viscosity that causes real dissipation in the energy dissipation length l_D.) The mean-free-path of atomic and molecular collisions in neutral fluid is usually much shorter than the characteristic 'wave length' of fluctuations. In contrast, Coulombic collisions between ions or electrons, which are the origins of the real dissipation, are rare in high-temperature plasmas. The mean-free-path can be longer than the typical wave length. In such cases, particles can traverse many wave periods along the magnetic field line without losing their memory by collisions. Wave–particle interaction occurs, and internal freedom of particle distribution can influence the plasma turbulence.

The objective of this monograph is to illustrate the methods which could be common to neutral fluids turbulence and to plasma turbulence. Therefore emphasis is made on the analyses based on fluid-like (moment) equations.

Nomenclature

a	radius of plasma column
$\boldsymbol{A}$	vector potential
$\boldsymbol{B}$	magnetic field
$\hat{\boldsymbol{b}}$	unit vector in the direction of magnetic field: $\hat{\boldsymbol{b}} = \boldsymbol{B}/B$
B_{p}	poloidal component of magnetic field
B_{t}	toroidal component of magnetic field
c	light velocity
c_{s}	ion sound velocity: equation (12.15)
D	diffusion coefficient
D_{M}	diffusion coefficient for magnetic field line: equation (18.6)
e	unit charge of electron
$\boldsymbol{E}$	electric field
E_{r}	radial component of electric field
$\mathcal{E}$	integrated fluctuating energy in a small volume: equation (23.9)
$\mathcal{E}_k$	wave energy density
$f(\boldsymbol{v})$	velocity distribution function of plasma particles
$\tilde{\boldsymbol{f}}$	fluctuation of fields and plasma parameters: equation (14.42)
g^2	magnitude of statistical source term: equation (23.10b)
G_0	normalized pressure gradient: equation (12.24c)
$I_{j,k}$	Intensity of jth element of fluctuation field of k mode
I_k	intensity of k mode fluctuations
$\boldsymbol{J}$	electric current
$\mathcal{K}$	Kubo number (ratio of correlation time to the eddy-turnover time): equation (14.13)
$\boldsymbol{k}$	wave vector
$k_{\parallel}$	wave vector component parallel to magnetic field
$k_{\perp}$	wave vector component perpendicular to magnetic field
$\boldsymbol{\mathcal{L}}$	dispersion operator: equation (14.46)
l_{cor}	correlation length of fluctuations
L_{M}	scale-length of magnetic field gradient: equation (12.24b)
L_n	scale-length of density gradient: equation (11.14a)
L_{p}	scale-length of pressure gradient: equation (12.24b)
L_{s}	magnetic shear length: Rqs^{-1}
L_{T}	scale-length of temperature gradient
m, n	poloidal and toroidal mode numbers
m_e	electron mass
m_i	ion mass
n_{s}	particle number density of species s (s: electron or ion)
p	plasma pressure
p_e	electron pressure
p_i	ion pressure
P_r	Prandtl number

q	inverse of the pitch of magnetic field line around a torus; safety factor: equation (13.9)
$\boldsymbol{q}$	heat flux
r	minor radius of toroidal plasma: Fig.13.1
R	major radius of toroidal plasma: Fig.13.1
R_a	Rayleigh number
R_e	Reynolds number
$r_{\mathrm{A}\to\mathrm{B}}$	transition rate from state A to state B
r_{s}	mode rational surface: $q(r_{\mathrm{s}}) = m/n$
s	magnetic shear parameter: equation (13.11)
$S(\mathcal{E})$	renormalized dissipation function: equation (23.13)
$\tilde{\boldsymbol{S}}_k$	turbulent noise source: equation (14.44)
$\tilde{\boldsymbol{S}}_{\mathrm{th}}$	thermal noise source
S_v	shear of cross-field plasma flow velocity: equation (20.8)
t	time
T	temperature
T_{n}	normalized temperature: equation (23.18)
$\boldsymbol{V}$	velocity
v_{A}	Alfvén velocity
v_{Ap}	Alfvén velocity at poloidal magnetic field
$v_{\mathrm{th},e}$	thermal velocity of electrons
$v_{\mathrm{th},i}$	thermal velocity of ions
$\boldsymbol{V}_{E\times B}$	$E \times B$ drift velocity: $\boldsymbol{v}_{E\times B} = \boldsymbol{E} \times \boldsymbol{B}B^{-2}$
V_{de}	drift velocity: $V_{\mathrm{de}} = T_e/eBL_n$
$w(\mathcal{E})$	flux of probability density function: equation (24.10)
$\tilde{w}(t)$	white noise fluctuation
W_{p}	internal energy of plasmas
w_{is}	width of magnetic island
α	normalized pressure gradient in tokamak: equation (17.7)
β	ratio of plasma pressure to magnetic pressure
β_{p}	ratio of plasma pressure to pressure of poloidal magnetic field
χ	thermal diffusivity
δ	collisionless skin depth: c/ω_{p}
$\boldsymbol{\varepsilon}$	dielectric tensor
$\varepsilon_1(k, \omega)$	linear dielectric constant
ε_0	vacuum susceptibility
$\varepsilon_\perp$	dielectric constant of magnetized plasma: equation (19.9)
ϕ	electrostatic potential
γ_{L}	linear growth rate
γ_{m}	mean decorrelation rate at thermodynamical equilibrium: equation (23.11)
γ_{ng}	nonlinear growth rate
γ_v	eddy-damping rate
$\gamma_{v,c}$	eddy-damping rate by collisional process: equation (23.5)

Γ_{dec}	decorrelation rate of test mode by fluctuations
Γ_r	particle flux in the radial direction
$\eta_{\parallel}$	resistivity for the parallel current
η_i	ratio of ion temperature gradient to density gradient: $\nabla(\ln \bar{T}_i)/\nabla(\ln \bar{n}_e)$
η_e	$\nabla(\ln \bar{T}_e)/\nabla(\ln \bar{n}_e)$
$\boldsymbol{\kappa}$	magnetic field gradient: equation (11.24)
λ	current diffusivity
λ_j	eigenvalue of the jth least-stable mode for renormalized operator $\mathcal{L}$: equation (14.50)
Λ	damping rate of amplitude of coarse-grained fluctuations
μ_0	magnetic permeability of vacuum
μ_e	electron viscosity (kinematic viscosity)
μ_i	ion viscosity (kinematic viscosity)
ν_{ei}	electron–ion collision frequency
ν_{ii}	ion–ion collision frequency
θ	poloidal angle: figure 13.1
ρ_e	electron gyroradius
ρ_i	ion gyroradius
ρ_{p}	gyroradius at poloidal magnetic field
ρ_{s}	ion gyroradius at electron temperature
σ	electric conductivity
τ_{A}	Alfvén-transit time: a/v_{A}
τ_{Ap}	transit time at poloidal Alfvén velocity: qR/v_{A}
τ_{cor}	correlation time of fluctuations
ω_{ce}	cyclotron frequency of electrons
ω_{ci}	cyclotron frequency of ions
ω_k	eigenfrequency of a mode
ω_{M}	drift frequency due to the magnetic field gradient: equation (12A.7)
ω_{p}	plasma frequency
ω_*	drift frequency: $\omega_* = V_{\text{de}}k_y$
ω_{*i}	ion drift frequency: equation (12.23)
ω_{*Ti}	ion-temperature gradient drift frequency: equation (12.26)
ζ	toroidal angle: figure 13.1

Chapter 11

Equations for Plasmas

One distinct feature of a plasma as a continuous medium is that the responses of the electrons and ions are not identical, hence they induce collective electromagnetic fields. In addition, inhomogeneities often exist in the plasma number density, pressure and velocity, and these have important roles in the evolution of turbulence. Therefore various plasma quantities together with the electromagnetic fields must be simultaneously solved to understand the turbulence dynamics in plasmas.

The dynamical equations of a plasma in the fluid limit are usually constructed by using a two-fluid picture. The number of relevant variables is much larger than in the case of neutral-fluid turbulence. The large number of independent variables is one of the reasons why a variety of processes can occur in turbulent plasmas. In some circumstances, motion of an individual particle becomes essential for the interaction between plasma and the fluctuating fields. In this case, kinetic equations are employed.

11.1 Fluid Equations

A representative set of equations is the Braginskii equations [11.1]. In this set of equations, variables are chosen as density n_s, velocity $\boldsymbol{V}_s$ and temperature T_s for each species ($s = e, i$). The higher-order moments, i.e., the fluxes of number density, momentum and energy ($\mathbf{\Gamma}_s$, $\mathbf{\Pi}_s^{\mathrm{d}}$ and $\boldsymbol{q}_s$, where the superscript d stands for deviatoric part (traceless part)) and the exchanges of momentum and energy between different species, $\boldsymbol{R}_{ei}$ and Q_s, are expressed in terms of n_s, $\boldsymbol{V}_s$ and T_s, i.e., the gradient–flux relations are given. (Explicit formulae for the gradient–flux relations are given in [11.1] and are not reproduced here.) These gradient–flux relations are the closure equations for deducing the fluid equation from the kinetic equations. The two-fluid equations are:

$$\frac{\partial}{\partial t} n_s + \nabla \cdot (n_s \boldsymbol{V}_s) = 0 \quad (s = i, e), \tag{11.1a}$$

$$n_i m_i \frac{\mathrm{d}}{\mathrm{d}t} \boldsymbol{V}_i - e_i n_i (\boldsymbol{E} + \boldsymbol{V}_i \times \boldsymbol{B}) + \nabla p_i + \nabla \cdot \boldsymbol{\Pi}_i^{\mathrm{d}} = -\boldsymbol{R}_{ei}, \tag{11.1b}$$

$$n_e m_e \frac{\mathrm{d}}{\mathrm{d}t} \boldsymbol{V}_e + e n_e (\boldsymbol{E} + \boldsymbol{V}_e \times \boldsymbol{B}) + \nabla p_e + \nabla \cdot \boldsymbol{\Pi}_e^{\mathrm{d}} = \boldsymbol{R}_{ei}, \tag{11.1c}$$

$$\frac{3}{2} n_s \frac{\mathrm{d}}{\mathrm{d}t} T_s + p_s \nabla \cdot \boldsymbol{V}_s = -\nabla \cdot \boldsymbol{q}_s - \boldsymbol{\Pi}_s^{\mathrm{d}} : \nabla \boldsymbol{V}_s + Q_s \quad (s = i, e). \tag{11.1d}$$

(See [2.24] for details.) By combining these equations with the Maxwell equations

$$\varepsilon_0 \nabla \cdot \boldsymbol{E} = \sum_{s=e,i} e_s n_s, \tag{11.2}$$

$$\nabla \times \boldsymbol{B} - \frac{1}{c^2} \frac{\partial}{\partial t} \boldsymbol{E} = \mu_0 \boldsymbol{J} = \mu_0 \sum_{s=e,i} e_s \boldsymbol{V}_s, \tag{11.3}$$

$$\frac{\partial}{\partial t} \boldsymbol{B} = -\nabla \times \boldsymbol{E}, \tag{11.4}$$

the dynamics of plasma and electromagnetic field are described.

As will be seen in the following, plasma dynamics can have different characteristic time scales and length scales. Therefore, the choice of normalization units (e.g., the Alfvén unit in the magnetohydrodynamic (MHD) equations, equation (6.25)) may not be unique for plasma turbulence. Instead, characteristic scales are chosen case by case in order to understand the complicated plasma dynamics more easily. Therefore the plasma equations in this monograph will be either expressed with explicit dimensions or normalized to convenient units depending on the problems.

11.2 Reduced Set of Equations

A set of two fluid equations is a highly reduced system to describe the plasma dynamics, because the differences in responses of each individual particle are neglected. Nevertheless, it is a set of nonlinear equations with 13 independent variables and is still very complicated. For analytic transparency, various efforts have been made for a further reduction in the number of variables. There are several characteristic waves in MHD equations: the compressional Alfvén wave, the shear Alfvén wave and the ion sound wave. The compressional Alfvén wave has a high frequency, so that it can be decoupled by use of a time scale separation. (The temporal change in magnitude of the strong magnetic field is neglected.) By eliminating the high-frequency oscillation associated with the compressional Alfvén wave, simplified models to analyze the nonlinear evolution of global MHD instabilities in tokamaks have been proposed [11.2–11.4]. This method has also been extended to cover the cases of pressure-gradient-driven turbulence, electrostatic perturbations or micro-fluctuations. Various reduced sets of equations have been derived. These include:

1. the Yagi–Horton seven-field model [11.5];
2. the Romaneli–Zonca six-field model [11.6];
3. the Drake–Antonsen five-field model [11.7];
4. the Hazeltine four-field model [§7.4 of 11.8];
5. the Strauss three-field model [11.9, 11.10];
6. the Itoh *et al* electron inertial three-field model [11.11];
7. the Rosenbluth *et al* MHD two-field model [11.4];
8. the Hasegawa–Wakatani two-field model [11.12];
9. the Hasegawa–Mima one-field model [11.13].

Examples of some of these reduced sets of equations are now briefly illustrated. The conservation properties of these equations are discussed in §11.3.

11.2.1 Yagi–Horton Equations

For the study of low frequency and pressure-gradient-driven turbulence, Yagi *et al* have proposed a seven-field model of plasma dynamics [11.5]. The density for which

$$n = n_e = e_i e^{-1} n_i$$

is assumed, the electrostatic potential ϕ, the stream function Φ, the ion velocity in the direction of magnetic field $v_\parallel$, the vector potential in the direction of magnetic field, $A_\parallel$, and the electron and ion pressures, p_e and p_i, are chosen as independent variables. A set of equations for these variables $(n, \phi, \Phi, v_\parallel, p_e, p_i, A_\parallel)$ follows:

Continuity

$$\frac{\mathrm{d}}{\mathrm{d}t} n + n\nabla \cdot \boldsymbol{V} = 0, \tag{11.5a}$$

Equation of motion

$$\nabla_\perp \cdot \left(\frac{nm_i}{B} \frac{\mathrm{d}}{\mathrm{d}t} \frac{1}{B} \nabla_\perp \Phi \right) = \nabla_\parallel J_\parallel + \nabla_\perp \cdot \{ \hat{\boldsymbol{b}} \times [\nabla_\perp (p_e + p_i) + \nabla_\perp \cdot \boldsymbol{\Pi}_i^{\mathrm{d}}] \}, \tag{11.5b}$$

Equation of motion

$$nm_i \frac{\mathrm{d}}{\mathrm{d}t} v_\parallel + \nabla_\parallel (p_e + p_i) + \hat{\boldsymbol{b}} \cdot \nabla_\perp \cdot \boldsymbol{\Pi}_i^{\mathrm{d}} = 0, \tag{11.5c}$$

Ohm's law

$$-\frac{\partial}{\partial t}A_{\parallel} - \nabla_{\parallel}\phi + \frac{1}{en}\nabla_{\parallel}p_e = \eta_{\parallel}J_{\parallel} - \frac{0.71}{e}\nabla_{\parallel}T_e, \tag{11.5d}$$

Drift motion

$$\boldsymbol{J}_{\perp} = \hat{\boldsymbol{b}} \times [\nabla(p_e + p_i) + \nabla\cdot\boldsymbol{\Pi}_i^{\mathrm{d}}] - \frac{nm_i}{B}\frac{\mathrm{d}}{\mathrm{d}t}\nabla_{\perp}\Phi, \tag{11.5e}$$

Ion energy balance

$$\frac{3}{2}\frac{\partial}{\partial t}p_i + \frac{3}{2}\nabla_{\perp}\cdot(p_i\boldsymbol{V}) + p_i\nabla\cdot\boldsymbol{V} = -\nabla\cdot\boldsymbol{q}_i + Q_i - \boldsymbol{\Pi}_i^{\mathrm{d}} : \nabla\boldsymbol{V}, \tag{11.5f}$$

Electron energy balance

$$\frac{3}{2}\frac{\partial}{\partial t}p_e + \frac{3}{2}\nabla_{\perp}\cdot\left[p_e\left(\boldsymbol{V} - \frac{1}{en}\boldsymbol{J}\right)\right] + p_e\nabla\cdot\left(\boldsymbol{V} - \frac{1}{en}\boldsymbol{J}\right) = -\nabla\cdot\boldsymbol{q}_e + Q_e, \tag{11.5g}$$

where the Lagrange derivative

$$\frac{\mathrm{d}}{\mathrm{d}t} = \frac{\partial}{\partial t} + \boldsymbol{V}\cdot\nabla \tag{11.6}$$

is used, and $\hat{\boldsymbol{b}} = \boldsymbol{B}/B$. In this set of equations, the velocity and current are expressed in terms of $(n, \phi, \Phi, v_{\parallel}, p_e, p_i, A_{\parallel})$ as follows:

$$\boldsymbol{V} = \frac{\hat{\boldsymbol{b}} \times \nabla_{\perp}\Phi}{B} + v_{\parallel}\hat{\boldsymbol{b}}, \tag{11.7a}$$

$$J_{\parallel} = -\nabla_{\perp}^2 A_{\parallel}, \tag{11.7b}$$

and the relation between the stream function Φ and potential ϕ is given by

Perpendicular Ohm's law

$$\nabla_{\perp}\Phi - \nabla_{\perp}\phi - \frac{1}{en}(\boldsymbol{J}_{\perp} \times \boldsymbol{B} - \nabla_{\perp}p_e) = \eta_{\perp}\boldsymbol{J}_{\perp} + \frac{3}{2}\frac{1}{e\omega_{ce}\tau_e}\hat{\boldsymbol{b}} \times \nabla_{\perp}T_e. \tag{11.7c}$$

With the help of the gradient–flux relations, that describe $\boldsymbol{\Pi}_i^{\mathrm{d}}$, $\boldsymbol{q}_e$, $\boldsymbol{q}_i$, Q_e, Q_i, $\eta_{\parallel}$ and $\eta_{\perp}$ [11.1], equations (11.5)–(11.7) form a closed set of equations that describes low-frequency turbulence in inhomogeneous plasmas. ($\eta_{\parallel}$ and $\eta_{\perp}$ are the resistivity in the direction of $\hat{\boldsymbol{b}}$ and perpendicular to $\hat{\boldsymbol{b}}$, respectively; ω_{ce} is the electron cyclotron frequency; τ_e is the electron collision frequency.)

The Yagi–Horton equations can describe plasma turbulence in a wider range of circumstances. However, depending on the subject of the analysis, much simpler versions are sometimes employed.

11.2.2 Hasegawa–Mima Equation

Investigating the nonlinearity of the $E \times B$ motion is the central subject of plasma turbulence in the presence of strong magnetic fields. This can be performed for electrostatic perturbations by keeping only one variable, which is usually chosen to be the electrostatic potential ϕ.

A simple model of a two-dimensional magnetized plasma is that of charged elements moving with the guiding center velocity

$$\boldsymbol{v}_{E\times B} = \frac{\boldsymbol{E} \times \boldsymbol{B}}{B^2} \tag{11.8}$$

in a constant magnetic field [11.14]. This is expressed by

$$\frac{\mathrm{d}}{\mathrm{d}t}\nabla_\perp^2\phi = 0, \tag{11.9}$$

which expresses the conservation of the z component of the vorticity $\boldsymbol{\omega} = \nabla \times \boldsymbol{v}_{E\times B}$. The convective derivative is given by the $E \times B$ drift velocity, $\boldsymbol{v}_{E\times B} = B^{-1}\boldsymbol{b} \times \nabla\phi$, and a total time derivative of a quantity F is expressed as

$$\frac{\mathrm{d}}{\mathrm{d}t}F = \frac{\partial}{\partial t}F + \boldsymbol{v}_{E\times B}\cdot\nabla F, \tag{11.10a}$$

i.e.,

$$\frac{\mathrm{d}}{\mathrm{d}t}F = \frac{\partial}{\partial t}F + \frac{1}{B}[\phi, F], \tag{11.10b}$$

where the Poisson bracket is defined as

$$[\phi, F] = (\nabla_\perp\phi \times \nabla_\perp F)\cdot\hat{\boldsymbol{b}}. \tag{11.11}$$

The nonlinear term is expressed in terms of the Poisson bracket. The model equation (11.9) describes the advection change of vortex, and does not include the plasma responses that generate the electric field. In plasmas, various mechanisms cause the screening (or anti-screening) of the electric field.

The simplest model equation that describes electrostatic turbulence in inhomogeneous plasmas is the Hasegawa–Mima (HM) equation (or Hasegawa–Mima–Charney (HMC) equation). In this model, the density response is assumed to be adiabatic, $n = e\phi/T_e$, and the pressure and magnetic perturbations are assumed to be small. The HM (or HMC) equation is a dissipationless system for two-dimensional $E \times B$ motion [11.13],

$$\frac{\mathrm{d}}{\mathrm{d}t}\rho_i^2\nabla_\perp^2\phi - \frac{\partial}{\partial t}\phi = 0, \tag{11.12}$$

where ρ_i is the ion gyroradius. ($\rho_i = v_{\mathrm{th},i}/\omega_{ci}$; $v_{\mathrm{th},i}$ is the ion thermal velocity; ω_{ci} is the ion cyclotron frequency.) The plasma property is included in the HM equation (11.12) through the polarization drift effect, which arises from the finite inertia of ions, and the adiabatic response of the electrons. In the presence of a density inhomogeneity in the x direction, the density perturbation is caused by $E \times B$ advection of the background profile. In this case, equation (11.12) takes the form

$$\frac{\mathrm{d}}{\mathrm{d}t}\rho_i^2\nabla_\perp^2\phi - \frac{\partial}{\partial t}\phi - V_{\mathrm{de}}\frac{\partial}{\partial y}\phi = 0, \tag{11.13}$$

where V_{de} is the electron drift velocity,

$$V_{\mathrm{de}} = \frac{T_e}{eBL_n} \quad \text{and} \quad \frac{1}{L_n} = \left|\frac{\nabla\bar{n}}{\bar{n}}\right|, \tag{11.14a}$$

and

$$\omega_* = V_{\mathrm{de}}k_y \tag{11.14b}$$

is the drift frequency. The HM equation is analogous to the vorticity equation which can be deduced from the two-dimensional Navier–Stokes equation. In neutral fluid dynamics, an equation with a similar structure has been derived [11.15]. Note that the HM equation can also be derived via gyrokinetics, which is sketched in [11.16].

11.2.3 Hasegawa–Wakatani Equations

Global scale transport can be described by cross-correlation functions (for instance, the cross-correlation function between the internal energy and velocity controls the fluctuation-driven energy flux). Analysis of a cross-correlation function requires at least two variables. The simplest set of model equations that allows the study of cross-field transport by electrostatic drift wave is that of Hasegawa and Wakatani (HW) [11.12]. This model describes $E \times B$ motion and density perturbations for collisional drift waves. A set of dynamical equations for the normalized potential and normalized density $(n \equiv \tilde{n}/n, \phi \equiv e\tilde{\phi}/T_e)$ of collisional drift waves (in electrostatic limit) is

$$\frac{\partial}{\partial t}\nabla_\perp^2\phi + [\phi, \nabla_\perp^2\phi] = d_\parallel(\phi - n) + \mu_c\nabla_\perp^4\phi, \tag{11.15a}$$

$$\frac{\partial}{\partial t}n + [\phi, n] = d_\parallel(\phi - n) - \frac{\partial\phi}{\partial y} + D_c\nabla_\perp^2 n, \tag{11.15b}$$

where

$$d_\parallel \equiv k_\parallel^2 D_\parallel \omega_{ci}^{-1} L_n \rho_s^{-1} \quad \text{and} \quad D_\parallel = v_{\mathrm{th},e}^2 \nu_{ei}^{-1}. \tag{11.15c}$$

$v_{\mathrm{th},e}$ is the electron thermal velocity; ν_{ei} is the electron-ion collision frequency; μ_c and D_c are the collisional viscosity and diffusivity, respectively. Time, length and electrostatic perturbation are normalized to $\rho_s/V_{\mathrm{de}} = L_n/c_s$, the ion gyroradius at the electron temperature ρ_s and $\rho_s L_n^{-1} e^{-1} T_e$, respectively.

The parameter $d_\parallel$, being the ratio between the parallel diffusion rate and the drift frequency, controls the phase difference between the potential and density. In the collisionless limit of

$$\nu_{ei}/\omega_* \to 0, \quad \text{i.e.,} \quad d_\parallel^{-1} \to 0, \tag{11.16}$$

the electron motion is not impeded and the plasma response tends to be adiabatic, i.e.,

$$n \simeq \phi. \tag{11.17}$$

In this limit with μ_c, $D_c \to 0$, equation (11.15b) gives

$$d_\parallel(\phi - n) = \frac{\partial}{\partial t}\phi + \frac{\partial \phi}{\partial y} + O(d_\parallel^{-1}). \tag{11.18}$$

Substitution of this relation into equation (11.15a) leads to the HM equation (11.13).

In the opposite limit,

$$d_\parallel \to 0, \tag{11.19}$$

equations (11.15a) and (11.15b) are decoupled. The density perturbation then becomes a passive scalar quantity. This set of equations is the simplest one that allows both the autocorrelation function and cross-correlation function to be studied. Not only the cascade in the turbulence spectrum, but also the transport, which is induced by the inhomogeneity-driven turbulence, can be studied. An intensive study has been performed.

11.2.4 Reduced MHD Equations

To study the evolution of the MHD instabilities, which are caused by inhomogeneities of plasma current and/or pressure gradient, another set of variables is chosen, i.e., the scalar and vector potentials in the direction of the magnetic field $\bar{\boldsymbol{B}}$, $(\phi, A_\parallel)$. A reduced set of equations follows:

$$\frac{\partial}{\partial t}\nabla_\perp^2\phi + [\phi, \nabla_\perp^2\phi] - \nabla_\parallel J_\parallel = 0, \tag{11.20a}$$

$$\frac{\partial}{\partial t}A_\parallel + \nabla_\parallel\phi = -\eta_\parallel J_\parallel, \tag{11.20b}$$

with $J_\parallel = -\nabla_\perp^2 A_\parallel$. In this set of equations, length, time, and the scalar and vector potentials are normalized to the plasma radius a, poloidal Alfvén transit time $\tau_{\mathrm{Ap}} = a/v_{\mathrm{Ap}} = qR/v_{\mathrm{A}}$, Ba^2/qR and $Bv_{\mathrm{A}}a^2/qR$, respectively. (Length in parallel direction: in qR.)

11.3 Reduced Set of Equations and Conservation Property

In order to analyze nonlinear interactions in strong plasma turbulence, various reduced sets of equations have been used. The representative ones are listed in §11.2. Some basic properties of these reduced sets of equations are explained here.

11.3.1 Hasegawa–Mima Equation

The simplest model equation that describes electrostatic turbulence in inhomogeneous plasmas, i.e., the HM equation (or the HMC equation), is derived for two-dimensional $E \times B$ motion as equation (11.13), and is given in dimensionless form as

$$\frac{\partial}{\partial t}\nabla_\perp^2 \phi + [\phi, \nabla_\perp^2 \phi] - \frac{\partial}{\partial t}\phi - \frac{\partial}{\partial y}\phi = 0. \tag{11.21}$$

Time, length and electrostatic potential are normalized using $\rho_s/V_{de} = L_n/c_s$, ρ_s, and $\rho_s L_n^{-1} e^{-1} T_e$ respectively. (It should be noted that if one employs the MHD normalization for equation (11.20), the same form is obtained for the HM equation (11.21).) The plasma properties are included in the HM equation in the form of the polarization drift effect due to the finite size of the ion-inertia (the first term), the electron adiabatic response (the third term) and the density inhomogeneity (the fourth term). The adiabatic limit of the response of dissipationless plasmas, equation (11.17),

$$\tilde{n}/\bar{n} = e\tilde{\phi}/T_e,$$

is studied in the HM equation.

When the Debye length becomes longer, an equation similar to equation (11.21) can be derived for fluctuations in the range of the Debye length. In this case, the length is regulated by the Debye length, not by the ion gyro-radius [11.17].

The HM equation conserves the 'energy', 'enstrophy' and 'mass' of the vortices which are defined as

$$\mathcal{E} = \frac{1}{2}\int d\boldsymbol{x}(\phi^2 + (\nabla\phi)^2), \tag{11.22a}$$

$$\mathcal{U} = \frac{1}{2}\int d\boldsymbol{x}((\nabla_\perp\phi)^2 + (\Delta_\perp\phi)^2), \tag{11.22b}$$

$$\mathcal{M} = \int d\boldsymbol{x}(\phi - \nabla_\perp^2\phi). \tag{11.22c}$$

Note that the integral $\int d\boldsymbol{x}\, f\,[f, g]$ vanishes for periodic boundary conditions. One can deduce that an inverse cascade is possible from this equation [11.18, 11.19]. This is one of the most important contributions of the HM equation for understanding the plasma turbulence. That is, micro-scale fluctuations can generate a structure which has a much longer scale length in comparison with the fluctuations. The structure with longer scales can also be induced by turbulence. The effects of turbulence on the formation and destruction of global structure can be recognized clearly by studying the HM equation. This aspect is discussed in chapter 19. The limit of thermodynamical equilibrium is discussed in chapter 26.

The HM equation is analogous to the vorticity equation in a neutral fluid, which can be applied to the case of atmospheric dynamics on a rotating planet. The atmospheric flow on the horizontal plane is governed by the equation

$$\frac{\partial}{\partial t}\boldsymbol{v} = -g\nabla h + R_{\mathrm{C}}\boldsymbol{v} \times \hat{z}, \tag{11.23}$$

where g is the gravitational acceleration, h is the atmospheric depth and $R_{\mathrm{C}}\boldsymbol{v} \times \hat{z}$ is the Coriolis force. The gradient (in the longitudinal direction) of the vertical component of the Coriolis force plays the role of a density gradient in equation (11.21). The Rossby wave in a neutral fluid corresponds to the drift wave (see, e.g., [11.19]). The inverse cascade in fluid dynamics has been discussed; see, e.g., the discussion in [11.20] and related references [11.21, 11.22].

11.3.2 Three-Field Equations

Plasma turbulence is often investigated in more general circumstances, i.e., for the case of electromagnetic turbulence or when there are multiple gradients (those of temperature, density, magnetic field, etc.). Fluctuations like the drift waves, the ion-temperature gradient mode, resistive MHD mode, interchange mode, ballooning mode, current-diffusive interchange mode, etc., are studied by three- or four-fields models.

As an example, a set of three-field equations is given by [2.28]

$$\frac{\partial}{\partial t}\nabla_\perp^2\tilde{\phi} + [\tilde{\phi}, \nabla_\perp^2\tilde{\phi}] = \nabla_\parallel \tilde{J}_\parallel + (\boldsymbol{b} \times \boldsymbol{\kappa})\cdot\nabla\tilde{p} + \mu_c\nabla_\perp^4\tilde{\phi}, \tag{11.24a}$$

$$\frac{\partial}{\partial t}\tilde{A}_\parallel + \frac{c^2}{\omega_{\mathrm{p}}^2 a^2}\left(\frac{\partial}{\partial t}\tilde{J}_\parallel + [\tilde{\phi}, \tilde{J}_\parallel]\right) = -\nabla_\parallel\tilde{\phi} - \eta_\parallel\tilde{J}_\parallel + \lambda_c\nabla_\perp^2\tilde{J}_\parallel, \tag{11.24b}$$

$$\frac{\partial}{\partial t}\tilde{p} + [\tilde{\phi}, \tilde{p}] + [\tilde{\phi}, \bar{p}] = \chi_c\nabla_\perp^2\tilde{p}, \tag{11.24c}$$

where $\boldsymbol{\kappa}$ denotes the gradient of the magnetic field strength, λ is the current diffusivity, χ is the thermal diffusivity, and ω_{p} is the plasma frequency. The suffix c in the transport coefficients stands for a collisional dissipation process. (In this set of equations, length, time, electrostatic potential, and vector potential are normalized to the plasma radius a, poloidal Alfvén transit time $\tau_{\mathrm{Ap}} = a/v_{\mathrm{Ap}} = qR/v_{\mathrm{A}}$, $Bv_{\mathrm{A}}a^2/qR$, and Ba^2/qR, respectively.)

A conservation relation is derived as follows.

$$\frac{\partial}{\partial t}\frac{1}{2}\left[\int \mathrm{d}V|\nabla_\perp\tilde{\phi}|^2 + \left(1+\frac{c^2}{\omega_p^2 a^2}\right)\int \mathrm{d}V|\tilde{J}_\parallel|^2 + \int \mathrm{d}V\,\tilde{p}^2\right]$$
$$= \left(\kappa_x + \frac{\mathrm{d}\bar{p}}{\mathrm{d}x}\right)\int \mathrm{d}V(\tilde{p}^*\nabla_y\tilde{\phi}) - \mu_c\int \mathrm{d}V|\nabla_\perp^2\tilde{\phi}|^2$$
$$-\eta_\parallel\int \mathrm{d}V|\tilde{J}_\parallel|^2 - \lambda_c\int \mathrm{d}V|\nabla_\perp\tilde{J}_\parallel|^2 - \chi_c\int \mathrm{d}V|\nabla_\perp\tilde{p}|^2. \tag{11.25}$$

11.3.3 Yagi–Horton Equations

In the reduced set of equations with a small number of fields (e.g., the HM equation, the HW equation, three-field models, etc.), 'energy' includes the quadratic terms of perturbations. The quadratic term of perturbed internal energy also appears. (For instance, the integrand of 'energy' in the HM equation, equation (11.22a), is essentially $\tilde{n}^2 + \tilde{v}_{E\times B}^2$ with a proper normalization. The perturbation of internal energy contributes to $\mathcal{E}$ in a quadratic form. The left-hand side of equation (11.25) consists of quadratic terms.) Conservation relations are not directly related to the conservation of energy. Such a difficulty is resolved in the reduced set of equations with a large number of fields [11.5, 11.6]. In the case of YH equations, the energy conservation relation is given by

$$\frac{\partial}{\partial t}\left(\langle\tfrac{3}{2}p_e\rangle + \langle\tfrac{3}{2}p_i\rangle + \left\langle\frac{nm_i}{2}\frac{|\nabla_\perp\Phi|^2}{B^2}\right\rangle + \left\langle\frac{nm_i}{2}v_\parallel^2\right\rangle + \langle\tfrac{1}{2}|\nabla_\perp A_\parallel|^2\rangle\right) = 0 \tag{11.26}$$

in the invicid limit (no collisional dissipation), where $\langle\ \rangle$ indicates the average within a volume, and the flow across the surface is assumed to vanish. The perturbation of internal energy appears in a linear form of $\frac{3}{2}p_e + \frac{3}{2}p_i$ in equation (11.26), as in the energy conservation relation. The conservation property of the gyro-averaged equations is discussed in [11.23].

11.3.4 Dissipation and Transport Flux

The fluctuation energy is induced by the energy release associated with the global turbulent-driven flux $-q_x = \langle\tilde{p}^*\nabla_y\tilde{\phi}\rangle$. This released energy is related to the dissipated energy. Nonlinear interactions transfer energy between modes with different scale lengths, and the dissipation finally takes place through collisional transport. The collisional transport coefficients (viscosity, resistivity, current diffusivity, thermal conductivity) are introduced. A small but finite value of collisional transport is essential for the presence of stationary turbulence. The role of collisional dissipation is illustrated by taking an example of the three-field model. In a stationary state, equation (11.25) gives the relation

$$-\left(\kappa_x + \frac{\mathrm{d}\bar{p}}{\mathrm{d}x}\right)\int \mathrm{d}V\, q_x = \mu_c \int \mathrm{d}V |\nabla_\perp^2 \tilde{\phi}|^2 + \eta_\parallel \int \mathrm{d}V |\tilde{J}_\parallel|^2$$
$$+ \lambda_c \int \mathrm{d}V |\nabla_\perp \tilde{J}_\parallel|^2 + \chi_c \int \mathrm{d}V |\nabla_\perp \tilde{p}|^2. \tag{11.27}$$

The right-hand side is positive semi-definite. This relation shows two features of turbulence and turbulent transport. First, non-trivial solutions, i.e., the finite amplitudes for fluctuations ($|\tilde{\phi}| \neq 0$, etc.), are induced and sustained by the turbulent-driven flux in the direction of the global inhomogeneity (left-hand side). Second, if collisional dissipation processes do not exist at all, the right-hand side vanishes. A stationary turbulent state in inhomogeneous plasma with cross-field flux is realized only if collisional dissipation exists. One must take care in studying the limit of $(\mu_c, \eta_\parallel, \lambda_c, \chi_c) \to 0$. In a strong turbulence state, the turbulent level and turbulent transport often do not depend on small values of $(\mu_c, \eta_\parallel, \lambda_c, \chi_c)$. One can consider a turbulent transport in the limit of $(\mu_c, \eta_\parallel, \lambda_c, \chi_c) \to 0$. Such a limit is different from that of the system with $(\mu_c, \eta_\parallel, \lambda_c, \chi_c) = 0$. The convergence at $(\mu_c, \eta_\parallel, \lambda_c, \chi_c) \to 0$ is not necessarily a uniform convergence.

This relation is a generalization of the relation between the collisional dissipation and transport. In the absence of turbulence, only a resistive dissipation of large scale current contributes to the right-hand side of equation (11.27). A formula has been derived in [11.24] as

$$\int \mathrm{d}S\, V_x = \int \mathrm{d}S\, \frac{\eta |J|^2}{|\mathrm{d}\bar{p}/\mathrm{d}x|}, \tag{11.28}$$

where V_x is the diffusion velocity in the direction of the gradient and surface-integral is taken on the magnetic surface. In inhomogeneous slab plasma, diamagnetic current exists. The dissipation of the diamagnetic current induces the collisional transport. In toroidal configuration, the diamagnetic current should be compensated by the current which flows along the magnetic field. This flow along the field line is called secondary current, or Pfirsch–Schlüter current. This symmetry-breaking current enhances the dissipation and increases the collisional transport in toroidal geometry. It is noted that the diamagnetic current and the secondary current are proportional to the gradient of plasma pressure. Therefore, the right-hand side of equation (11.28) is a linear function of the pressure gradient. A linear relation between the gradient and flux holds in the case of collisional transport.

As is shown by equation (11.27), the microscopic fluctuations enhance dissipation, so that the cross-field transport is increased by turbulence in the stationary state.

11.4 Kinetic Equation

11.4.1 Vlasov Equation

When the motions of individual particles are essential, a kinetic equation must be used and

$$\left(\frac{\partial}{\partial t}+\boldsymbol{v}\cdot\nabla+\frac{e_s}{m_s}(\boldsymbol{E}+\boldsymbol{v}\times\boldsymbol{B})\cdot\nabla_v\right)f_s(\boldsymbol{v}:t)=\mathcal{C}, \tag{11.29}$$

where $\mathcal{C}$ is the collision operator is a typical example. In the case of

$$\mathcal{C}=0, \tag{11.30}$$

the Vlasov equation is obtained. Upon solving equation (11.29) for the plasma density and current, and together with Maxwell's equations (11.2)–(11.4), plasma turbulence can be investigated.

11.4.2 Gyro-Averaged Equations

In many cases, the scale length of the fluctuation is in the range of the gyroradius, but the time scale is much longer than the inverse cyclotron frequency. In such a case, the resonances associated with the gyromotion are unimportant, and the distribution function becomes a function of $(v_\perp, v_\parallel)$,

$$f(\boldsymbol{v}:t)=\hat{f}(v_\perp, v_\parallel:t). \tag{11.31}$$

Then the kinetic equation is reduced to

$$\frac{\partial}{\partial t}\hat{f}_s+(v_\parallel\hat{\boldsymbol{b}}+\boldsymbol{v}_{\mathrm{D}})\cdot\nabla\hat{f}_s+\left(e_s\frac{\partial\phi}{\partial t}+\mu_s\frac{\partial B}{\partial t}-e_s\boldsymbol{v}_\parallel\cdot\frac{\partial\boldsymbol{A}}{\partial t}\right)\frac{\partial}{\partial U}\hat{f}_s=\mathcal{C}, \tag{11.32}$$

where $\boldsymbol{v}_\parallel=v_\parallel\hat{\boldsymbol{b}}$, $\boldsymbol{v}_{\mathrm{D}}$ is the particle drift velocity,

$$\boldsymbol{v}_{\mathrm{D}}=B^{-1}\boldsymbol{E}\times\hat{\boldsymbol{b}}+\omega_{cs}^{-1}\hat{\boldsymbol{b}}\times(m_s^{-1}\mu_s\nabla B+v_\parallel^2\boldsymbol{\kappa}+v_\parallel\partial\hat{\boldsymbol{b}}/\partial t), \tag{11.33a}$$

$$U=m_s v_\parallel^2/2+e_s\phi+\mu_s B, \tag{11.33b}$$

μ_s is the magnetic moment, and $\boldsymbol{\kappa}$ is the curvature of the magnetic field. An explicit form for non-uniform plasmas can be seen in, e.g., [11.8, 11.25, 11.26].

Appendix 11A

Relations in Thermodynamics and Mean-Field Equation

In fluid-like equations of plasmas, equations (11.1) or (11.5), there appear many terms which describe various fluxes. They are extensions of thermodynamical relations, in which interference between gradients and fluxes have been given. Ohm's law and heat flux equation have been given in classical thermodynamics as [11.27]

$$\boldsymbol{E} + \boldsymbol{V} \times \boldsymbol{B} = \eta \boldsymbol{J} + a\nabla T + \mathcal{R}\boldsymbol{B} \times \boldsymbol{J} + \mathcal{N}\boldsymbol{B} \times \nabla T \tag{11A.1}$$

and

$$\boldsymbol{q} - \phi \boldsymbol{J} = aT\boldsymbol{J} - \kappa \nabla T + \mathcal{N}T\boldsymbol{B} \times \boldsymbol{J} + \mathcal{L}\boldsymbol{B} \times \nabla T, \tag{11A.2}$$

where $\boldsymbol{q}$ is the heat flux. Some fluxes are perpendicular to the magnetic field and gradients. The term $\mathcal{R}\boldsymbol{B} \times \boldsymbol{J}$ is called the Hall term, $\mathcal{N}\boldsymbol{B} \times \nabla T$ is the Nernst effect, $\mathcal{N}T\boldsymbol{B} \times \boldsymbol{J}$ is the Ettingshausen effect, and $\mathcal{L}\boldsymbol{B} \times \nabla T$ is the Leduc–Righi effect. All these effects appear in plasma equations. The interference is also important. The terms $a\nabla T$ in Ohm's law and $aT\boldsymbol{J}$ in the heat flux equation are reciprocal, and the common coefficient a reflects the symmetry in the transport coefficients. The Nernst effect and Ettingshausen effect also form a pair of interferences in transport, and have a common coefficient $\mathcal{N}$ in classical thermodynamics. It is noted that the interference of transport does not appear, in classical thermodynamics, between those of scalar quantity and vector quantity. For example, the cross-field flux of momentum does not interfere with those of energy and number density. This law in classical thermodynamics is know as 'Curie's principle'.

In turbulent plasmas, one might consider a mean-field Ohm's law which describes the relations between averaged quantities:

$$\bar{\boldsymbol{E}} + \bar{\boldsymbol{V}} \times \bar{\boldsymbol{B}} = \eta \bar{\boldsymbol{J}} + a_t \nabla \bar{T} + \mathcal{R}_t \bar{\boldsymbol{B}} \times \bar{\boldsymbol{J}} + \mathcal{N}_t \bar{\boldsymbol{B}} \times \nabla \bar{T} - \alpha \bar{\boldsymbol{B}} + \beta \bar{\boldsymbol{J}} - \gamma \nabla \times \bar{\boldsymbol{V}}.$$

The last three terms are known as dynamo-terms. The term $\alpha\bar{\boldsymbol{B}}$ is called the helicity dynamo, $\beta\bar{\boldsymbol{J}}$ the turbulent resistivity, and $\gamma\nabla \times \bar{\boldsymbol{V}}$ the cross-helicity dynamo. Dynamo terms are discussed in §8.2. Derivation of an appropriate form of mean-field transport coefficient (such as turbulent thermal conductivity) is one of the main subjects of plasma turbulence. It is also shown that Curie's principle does not hold in turbulent systems.

References

[11.1] Braginskii S I 1965 in *Reviews of Plasma Physics* ed M A Leontovich (New York: Consultants Bureau) vol 1, p 205

[11.2] Kadomtsev B B and Pogutse O P 1974 *Sov. Phys.-JETP* **38** 283 [1973 *Zh. Eksp. Teor. Fiz.* **65** 575]
[11.3] Sykes A and Wesson J A 1976 *Phys. Rev. Lett.* **37** 140
[11.4] Rosenbluth M N, Monticello D A, Strauss H R and White R B 1976 *Phys. Fluids* **19** 198
[11.5] Yagi M and Horton C W 1994 *Phys. Plasmas* **1** 2135
[11.6] Romaneli F and Zonca F 1989 *Plasma Phys. Contr. Fusion* **31** 1365
[11.7] Drake J F and Antonsen T M Jr 1984 *Phys. Fluids* **27** 898
[11.8] Hazeltine R D and Meiss J D 1992 *Plasma Confinement* (Addison Wesley)
[11.9] Strauss H R 1976 *Phys. Fluids* **19** 134; 1977 *Phys. Fluids* **20** 1354
[11.10] Hazeltine R D 1983 *Phys. Fluids* **26** 3242
[11.11] Itoh K, Yagi M, Itoh S-I, Fukuyama A and Azumi M 1993 *Plasma Phys. Contr. Fusion* **35** 543
[11.12] Hasegawa A and Wakatani M 1983 *Phys. Rev. Lett.* **50** 682
[11.13] Hasegawa A and Mima K 1977 *Phys. Rev. Lett.* **39** 205
[11.14] Taylor J B and McNamara B 1971 *Phys. Fluids* **14** 1492
[11.15] Charney J G 1948 *Geophys. Public. Kosjones Nors. Videnshap.-Akad. Oslo* **17** 3
[11.16] Hu G, Krommes J A and Bowman J C 1997 *Phys. Plasmas* **4** 2116
[11.17] Taylor J B 1997 *Plasma Phys. Contr. Fusion* **39** A1
[11.18] Horton C W 1984 in *Basic Plasma Physics II* ed A A Galeev and R N Sudan (Amsterdam: North Holland) chapter 6.4
[11.19] Hasegawa A 1985 *Adv. Phys.* **34** 1
[11.20] Sagdeev R Z, Shapiro V D and Shevchenko V I 1978 *Sov. J. Plasma Phys.* **4** 306 [1978 *Fiz. Plasmy* **4** 551]
[11.21] Onsager L 1949 *Nuovo Cimento Suppl.* **6** 279
[11.22] Batchelor G K 1959 *Theory of Homogeneous Turbulence* (Cambridge: Cambridge Univeristy Press)
[11.23] Sugama H 2000 *Phys. Plasmas* **7** 466
[11.24] Kruskal M D and Kulsrud R M 1958 *Phys. Fluids* **1** 265
[11.25] Frieman E A and Chen L 1982 *Phys. Fluids* **25** 502
[11.26] Antonsen T M Jr and Lane B 1980 *Phys. Fluids* **23** 1205
[11.27] Landau L D and Lifshitz E M 1960 *Electrodynamics of Continuous Media* (Oxford: Pergamon Press) §25

Chapter 12

Inhomogeneity and Modes in Plasmas

In this chapter, linear waves are explained first. A few characteristic examples of the linear modes are introduced. Then, methods for studying weak turbulence are explained and the conditions for strong turbulence are given. Some cases are discussed on the basis of reduced set of equations.

12.1 Linear Mode

There are many characteristic waves in magnetized plasmas. The discussion here is limited to these perturbations with the frequencies lower than the ion cyclotron frequency.

12.1.1 Dispersion Relation

The spatio-temporal patterns to appear with periodicity (either stationary or propagating) are called *modes*. The property of a mode is the characterizing information of a continuous medium. If one imposes an external perturbation of a form

$$\tilde{\boldsymbol{E}}_{\mathrm{ext}} \exp(\mathrm{i}\boldsymbol{k}\cdot\boldsymbol{x} - \mathrm{i}\omega t), \tag{12.1}$$

where $\boldsymbol{k}$ and ω can be complex, then the charged elements of plasma are displaced so as to generate a response field

$$\tilde{\boldsymbol{E}}_{\mathrm{induced}} \exp(\mathrm{i}\boldsymbol{k}\cdot\boldsymbol{x} - \mathrm{i}\omega t).$$

If there is a relation of $(\boldsymbol{k},\omega)$ as

$$\omega = \omega(\boldsymbol{k}), \tag{12.2}$$

for which the ratio $\tilde{E}_{\mathrm{induced}}/\tilde{E}_{\mathrm{ext}}$ becomes very large, then a pattern with the form $\exp\{\mathrm{i}\boldsymbol{k}\cdot\boldsymbol{x} - \mathrm{i}\omega(\boldsymbol{k})t\}$ is expected to have a large amplitude and to be selectively observed. Such a pattern is called a *mode*, and the relation equation (12.2) is called the *dispersion relation*.

The dispersion relation is determined by the electric conductivity tensor, $\boldsymbol{\sigma}_{s,k\omega}$, where the suffix s stands for the species of plasma elements, i.e., the electrons and ions ($s = e,\ i$). The perturbed current which is carried by sth component $\tilde{\boldsymbol{J}}_{s,k\omega}$ is expressed as

$$\tilde{\boldsymbol{J}}_{s,k\omega} = \boldsymbol{\sigma}_{s,k\omega}\tilde{\boldsymbol{E}}_{k\omega}. \tag{12.3}$$

The perturbed current is given as

$$\tilde{\boldsymbol{J}}_{k\omega} = \left(\sum_s \boldsymbol{\sigma}_{s,k\omega}\right)\tilde{\boldsymbol{E}}_{k\omega}, \tag{12.4}$$

and the dielectric tensor is introduced (according to convention) as

$$\varepsilon(k,\omega) = \boldsymbol{I} - \frac{\mathrm{i}c^2\mu_0}{\omega}\left(\sum_s \boldsymbol{\sigma}_{s,k\omega}\right), \tag{12.5}$$

where $\boldsymbol{I}$ is the unit tensor, μ_0 is the magnetic permeability of vacuum and c is the velocity of light. By substituting equation (12.5) into the Maxwell equation, i.e., equation (11.3) and equation (11.4), one has the relation

$$\left[\varepsilon(k,\omega) - \left(\frac{kc}{\omega}\right)^2\left(\boldsymbol{I} - \frac{\boldsymbol{kk}}{k^2}\right)\right]\boldsymbol{E}_{k,\omega} = \frac{c\mu_0}{\mathrm{i}\omega}\boldsymbol{j}_{\mathrm{ext};k,\omega} \tag{12.6}$$

where $\boldsymbol{j}_{\mathrm{ext};k,\omega}$ is an externally-imposed current perturbation of the Fourier component $(\boldsymbol{k},\omega)$. Equation (12.5) predicts that the perturbed field $\boldsymbol{E}_{k,\omega}$ can take a finite amplitude even without the external perturbation, if the condition

$$\det\left|\varepsilon(\boldsymbol{k},\omega) - \left(\frac{kc}{\omega}\right)^2\left(\boldsymbol{I} - \frac{\boldsymbol{kk}}{k^2}\right)\right| = 0 \tag{12.7}$$

is satisfied. The dispersion relation equation (12.2) is given as a solution of equation (12.7).

The linear plasma dielectric tensor has been discussed in literature [2.21, 2.40]. Linear-instability analysis is reduced to obtaining the conductivity tensor and solving the dispersion relation. A variety of low-frequency instabilities have been reviewed in literature [2.22, 2.25, 2.39, 11.18, 12.1–12.8].

12.1.2 Vlasov Equation and Linear Dielectric Tensor

The perturbed distribution function is calculated within a framework of linear response theory. The distribution function is written as the sum of the average and the fluctuating parts as

$$f_s(\boldsymbol{x},\boldsymbol{v},t) = \bar{f}_s(\boldsymbol{x},\boldsymbol{v},t) + \tilde{f}_s(\boldsymbol{x},\boldsymbol{v},t), \tag{12.8}$$

where the time dependence is slow for $\bar{f}_s$ and fast for $\tilde{f}_s$. The fluctuating part is given by a path integral of the Vlasov equation (equation (11.29) with

$\mathcal{C} = 0$) as

$$\tilde{f}_s(\boldsymbol{r}, \boldsymbol{v}, t) = \frac{e_s}{m_s} \int_{-\infty}^{t} \mathrm{d}t' [\tilde{\boldsymbol{E}}(\boldsymbol{x}', t') + \boldsymbol{v}' \times \tilde{\boldsymbol{B}}(\boldsymbol{x}', t')] \frac{\partial}{\partial \boldsymbol{v}'} \bar{f}_s(\boldsymbol{x}', \boldsymbol{v}', t), \quad (12.9)$$

where $\boldsymbol{x}'(t')$ and $\boldsymbol{v}'(t')$ are the particle position and velocity along the *unperturbed orbit* of particles with the boundary condition $\boldsymbol{x} = \boldsymbol{x}'(t)$ and $\boldsymbol{v} = \boldsymbol{v}'(t)$ at $t' = t$. The $(\boldsymbol{k}, \omega)$-Fourier component of the perturbed distribution function, $\tilde{f}_{k,\omega}(\boldsymbol{v}) \exp(\mathrm{i}\boldsymbol{k} \cdot \boldsymbol{r} - \mathrm{i}\omega t)$, is given from equation (12.9) as

$$\tilde{f}_{s,k\omega}(\boldsymbol{v}) = \frac{e_s}{m_s} \int_{-\infty}^{t} \mathrm{d}t' \exp\{\mathrm{i}\boldsymbol{k} \cdot (\boldsymbol{x}' - \boldsymbol{x}) - \mathrm{i}\omega(t' - t)\}$$
$$\times \frac{\partial}{\partial \boldsymbol{v}'} \bar{f}_s(\boldsymbol{x}', \boldsymbol{v}') \cdot \left(\boldsymbol{I} + \frac{\boldsymbol{v}'}{\omega} \times \boldsymbol{k} \right) \tilde{\boldsymbol{E}}_{k\omega}. \quad (12.10)$$

The perturbed current $\tilde{\boldsymbol{J}}_{s,k\omega}$ is calculated as

$$\tilde{\boldsymbol{J}}_{s,k\omega} = e_s \int \boldsymbol{v} \tilde{f}_{s,k\omega}(\boldsymbol{v})\, \mathrm{d}\boldsymbol{v} \quad (12.11)$$

Once the distribution function of the average part $\bar{f}_s(\boldsymbol{x}', \boldsymbol{v}')$ is given, then the dispersion relation is obtained by direct manipulation.

Before illustrating some typical examples, we briefly discuss the relation between the analyses based on the fluid equations (like MHD equations) and those based on the kinetic equations.

Figure 12.1 illustrates the difference between the fluid and kinetic approaches. The dynamical equations can be given by fluid equations as in Part II. In the fluid approach, the kinetic equation (the Boltzmann equation or the Vlasov equation or others) is first integrated over the velocity space, and higher-order moments modelled (e.g., mass density or number density, velocity, pressure, etc.). Then the fluid equations are solved in the presence of the perturbations, and the perturbed current calculated. In the kinetic

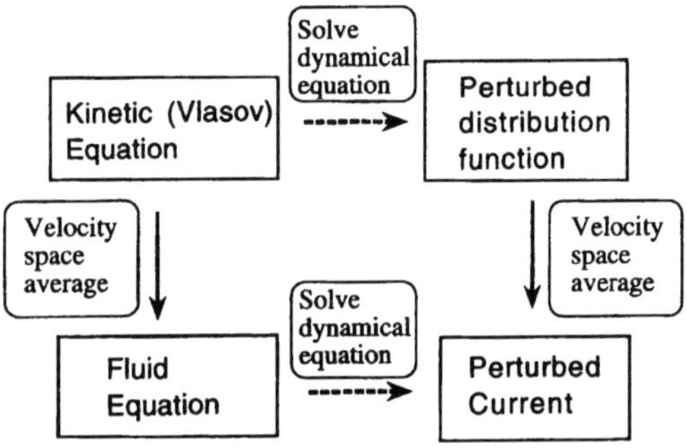

Figure 12.1. Comparison between fluid and kinetic approaches. The solution of dynamical equations in the presence of perturbing electromagnetic fields and the velocity–space average are shown by broken and full arrows, respectively.

approach, the dynamical equation is solved first and the perturbed distribution is obtained. Then the velocity space integral is performed to yield the perturbed current.

The processes of 'solution of dynamical equation' and 'velocity space average' are not necessarily commutable. The propagator in equation (12.10) contains the term with

$$(k_\parallel v_\parallel - \omega)^{-1}. \tag{12.12}$$

The weighted average differs, if the velocity space average is taken before solving the dynamical equation. The difference in the averages with $(k_\parallel v_\parallel - \omega)^{-1}$ comes from the resonant particles, the velocity of which satisfies the relation

$$k_\parallel v_\parallel \simeq \omega. \tag{12.13}$$

The fluid approach ignores the information about the resonant particles. (The Braginskii equations, equations (11.1a–d), which use the lowest order cut-off for the gradient–flux relation, are derived in a collisional limit. In order to extend the range of applicability of the fluid equations to less collisional cases, the higher-order closure models have been employed. See, e.g., discussions in [12.9–12.13]. Such approaches are called 'gyrofluid models'.) As long as the linear response of the plasma is being studied, the kinetic approach is more precise. Nevertheless, the fluid approach is relevant and useful on many occasions when studying turbulent plasmas. A basis of the relevance is discussed later.

12.2 Examples of Modes

We consider a strongly magnetized plasma slab and take the z axis in the direction of the main magnetic field. The plasma parameters (number density, temperature) are assumed to be inhomogeneous in the x direction.

12.2.1 Ion Sound Wave, Drift Wave and Convective Cell

The sound wave in the plasma

$$\omega^2 = c_s^2 k_\parallel^2, \tag{12.14}$$

with the ion sound speed

$$c_s^2 = (Z_i T_e + \gamma T_i) m_i^{-1}, \tag{12.15}$$

where γ is the specific heat ratio of ions, is modified by the density gradient. The drift wave appears with the dispersion relation

$$\omega^2 - \omega\omega_* - k_\parallel^2 c_s^2 = 0, \tag{12.16}$$

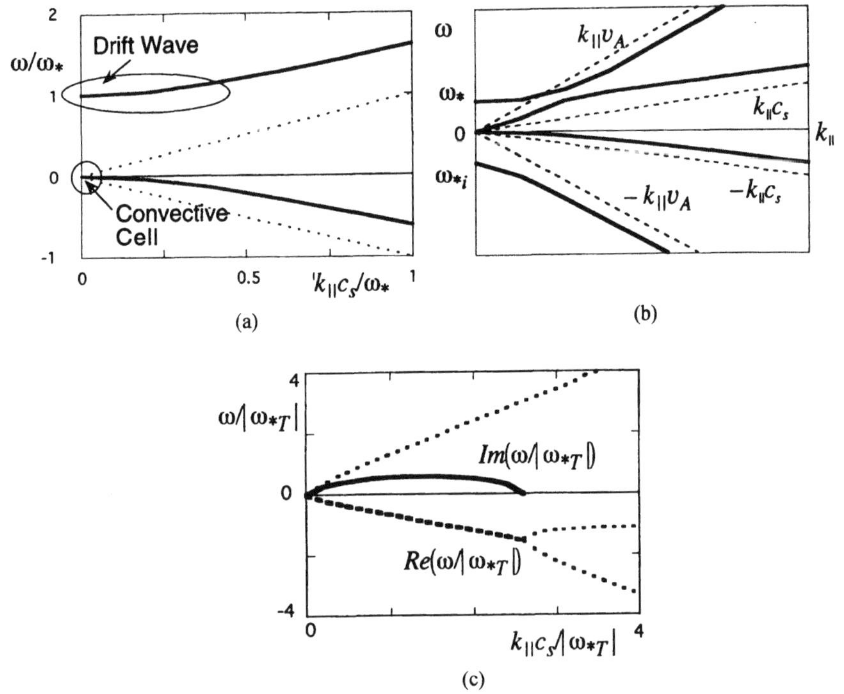

Figure 12.2. Dispersion relations for the drift wave and the convective cell (a), the drift Alfvén waves together with the drift wave (b) and the ion temperature gradient mode (c).

where ω_* is the drift frequency defined in equation (11.14b). The drift mode,

$$\omega \simeq \omega_*, \tag{12.17}$$

exists under the condition of

$$|k_\parallel v_{\mathrm{th},i}| \ll \omega \ll |k_\parallel v_{\mathrm{th},e}|. \tag{12.18}$$

The dispersion relation is shown in figure 12.2(a).

There is another low frequency branch satisfying

$$\omega \simeq -\frac{k_\parallel^2 c_s^2}{\omega_*}. \tag{12.19}$$

(see figure 12.2(a)). In the long wavelength limit, this real frequency vanishes, i.e., this mode does not propagate. This perturbation branch is called a 'convective cell' [2.22, 11.20, 12.14]. In the linear theory, the convective cell is a damped mode with the damping rate of $\mu_\perp k_\perp^2$ where $\mu_\perp$ is the viscosity for the motion in the perpendicular direction. (It is discussed in chapter 19.) The convective cell for electromagnetic perturbations is discussed in [12.15].

12.2.2 Shear Alfvén Wave and Drift Alfvén Mode

The shear Alfvén wave

$$\omega^2 = k_\parallel^2 v_A^2, \tag{12.20}$$

with the Alfvén velocity

$$v_A^2 = \bar{B}^2 / m_i \bar{n}_i \mu_0, \tag{12.21}$$

is also modified by a density gradient. The drift Alfvén mode is generated. Its dispersion relation is

$$\omega^2 - \omega\omega_* - k_\parallel^2 c_s^2 = \frac{\omega^2(\omega - \omega_*)(\omega + |\omega_{*i}|)}{k_\parallel^2 v_A^2}, \tag{12.22}$$

where

$$\omega_{*i} = -(T_i/T_e)\omega_* \tag{12.23}$$

is the ion diamagnetic drift frequency. The dispersion relation is illustrated in figure 12.2(b).

12.2.3 Interchange Mode

When both the pressure and magnetic field strength are inhomogeneous in the x direction, the drift Alfvén mode is modified to yield the interchange instability. The dispersion relation is given as

$$-\omega^2 = \gamma_0^2 \equiv c_s^2 \frac{\nabla \bar{p}}{\bar{p}} \cdot \frac{\nabla \bar{B}}{\bar{B}}, \tag{12.24a}$$

and the scale lengths of pressure gradient and magnetic field gradient are introduced as

$$L_p = |\nabla \bar{p}/\bar{p}|^{-1} \quad \text{and} \quad L_M = |\nabla \bar{B}/\bar{B}|^{-1}. \tag{12.24b}$$

For convenience, a notation

$$G_0 = a^2 \nabla \ln \bar{p} \cdot \nabla \ln \bar{B}, \tag{12.24c}$$

where a is the size of plasma in the direction of the gradient (e.g., radius of plasma column), is also used to denote the magnitude of the driving parameter. When G_0 is positive, the mode becomes unstable as is shown by equation (12.24a). The combination of the inhomogeneities of the plasma pressure and magnetic pressure causes the instability [11.8, 12.1]. This instability is an MHD analogue to the Rayleigh–Benard instability. In toroidal geometry, its counterpart is the ballooning instability [12.16–12.18].

12.2.4 Ion Temperature Gradient Mode

The influence of the temperature gradient of ions is coupled with the parallel ion motion. The ion-sound term $k_\parallel^2 v_{\mathrm{th},i}^2/\omega^2$ is modified by the factor of $(1 - \omega_{*Ti}/\omega)$ in the ion response as [12.1]

$$\frac{\tilde{n}_i}{\bar{n}} \cong \frac{k_\parallel^2 c_\mathrm{s}^2}{\omega^2}\left(1 - \frac{\omega_{*Ti}}{\omega}\right)\frac{e\tilde{\phi}}{T_e}, \tag{12.25}$$

where

$$\omega_{*Ti} = \frac{k_y T_i}{eBL_{Ti}} \tag{12.26}$$

is the ion temperature gradient drift frequency and $L_{Ti} = |\nabla T_i/T_i|^{-1}$ is the scale length of the ion temperature gradient. (The sign of ω_{*Ti} is chosen according to the choice that ∇T_i and ∇n direct in the same direction.) The electrons have the Boltzmann response, $\tilde{n}_e/\bar{n} \cong e\tilde{\phi}/T_e$. The dispersion relation of the ion sound wave is modified to

$$1 = \frac{k_\parallel^2 c_\mathrm{s}^2}{\omega^2}\left(1 - \frac{\omega_{*Ti}}{\omega}\right). \tag{12.27}$$

The motion of ions may drive an instability in the long-parallel-wavelength region as shown in figure 12.2(c), and this instability is called the ion-temperature gradient (ITG) mode.

12.2.5 Dissipative Drift Mode

A drift wave becomes unstable if electron dissipation exists. The electron response to the electrostatic perturbation is usually close to the Boltzmann response, i.e., $\tilde{n}_e/\bar{n}_e \simeq e\tilde{\phi}/T_e$, but is not always so. If there is a small but finite phase difference in the electron response

$$\frac{\tilde{n}_e}{\bar{n}_e} = (1 - \mathrm{i}\delta_\mathrm{d})\frac{e\tilde{\phi}}{T_e}, \tag{12.28}$$

equation (12.16) is modified, and the drift wave becomes unstable as

$$\omega = \omega_*(1 + \mathrm{i}\delta_\mathrm{d}) \tag{12.29}$$

in the limit of $k_\parallel^2 v_{\mathrm{th},i}^2 \ll \omega^2$.

12.3 Weak Turbulence Theory

12.3.1 Ansatz of Weak Turbulence

Weak turbulence theory assumes that the nonlinearity is weak so that the excited fluctuations are composed of the linear eigenmodes, e.g.,

$$\langle \tilde{E}_y(k,\omega)\tilde{E}_y(k',\omega')\rangle = 2\pi\delta_{k,-k'}\delta(\omega+\omega')I_k \tag{12.30}$$

where ω satisfies the dispersion relation $\omega = \omega(\boldsymbol{k})$. The spectral function (autocorrelation function) is non-zero on a hyper-surface, $\omega = \omega(\boldsymbol{k})$, in the four-dimensional space $(\boldsymbol{k},\omega)$. In other words, the nonlinear terms in the dynamical equation (e.g. the $(\partial/\partial x_j)(u_i'u_j' - R_{ij})$ term in equation (3.31)) are assumed to be much smaller than the linear terms. Nonlinear interactions are often truncated at the lowest order correction. In the dynamical equation for the $(\boldsymbol{k},\omega(\boldsymbol{k}))$ mode, the nonlinear interactions from the $(\boldsymbol{k}',\omega(\boldsymbol{k}'))$ and $(\boldsymbol{k}'',\omega(\boldsymbol{k}''))$ linear modes are retained only if

$$\boldsymbol{k} = \boldsymbol{k}' + \boldsymbol{k}'' \quad \text{and} \quad \omega(\boldsymbol{k}) = \omega(\boldsymbol{k}') + \omega(\boldsymbol{k}'') \tag{12.31}$$

hold. This method belongs to those of the truncation of nonlinearities.

12.3.2 Wave Kinetic Equation

The wave kinetic equation is derived by the ansatz of weak plasma turbulence, and takes the form [2.39, 12.19, 12.20]

$$\frac{1}{2}\left[\frac{\partial}{\partial\omega}\Re\varepsilon_1(\boldsymbol{k},\omega)\right]\frac{\mathrm{d}}{\mathrm{d}t}I_k = -\Im\varepsilon_1(\boldsymbol{k},\omega)I_k + \sum_{k'} A_{kk'}I_{k'}I_k + \sum_{\substack{k'+k''=k\\ \omega'+\omega''=\omega}} B_{k'k''}I_{k'}I_{k''} \tag{12.32}$$

where $\varepsilon_1(\boldsymbol{k},\omega)$ is the dielectric constant for linear response, and the linear growth rate is given as

$$\gamma_k = -\Im\varepsilon_1(\boldsymbol{k},\omega)\left[\frac{\partial}{\partial\omega}\Re\varepsilon_1(\boldsymbol{k},\omega)\right]^{-1}. \tag{12.33}$$

The relation $\omega(\boldsymbol{k}) = \omega(\boldsymbol{k}') + \omega(\boldsymbol{k}'')$ is abbreviated as $\omega = \omega' + \omega''$.

Equation (12.32), derived by the weak turbulence theory, has been used to study the statistical property of plasma turbulence. If one solves the stationary state, $\mathrm{d}I_k/\mathrm{d}t = 0$, an order of the magnitude of the turbulence level is

$$I_k \approx O(A_{kk'}^{-1}\Im\varepsilon_1(\boldsymbol{k}',\omega')). \tag{12.34}$$

The weak turbulence ansatz means that the nonlinear interactions are weak; that is, the condition

$$\left|\sum_{k'} A_{kk'}I_{k'}\right| \ll |\omega(\boldsymbol{k})| \tag{12.35}$$

must be satisfied. Therefore, one of the necessary conditions is

$$\gamma_k \ll |\omega(\boldsymbol{k})|. \tag{12.36}$$

12.3.3 Integral, Lyapunov Function and Thermodynamics

The weak turbulence theory has been applied to such problems as one-dimensional turbulence [12.21] or the excitation of toroidal Alfvén eigenmodes (TAE modes) [12.22]. It also gives some perspectives on plasma turbulence [12.19, 12.23, 12.24].

The second term in the right-hand side of equation (12.32) is linear in I_k, and acts on the growth rate of the mode (like an eddy viscosity term). The anti-symmetry relation holds as

$$A_{kk'} = -A_{k'k}, \tag{12.37}$$

reflecting the detailed balance, e.g., in the induced scattering process. (Induced wave scattering is the process in which $\omega(\boldsymbol{k}) - \omega(\boldsymbol{k}') = (\boldsymbol{k} - \boldsymbol{k}') \cdot \boldsymbol{v}$, where $\boldsymbol{v}$ is the particle velocity. The details are explained in [12.19].) The third term $B_{k'k''} I_{k'} I_{k''}$, is considered to be the 'noise source term', when many linear modes are independently excited. An integral Lyapunov function can be constructed within the framework of quasi-linear theory [12.23, 12.24].

If one first neglects the $B_{k'k''} I_{k'} I_{k''}$ term (noise source), equation (12.32) has an integral

$$\frac{\mathrm{d}}{\mathrm{d}t} \hat{\mathfrak{F}} = 0 \tag{12.38}$$

from the anti-symmetry of the coefficient $A_{kk'}$, where $\hat{\mathfrak{F}}$ is defined as

$$\hat{\mathfrak{F}} = \frac{1}{2} \sum_k \left[\frac{\partial}{\partial \omega} \Re \varepsilon_1(\boldsymbol{k}, \omega) \right] (I_k - \hat{I}_{0,k} \ln I_k) \tag{12.39}$$

and $\hat{I}_{0,k}$ is the solution of stationary equation

$$\sum_{k'} A_{kk'} \hat{I}_{0,k'} = \Im \varepsilon_1(\boldsymbol{k}, \omega). \tag{12.40}$$

The trajectory of $\{I_k\}$ in the phase space is confined on the hypersurface

$$\hat{\mathfrak{F}} = \text{const.} \tag{12.41}$$

Figure 12.3 illustrates the trajectory of spectra of linearly unstable and stable modes.

The 'noise source term' $B_{k'k''} I_{k'} I_{k''}$ modifies the trajectory of $\{I_k\}$ in the phase space. Writing the source term symbolically, $\sum_{k'} B_{k'k''} I_{k'} I_{k''} \to \varepsilon_{\text{noise},k}$, one defines

$$\mathfrak{F} = \frac{1}{2} \sum_k \left[\frac{\partial}{\partial \omega} \Re \varepsilon_1(\boldsymbol{k}, \omega) \right] (I_k - I_{0,k} \ln I_k) \tag{12.42}$$

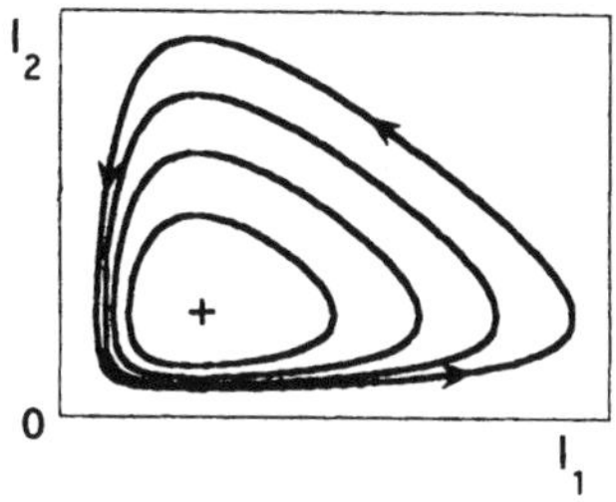

Figure 12.3. Quasi-linear coupling of unstable and stable modes, I_1 and I_2, respectively, in the absence of noise. Arrows indicate the evolution of $I_1(t)$, $I_2(t)$. Contours show those for $\hat{\mathfrak{F}} = \text{const}$. In the presence of small but finite noise, the trajectory approaches the stationary state, denoted by a symbol $+$.

with the stationary solution $I_{0,k}$ in the presence of the source term changing

$$-\Im \varepsilon_1(\boldsymbol{k}, \omega) I_{0,k} + \sum_{k'} A_{kk'} I_{0,k'} I_{0,k} + \varepsilon_{\text{noise},k} = 0. \tag{12.43}$$

This $\mathfrak{F}$ is shown to be the Lyapunov function of the system. From equation (12.43), an imaginary part of dielectric constant satisfies the relation as

$$\Im \varepsilon_1(\boldsymbol{k}, \omega) = \sum_{k'} A_{kk'} I_{0,k'} + \frac{\varepsilon_{\text{noise},k}}{I_{0,k}}. \tag{12.44}$$

Substitution of equation (12.44) into equation (12.32) provides relations

$$\frac{1}{2}\left[\frac{\partial}{\partial \omega} \Re \varepsilon_1(\boldsymbol{k}, \omega)\right] \frac{\mathrm{d}}{\mathrm{d}t} I_k = \sum_{k'} A_{kk'} (I_{k'} - I_{0,k'}) I_k + \frac{\varepsilon_{\text{noise},k}(I_{0,k} - I_k)}{I_{0,k}} \tag{12.45a}$$

and

$$\frac{1}{2}\left[\frac{\partial}{\partial \omega} \Re \varepsilon_1(\boldsymbol{k}, \omega)\right] I_{0,k} \frac{\mathrm{d}}{\mathrm{d}t} \ln I_k = \sum_{k'} A_{kk'} (I_{k'} - I_{0,k'}) I_{0,k} + \frac{\varepsilon_{\text{noise},k}(I_{0,k} - I_k)}{I_k}. \tag{12.45b}$$

Combining equations (12.45a) and (12.45b), one has

$$\frac{\mathrm{d}}{\mathrm{d}t} \mathfrak{F} = \sum_{k,k'} A_{kk'} (I_{k'} - I_{0,k'})(I_k - I_{0,k}) - \sum_k \frac{\varepsilon_{\text{noise},k}(I_{0,k} - I_k)^2}{I_{0,k} I_k}. \tag{12.46}$$

The first term of the right-hand side vanishes, because the matrix A is anti-symmetric, i.e., $A_{k,k'} = -A_{k',k}$ holds. One has

$$\frac{\mathrm{d}}{\mathrm{d}t} \mathfrak{F} = -\sum_k \frac{\varepsilon_{\text{noise},k}(I_{0,k} - I_k)^2}{I_{0,k} I_k} \leq 0 \tag{12.47}$$

for positive or zero noise sources. Now the functional $\mathfrak{F}$ is the Lyapunov function of the dynamics of $\{I_k\}$, and $\mathcal{F}$ is minimum if $I_k = I_{0,k}$ holds for all k. The noise source term $\varepsilon_{\text{noise},k}$ acts as friction, and, as the average, all trajectories finally approach

$$\{I_k\} \to \{I_{0,k}\}. \tag{12.48}$$

The thermodynamic implications are now explained. The number density of the mode, N_k, is introduced as

$$N_k = \frac{1}{8\pi}\frac{\partial}{\partial\omega}\Re\,\varepsilon_1(\boldsymbol{k},\omega)I_k. \tag{12.49}$$

Adding a constant value $-\sum_k \omega_k N_{0,k}$ to equation (12.42), $\mathcal{F}$ is transformed to $\mathcal{F}_{\text{H}}$ as

$$\mathcal{F}_{\text{H}} = \mathcal{F} - \sum_k \omega_k N_{0,k} = \sum_k \omega_k [N_k - N_{0,k}(\ln N_k + 1)]. \tag{12.50}$$

The internal energy of the wave is

$$\mathcal{E} = \sum_k \omega_k N_k, \tag{12.51}$$

and the wave entropy is defined as

$$S_{\text{wave}} \approx k_{\text{B}} \sum_k (\ln N_k + 1) \tag{12.52}$$

in the large N_k limit.

If $\omega_k N_{0,k}$ is common to all k (i.e., equipartition holds), it can be written as a 'temperature',

$$\omega_k N_{0,k} = k_{\text{B}} T_{\text{wave}} \quad \text{(for all } \boldsymbol{k}\text{)} \tag{12.53}$$

and $\mathcal{F}_{\text{H}}$ is written as

$$\mathcal{F}_{\text{H}} = \mathcal{E} - k_{\text{B}} T_{\text{wave}} S_{\text{wave}}. \tag{12.54}$$

In this case, the quantity $\mathcal{F}_{\text{H}}$ corresponds to the Helmholtz free energy [12.24]. Equation (12.47) is considered to be the approach to finding the minimum of the Helmholtz free energy of fluctuation state, if the equipartition (12.53) holds. In realistic cases, however, the stationary solution equation (12.43) does not satisfy the equipartition law equation (12.53).

12.4 Transport Matrix and Symmetry

The distribution function evolves by fluctuation feedback. Quasi-linear diffusion of the distribution function is induced. Cross-correlation functions are related to the fluxes of global quantities, which are expressed in terms of the gradient and fluctuation spectrum $\{I_k\}$. The transport matrix of fluxes and gradients is symmetric in the framework of quasi-linear theory [12.25–12.29]. The details are explained in Appendix 12A.

Appendix 12A

Quasi-Linear Theory of Transport

Within the framework of the quasi-linear theory, which is briefly explained here, the reactions of turbulence with a mean distribution and global profile can be calculated [12.19]. The fluctuation-driven flux can be characterized by a transport matrix.

12A.1 Reaction on Mean Distribution

The back-reaction of the linear response affects the average distribution function as

$$\frac{\partial}{\partial t}\bar{f} = \frac{-e_s}{m_s}\left\langle (\tilde{\boldsymbol{E}} + \boldsymbol{v}\times\tilde{\boldsymbol{B}})\cdot\frac{\partial}{\partial \boldsymbol{v}}\tilde{f}\right\rangle. \tag{12A.1}$$

Upon substitution of the linear response function $\tilde{f}$ into equation (12A.1), the evolution equation is obtained. The electrostatic limit, $\tilde{\boldsymbol{E}} = -\nabla\tilde{\phi}$, is shown for an illustration as

$$\frac{\partial}{\partial t}\bar{f} = -\left(\frac{e}{m}\right)^2 \sum_k |\phi_k|^2 \boldsymbol{k}\cdot\frac{\partial}{\partial \boldsymbol{v}}\Im\left(\frac{1}{\omega - \boldsymbol{k}\cdot\boldsymbol{v}}\right)\boldsymbol{k}\cdot\frac{\partial}{\partial \boldsymbol{v}}\bar{f}. \tag{12A.2}$$

A diffusion equation in the phase space is obtained, with the diffusion coefficient being a quadratic function of the perturbation amplitude in the quasi-linear model. The influence of a strong magnetic field and finite-gyroradius effects has been extensively discussed in the literature.

12A.2 Influence on the Global Profile and Transport Matrix

In inhomogeneous plasmas, diffusion in the phase space is also associated with transport in the real space.

The fluctuation-driven fluxes of particle, energy, and current have been calculated by use of the quasi-linear theory. Fluxes are expressed in terms of a transport matrix. Similarly to neoclassical transport, the transport matrix has been obtained for electron fluxes. In this appendix, by convention Γ denotes the particle flux, not the decorrelation rate. The following is an example of electrons from [12.26]:

$$\begin{pmatrix} \Gamma_r \\ q_{e,r}/T_e \\ \bar{J}_\parallel/T_e \end{pmatrix} = -\begin{pmatrix} D & -\frac{3}{2}D & D_{31} \\ -\frac{3}{2}D & \frac{13}{4}D & -0.87D_{31} \\ D_{31} & -0.87D_{31} & -\sigma_{e\varsigma}/T_e \end{pmatrix}\begin{pmatrix} X_1 \\ X_2 \\ \bar{E}_\parallel \end{pmatrix}, \tag{12A.3}$$

where the driving forces are given by

$$X_1 = -\left\langle \frac{\omega}{m} \right\rangle \frac{eBr}{T_e} - \frac{T_i}{T_e}\frac{1}{p_i}\frac{\mathrm{d}p_i}{\mathrm{d}r} - \frac{e}{T_e}\frac{\mathrm{d}\phi}{\mathrm{d}r} + \frac{T_e + T_i}{T_e}\frac{1}{p}\frac{\mathrm{d}p}{\mathrm{d}r}, \tag{12A.4a}$$

$$X_2 = \frac{1}{T_e}\frac{\mathrm{d}T_e}{\mathrm{d}r}, \tag{12A.4b}$$

and the matrix elements are given by

$$D = \frac{\sqrt{\pi}}{4} n_e \rho_{pe}^2 \frac{v_{\mathrm{th},e}}{qR} \sum_{m,n,\omega} \frac{m^2}{|m - nq|} \left(\frac{e\tilde{\phi}_{m,n,\omega}}{\sqrt{2}T_e} \right)^2, \tag{12A.5a}$$

$$D_{31} = \frac{\sqrt{\pi}}{2} \frac{n_e}{\nu_{ei} B_{\mathrm{p}}} \frac{v_{\mathrm{th},e}}{qR} \sum_{m,n,\omega} \frac{m(m - nq)}{|m - nq|} \left(\frac{e\tilde{\phi}_{m,n,\omega}}{\sqrt{2}T_e} \right)^2, \tag{12A.5b}$$

and

$$\sigma_{e\zeta} = \sigma_c \left\{ 1 - 0.175 \frac{\sqrt{\pi} v_{\mathrm{th},e}}{\nu_{ei} qR} \sum_{m,n,\omega} |m - nq| \left(\frac{e\tilde{\phi}_{m,n,\omega}}{\sqrt{2}T_e} \right)^2 \right\}, \tag{12A.5c}$$

where the over-bar, which denotes the global (averaged) quantity, is suppressed for simplicity. In this expression, m and n are the poloidal and toroidal mode numbers, respectively, ρ_{pe} is the electron gyroradius in the poloidal magnetic field, and σ_c is the electrical conductivity (determined by the collisional processes). There are other anomalous fluxes, such as the ion energy flux $q_{i,r}$, electron-ion energy equipartition Q_{ei}, ion shear viscosity, etc. More complete transport matrices for the sets of fluxes $(\Gamma, q_{e,r}, q_{i,r}, J_{\|})$ or $(\Gamma, q_{e,r}, q_{i,r}, Q_{ei})$ are explicitly given in [12.28]. A transport matrix that includes the ion shear viscosity is discussed in [12.27].

The transport matrix is a quadratic function of the fluctuation amplitude. The fluctuation level is not determined in this framework, and must be found from other theoretical considerations. An order-of-magnitude estimate of the flux is often made by employing the fluctuation level of $e\tilde{\phi}/T \sim 1/kL_n$. However, that level is sufficiently high that the assumption of quasi-linear response is violated, as is noted in §13.1.

12A.3 On the Inward Pinch

In addition to the off-diagonal terms of the transport matrix, which are associated with pinch terms, an additional mechanism exists that causes particle and energy pinches. Specifically, the fluctuation-driven flux caused by the inhomogeneity of the magnetic field has been analyzed. This flux is seen as a pinch.

The linear response of the distribution function to the electric field perturbation is calculated in a toroidal geometry. By use of a simplified equation, i.e., the gyrokinetic equation (see, e.g., equation (11.32)), one has the response as

$$\tilde{f} = \frac{-i}{\omega - k_{\parallel}v_{\parallel} - \omega_{\mathrm{M}}}\left(\tilde{\boldsymbol{V}}_{E\times B}\cdot\nabla\bar{f} - \frac{e}{T}v_{\parallel}\tilde{E}_{\parallel}\bar{f} - \frac{v_{\perp}^2 + 2v_{\parallel}^2}{v_{\mathrm{th}}^2}\tilde{\boldsymbol{V}}_{E\times B}\cdot\frac{\nabla\bar{B}}{\bar{B}}\bar{f}\right), \tag{12A.6}$$

where $\tilde{\boldsymbol{V}}_{E\times B} = \tilde{\boldsymbol{E}} \times \hat{\boldsymbol{b}}/\bar{B}$ and

$$\omega_{\mathrm{M}} = (v_{\perp}^2 + 2v_{\parallel}^2)\omega_c^{-1}\boldsymbol{k}\cdot(\hat{\boldsymbol{b}} \times \nabla\bar{B})\bar{B}^{-1} \tag{12A.7}$$

is the magnetic drift frequency [12.30]. The particle flux, which is caused by the fluctuating $E \times B$ drift, is given by

$$\Gamma = \langle\tilde{\boldsymbol{V}}_{E\times B}\tilde{n}\rangle = \left\langle\tilde{\boldsymbol{V}}_{E\times B}\int \mathrm{d}\boldsymbol{v}\,\tilde{f}(\boldsymbol{v})\right\rangle. \tag{12A.8}$$

There thus appears a term proportional to the gradient of the magnetic field (the third term on the right-hand side of equation (12A.6)). The third term, which is a diffusion in phase space, is associated with a plasma flux directed to the region of higher magnetic field. Note that the origin of this third term in equation (12A.6) is the gradient-B drift term in the gyrokinetic equation (11.32) (the second term of $\boldsymbol{v}_D$, which is explicitly given in equation (11.33a)).

This flux can cause the pinch in tokamaks. It is in the direction of the major radius for tokamak configurations, so the component of this flux across the magnetic surface has a poloidal dependence of $\cos\theta$, where θ is the poloidal angle. As has been discussed in the case of collisional transport (Pfirsch–Schlüter diffusion), the surface area has a dependence as $1 + (r/R)\cos\theta$, and is larger in the lower-field side. As a result, an inward pinch is induced. Upon comparing the first term and the third term on the right-hand side, one has an order-of-magnitude estimate

$$\frac{\text{flux by the third term}}{\text{flux by the first term}} \simeq \frac{-D_{\mathrm{QL}}rR^{-2}n}{-D_{\mathrm{QL}}\nabla n}. \tag{12A.9}$$

From this estimate one has the particle and heat fluxes, Γ and q_r, as

$$\Gamma \simeq -D_{\mathrm{QL}}\nabla n - D_{\mathrm{QL}}\frac{r}{R^2}n \tag{12A.10a}$$

and

$$q_r \simeq -\chi_{\mathrm{QL}}\nabla\bar{p} - \chi_{\mathrm{QL}}\frac{r}{R^2}\bar{p}, \tag{12A.10b}$$

where D_{QL} and χ_{QL} are the quasi-linear diffusion coefficient and quasi-linear thermal diffusion coefficient, respectively. If the gradient is weak, the pinch term can be effective. The pinch term vanishes near the magnetic axis $r = 0$, because the poloidal angle dependence of the surface element vanishes there. If the fluctuation amplitude is not constant on the magnetic surface, like the case of ballooning instabilities, the pinch term could be larger.

References

[12.1] Mikailowski A B 1974 *Theory of Plasma Instabilities* transl. J B Barbour (New York: Consultants Bureau); 1992 *Electromagnetic Instabilities in an Inhomogeneous Plasma* (Bristol: IOP Publishing)

[12.2] Hasegawa A 1975 *Plasma Instabilities and Nonlinear Effects* (Berlin: Springer)

[12.3] Tang W M 1978 *Nucl. Fusion* **18** 1089

[12.4] Kadomtsev B B and Pogutse O P 1971 *Nucl. Fusion* **11** 67

[12.5] Connor J W *et al* 1993 *Plasma Phys. Contr. Fusion* **35** 319

[12.6] Connor J W and Wilson H R 1994 *Plasma Phys. Contr. Fusion* **36** 719

[12.7] Horton C W 1999 *Rev. Mod. Phys.* **71** 735

[12.8] Weiland J 2000 *Collective Modes in Inhomogeneous Plasma* (Bristol: IOP Publishing)

[12.9] Hammett G W and Perkins F W 1990 *Phys. Rev. Lett.* **64** 3019

[12.10] Chang Z and Callen J D 1992 *Phys. Fluids B* **4** 1167

[12.11] Snyder P B, Hammett G W, and Dorland W 1997 *Phys. Plasmas* **4** 2687

[12.12] Mattor N 1998 *Phys. Plasmas* **5** 1822

[12.13] Sugama H, Watanae T-H and Horton W 2001 *Phys. Plasmas* **8** 2617

[12.14] Okuda H and Dawson J M 1973 *Phys. Fluids* **16** 408

[12.15] Sagdeev R Z, Shapiro V D and Shevchenko V I 1978 *JETP Lett.* **27** 340 [1978 *Pis'ma Zh. Eksp. Teor. Fiz.* **27** 361]

[12.16] Connor J W, Hastie R J and Taylor J B 1979 *Proc. R. Soc. A* **365** 1

[12.17] Lortz D and Nuhrenberg J 1978 *Phys. Lett.* **68**A 49

[12.18] Coppi B *et al* 1979 *Comments Plasma Phys. Contr. Fusion* **5** 1

[12.19] Sagdeev R Z and Galeev A A 1969 *Nonlinear Plasma Theory* ed T M O'Neil and D L Book (New York: Benjamin)

[12.20] Elsässer K and Gräff P 1971 *Ann. Phys. (NY)* **68** 305

[12.21] Sagdeev R Z 1979 *Rev. Mod. Phys.* **51** 1

[12.22] Berk H and Breizman B N 1995 *Phys. Fluids B* **2** 2226, 2235, 2246

[12.23] Breizman B N 1978 *Sov. Phys.–JETP* **45** 271 [1977 *Zh. Eksp. Teor. Fiz.* **72** 518]

[12.24] Itoh K and Itoh S-I 1985 *J. Phys. Soc. Jpn.* **54** 1228

[12.25] Hazeltine R D, Mahajan S M and Hitchcock D A 1981 *Phys. Fluids* **24** 1164

[12.26] Shaing K C 1988 *Phys. Fluids* **31** 2249

[12.27] Itoh S-I 1992 *Phys. Fluids B* **4** 796

[12.28] Sugama H and Horton W 1995 *Phys. Plasmas* **2** 2989

[12.29] Balescu R 1991 *Phys. Fluids B* **3** 564

[12.30] Smolyakov A I, Callen J D and Hirose A 1993 in *Local Transport Studies in Fusion Plasmas* ed J D Callen, G Gorini and E Sindoni (Bologna: SIF) p 87

Chapter 13

Inhomogeneous Strong Turbulence

13.1 Regime of Strong Plasma Turbulence

In the presence of fluctuating fields, plasma elements (particles) are subject to $E \times B$ motion. The Doppler shift of the $\boldsymbol{k}$ mode by the fluctuating velocity $\tilde{v}_{E \times B}$ of equation (11.8),

$$\omega_{E \times B} = \tilde{\boldsymbol{v}}_{E \times B} \cdot \boldsymbol{k}_{\perp}, \tag{13.1}$$

can substantially influence the mode characteristics if the condition

$$\omega_{E \times B} \simeq |\omega_k| \tag{13.2}$$

holds, where $\boldsymbol{k}_{\perp}$ is the wave vector perpendicular to the main magnetic field. When the fluctuations are quasi-electrostatic (which is a relevant assumption for the low-β plasmas, β being the ratio of the plasma pressure to the magnetic pressure), the perturbed velocity is given in terms of an electrostatic potential perturbation as $\tilde{v}_{E \times B} = -i\boldsymbol{k}' \times \hat{\boldsymbol{b}}\tilde{\phi}/B$ ($\boldsymbol{k}'_{\perp}$ and $\tilde{\phi}$ being the wave vector and the electrostatic potential of the background fluctuations, respectively) and condition (13.2) is rewritten as

$$\frac{e\tilde{\phi}}{T} \approx \frac{1}{kL_n} \tag{13.3}$$

for the range of drift frequency. Condition (13.3) is easily satisfied for the instabilities listed in §12.2 as will be discussed later. In this section, 'strong turbulence' is used for the case where equation (13.2) is satisfied.

It is noted that a strong plasma turbulence occurs even if the 'Reynolds number' remains of order unity. The Reynolds number R_e might be evaluated by the ratio between the convective nonlinear term $\boldsymbol{v} \cdot \nabla \boldsymbol{v}$ and the viscous damping term. The molecular viscosity μ_c is of the order of

$$\mu_c \approx \nu_{ii}\rho_i^2, \tag{13.4}$$

where ν_{ii} is the ion–ion collision frequency and ρ_i is the ion gyroradius. (Note that μ_c can be enhanced by an order of magnitude in toroidal plasmas [13.1–13.3]. Such details are, however, not the subject of this heuristic argument.) For a turbulence which is induced by the instabilities in the range of $k\rho_i \approx 1$, one has an estimate for R_e as

$$R_e = \frac{v_{E\times B}k}{\nu_{ii}\rho_i^2 k^2} \simeq \frac{\omega_*}{\nu_{ii}} \tag{13.5}$$

where the fluctuating $E \times B$ velocity is estimated for the fluctuation level of equation (13.3) as $k\tilde{\phi}/B \approx T/eL_nB$. The value of R_e can remain of the order of unity. (An example of resistive drift wave turbulence is given in [13.4].) The strong plasma turbulence, characterized by condition (13.2), develops without requiring the large Reynolds number. For fluctuations with global scale length, $k \approx 1/L_n$, one has

$$R_e \simeq \frac{\omega_*}{\nu_{ii}} \frac{L_n^2}{\rho_i^2} \tag{13.6}$$

for the level of (13.3). For such a case, the Reynolds number becomes greater than unity.

In the regime of the strong turbulence, nonlinear interactions of various gradients become noticeable. The flux in a transport matrix is no longer linear with respect to the wave spectrum. The symmetry of the transport matrix does not necessarily hold.

13.2 Concepts to Describe Inhomogeneous Turbulent Plasmas

Common features are seen between the neutral fluid turbulence and plasma turbulence: however, there are specific aspects in the turbulence of magnetized plasmas. Concepts that characterize turbulence in inhomogeneous plasmas are briefly surveyed here.

13.2.1 Gradients (Magnetic Surface, Shear, etc.)

Magnetized plasmas have a strong anisotropy between the motions along and perpendicular to the magnetic field. The global plasma parameters, say the temperature $\bar{T}$, tend to be uniform along the magnetic field in a short time. A plasma confinement scheme is constructed so as to insulate the high temperature plasma from cold surrounding materials, forming the nested and closed magnetic surfaces (as an average). One example is the toroidal plasma confinement, for which one uses quasi-toroidal coordinates, r, θ, and ζ, where r is the minor radius, θ is the poloidal angle, and ζ is the toroidal angle (see figure 13.1). An overview of toroidal plasma is given in, e.g., [2.22, 2.23, 2.26].

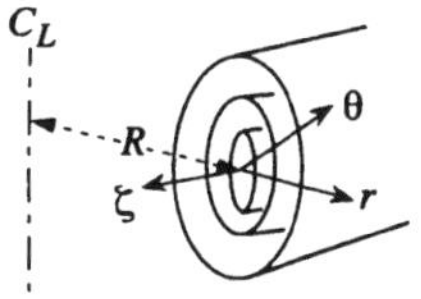

Figure 13.1. Geometry of toroidal plasma and quasi-toroidal coordinates (r, θ, ζ). Local Cartesian coordinates (x, y, z) are taken where the x axis is in the r direction and the z axis is in the direction of the magnetic field.

Inhomogeneities that drive (or suppress) instabilities and turbulence exist primarily perpendicular to the magnetic field. The gradients in the number density, pressure and/or temperature are across the magnetic surfaces. The global plasma velocity, $\bar{\boldsymbol{V}}$, flows on the magnetic surface. The gradient of $\bar{\boldsymbol{V}}$ across the magnetic surface, $\mathrm{d}\bar{\boldsymbol{V}}/\mathrm{d}r$, strongly influences the characteristics of the turbulence.

Inhomogeneity in the electromagnetic field is also important for plasmas. The global plasma velocity $\bar{\boldsymbol{V}}$ across the magnetic field is related to the radial electric field through the relation

$$-e\bar{n}_i(\bar{\boldsymbol{E}} + \bar{\boldsymbol{V}} \times \bar{\boldsymbol{B}}) + \nabla \bar{p}_i + \nabla \cdot \bar{\boldsymbol{\Pi}}_i^{\mathrm{d}} = 0 \tag{13.7}$$

where $\boldsymbol{\Pi}^{\mathrm{d}}$ is the traceless part (deviatoric part) of the stress tensor. The inhomogeneity of $\bar{\boldsymbol{E}}$ is related to that of $\bar{\boldsymbol{V}}$ and affects the turbulence. (Details about the effects on turbulence are discussed in chapter 20.) The gradient of the magnetic field structure $\nabla|\bar{\boldsymbol{B}}|$ also influences the free energy, which generates instabilities and turbulence as is shown in equation (12.24). $\nabla|\bar{\boldsymbol{B}}|$ plays the role of effective gravity, being analogous to a buoyancy force in fluid dynamics. It is a convention to name a 'magnetic hill' if

$$\nabla|\bar{\boldsymbol{B}}|\nabla\bar{p} > 0 \tag{13.8}$$

holds so that the $\nabla|\bar{\boldsymbol{B}}|$ term contributes to cause an instability. If $\nabla|\bar{\boldsymbol{B}}|\nabla\bar{p} < 0$, it is called a 'magnetic well'. The pitch of the magnetic field line,

$$1/q = RB_\theta/rB_\zeta, \tag{13.9}$$

selects the wave vector; R is the major radius of the torus. The wavevector in the direction of $\bar{\boldsymbol{B}}$, $k_\| = \boldsymbol{k} \cdot \boldsymbol{B}B^{-1}$ has the form $k_\| \propto (k_\zeta + k_\theta r/qR)$. For a combination with

$$k_\theta/k_\zeta = -qR/r, \tag{13.10a}$$

it vanishes,

$$k_\| = 0. \tag{13.10b}$$

A surface on which $k_\| = 0$ holds is called a mode rational surface. The variation of the magnetic field direction (in the r direction) is called the 'magnetic shear' and is measured by

$$s = rq^{-1}\frac{\mathrm{d}q}{\mathrm{d}r} \tag{13.11}$$

called the magnetic shear parameter. In the presence of magnetic shear, the condition $k_{\|} \approx 0$ holds near the mode rational surface, in a narrow region with the width in proportion to s^{-1}.

These geometrical and inhomogeneous structures play an important role in the turbulence of confined plasmas.

13.2.2 Mode, Wave, and Vortex

The wavelength of plasma instabilities can be much shorter than the system size. Therefore, fluctuations could be observed as a 'mode' or a 'wave', if the spatial correlation length is longer than the wavelength. Similar arguments apply to the temporal evolution.

An example of the *weak turbulence* picture is seen in §12.3. In the limit of equation (12.30), the observed correlation function takes the form

$$\langle \tilde{E}(t')^* \tilde{E}(t) \rangle = \sum_k I(\boldsymbol{k}, \omega_k)\, \mathrm{e}^{-\mathrm{i}\omega_k (t-t')}, \tag{13.12}$$

i.e., a wave feature (a pattern that satisfies $\omega = \omega_k$) is observed. In the case where equation (12.36) is satisfied, fluctuations might be observed as waves.

A wave-like feature is barely visible, if the correlation length becomes shorter. When the spectrum becomes broader, so that

$$\langle \tilde{E}(\omega')^* \tilde{E}(\omega) \rangle : \; \delta(\omega - \omega') \rightarrow \frac{\Delta\omega}{(\omega - \omega')^2 + \Delta\omega^2}, \tag{13.13}$$

the autocorrelation time becomes shorter:

$$\langle \tilde{E}(t')^* \tilde{E}(t) \rangle \rightarrow \exp[-\mathrm{i}\omega_k (t - t') - \Delta\omega |t - t'|]. \tag{13.14}$$

If the autocorrelation time is comparable with or shorter than ω_k^{-1}, the wave-like feature is not observed. When one measures, say, the electrostatic potential $\tilde{\phi}$, the transient peak of $\tilde{\phi}$ could disappear in a short time $\Delta\omega^{-1}$, so that the perturbation is no longer seen as a wave. When the width of spectrum in k space broadens and the width Δk becomes of the order of the representative mode number, the perturbation is no longer considered as a wave but as a small-scale vortex (with short life time). When the turbulence becomes strong, the perturbation can be seen to be composed of vortices.

On some occasions, a solitary vortex appears. The life time of an isolated potential structure can be much longer than the eddy turnover time, as discussed in the next subsection.

13.2.3 Propagating Solitary Structure

The other solution of interest is a solitary structure which moves with a constant velocity. In inhomogeneous plasmas, a localized propagating perturbation has been obtained from a reduced set of equations in a form like

$$\phi(x, y, t) = \phi(x, \hat{y}) \quad \text{and} \quad \hat{y} = y - v_{**}t. \tag{13.15}$$

(x and z coordinates are taken in the direction of decreasing density and in the direction of magnetic field, respectively.) The velocity v_{**} stands for the propagating velocity in the direction of diamagnetic drift. An example is

$$\phi(x, \hat{y}) = \begin{cases} v_{**}a_{\mathrm{m}}K_1^{-1}(b)K_1(bra_{\mathrm{m}}^{-1})\cos\theta & \text{for } r > a_{\mathrm{m}} \\ v_{**}[r + b^2\gamma^{-2}(r - a_{\mathrm{m}}J_1^{-1}(\gamma)J_1(\gamma r a_{\mathrm{m}}^{-1}))]\cos\theta & \text{for } r < a_{\mathrm{m}} \end{cases} \tag{13.16a}$$

where $r^2 = x^2 + \hat{y}^2$, $x = r\cos\theta$, $b = a_{\mathrm{m}}\sqrt{1 - V_{\mathrm{de}}/v_{**}}$, γ is a solution of the equation $\gamma J_1(\gamma)J_2^{-1}(\gamma) = -bK_1(b)K_2^{-1}(b)$, and functions J and K are the Bessel function and modified Bessel function, respectively. (In this form, normalization follows that of drift wave ordering, i.e., that of equation (11.15).) This propagating solitary structure is a pair of oppositely-rotating vortices. It is often called drift wave *modon*. It is anti-symmetric in the x direction (in the direction of the density gradient) and symmetric in the $\hat{y}$ direction. Away from the vortex, the perturbation decays with the asymptotic form

$$|\phi(x, \hat{y})| \to \exp(-r\sqrt{1 - V_{\mathrm{de}}v_{**}^{-1}}). \tag{13.16b}$$

Thus this structure is localized in the region of $r \approx a_{\mathrm{m}}/b$.

This solitary solution is characterized by the velocity v_{**} and the size a_{m}. The magnitude of the potential perturbation, velocity and size are coupled to each other. For the same size a_{m}, the stronger vortex (with larger $|\phi|$) has the larger velocity v_{**}. The vortex is propagating either in the direction of the ion diamagnetic drift $v_{**} < 0$ (ion modon) or of the electron diamagnetic drift $v_{**} > 0$ (electron modon). The velocity of the drift wave vortex (electron modon) is faster than the drift velocity V_{de}. The dynamics and life time have been discussed [13.5, 13.6].

13.2.4 Convective Cell, Zonal Flow and Streamer

By the words 'mode' and 'wave', one describes patterns that change in space and time. There is a mode that does not propagate but which is subject to growth or damping. A typical example is the *convective cell* shown in figure 12.2(a). This has a very small wavenumber along the magnetic field line and is associated with the electric field across the magnetic field line. Two limiting cases have special importance in the turbulence of inhomogeneous plasmas.

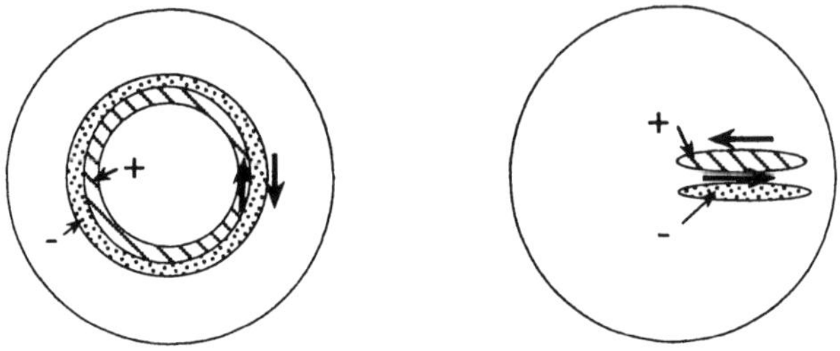

Figure 13.2. Examples of convective cell formation are illustrated in a poloidal cross-section of toroidal plasmas. Zonal flow (left) and streamer (right). In these examples plasma is charged positive in the hatched region and negative in the dotted region, respectively. Radially-inhomogeneous poloidal flow occurs for the zonal flow (a), and poloidally-inhomogeneous radial flow happens for the streamer (b).

One limit is that the perturbation is almost constant on a magnetic surface and is localized in the vicinity of the particular magnetic surface. The perturbed electric field is in the radial direction:

$$\boldsymbol{k} \simeq (k_r, 0, 0) \quad \text{and} \quad \tilde{\boldsymbol{E}} \simeq (\tilde{E}_r, 0, 0) \tag{13.17}$$

(see figure 13.2(a)). The perturbed flow on the magnetic surface is associated with the field. The flow is strongly localized in the vicinity of a particular magnetic surface. This perturbation is called a *zonal flow* [13.7]. The presence of the zonal flow has a strong impact on the evolution of micro-scale turbulence as will be explained later in chapter 20.

The other limit is that the perturbation is rapidly changing in the poloidal direction, but is almost constant in the radial direction,

$$\boldsymbol{k} \simeq (0, k_\theta, 0) \quad \text{and} \quad \tilde{\boldsymbol{E}} \simeq (0, \tilde{E}_\theta, 0) \tag{13.18}$$

(see figure 13.2(b)). The perturbed flow due to this electric field perturbation is in the radial direction in the vicinity of a particular poloidal angle. This perturbation is called *streamer*. (An analogy holds between the 'streamer' in plasma modelling and the 'streak' in fluid dynamics. Both of them are caused by advection in the direction of the gradients of global parameters.) Once a streamer is generated, the radial flow of plasma (energy, etc.) can be enhanced. This has also a strong impact on the transport phenomena in plasmas.

13.2.5 Reconnection, Island Overlapping, Braiding, and Mixing

A topological change in the magnetic field is a key to understanding plasma turbulence, because plasma particles move almost freely along magnetic field lines. The motion of electrons is much faster than that of ions. Once the topology of the magnetic field lines changes, a global variation in the plasma parameters suddenly arises. This global change becomes complicated if there is a substantial difference in the velocity of the electrons and ions: not only the plasma parameters, but also the electromagnetic fields, are modified rapidly.

A topological change takes place owing to reconnection of the magnetic field lines. If a system with nested closed magnetic surfaces is subject to symmetry breaking perturbations, the magnetic surfaces are deformed. A high electric conductivity of plasmas tends to preserve the topology. The deformation generates strongly localized current near the mode rational surface. Due to the collisional resistivity, electron inertial impedance, or current diffusion within a collisionless skin depth, a small but finite perturbed electric field is induced in the direction of the magnetic field line, and induces the perturbed magnetic field that causes reconnection. Through this reconnection process, the topology of magnetic field lines and magnetic surfaces changes. Magnetic surfaces are no longer nested or isolated, but have many islands in between isolated surfaces.

In the case where many magnetic islands are generated on various magnetic surfaces, the interactions between islands induce the global variation of the plasma. If the width of the magnetic islands w_{is} becomes larger than the radial distance of magnetic islands d_{is}, i.e.,

$$\frac{w_{\text{is}}}{d_{\text{is}}} \geq 1 \tag{13.19}$$

a global stochastization of the magnetic field can take place [13.8, 13.9]. Field lines become stochastic and meander diffusively over a wide area. Phase space mixing and much transport may occur [13.10]. Under such circumstances, the electron temperature and pressure tend to become uniform in this region of stochasticity. Ions tend to move as well. There could arise the difference between the electron and ion responses, so that a selective loss of electron energy and momentum can be induced. See also [13.11–13.13] for the application to plasma physics.

13.2.6 Plume and Avalanche (Time Intermittence)

A perturbation like a streamer might not be constant in time but could be intermittent. The heat flux associated with the perturbations happens abruptly and the flow across the magnetic surfaces can occur like an avalanche. Non-stationary features of the flux are seen as plumes or avalanches.

13.2.7 Clumps

The Balescu–Lenard collision operator includes the collective nature of the plasma dynamics through the form of the dielectric constant [2.40]. Thus, important plasma characteristics, which arise from the long-range interactions mediated by the electromagnetic field, could be described by this collision operator. Under some circumstances, the initial condition (ballistic term) can affect the relaxation. If the deviation of the initial condition from the equilibrium distribution is large, it can influence (accelerate) the relaxation. The ballistic effect is usually analyzed in terms of the concept of a *clump*, because a clump (cluster) of particles behaves together as a macroparticle with a finite life time [13.14, 13.15]. An explanation of the clump concept is briefly given in Appendix 13A. The description of clumps is limited in this article: here we stress the fluid (moment) description of plasma turbulence, because we emphasize the mutual relationship between the plasma and a neutral fluid. More detailed discussion of the clump formalism, with particular emphasis on its application to nonlinear fluid theory, can be found in, e.g., [2.52] and references therein.

13.3 Microscale and Mesoscale Structures and Competition

A 'mesoscale' is a scale length intermediate between that of the global inhomogeneity and microscopic fluctuations [2.28]. A typical example is the electric domain interface, which is illustrated in figure 13.3. Two radial regions of different radial electric field touch across a thin layer, and a steep gradient is established. This is called a domain interface. One of the central themes of recent turbulence theory in plasmas is the mutual interaction between macroscale and mesoscale inhomogeneities and the turbulent fluctuations that regulate them, leading to anomalous transport, transitions and improved confinement [2.28, 2.41–2.52]. Microscopic fluctuations generate the convective cells and mesoscale structures. The convective cells, including the electric field domain interface, zonal flow and streamers, cause the suppression and excitation of micro-instabilities (figure 13.4). The stabilizing and destabilizing effects of mesoscale structure on the turbulence cause the complex dynamics, including a subcritical excitation of turbulence. These nonlinear features are surveyed later.

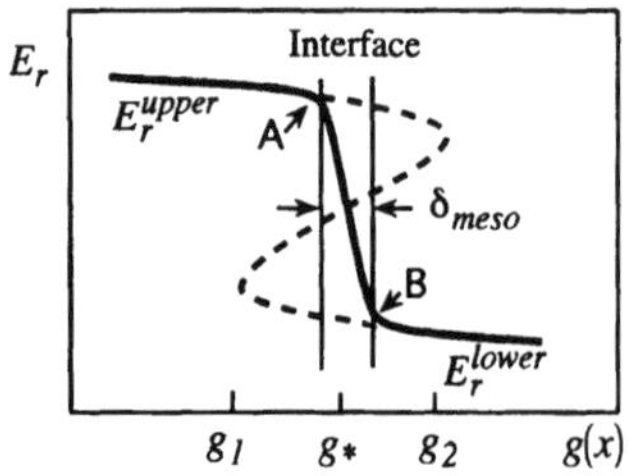

Figure 13.3. An electric field domain interface as a typical example of mesoscale structure. Radial electric field as a function of radial parameter is illustrated.

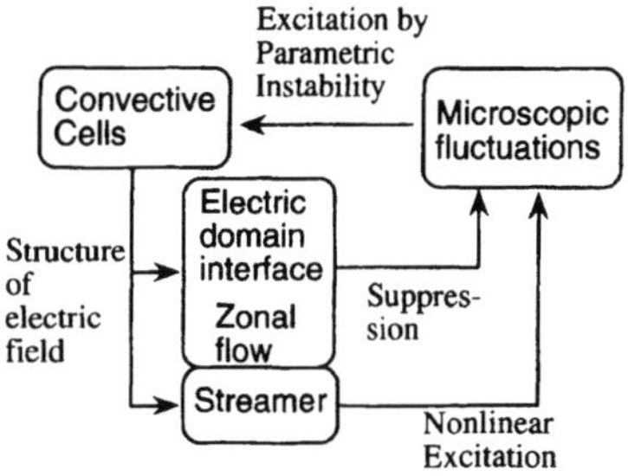

Figure 13.4. Generation–destruction mechanisms between the microscopic turbulence and mesoscale structures (like the electric domain interface, convective cells, zonal flow and streamers).

Appendix 13A

Clumps

The contribution of ballistic terms in the response function is discussed in [13.14–13.17]. A brief explanation of the concept is made after [13.15]. A more thorough survey of the clump theories is given in [2.52].

In deriving equation (12A.2) from (12A.1), the wave-like solution of the linearlized Vlasov equation has been substituted. However, the solution of the linearlized equation of equation (11.29), for electrons

$$\frac{\partial}{\partial t}\tilde{f} + \boldsymbol{v}\cdot\nabla\tilde{f} = \frac{-\mathrm{i}e}{m_e}\tilde{\phi}\boldsymbol{k}\cdot\frac{\partial}{\partial \boldsymbol{v}}\bar{f},$$

is a sum of the wave-like solution and that of the homogeneous equation as

$$\tilde{f} = \frac{e}{m_e}\frac{\tilde{\phi}}{\omega - \boldsymbol{k}\cdot\boldsymbol{v}}\boldsymbol{k}\cdot\frac{\partial}{\partial \boldsymbol{v}}\bar{f} + \tilde{f}_{\mathrm{cl}}, \tag{13A.1}$$

where $\tilde{f}_{\mathrm{cl}}$ is a solution of the homogeneous equation $(\omega - \boldsymbol{k}\cdot\boldsymbol{v})\tilde{f}_{\mathrm{cl}} = 0$. (The suffix cl denotes clump here.) The self-consistent electric field that is induced by this clump is

$$\tilde{\phi} = \frac{-1}{\varepsilon}\frac{e}{\varepsilon_0 k^2}\int \mathrm{d}v\,\tilde{f}_{\mathrm{cl}}, \tag{13A.2}$$

where ε is the dielectric constant (see equation (12.32)). The clump in the phase space $\tilde{f}_{\mathrm{cl}}$ and this induced electric field cooperatively introduce the evolution of the averaged distribution function. Substituting equations (13A.1) and (13A.2) into equation (12A.1), one has

$$\frac{\partial}{\partial t}\bar{f} = \frac{e^2}{m^2}\int\left(\frac{e}{\varepsilon\varepsilon_0 k}\right)^2\boldsymbol{k}\cdot\frac{\partial}{\partial \boldsymbol{v}}H[\bar{f}(\boldsymbol{v}),\bar{f}(\boldsymbol{v}')]\,\mathrm{d}\boldsymbol{v}'\,\mathrm{d}\boldsymbol{v}''\,\mathrm{d}\boldsymbol{k}\,\mathrm{d}\omega, \tag{13A.3}$$

$$H[\bar{f}(\boldsymbol{v}), \bar{f}(\boldsymbol{v}')] = \left\{ \langle \tilde{f}_{\text{cl}}(\boldsymbol{v}')\tilde{f}^*_{\text{cl}}(\boldsymbol{v}'')\rangle_{k\omega}\delta(\omega - \boldsymbol{k}\cdot\boldsymbol{v})\boldsymbol{k}\cdot\frac{\partial}{\partial\boldsymbol{v}}\bar{f}(\boldsymbol{v}) \right.$$

$$\left. - \langle \tilde{f}_{\text{cl}}(\boldsymbol{v}'')\tilde{f}^*_{\text{cl}}(\boldsymbol{v})\rangle_{k\omega}\delta(\omega - \boldsymbol{k}\cdot\boldsymbol{v}')\boldsymbol{k}\cdot\frac{\partial}{\partial\boldsymbol{v}'}\bar{f}(\boldsymbol{v}') \right\} \quad (13A.4)$$

with the two-point correlation function

$$\langle \tilde{f}_{\text{cl}}(\boldsymbol{r}+\boldsymbol{l}, \boldsymbol{v}', t+\tau)\tilde{f}_{\text{cl}}(\boldsymbol{r}, \boldsymbol{v}, t)\rangle = \int \langle \tilde{f}_{\text{cl}}(\boldsymbol{v})\tilde{f}_{\text{cl}}(\boldsymbol{v}')\rangle_{k\omega} \exp(\mathrm{i}\boldsymbol{k}\cdot\boldsymbol{l} - i\omega\tau)\,\mathrm{d}\boldsymbol{k}\,\mathrm{d}\omega. \quad (13A.5)$$

The clump contribution in equation (13A.3) could cause the relaxation in the non-wave region where the dielectric constant ε does not vanish. The correlation of the clumps is proportional to $\delta(\omega - \boldsymbol{k}\cdot\boldsymbol{v})$, and remains finite in a small interval $\Delta r\,\Delta v$ in the phase space. The interval must satisfy $\Delta r \ll v_{\text{th}}/\omega_p$ and $\Delta v \ll v_{\text{th}}$. The correlation relation is approximated as

$$\langle \tilde{f}_{\text{cl}}(\boldsymbol{v})\tilde{f}_{\text{cl}}(\boldsymbol{v}')\rangle_{k\omega} = A(\Delta r\,\Delta v)^3\delta(\omega - \boldsymbol{k}\cdot\boldsymbol{v})\delta(\boldsymbol{v}-\boldsymbol{v}'), \quad (13A.6)$$

where A is an amplitude given by the initial condition. Notice that the enhancement factor $Q^2 \equiv (\Delta r\,\Delta v)^3$ indicates that the contribution of the relaxation is due to a macro particle with charge Qe. Since the binary collision frequency is proportional to the square of the charge of particle, the clump formation is influential in accelerating the relaxation.

An example is the case where an initial distribution function is given by such a model that $\bar{f} = 1$ if $\bar{f} \neq 0$ (like a spinodal decomposition) [13.15]. With this initial condition, the coefficient A is approximated as $A = (1-\bar{f})\bar{f}$. An accelerated relaxation to the Fermi distribution is shown. (It has been pointed out that the interactions with collective modes can give rise to a high-energy tail.) Also note that the initial memory of the clump decays in time. Therefore the acceleration of the relaxation is effective when the initial distribution function deviates substantially from the equilibrium distribution function.

A quantitative analysis has been developed for the study of the clump correlation function in, e.g., [13.16–13.19]. The method of deriving the two-point correlation has been extended to pressure-driven turbulence, e.g., [13.4, 13.20, 13.21]. More extensive discussion of the clump algorithm as applied to fluid problems is given in [2.52].

References

[13.1] Galeev A A and Sagdeev R Z 1973 in *Review of Plasma Physics* ed M A Leontovich (New York: Consultants Bureau) vol 7, p 257

[13.2] Rosenbluth M N, Hinton F L and Hazeltine R D 1972 *Phys. Fluids* **15** 116

[13.3] Balescu R 1988 *Transport Processes in Plasmas. 2. Neoclassical Transport Theory* (Amsterdam: North-Holland)
[13.4] Terry P W and Diamond P H 1985 *Phys. Fluids* **28** 1419
[13.5] Makino M, Kamimura T and Taniuti T 1981 *J. Phys. Soc. Jpn.* **50** 980
[13.6] Meiss J D and Horton C W 1983 *Phys. Fluids* **26** 990
[13.7] Hasegawa A, Maclennan C G and Kodama Y 1979 *Phys. Fluids* **22** 2122
[13.8] Chirikov B V 1979 *Phys. Reports* **52** 263
[13.9] Lichtenberg A J and Liebermann M A 1984 *Regular and Stochastic Motion* (New York: Springer)
[13.10] Balescu R 1975 *Equilibrium and Non-Equilibrium Statistical Mechanics* (New York: Wiley)
[13.11] Rosenbluth M N, Sagdeev R Z, Taylor J B and Zaslavsky G M 1966 *Nucl. Fusion* **6** 297
[13.12] Filonenko N N, Sagdeev R Z and Zaslavsky G M 1967 *Nucl. Fusion* **7** 253
[13.13] Fukuyama A, Momota H and Itatani R 1977 *Phys. Rev. Lett.* **38** 701
[13.14] Dupree T H 1970 *Phys. Rev. Lett.* **25** 789
[13.15] Kadomtsev B B and Pogutse O P 1970 *Phys. Rev. Lett.* **25** 1155
[13.16] Dupree T H 1972 *Phys. Fluids* **15** 334
[13.17] Dupree T H 1983 *Phys. Fluids* **26** 2460
[13.18] Boutros-Ghali T and Dupree T H 1981 *Phys. Fluids* **24** 1839
[13.19] Berman R H, Tetreault D J and Dupree T H 1983 *Phys. Fluids* **26** 2437
[13.20] Krommes J A 1997 *Phys. Plasmas* **4** 655
[13.21] Terry P W, Diamond P H, Shaing K C, Garcia L and Carreras B A 1986 *Phys. Fluids* **29** 2501

Chapter 14

Method for Strong Turbulence I. Renormalization and Statistical Method

The perturbed electromagnetic field in plasmas is coupled to the perturbed current as equation (11.3), i.e.,

$$\frac{1}{c^2}\frac{\partial^2}{\partial t^2}\tilde{\boldsymbol{E}} + \nabla \times \nabla \times \tilde{\boldsymbol{E}} = -\mu_0 \frac{\partial}{\partial t}\tilde{\boldsymbol{J}}.$$

The perturbed current, as in equation (12.11),

$$\tilde{\boldsymbol{J}} = \sum_{s:k,\omega} \exp(\mathrm{i}\boldsymbol{k}\cdot\boldsymbol{x} - \mathrm{i}\omega t) e_s \int \boldsymbol{v}\tilde{f}_{s,k\omega}(\boldsymbol{v})\,\mathrm{d}\boldsymbol{v} \tag{14.1}$$

is determined by the dynamical equation like the Vlasov equation. The properties of the fluctuations, particularly their nonlinear behaviour, are solely determined by the response of plasma current perturbation to the perturbed electromagnetic field, $\tilde{\boldsymbol{J}}[\tilde{\boldsymbol{E}}, \tilde{\boldsymbol{B}}]$. The study of nonlinear dielectric tensors has been a central issue for plasma turbulence.

14.1 Resonance Broadening and Renormalization in the Kinetic Propagator

14.1.1 Renormalization of the Propagator

When one describes plasma dynamics through kinetic equations, the perturbed velocity distribution function is obtained by solving the Vlasov equation. In the framework of the linear response, the perturbed distribution function to the electric perturbation $\tilde{E} \propto \tilde{E}_{k,\omega} \exp(\mathrm{i}\boldsymbol{k}\cdot\boldsymbol{x} - \mathrm{i}\omega t)$ has the form

$$\tilde{f} \propto \frac{1}{\omega - k_{\parallel}v_{\parallel} + \mathrm{i}\nu}\tilde{E}_{k,\omega}, \tag{14.2}$$

where ν is the time rate at which the particles lose the phase memory through collisions. The particles satisfying the condition $v_{\parallel} \simeq \omega/k_{\parallel}$, i.e., the resonant particles, contribute to the process of the energy exchange between waves. In the presence of the background fluctuations, resonant particles are scattered by fluctuations. The mean free time Γ^{-1} can be much shorter than the Coulombic collision time. Then the resonance in the response function is broadened as

$$\tilde{f} \propto \frac{1}{\omega - k_{\parallel}v_{\parallel} + \mathrm{i}\Gamma}\tilde{E}_{k,\omega}. \tag{14.3}$$

The resonance condition can be rewritten as

$$|\omega - k_{\parallel}v_{\parallel}| < \Gamma. \tag{14.4}$$

The rate at which resonant particles are scattered depends on the magnitude of the fluctuations. One can calculate the resonance broadening by means of the renormalization given in, e.g., [2.52]. When one perturbatively solves the Vlasov equation

$$\begin{aligned}\tilde{f}_k(\boldsymbol{v}) &= \frac{e_s}{m_s}\int_{-\infty}^{t} \mathrm{d}t' \exp\{\mathrm{i}\boldsymbol{k}\cdot(\boldsymbol{x}'-\boldsymbol{x}) - \mathrm{i}\omega(t'-t)\}\frac{\partial}{\partial \boldsymbol{v}'}\bar{f}(\boldsymbol{x}',\boldsymbol{v}')\cdot\tilde{\boldsymbol{E}}_k \\ &+ \frac{e_s}{m_s}\int_{-\infty}^{t} \mathrm{d}t' \exp\{\mathrm{i}\boldsymbol{k}\cdot(\boldsymbol{x}'-\boldsymbol{x}) - \mathrm{i}\omega(t'-t)\}\frac{\partial}{\partial \boldsymbol{v}'}\tilde{f}_p(\boldsymbol{x}',\boldsymbol{v}')\cdot\tilde{\boldsymbol{E}}_{k-p} \\ &+ \cdots \end{aligned} \tag{14.5}$$

one obtains $\tilde{f}_k(\boldsymbol{v})$ in a series of $\tilde{E}^n$. The nth order terms with $\tilde{E}^n$ have a divergence like $(\omega - k_{\parallel}v_{\parallel})^{-n}$. If one collects the most secular terms among these divergent terms at each order, one obtains the renormalized dielectric tensor. In the case of electrostatic fluctuations, $\tilde{\boldsymbol{E}} = -\nabla\tilde{\phi}$, an explicit formula for the dielectric constant is given by

$$\begin{aligned}\varepsilon_{k,\omega} &= 1 - \sum_{s=e,i}\frac{e_s^2}{k^2 m_s} \\ &\times \int \mathrm{d}v\left(\frac{\partial \bar{f}_s}{\partial W} - \frac{J_0\left(\frac{k'_{\perp}v_{\perp}}{\omega_{cs}}\right)}{\omega - k_{\parallel}v_{\parallel} + \mathrm{i}\Gamma_{k,\omega}}\left(\omega\frac{\partial \bar{f}_s}{\partial W} - \frac{\hat{\boldsymbol{b}}\cdot(\boldsymbol{k}\times\nabla)\bar{f}_s}{\omega_{cs}}\right)\right)\end{aligned} \tag{14.6}$$

where $W = \frac{1}{2}m_s(v_{\perp}^2 + v_{\parallel}^2)$ is the kinetic energy of particle, $\hat{\boldsymbol{b}}$ is the unit vector in the direction of the magnetic field, ω_{cs} is the cyclotron frequency and J_0 is the zeroth-order Bessel function of the first kind. The nonlinear turbulent decorrelation operator Γ is determined by the recurrence formula [14.1–14.5]. (The formal theory and a more complete form of the renormalized dielectric function than equation (14.6) can be found in [14.2].) The recurrent formula is explicitly given as

$$\Gamma_{k,\omega}(v_\perp, v_\parallel) = -\Im \sum_{k',\omega'} \frac{J_0^2\left(\dfrac{k'_\perp v_\perp}{\omega_{ci}}\right)}{\omega - \omega' - (k_\parallel - k'_\parallel)v_\parallel + \mathrm{i}\Gamma_{k',\omega'}} \frac{(\boldsymbol{k} \times \boldsymbol{k}' \cdot \hat{b})^2 |\phi_{k',\omega'}|^2}{B^2} \tag{14.7}$$

When the scales of the test mode (k, ω) and the background modes (k', ω') are separated,

$$|k, \omega| \ll |k', \omega'|, \tag{14.8}$$

the damping rate of the (k, ω) component is expressed through a diffusion operator $\boldsymbol{D}$ as

$$\Gamma_{k,\omega} = \boldsymbol{k} \cdot \boldsymbol{D} \cdot \boldsymbol{k}. \tag{14.9}$$

The recurrence formula is then given in terms of the diffusion operator [14.6]:

$$\boldsymbol{D}(v_\perp, v_\parallel) = -\Im \sum_{k',\omega'} \frac{J_0^2\left(\dfrac{k'_\perp v_\perp}{\omega_{ci}}\right)}{\omega' - k'_\parallel v_\parallel + \mathrm{i}\boldsymbol{k}' \cdot \boldsymbol{D} \cdot \boldsymbol{k}'} \frac{(\boldsymbol{k}' \times \hat{b})(\boldsymbol{k}' \times \hat{b})|\phi_{k',\omega'}|^2}{B^2}. \tag{14.10}$$

In the absence of turbulence, the right-hand side of equation (14.10) reduces to the quasi-linear formula as

$$\boldsymbol{D}(v_\perp, v_\parallel) = \sum_{k',\omega'} J_0^2\left(\frac{k'_\perp v_\perp}{\omega_{ci}}\right) \Im\left(\frac{-1}{\omega' - k'_\parallel v_\parallel + \mathrm{i}\nu}\right) \frac{|\tilde{E}_{\perp:k',\omega'}|^2}{B^2}. \tag{14.11}$$

Kinetic equations can describe the wave–particle interaction. Substitution of the renormalized operator (14.7) into formula (14.6) gives the renormalized dielectric constant. Combining the dielectric constant with the Maxwell equation (or charge neutrality condition for the case of long-wavelength and electrostatic perturbations) would provide a solution for the turbulent state. In this turbulent state, the orbits of plasma particles are modified by the turbulence. The analysis of the renormalized propagator has been put upon the calculation of the dielectric screened by the background turbulence. The incoherent noise has often been put aside. Modelling of the incoherent random noise is explained in §14.4.

14.1.2 Strong Turbulence Limit and Fluid Model

The broadening of the resonance in the response function is one reason for using fluid equations for plasma turbulence. If the turbulent scattering becomes stronger and the parallel wavenumber is small, $|k_\parallel v_{\mathrm{th}}| < \Gamma$, all of the plasma particles respond in the same way with one decorrelation rate Γ ($v_{\rm th}$ is the thermal velocity) in equation (14.7) or (14.10). In this limit, the individuality of the particle has no influence on the dielectric tensor. A fluid description is valid in this limit. This approximation is a basis to employ a fluid approach for the problems of plasma turbulence.

14.1.3 Strong Turbulence Limit and Kubo Number

Corrsin approximation [14.7] in strong turbulence is also related with this simplification. In the limit of strong turbulence,

$$|\omega - k_{\parallel} v_{\parallel}| < \Gamma,$$

the $\boldsymbol{k}' \cdot \boldsymbol{D} \cdot \boldsymbol{k}'$ term dominates the denominator of equation (14.10). Then equation (14.10) gives an approximation called Corrsin approximation as

$$D \simeq \frac{|\tilde{\phi}|}{B}. \tag{14.12}$$

In this expression, diffusivity is a linear function of the Kubo number $\mathcal{K}$ if the decorrelation rate and wavelength are specified as being independent of the diffusivity. Here $\mathcal{K}$ is the ratio of the $E \times B$ velocity to the wavelength divided by the correlation time (i.e., correlation time of the vortex divided by eddy turnover time) as [14.8, 14.9]

$$\mathcal{K} = \frac{k^2 |\tilde{\phi}|}{\Gamma B}. \tag{14.13}$$

The assumption that nonlinear decorrelation dominates for all particles, $|\omega - k_{\parallel} v_{\parallel}| < \Gamma$, is a simplified limit. Equation (14.10), which is a closed equation in D for given fluctuations, is derived by use of the Markovian approximation. These analytic simplifications (i.e., Markovian approximation as well as neglect of wave–particle resonance) do not necessarily hold in all circumstances. For example, when particles can be trapped, the $\mathcal{K}$-dependence on D becomes weaker [14.10, 14.11]. Weak dependence on $\mathcal{K}$ is also recovered in transport due to a percolation process [14.12, 14.15].

When the nonlinear decorrelation rate Γ is caused by damping through diffusivity (or viscosity of the same magnitude), $\Gamma \simeq Dk^2$, equations (14.12) and (14.13) mean that

$$\mathcal{K} \simeq 1. \tag{14.14}$$

The strong turbulence state, which is governed by the fluctuating $E \times B$ motion, is specified by equation (14.14). Geometrical effects can modify the Kubo number, even though the plasma turbulence level is given by equation (13.3). This is discussed in §18.2.1 and §25.2.

14.2 Nonlinear Response in Fluid-Like Equations

A system of renormalized kinetic equations is usually too complicated for the study of plasma turbulence. Therefore, renormalization approaches to the moment equations (fluid equations) are applied. By using the fluid (moment) equations, one can study the properties of interactions through convective nonlinearity by sacrificing wave–particle interactions. Here, the analysis of convective nonlinearity is briefly discussed in the following [14.16]. The statistical reduction of the dynamical equation for plasmas is discussed in the next subsection.

A model equation which describes the passive advection of a scalar quantity X in the presence of a statistically specified background velocity $\tilde{\boldsymbol{v}}$ with an external statistical source is given by

$$\frac{\partial \tilde{X}}{\partial t} + (\bar{\boldsymbol{V}} + \tilde{\boldsymbol{v}}) \cdot \nabla \tilde{X} - D_c \nabla^2 \tilde{X} = \tilde{S}^{\mathrm{ext}} \tag{14.15}$$

where D_c is the collisional diffusion coefficient. The fluctuating field quantity $\tilde{X}$ has its own characteristic correlation length l_{cor}. Two categories of background fluctuation fields are considered in $\tilde{v}$, i.e.,

$$\tilde{\boldsymbol{v}} = \tilde{\boldsymbol{v}}_s + \tilde{\boldsymbol{v}}_l, \tag{14.16}$$

where $\tilde{\boldsymbol{v}}_s$ has a much shorter wavelength than l_{cor}, and $\tilde{\boldsymbol{v}}_l$ has a much longer wavelength than l_{cor}. A static and inhomogeneous flow $\bar{\boldsymbol{V}}$ is a special limit of $\tilde{\boldsymbol{v}}_l$. The separation of background flows is shown in figure 14.1.

14.2.1 Short-Wavelength Fluctuations

First, the shorter wavelength components are studied. When the background fluctuation velocity changes very rapidly, one takes the rapid change model (RCM) [14.17] in which the correlation functions are assumed to have the form

$$\langle \tilde{v}_s(x + l, t + \tau) \tilde{v}_s(x, t) \rangle = 2 \langle \tilde{v}_s, \tilde{v}_s \rangle (l) \tau_{\mathrm{ac}} \delta(\tau), \tag{14.17a}$$

$$\langle \tilde{S}^{\mathrm{ext}}(x + l, t + \tau) \tilde{S}^{\mathrm{ext}}(x, t) \rangle = \mathcal{S}^{\mathrm{ext}}(l) \delta(\tau), \tag{14.17b}$$

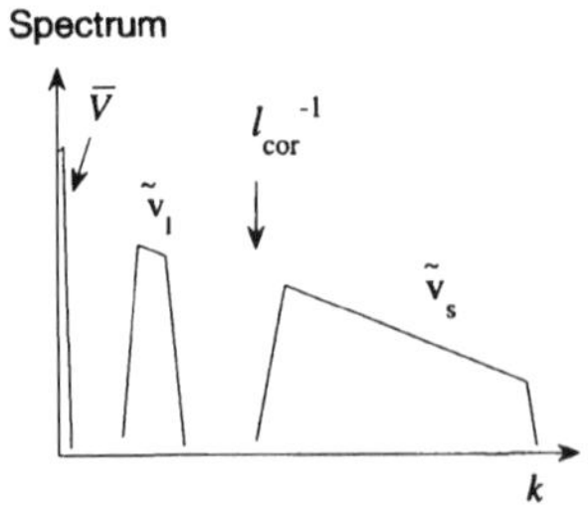

Figure 14.1. Schematic illustration of background fluctuations, $\tilde{v}_s$ and $\tilde{v}_l$, and an inhomogeneous flow $\bar{V}$. They are assumed to be statistically independent.

The autocorrelation time of $\tilde{v}_s$, τ_{ac}, is assumed to be much shorter than the eddy turnover time of the perturbation $\tilde{X}$. With the help of this model, a statistical equation for the passive scalar perturbation can be derived. The correlation function

$$C(l, \tau) = \langle \tilde{X}(x + l, t + \tau)\tilde{X}(x, t)\rangle \tag{14.18}$$

is then governed by the equation

$$\frac{\partial}{\partial t} C - \frac{2}{l^{d-1}} \frac{\partial}{\partial l} l^{d-1} D_{-}(l) \frac{\partial}{\partial l} C = \mathcal{S}^{\text{ext}}(l) \tag{14.19}$$

with

$$D_{-}(l) = D_c + \tau_{ac} \sum_k (1 - e^{ik \cdot l})\langle|\tilde{v}_s|^2\rangle, \tag{14.20}$$

and where d stands for the space dimension.

From the relations (14.19) and (14.20), an order of estimate for the intensity of fluctuation, $I = \lim_{l \to 0} C(l)$, can be made:

$$I \approx \tau_D \left\{ \lim_{l \to 0} \mathcal{S}^{\text{ext}}(l) \right\}, \tag{14.21}$$

$$\tau_D \approx \frac{l_{\text{cor}}^2}{D}, \tag{14.22}$$

and

$$D = \lim_{l \to \infty} D_{-}(l) = D_c + \tau_{ac} \sum_k \langle|\tilde{v}_{s,k}|^2\rangle. \tag{14.23}$$

The diffusion time τ_D over the scale of l_{cor} is a time scale that determines the perturbation lifetime of the passive field [14.16]. Note that equation (14.21) is one form of the fluctuation–dissipation (FD) relation in turbulent plasmas. This relation, which is discussed in several places in this monograph, is an extension, to nonequilibrium plasmas, of the fluctuation–dissipation theorem (FDT) which has been established in the limit of thermodynamical equilibrium [14.18]. It also called the *extended FDT*.

Equation (14.23) agrees with a limiting form of equation (14.10) which is derived by a kinetic approach: in the limit of $k'_{\perp}\rho_i/\omega_{ci} \to 0$ and $|\omega - k_{\parallel}v_{\parallel}| \ll \Gamma$, equation (14.10) with

$$\tau_{ac} = \Gamma^{-1} \tag{14.24}$$

reduces to equation (14.23).

14.2.2 Rapidly-Changing, Long-Wavelength Components

Next the influence of longer wavelength fluctuations is studied. As for $\tilde{\boldsymbol{v}}_s$, the fluctuation velocity $\tilde{\boldsymbol{v}}_l$ is also assumed to be rapidly changing in time. The influence of $\tilde{\boldsymbol{v}}_l$ is introduced as a rapidly changing Doppler-shift,

$\mathrm{i}\tilde{\omega}_k \tilde{X}_k = \tilde{\boldsymbol{v}}_l \cdot \nabla \tilde{X}_k$. Then, from equation (14.15), the forced stochastic oscillator equation

$$\frac{\partial}{\partial t}\tilde{X}_k + \mathrm{i}\tilde{\omega}_k \tilde{X}_k - D_k \nabla^2 \tilde{X}_k = \tilde{S}_k^{\mathrm{ext}} \tag{14.25}$$

is obtained for each Fourier component. The diffusion coefficient D_k is obtained from the contribution of random shorter wavelength components, $\tilde{\boldsymbol{v}}_s$ in equation (14.23). Its contribution is evaluated by the damping rate

$$\Gamma_s = D_k k_\perp^2. \tag{14.26}$$

The impact of stochastic frequency shift is characterized by the parameter

$$\Gamma_l = \tau_{\mathrm{ac},l}\langle \tilde{\omega}_k^2 \rangle \tag{14.27}$$

where $\tau_{\mathrm{ac},l}$ is the autocorrelation time of the longer wavelength fluctuations $\tilde{v}_l$. The statistical property of the response function of equation (14.25),

$$G(t;t') = \exp\left(-\int_{t'}^{t} \mathrm{d}\tau(\mathrm{i}\tilde{\omega}_k + \Gamma_s)\right), \tag{14.28}$$

has been studied when the correlation $\langle \tilde{\omega}_k(t)\tilde{\omega}_k(0)\rangle$ decays as [14.18]

$$\langle \tilde{\omega}_k(t)\tilde{\omega}_k(0)\rangle = \langle \tilde{\omega}_k^2 \rangle \exp(-t/\tau_{\mathrm{ac},l}). \tag{14.29}$$

In the limit of rapidly changing background fluctuations,

$$\tau_{\mathrm{ac},l} \ll \tau_D, \tag{14.30}$$

one obtains

$$\langle G(t;0)G(t';0)\rangle = \exp(-2(\Gamma_l + \Gamma_s)|t - t'|) \tag{14.31}$$

with $\Gamma_s = \tau_D^{-1}$, where τ_D is given in equation (14.22). The decorrelation of the test field occurs with the rate of

$$\Gamma_s + \Gamma_l = \tau_D^{-1} + \Gamma_l. \tag{14.32}$$

The statistical average of fluctuating field $\tilde{X}$ is given as

$$I \sim \frac{\lim_{l\to 0} \mathcal{S}^{\mathrm{ext}}(l)}{\tau_D^{-1} + \Gamma_l} = \frac{1}{1 + \tau_D \Gamma_l}(\tau_D \lim_{l\to 0} \mathcal{S}^{\mathrm{ext}}(l)). \tag{14.33}$$

Comparing equations (14.21) and (14.33), one finds that the fluctuation level is suppressed by the stochastic Doppler shift due to the longer wavelength fluctuations by the factor of

$$\frac{1}{1 + \tau_D \Gamma_l} = \frac{1}{1 + \tau_D \tau_{\mathrm{ac},l}\langle \tilde{\omega}_k^2 \rangle} \quad (\text{for } \tau_{\mathrm{ac},l} \ll \tau_D) \tag{14.34}$$

assuming that the source $\lim_{l\to 0} \mathcal{S}^{\mathrm{ext}}(l)$ is unchanged.

In the large-amplitude limit of random oscillation (or long correlation time $\tau_{\mathrm{ac},l}$),

$$\tau_{\mathrm{ac},l}^2 \langle \tilde{\omega}_k^2 \rangle > 1, \tag{14.35}$$

the Gaussian response is given by

$$\langle G(t;0)G(t';0) \rangle \propto \exp(-2\langle \tilde{\omega}_k^2 \rangle (t-t')^2). \tag{14.36}$$

(Equation (14.36), when compared with equation (14.31), is called motional narrowing.) This results in

$$I \sim \sqrt{\frac{\pi}{2}} \frac{1}{\sqrt{\langle \tilde{\omega}_k^2 \rangle}} \lim_{l \to 0} \mathcal{S}^{\mathrm{ext}}(l). \tag{14.37}$$

Comparing equation (14.37) with equation (14.21), a reduction by the factor

$$\frac{1}{\tau_D \sqrt{\langle \tilde{\omega}_k^2 \rangle}} \tag{14.38}$$

is obtained.

Figure 14.2 illustrates the suppression factor of fluctuation level, equations (14.34) and (14.38), as a function of an average Doppler-shift frequency $\sqrt{\langle \tilde{\omega}_k^2 \rangle}$.

The statistical quantities $\tilde{\omega}_k$ and $\tilde{S}_k^{\mathrm{ext}}$ might be either independent or correlated. The influence of a cross-correlation between $\tilde{\omega}_k$ and $\tilde{S}_k^{\mathrm{ext}}$ has also been studied [14.16]. An example of the result is

$$I \propto 1 - \alpha_1 Z^2, \tag{14.39}$$

where α_1 is a numerical coefficient, and Z is the ratio of cross-correlation to autocorrelation, $Z = \langle \tilde{\omega}_k \tilde{S}_k^{\mathrm{ext}} \rangle / \sqrt{\langle \tilde{\omega}_k^2 \rangle \langle \tilde{S}_k^{\mathrm{ext}} \rangle}$. The cross-correlation between the Doppler shift and the external noise source suppresses the fluctuation level.

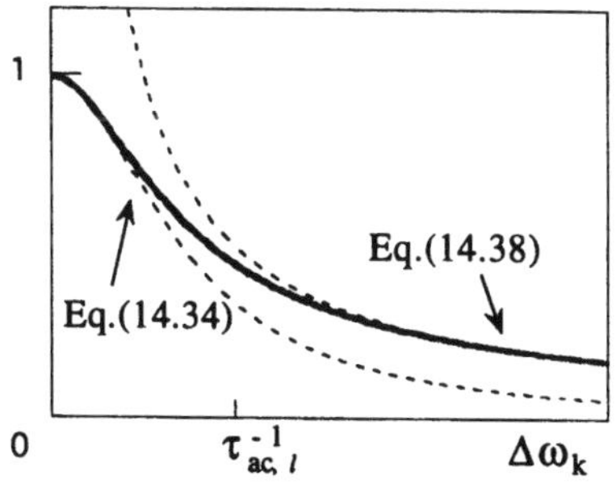

Figure 14.2. A schematic illustration of the suppression factor of the fluctuation level owing to the rapidly changing long-wavelength components. In this figure τ_D and $\tau_{\mathrm{ac},l}$ are fixed and $\Delta\omega_k \equiv \sqrt{\langle \tilde{\omega}_k^2 \rangle}$ is varied. The limiting formulae are shown by dotted curves.

14.2.3 Static but Sheared Flow

A Doppler shift by the sheared static velocity $\bar{V} \cdot \nabla$ also influences the perturbation level of the passive field. For a case of sheared flow, one takes

$$V = (0, V_y(x), 0).$$

The contribution of short-wavelength turbulence is modelled by the turbulent transport coefficient D, and equation (14.15) can be rewritten as

$$\frac{\partial}{\partial t}\tilde{X} + \bar{V}_y(x)\frac{\partial}{\partial y}\tilde{X} - D\nabla^2\tilde{X} = \tilde{S}^{\mathrm{ext}}. \tag{14.40}$$

The sheared static flow induces a stretched vortex, and the correlation time of the test field $\tilde{X}$ is modified. The influences on statistical properties are discussed in detail later (see chapter 20).

14.2.4 On Rigorous Upper Bound

Equations in this subsection are a useful basis for the analysis of plasma turbulence. For some cases, rigorous upper bounds for turbulence-driven transport can be calculated [14.9]. A rigorous upper bound might have a different parameter dependence compared with the true value. However, it provides a firm basis for the understanding of turbulence. For instance, one could find a boundary in a parameter space, below which the upper bound is zero for steady states. Such a criterion might be called the 'energy-stability criterion' for turbulence: perturbation of any amplitude (either subcritical or supercritical ones) decays in time [2.49, 2.52, 14.19]. A brief description on the rigorous upper bound is given in Appendix 14A.

14.3 Renormalization in a Reduced Set of (Fluid-Like) Equations

The nonlinear dispersion relation is derived by use of the renormalized dielectric tensor. This method is illustrated by the example of a reduced set of equations. The reduced set of equations (see §11.2) has the form

$$\frac{\partial}{\partial t}\boldsymbol{f} + \mathcal{L}^{(0)}\boldsymbol{f} = \mathcal{N}(\boldsymbol{f},\boldsymbol{f}) + \tilde{\boldsymbol{S}}_{\mathrm{th}} \tag{14.41}$$

where $\boldsymbol{f}$ denotes the set of variables:

$$\begin{gathered}\boldsymbol{f}^{\mathrm{T}} = (\phi, n), \qquad \boldsymbol{f}^{\mathrm{T}} = (\phi, J, p), \qquad \boldsymbol{f}^{\mathrm{T}} = (\phi, J, v_{\parallel}, p) \\ \boldsymbol{f}^{\mathrm{T}} = (n, \phi, \Phi, v_{\parallel}, p_e, p_i, A_{\parallel})\end{gathered} \tag{14.42}$$

for the two-field Hasegawa–Wakatani (HW) model, three-field model, four-field model, and seven-field Yagi–Horton (YH) model, respectively. The linear operator $\mathcal{L}^{(0)}$ is an $N \times N$ matrix for the N-field model and controls the linear modes. $\mathcal{N}(\boldsymbol{f},\boldsymbol{f})$ are the nonlinear terms, e.g.,

$$\mathcal{N}(\boldsymbol{f},\boldsymbol{f}) = -\begin{pmatrix} \nabla_\perp^{-2}[\phi, \nabla_\perp^2 \phi] \\ [\phi, J] \\ [\phi, p] \end{pmatrix} \tag{14.43}$$

for the case of $\boldsymbol{f}^{\mathrm{T}} = (\phi, J, p)$. The term $\tilde{\boldsymbol{S}}_{\mathrm{th}}$ stands for the thermodynamical excitations induced by the interaction with a heat bath.*

Theoretical models have been developed to separate the nonlinear interaction term into two terms:

$$\mathcal{N}(\boldsymbol{f},\boldsymbol{f}) = \mathcal{N}_{\mathrm{coherent}}(\boldsymbol{f},\boldsymbol{f}) + \tilde{\boldsymbol{S}}, \tag{14.44}$$

where $\mathcal{N}_{\mathrm{coherent}}(\boldsymbol{f},\boldsymbol{f})$ is the coherent part, which changes with the phase of the test mode, f, and $\tilde{\boldsymbol{S}}$ is the incoherent part (noise part). ($\tilde{\boldsymbol{S}}$ can change very rapidly in time. For a description of rapid change, see the discussion in §14.4.1.) Explicit forms of $\mathcal{N}_{\mathrm{coherent}}(\boldsymbol{f},\boldsymbol{f})$ and $\tilde{\boldsymbol{S}}_k$ are given by modelling. The various models for the coherent and incoherent parts have been analyzed (for a detailed discussion, see, e.g., [2.52]). A sample is explained in the following.

A test mode $\boldsymbol{f}_k$ is chosen, and the terms $\mathcal{N}_{\mathrm{coherent}}(\boldsymbol{f},\boldsymbol{f})$ and $\tilde{\boldsymbol{S}}_k$ for $\boldsymbol{f}_k$ are estimated. The coherent part can be estimated by renormalization when the background fluctuations have a smaller scale length than the test perturbation. The coherent part $\mathcal{N}_{\mathrm{coherent}}(\boldsymbol{f},\boldsymbol{f})$ is modelled by a matrix form for the N-field models. In the case of two-field model (HW equation), an explicit form of the matrix is given in [11.16]. (The three-field model is discussed in [2.28].) A diagonalization approximation is often used for analytic insight, based on the view that the parameter dependence of the obtained result is little influenced by off-diagonal terms. (The contribution of the off-diagonal elements must be included for the quantitative conclusion.) The diagonal terms are approximated by the diffusion terms with the turbulent viscosity (μ_N for ion viscosity, μ_{Ne} for electron viscosity, and χ_N for thermal diffusivity), or by the eddy-damping coefficients (γ_1 for ion momentum, γ_2 for parallel electron momentum, and γ_3 for thermal energy), as

$$\mathcal{N}_{\mathrm{coherent}}(\boldsymbol{f},\boldsymbol{f})_k = \begin{pmatrix} \mu_N \nabla_\perp^2 f_1 \\ \mu_{Ne} \nabla_\perp^2 f_2 \\ \chi_N \nabla_\perp^2 f_3 \end{pmatrix}_k = -\begin{pmatrix} \gamma_1 f_1 \\ \gamma_2 f_2 \\ \gamma_3 f_3 \end{pmatrix}_k. \tag{14.45}$$

* The phrase 'thermal noise source' is often used for the noise source of the system in thermal equilibrium. This is not used in the following: the fluctuations and noises of energy or temperature in nonuniform plasmas are the key subject, so that the word 'thermal' might be confusing. Instead, 'thermodynamical excitation' is used.

(Note that nonlinear damping rates γ_1, γ_2 and γ_3 could be different.) Within this diagonal approximation, the renormalized operator $\mathfrak{L}$ is given by

$$\mathfrak{L}_{ij} = \mathfrak{L}_{ij}^{(0)} + \gamma_i \delta_{ij} \tag{14.46}$$

and one has a renormalized reduced set of equations (with a thermodynamical noise source) as

$$\frac{\partial}{\partial t} \boldsymbol{f}_k + \mathfrak{L} \boldsymbol{f}_k = \tilde{\boldsymbol{S}}_k + \tilde{\boldsymbol{S}}_{\mathrm{th},k}, \tag{14.47}$$

where $\boldsymbol{k}$ denotes the test mode.

The renormalized transfer rates are given by [2.49]

$$\gamma_{i,k} = -\sum_{\Delta} M_{i,kpq} M_{i,qkp}^{*} \tau_{i,qkp}^{*} |\tilde{f}_{1,p}^{2}|. \tag{14.48}$$

(The suffix i denotes the field component, $i = 1, 2, 3$ for three-field models, and k, p, q stand for interacting Fourier modes.) In these expressions, the summation Δ indicates the constraint $\boldsymbol{k} + \boldsymbol{p} + \boldsymbol{q} = 0$. The triad interaction time $\tau_{i,kpq}$ obeys the same linear and nonlinear physics as the test mode, as is explained in [2.49]. By employing the Markovian approximation, the triad interaction time is approximately estimated by the relation

$$(\tau_{j,kpq}) \simeq [\mathfrak{L}(k) + \mathfrak{L}(p) + \mathfrak{L}(q)]^{-1} \boldsymbol{I} \tag{14.49}$$

where $\boldsymbol{I}$ is a unit tensor. ($\tau_{j,kpq}$ might be different for different j, and is often approximately estimated by $\tau_{kpq} = \tau_{1,kpq}$. A more thorough discussion of the full matrix model of $\mathfrak{N}_{\mathrm{coherent}}(\boldsymbol{f}, \boldsymbol{f})$ is given in [11.16].) The explicit form of the nonlinear interaction matrix is given as, e.g., $M_{(2,3),kpq} = (\boldsymbol{p} \times \boldsymbol{q}) \cdot \boldsymbol{b}$.

Equation (14.47) shows that the amplitude of the fluctuation $|\boldsymbol{f}_k|$ becomes large in the vicinity of the pole of the renormalized operator $\mathfrak{L}$. Thus the nonlinear dispersion relation

$$\det(\lambda \boldsymbol{I} + \mathfrak{L}) = 0 \tag{14.50}$$

describes the characteristic feature of the turbulence, where $\boldsymbol{I}$ is a unit tensor, and $-\lambda$ is the eigenvalue of the operator $\mathfrak{L}$. (The sign is chosen so that $\lambda < 0$ holds for unstable cases.) In a steady state, the fluctuation level does not grow. This state could be evaluated by the condition,

$$\Re \lambda \simeq 0. \tag{14.51}$$

For purely growing/damping modes, one has $\lambda = 0$, i.e.,

$$\det(\mathfrak{L}) = 0. \tag{14.52}$$

Equation (14.50) is a simplified nonlinear dispersion relation, in which the effects of the noise source are neglected. This describes a dispersion relation of a test mode which is screened by the eddy-damping-like effects of the background turbulence [2.28]. The screened test mode is also called *dressed test mode*. Equation (14.52) gives a rough estimate of the level of turbulence, because $\mathfrak{L}$ is a function of the turbulence spectrum through the renormalization relation, e.g., equation (14.48). It is used to understand the role of various nonequilibrium parameters (e.g., the gradient, magnetic field configuration, etc.) on plasma turbulence. This analytic simplicity is given at a sacrifice of mathematical rigour. This simplified relation neglects the noise source and thus the nonlinear conservation of the quadratic invariants like energy is not guaranteed. The validity of the solution of equation (14.51) must be examined, *a posteriori*, by more accurate modelling in which the renormalized operator $\mathfrak{L}$ and the noise source $\tilde{\boldsymbol{S}}$ are determined self-consistently. The modelling including the noise sources is explained in §14.4, and the relevance of this solution is discussed there.

14.4 Randomness and the Statistical Picture

A fundamental origin of randomness in plasmas is collisions. Parts of nonlinear interactions of the background fluctuations are treated as a random noise source for a test mode. Modelling and the effects of noise source are discussed.

14.4.1 Estimate of Random Source Term

Consider a generic form of equation [14.20]

$$\left(\frac{\partial}{\partial t}+\mathrm{i}\omega_k\right)\tilde{X}_k=\frac{1}{2}\sum_{\Delta}M_{kpq}\tilde{X}_p^*\tilde{X}_q^*, \tag{14.53}$$

where $\tilde{X}_k$ is the fluctuation quantity, ω_k is a complex linear wave frequency, M_{kpq} is the coefficient of nonlinear interactions, $(\boldsymbol{k},\boldsymbol{p},\boldsymbol{q})$ are the wave vectors of interacting modes and $\sum_{\Delta}$ means the summation under the condition $\boldsymbol{k}+\boldsymbol{p}+\boldsymbol{q}=0$. Its counterpart in fluid dynamics is given by equation (3.120). This equation can be modelled by the Langevin equation

$$\left(\frac{\partial}{\partial t}+\mathrm{i}\omega_k+\Gamma_k(t)\right)\tilde{X}_k(t)=\tilde{S}_k(t), \tag{14.54}$$

where $\Gamma_k(t)$ is the nonlinear damping rate discussed in §14.3 and $\tilde{S}_k(t)$ is the nonlinear noise. The assumption

$$\langle\tilde{S}(t)\tilde{S}(t')\rangle\propto\delta(t-t')$$

is often employed in order to ensure the Markovian approximation. The term is treated by use of a white Gaussian model

$$\tilde{S}_k(t)=\tilde{w}(t)\sum_{\Delta}\sqrt{\tau_{kpq}}g_{kpq}, \tag{14.55}$$

where $\tilde{w}(t)$ denotes the white-noise function, τ_{kpq} is the Markovianized nonlinear propagator (long-time average of τ_{kpq} gives the interaction time, that plays the role of autocorrelation time), and $\tilde{g}_{kpq}$ is the average amplitude of instantaneous noise source,

$$\langle \tilde{S}(t)\tilde{S}(t')^*\rangle = \delta(t-t')\tau_{kpq}\langle \tilde{g}_{kpq}\tilde{g}^*_{kpq}\rangle. \tag{14.56}$$

When the noise amplitude $\tilde{g}_{kpq}$ is expressed by the use of the statistically-independent variable ζ (virtual field) as

$$\tilde{g}_{kpq} = M_{kpq}\zeta^*_p\zeta^*_q \tag{14.57}$$

the correlation functions for the virtual field ζ_p is identical to the one for the fluctuating field

$$\langle \zeta_p\zeta^*_q\rangle = \langle \tilde{X}_p\tilde{X}^*_q\rangle. \tag{14.58}$$

14.4.2 Dynamical Equations for Correlation Functions

A set of dynamical equations for correlation functions, called the realizable Markovian closure (RMC), has been discussed [14.21, 14.22]. A two-time correlation function

$$C_k(t,t') = \langle \tilde{X}_k(t)\tilde{X}^*_k(t')\rangle \tag{14.59}$$

can be written in the form of

$$C_k(t,t') = C_k^{1/2}(t)\bar{r}_k(t,t')C_k^{1/2}(t'). \tag{14.60}$$

Here $C_k(t)$ in the one-time correlation function, and $\bar{r}_k(t,t')$ is related to the Markovianized nonlinear damping Γ_k as

$$\bar{r}_k(t,t') = \begin{cases} \exp\left(-\int_{t'}^{t} \mathcal{P}\Gamma(\tau)\,\mathrm{d}\tau\right) & \text{for } t > t' \\ \exp\left(-\int_{t'}^{t} \mathcal{P}\Gamma(\tau)^*\,\mathrm{d}\tau\right) & \text{for } t \le t' \end{cases} \tag{14.61}$$

where the operator $\mathcal{P}$ is defined as

$$\mathcal{P}\Gamma = \Re\Gamma H(\Re\Gamma) + \mathrm{i}\Im\Gamma \tag{14.62}$$

and $H(\Re\Gamma)$ is a Heaviside step function. As discussed in [14.21], the role of $\mathcal{P}$ is to ensure realizability, i.e., to guarantee that the correlation function remains positive.

With the help of this Markovianization, a set of dynamical equations for the correlation function C and the normalized triad interaction time $\hat{\tau}_{kpq}$ can be derived as

$$\left(\frac{\partial}{\partial t} + 2\Re\Gamma_k(t)\right)C_k(t) = 2F_k(t) \tag{14.63}$$

and

$$\left(\frac{\partial}{\partial t}+\Gamma_k(t)+\mathcal{P}\Gamma_p(t)+\mathcal{P}\Gamma_q(t)\right)\hat{\tau}_{kpq}(t)=C_p^{1/2}(t)C_q^{1/2}(t), \tag{14.64}$$

where the damping rate Γ_k and the source term are given as functions of C and τ_{kpq}:

$$\Gamma_k=\gamma_{c,k}-\sum_{\Delta}M_{kpq}M^*_{pqk}\hat{\tau}^*_{kpq}C_q^{1/2}C_k^{-1/2}, \tag{14.65}$$

$$F_k=\tfrac{1}{2}\Re\sum_{\Delta}|M_{kpq}|^2\hat{\tau}_{kpq}C_p^{1/2}C_q^{1/2}, \tag{14.66}$$

with $\gamma_{c,k}$ being the collisional damping rate. This model equation is realizable, i.e., the correlation function C_k remains positive [14.22].

In a stationary state, one has,

$$C_k(t)=\frac{F_k(t)}{\Re\Gamma_k(t)}, \tag{14.67}$$

which is a fluctuation–dissipation relation (extended FDT) for the turbulent state.

14.4.3 Langevin Equations

The Langevin equation (14.47)

$$\frac{\partial}{\partial t}\boldsymbol{f}+\mathcal{L}\boldsymbol{f}=\tilde{\boldsymbol{S}}+\tilde{\boldsymbol{S}}_{\text{th}}$$

is constructed with equations (14.56) and (14.57) as the model form of the nonlinear noise functions [14.23, 14.24].

In order to solve the Langevin equation (14.47), an ansatz for a large number of degrees of freedom in the random modes, N, is introduced. The renormalized term γ_j in $\mathcal{L}$ arises from the statistical sum of N components, so that its variation in time becomes $O(N^{-1/2})$ less than that of f_k. Therefore, in solving f_k, $\mathcal{L}$ is approximated to be constant in time in the limit of $N\to\infty$. The general solution is formally given as

$$\boldsymbol{f}(t)=\sum_m\exp(-\lambda_m t)\boldsymbol{f}^{(m)}(0)+\int_0^t\exp[-\mathcal{L}(t-\tau)]\tilde{\boldsymbol{S}}(\tau)\,\mathrm{d}\tau \tag{14.68}$$

where $-\lambda_m$, $m=1,2,3,\ldots$, and

$$\Re\lambda_1<\Re\lambda_2<\Re\lambda_3<\cdots \tag{14.69}$$

represent the eigenvalues of the renormalized matrix $\mathcal{L}$, and are determined by equation (14.50). ($\boldsymbol{f}^{(m)}(0)$ represents the initial value which is transformed into a diagonal basis.) The sign of λ_m is defined so that $\Re\lambda_m$ is positive when the test mode perturbation does not increase. The decorrelation rate is given by $\Re\lambda_m$. The eigenvector with the eigenvalue $-\lambda_1$ corresponds to the least stable branch (mode) with the longest decay time. Others with $-\lambda_2, -\lambda_3, \ldots$ denote more stable branches, which decay much faster. For the N-field model, equation (14.50) is an Nth-order equation of λ. Equation (14.50) provides a relation between λ_m, γ_j, and global parameters such as

$$G_0 \equiv \nabla(\ln \bar{p}) \cdot \nabla(\ln \bar{B}), \qquad \eta_i \equiv \frac{\nabla(\ln \bar{T}_i)}{\nabla(\ln \bar{n}_e)},$$

etc.

14.4.4 Example of Three-Field Model

The matrix $\exp[-\mathcal{L}(t-\tau)]$ in equation (14.68) can be given explicitly. Matrix-elements of $\exp[-\mathcal{L}(t-\tau)]$ for the operator $\mathcal{L}$ of equation (11.24) are expressed in terms of three discrete eigenvalues of equation (14.50), λ_1, λ_2, and λ_3. They are given as

$$\{\exp[-\mathcal{L}(t-\tau)]\}_{ij} = A_{ij}^{(1)} \exp[-\lambda_1(t-\tau)] + A_{ij}^{(2)} \exp[-\lambda_2(t-\tau)] + A_{ij}^{(3)} \exp[-\lambda_3(t-\tau)] \tag{14.70}$$

where the matrices $\boldsymbol{A}$ are given in [14.23]. By introducing a projected amplitude of noise source,

$$\mathcal{S}^{(m)}(\tau) = \left(1, \frac{-\mathrm{i}k_\parallel}{k_\perp^2(\bar{\gamma}_e - \lambda_m)}, \frac{-\mathrm{i}k_y\kappa}{k_\perp^2(\bar{\gamma}_p - \lambda_m)}\right) \cdot \{\tilde{\boldsymbol{S}}(\tau) + \tilde{\boldsymbol{S}}_{\mathrm{th}}(\tau)\}, \tag{14.71}$$

where γ_e and γ_p stand for γ_2 and γ_3 in equation (14.45), respectively, and κ is the magnetic field gradient in equation (11.24); the fluctuation field is given by

$$\boldsymbol{f}^{(m)}(t) = A_{11}^{(m)} \begin{pmatrix} 1 \\ \dfrac{-\mathrm{i}k_\parallel \xi}{\bar{\gamma}_e - \lambda_m} \\ \dfrac{\mathrm{i}k_y p_0'}{\bar{\gamma}_p - \lambda_m} \end{pmatrix} \int_0^t \exp(-\lambda_m(t-\tau))\mathcal{S}^{(m)}(\tau)\,\mathrm{d}\tau \tag{14.72}$$

upon neglecting the initial condition, which is ineffective in determining the statistical average [14.25]. The superscript '(m)' denotes the mth branch of equation (14.50).

The long-time average of the decomposed amplitude is obtained from equation (14.72) as

$$\langle f_1^{(m)*} f_1^{(m)} \rangle = \frac{1}{2\Re(\lambda_m)} |A_{11}^{(m)}|^2 \langle \mathcal{S}^{(m)*}\mathcal{S}^{(m)} \rangle. \tag{14.73}$$

This is the fluctuation–dissipation relation (extended FDT) in the case of strongly unstable plasmas. One can estimate the noise source in equation

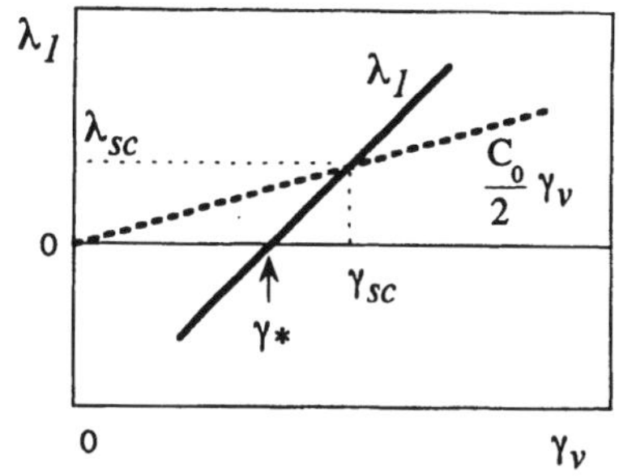

Figure 14.3. The balance between the nonlinear decorrelation rate λ_1 (full line) and the noise source rate $C_0\gamma_v/2$ (broken line). Both written as a function of γ_v. The crossover point determines the statistical averages of the nonlinear decorrelation rate and eddy damping rate. λ_{sc} indicates the self-consistent solution for the turbulence decorrelation rate. γ_{sc} is the eddy damping rate in the self-consistent solution and γ_* denotes the solution of the nonlinear marginal stability condition.

(14.73) as

$$\langle \mathcal{S}^{(1)*}\mathcal{S}^{(1)} \rangle \simeq C_0\gamma_v A_{11}^{-2} \langle f_{1,k}^{(1)*} f_{1,k}^{(1)} \rangle + \text{thermal excitations}, \tag{14.74}$$

where γ_v is the nonlinear damping rate of the vorticity (eddy damping rate), and C_0 is a numerical factor of the order of unity. (This is a crude estimate of the noise level, and as yet there is no modelling of the C_0 that can preserve a conservation property of equation (14.74).) With this estimate, the extended FDT equation (14.74) can be rewritten as

$$\langle f_{1,k}^{(1)*} f_{1,k}^{(1)} \rangle = \frac{C_0\gamma_v}{2\,\Re\,\lambda_1} \langle f_{1,k}^{(1)*} f_{1,k}^{(1)} \rangle + \text{thermal excitations}. \tag{14.75}$$

In the limit of strong turbulence, i.e., fluctuation level is much larger than the thermodynamical fluctuations,* equation (14.75) is simplified as

$$\Re\,\lambda_1 = \frac{C_0}{2}\gamma_v. \tag{14.76}$$

Both the nonlinear decorrelation rate λ_1 and the eddy damping rate γ_v depend on the fluctuation spectrum according to equations (14.48), (14.50) and (14.75); these are the equations that determine the fluctuation level together with the decorrelation rate. A schematic explanation of the foregoing mathematical procedure is given in figure 14.3. The fluctuation level that is determined by the extended FDT equation (14.76) is higher than that determined by the nonlinear marginal stability condition $\Re\,\lambda_1 = 0$.

* The phrase 'thermal fluctation' is often used for the system in thermal equilibrium. Instead of this, 'thermodynamical fluctuation' is used here: the word 'thermal' might be confusing, because the fluctuations of energy and temperature are induced by instabilities as well.

14.5 Fokker–Planck Equation

14.5.1 Projected Variable

By use of the decomposed elements, a Fokker–Planck equation may be derived. As an example of the three-field model a case of equation (14.70) is illustrated. A projected variable for an eigenmode with $\lambda = \lambda_1$

$$F^{(1)}(t) = \left(1, \frac{-ik_{\parallel}}{(k_\perp^2 + \xi)(\bar{\gamma}_e - \lambda_1)}, \frac{-ik_y\kappa}{k_\perp^2(\bar{\gamma}_p - \lambda_1)}\right) \cdot f(t) \qquad (14.77)$$

is introduced. ($\xi = a^2\omega_p^2 c^{-2}$ for the MHD normalization of equation (11.24).) Then the Langevin equation is decomposed as

$$\frac{\partial}{\partial t} F_k^{(1)} + \lambda_1 F_k^{(1)} = \mathcal{S}_k^{(1)}, \qquad (14.78)$$

where the projected noise source $\mathcal{S}_k^{(1)}$ is modelled by white noise. Statistical independence between the different k components holds, and we write

$$\langle \mathcal{S}_k^{(1)}(t)\mathcal{S}_{k'}^{(1)}(t')\rangle \propto g_k^{(1)^2} \delta_{k,k'}\delta(t - t'). \qquad (14.79)$$

14.5.2 Fokker–Planck Equation

Based on the property equation (14.79), the Fokker–Planck equation for the probability density function (PDF) $P(F^{(1)})$ is derived [14.26, 14.27] from equation (14.78):

$$\frac{\partial}{\partial t} P = \sum_k \frac{\partial}{\partial F_k^{(1)}} \left(\lambda_1 F_k^{(1)} + \tfrac{1}{2} g_k^{(1)} \frac{\partial}{\partial F_k^{(1)}} g_k^{(1)}\right) P. \qquad (14.80)$$

14.5.3 Equilibrium Probability Density Function

The equilibrium PDF is given by

$$P_{eq}(\{F_k^{(1)}\}) = \bar{P} \prod_k \frac{1}{g_k^{(1)}} \exp\left\{ - \int^{F_k^{(1)}} \frac{2\lambda_{1,k} F_k^{(1)}}{g_k^{(1)^2}} \, \mathrm{d}F_k^{(1)} \right\}. \qquad (14.81)$$

The presence of the nonlinear noise term, $g_k^{(1)} \neq 0$, allows access to the equilibrium probability density function.

14.5.4 H-Theorem

We define the $\mathcal{H}$ function as

$$\mathcal{H}^{(1)}(t) = \int \mathrm{d}F^{(1)} P(F^{(1)}; t) \ln\left(\frac{P(F^{(1)}; t)}{P_{eq}(F^{(1)})}\right) \qquad (14.82)$$

and obtain the inequality

$$\frac{\mathrm{d}}{\mathrm{d}t}\mathcal{H}^{(1)}(t) \le 0, \tag{14.83}$$

showing that the probability density function relaxes to the equilibrium one in the long-time limit. The irreversibility of evolution, equation (14.83), is due to the nonlinear noise term, $g_k^{(1)} \neq 0$, which is enhanced by the turbulence. The microscopic origin of the irreversibility is the Coulomb collisions, and the Lyapunov exponent of individual particles has been analyzed [14.28]. The irreversibility associated with the renormalized diffusion coefficient $g_k^{(1)^2}$ includes the cascade of energy, which leads to an enhanced irreversible dissipation by Coulombic collisions.

14.5.5 Tail in Probability Density

The tail component in the probability density function is another feature of nonlinear plasma turbulence. The 'effective temperature' that determines the width of $P_{eq}(F_k^{(1)})$ is given by $g_k^{(1)^2}$. If the fluctuation amplitude becomes high, the turbulent noise amplitude $g_k^{(1)^2}$ increases. As a result, $2\lambda_{1,k}F_k^{(1)}g_k^{(1)^{-2}}$ in the integrand of equation (14.81) has an $F_k^{(1)}$ dependence

$$\frac{2\lambda_{1,k}F_k^{(1)}}{g_k^{(1)^2}} \propto \frac{1}{F^{(1)}} \tag{14.84}$$

in the large-amplitude limit. The integral in (14.81) has a logarithmic dependence on $F^{(1)}$, and equation (14.81) predicts a power law distribution in the large-amplitude limit. This power law originates from the fact that the diffusion coefficient $g_k^{(1)^2}$ is renormalized [14.23].

One may compare the analysis with those in fluid theory. In [14.29], however, the integrand of equation (14.81) was expanded with respect to the fluctuation amplitude. Once the integrand is expanded, exponential forms are deduced.

It is also noted that equations (14.80) and (14.81) are not necessarily tractable for the analysis of turbulence in inhomogeneous plasmas. A Fokker–Planck equation for a coarse-grained variable is discussed in chapters 23–25.

14.6 Memory Effects and Non-Markovian Property

The memory effect could be important for plasma turbulence similar to a fluid turbulence. The statistical procedure used for the reduction of variables results in equations with a non-Markovian nature. A proper treatment of memory effect is necessary [14.30, 14.31]. (For instance, the quasi-normal approximation (QNA) retained only the decorrelation due to the molecular viscosity and turned out to be inadequate owing to the memory time being too long.)

Memory effects can be important when the Kubo number is large. Examples are given in §14.1.3. If the fluctuation amplitude $\tilde{A}$ becomes large, fluid elements circumnavigate the fluctuation eddy many times during the decorrelation time τ_{cor} of the fluctuations. The correlation function along the path of fluid elements (the Lagrangian correlation function) becomes different from that at one-time slice (the Eulerian correlation function). The 'decorrelation path method' can be applied to obtain the Lagrangian correlation function [14.10]

$$C_{ij}(t_1, t_2) = \langle v_i(x(t_1); t_1) v_j(x(t_2); t_2) \rangle, \quad i, j = (x, y). \tag{14.85}$$

In stationary homogeneous turbulence, this can be written as

$$C_{ij}(t_1, t_2) = C_L(t_1 - t_2)\delta_{ij}. \tag{14.86}$$

For small $\mathcal{K}$ limit, the quasi-linear limit is recovered as

$$C_L(\Delta t) \simeq \left(\frac{\tilde{A}}{l_{cor}}\right)^2 \exp\left(-\frac{|\Delta t|}{\tau_{ac}}\right), \quad \mathcal{K} \ll 1 \tag{14.87}$$

where l_{cor} and τ_{ac} are the correlation length and autocorrelation time of the fluctuating field, respectively, $\tilde{A}$ being an appropriately normalized amplitude, and the Kubo number is defined as

$$\mathcal{K} = \tilde{A}\tau_{ac}/l_{cor}^2$$

in this case. In the large-$\mathcal{K}$ limit, an asymptotic formula for $C_L(\Delta t)$ can be given by [14.30]

$$C_L(\Delta t) \approx -0.32\left(\frac{\tilde{A}}{l_{cor}}\right)^2 \left(\mathcal{K}\frac{|\Delta t|}{\tau_{ac}}\right)^{-1.42}, \tag{14.88}$$

where $\mathcal{K}|\Delta t/\tau_{ac}| > 3$ and $\mathcal{K} \gg 1$. A slow power-law decay is demonstrated.

This power-law decay of the correlation function demonstrates the limitation of theories that use the Markovian approximation. This approximation is valid in a certain range of Kubo number; investigations which do not use the Markovian property are surveyed in [14.30]. The appearance of a 'subdiffusive process' and a related discussion of transient responses are given in §25.4. The dispersion evolves in time as [14.32]

$$\langle (x(t) - x(0))^2 \rangle \propto t^{\alpha_{tail}} \tag{14.89}$$

with the index of

$$\alpha_{tail} = 0.58. \tag{14.90}$$

The index α_{tail} is smaller than unity, and this process is subdiffusive. The initially localized perturbation, however, develops a long tail in space. An application to drift-wave turbulence has been recently discussed [14.33].

The non-Markovian effect could be important when one studies the relaxation of the velocity distribution in the presence of turbulence and external drive (e.g., current drive). Analysis has been discussed in, e.g., [14.34]. An alternate modelling of statistical theory, putting an emphasis on application to atmospheric and oceanic problems, has been proposed [14.35].

Appendix 14A

Rigorous Upper Bounds for Transport

In some cases, a rigorous upper bound for turbulent transport can be obtained. Analysis of an upper bound has been explored in, e.g., [14.9, 14.19, 14.36]. A brief survey of the concept is explained here; details are given in [14.9]. Let us take an equation of, say, temperature

$$\left(\frac{\partial}{\partial t}+\tilde{u}\frac{\partial}{\partial x}-\chi_c\frac{\partial^2}{\partial x^2}\right)T(x,t)=0 \tag{14A.1}$$

where $\tilde{u}$ is a turbulent velocity field and χ_c stands for the molecular thermal diffusivity that is caused by binary particle collisions. The total flux Q_{tot} is composed of the turbulent transport Q and by the collisional transport Q_c as

$$Q_{\text{tot}}=Q+Q_c, \tag{14A.2}$$

where

$$Q=\langle\tilde{u}T(x,t)\rangle \quad \text{and} \quad Q_c=-\chi_c\frac{\partial}{\partial x}T \tag{14A.3}$$

hold.

When the statistics of the turbulent velocity field $\tilde{u}$ are specified, and are independent of the response of the temperature fluctuation, the problem is called passive. When the dynamics of $\tilde{u}$ is determined by the dynamics of plasmas or fluids (e.g., through the MHD equations or the Navier–Stokes equation), the problem is called self-consistent.

The structure of the theory is explained by taking an example from the passive problem, following [14.9]. The system of equation (14A.1) may be normalized as follows: length to L (the distance between the two boundaries), temperature to ΔT (the temperature difference between the two boundaries), fluctuating velocity to the average

$$\bar{\tilde{u}}=\langle\tilde{u}^2\rangle^{1/2}, \tag{14A.4a}$$

and time to

$$\tau_L=L/\bar{\tilde{u}} \tag{14A.4b}$$

(the macroscopic eddy turnover time). Under this normalization, the characteristic parameters are the Reynolds number R_e and the Kubo number $\mathcal{K}$, which are given as

$$R_e = \bar{\tilde{u}}L/\chi_c \quad \text{and} \quad \mathcal{K} = \bar{\tilde{u}}\tau_{\rm ac}/L, \tag{14A.5}$$

where $\tau_{\rm ac}$ is the autocorrelation time of the fluctuating field $\tilde{u}$. In a passive problem, R_e and $\mathcal{K}$ are given parameters.

The maximum of the possible turbulent flux is discussed with the boundary condition $Q(0) = Q(1) = 0$. As discussed in §11.3 (e.g., equations (11.25) and (11.27)), a quadratic form in perturbed quantities can be obtained from equation (14A.1). A space average such as

$$\bar{\bar{Q}} = \int_0^1 \mathrm{d}x\, Q(x)$$

is employed. In a stationary state, one has

$$\overline{\overline{-Q(x)\frac{\partial}{\partial x}\langle T(x)\rangle}} = R_e^{-1}\overline{\overline{\left(d\frac{\delta T}{\mathrm{d}x}\right)^2}}, \tag{14A.6}$$

where the deviation of the temperature profile is denoted as $\delta T(x)$, with respect to which the turbulent driven flux Q is maximized. Equation (14A.6) can be rewritten as

$$\bar{\bar{Q}} = R_e\overline{\overline{(Q-\bar{\bar{Q}})^2}} + R_e^{-1}\overline{\overline{\left(d\frac{\delta T}{\mathrm{d}x}\right)^2}}. \tag{14A.7}$$

The variational principle is stated as

Maximize the functional $\bar{\bar{Q}}[\delta T]$ subject to the constraint (14A.7). (14A.8)

This maximum principle is well-posed. One has the Schwartz inequality

$$\bar{\bar{Q}} = \overline{\overline{\langle \delta u\, \delta T\rangle}} \le \overline{\overline{\langle \delta u^2\rangle^{1/2}}}\;\overline{\overline{\langle \delta T^2\rangle^{1/2}}}. \tag{14A.9}$$

That is, the left-hand side of equation (14A.7) is linear (at most) in the temperature perturbation, while the right-hand side of equation (14A.7) is quadratic. The constraint equation (14A.7) bounds the domain of δT, and confirms the presence of the upper-bound of the left-hand side, $\bar{\bar{Q}}[\delta T]$.

Several cases have been studied. In the limit of large Kubo number, $\mathcal{K} \to \infty$, i.e., the perturbation is static, the upper bound of the flux is given as

$$\bar{\bar{Q}}_{\rm ub} = \bar{\bar{Q}}_{\infty}(R_e) = R_e^{-1}\{\tfrac{1}{2}R_e\coth(\tfrac{1}{2}R_e) - 1\}. \tag{14A.10}$$

In the limit of small Kubo number (i.e., very short autocorrelation time of fluctuations), the result is obtained as $\bar{\bar{Q}} = \bar{\bar{Q}}_{\rm ql}(\mathcal{K})$. It was pointed out that in order to achieve proper quasi-linear scaling, a two-time constraint is necessary in addition to the constraint equation (14A.7). (See [14.19] for details.) An upper bound is given for an arbitrary Kubo number:

$$\bar{\bar{Q}}_{\rm ub} = \frac{\bar{\bar{Q}}_{\rm ql}(\mathcal{K})\bar{\bar{Q}}_{\infty}(R_e)}{\bar{\bar{Q}}_{\rm ql}(\mathcal{K}) + \bar{\bar{Q}}_{\infty}(R_e)}. \tag{14A.11}$$

Application has also been made to a self-consistent problem in plasma turbulence theories. The MHD equation has been analyzed for the turbulent resistivity of the reversed-field pinch (RFP), and an upper bound for the volume-averaged turbulent resistivity $\eta_{\rm turb}^{\rm ub}$ has been obtained [14.37]. An energy-stability criterion for the turbulence, below which $\eta_{\rm turb}^{\rm ub} = 0$ holds, i.e., any (subcritical as well as supercritical excitation) perturbation is predicted to decay in time, is also obtained.

References

[14.1] Horton C W and Choi D 1979 *Phys. Reports* **49** 273

[14.2] Krommes J A 1984 in *Basic Plasma Physics II* ed A A Galeev and R N Sudan (Amsterdam: North Holland) chapter 5.5

[14.3] Cook I and Taylor J B 1973 *J. Plasma Phys.* **9** 131

[14.4] Dupree T H 1966 *Phys. Fluids* **9** 1773

[14.5] Kono M, Sanuki H and Todoroki J 1975 *Phys. Lett.* **51A** 247

[14.6] Dupree T H 1967 *Phys. Fluids* **10** 1049

[14.7] Corrsin S 1959 in *Atmospheric Diffusion and Air Pollution* ed F N Frenkiel and P A Sheppard (New York: Academic)

[14.8] Kubo R 1962 *J. Phys. Soc. Jpn.* **17** 1100

[14.9] Krommes J A and Smith R A 1987 *Ann. Phys.* **177** 246

[14.10] Vlad M, Spineanu F, Misguich J H and Balescu R 1998 *Phys. Rev. E* **53** 7359

[14.11] Isichenko M B 1992 *Rev. Mod. Phys.* **64** 961

[14.12] Kleva R G and Drake J F 1984 *Phys. Fluids* **27** 1686

[14.13] Chernikov A A, Neishtadt A I, Rogalsky A V and Yakhnin V Z 1991 *Chaos* **1** 206

[14.14] Isichenko M B, Horton W, Kim D E, Heo E G and Choi D-I 1992 *Phys. Fluids B* **4** 3973

[14.15] Ottaviani M 1992 *Euro Phys. Lett.* **20** 111

[14.16] Krommes J A 2000 *Phys. Plasmas* **7** 1148

[14.17] Kraichnan R H 1994 *Phys. Rev. Lett.* **72** 1016

[14.18] Kubo R, Toda M and Hashitsume N 1985 *Statistical Physics II* (Berlin: Springer)

[14.19] Howard L N 1963 *J. Fluid Mech.* **17** 405

[14.20] Krommes J A 1996 *Phys. Rev. E* **53** 4865

[14.21] Bowman J C, Krommes J A and Ottaviani M 1993 *Phys. Fluids B* **5** 3558

[14.22] Bowman J C and Krommes J A 1997 *Phys. Plasmas* **4** 3895

[14.23] Itoh S-I and Itoh K 1999 *J. Phys. Soc. Jpn.* **68** 1891

[14.24] Itoh S-I and Itoh K 1999 *J. Phys. Soc. Jpn.* **68** 2611

[14.25] Itoh S-I and Itoh K 2000 *J. Phys. Soc. Jpn.* **69** 3253

[14.26] Kubo R 1963 *J. Math. Phys.* **4** 174

[14.27] Kitahara K and Yoshikawa K 1994 *Science of Nonequilibrium Systems I* (Tokyo: Kodansha Scientific) [in Japanese] §3.2

[14.28] Ueshima Y, Nishihara K, Barnett D M, Tajima T and Furukawa H 1997 *Phys. Rev. E* **55** 3439

[14.29] Quian J 1983 *Phys. Fluids* **26** 2098

[14.30] Balescu R 2000 *Plasma Phys. Contr. Fusion* **42** B1

[14.31] Balescu R 1995 *Phys. Rev. E* **51** 4807

[14.32] Balescu R 1997 *Statistical Dynamics: Matter Out of Equilibrium* (London: Imperial College Press)

[14.33] Zagorodny A and Weiland J 1999 *Phys. Plasmas* **6** 2359

[14.34] Taguchi M 2001 private communications

[14.35] Frederiksen J S 1999 *J. Atmos. Sci.* **56** 1481; Frederiksen J S and Davies G 1977 *J. Atmos. Sci.* **54** 2475

[14.36] Busse F H 1978 *Adv. Appl. Mech.* **18** 77

[14.37] Krommes J A 1990 *Phys. Fluids B* **2** 1331

Chapter 15

Methods for Strong Turbulence II. Scale Invariance Method

Scale invariance is an alternative approach to understanding turbulence properties. The method and applications for plasma turbulence are reviewed in [15.1].

15.1 Fluid Models

15.1.1 Reynolds Number and Drag

There are two well known examples in fluid dynamics: one is the analysis of the drag force acting on a body immersed in a flow [15.2, 15.3]; the other is Kolmogorov spectrum [15.4].

The drag force is discussed in terms of the scale invariance property. It can be shown that the Navier–Stokes equation for an incompressible fluid

$$\rho\left(\frac{\partial \boldsymbol{V}}{\partial t} + \boldsymbol{V} \cdot \nabla \boldsymbol{V}\right) = -\nabla p + \nu \, \Delta \boldsymbol{V}, \tag{15.1}$$

where ρ is the mass density and ν is the fluid viscosity, is invariant under the following three transformations:

$$\mathcal{T}_1: \quad V \to a_1 V, \quad t \to a_1^{-1} t, \quad p \to a_1^2 p, \quad \nu \to a_1 \nu \tag{15.2a}$$

$$\mathcal{T}_2: \quad \boldsymbol{x} \to a_2 \boldsymbol{x}, \quad t \to a_2 t, \quad \nu \to a_2 \nu, \tag{15.2b}$$

$$\mathcal{T}_3: \quad \rho \to a_3 \rho, \quad p \to a_3 p, \quad \nu \to a_3 \nu. \tag{15.2c}$$

The drag force F, acting on a sphere of radius L in a fluid with velocity V, has the dimension of $\rho V x^3 t^{-1}$, so that $F/\rho V^2 L^2$ is invariant under transformations $\mathcal{T}_1$, $\mathcal{T}_2$ and $\mathcal{T}_3$ (L system size). The Reynolds number $R_e = \rho L V \nu^{-1}$ is also invariant under these transformations, so is an arbitrary function of R_e, $F_a(R_e)$. Therefore the ratio $(\rho V^2 L^2 F_a(R_e))^{-1} F$ is invariant. The drag force must therefore take the form

$$F = \rho V^2 L^2 F_a(R_e). \tag{15.3}$$

15.1.2 Spectrum

The scale invariance method is also useful in determining the turbulent spectrum. Let us consider a homogeneous system and assume that dissipated power per unit volume ε is given. The change of scales is applied as

$$\mathcal{T}_K: \quad x \to a_1 x \quad \text{and} \quad t \to a_2 t. \tag{15.4}$$

Noting the fact that the viscosity μ, the dissipation ε, the energy spectrum of the fluctuations $E(k)$, and the wavenumber are transformed as

$$\mu \to a_1^2 a_2^{-1} \mu, \quad \varepsilon \to a_1^2 a_2^{-3} \varepsilon, \quad E(k) \to a_1^3 a_2^{-2} E(k), \quad k \to a_1^{-1} k, \tag{15.5}$$

one finds that

$$\frac{E(k)}{\mu^{5/4} \varepsilon^{1/4}} \quad \text{and} \quad \frac{k}{\mu^{-3/2} \varepsilon^{1/4}} \tag{15.6}$$

are invariant under this transformation $\mathcal{T}_K$ equation (15.4). From this result, the spectrum is independent of the choice of coordinates if it is expressed as

$$E(k) = \mu^{5/4} \varepsilon^{1/4} F_a(k \mu^{3/4} \varepsilon^{-1/4}) \tag{15.7}$$

where $F_a(z)$ is an arbitrary function. The essence of the inertial range is that the fluctuation energy is independent of viscosity. The factor $\mu^{5/4} F_a(k \mu^{3/4} \varepsilon^{-1/4})$ is chosen as being independent of μ, i.e.,

$$F_a(z) \propto \mu^{-5/4}. \tag{15.8}$$

By use of this form factor, one finally has the Kolmogorov spectrum

$$E(k) \propto k^{-5/3} \varepsilon^{2/3}. \tag{15.9}$$

In the presence of external forces, the counterparts of equation (15.9) are also derived. The surface waves of water are induced by gravity and by surface tension. In the long wave length limit, it is the gravity wave satisfying

$$\omega_k \propto k^{1/2}. \tag{15.10}$$

In the short wave length limit, it is the capillary wave:

$$\omega_k \propto k^{3/2}. \tag{15.11}$$

The counterpart of equation (15.9) has been derived as

$$E(k) \propto k^{-5/2} \quad \text{(gravity waves)} \tag{15.12}$$

and

$$E(k) \propto k^{-3} \quad \text{(capillary waves)} \tag{15.13}$$

respectively [12.21].

15.2 Plasma Models

15.2.1 Transport Coefficient

This scale invariance method is applied to plasma turbulence. The representative examples are shown [15.1].

In the case of electrostatic perturbations, which is valid in the limit that the plasma pressure $\bar{p}$ is low compared with the magnetic pressure $\bar{B}^2/2\mu_0$, the Vlasov equation takes the form

$$\left(\frac{\partial}{\partial t}+\boldsymbol{v}\cdot\nabla+\frac{e_s}{m_s}(\boldsymbol{e}+\boldsymbol{v}\times\bar{\boldsymbol{B}})\frac{\partial}{\partial \boldsymbol{v}}\right)f_s(\boldsymbol{x},\boldsymbol{v};t)=0,\quad s=i,e, \tag{15.14}$$

where $\bar{B}$ is a constant magnetic field, and the distribution function f_s and the electric field contain fluctuating components. In the limit that the scale length of the perturbation is longer than the Debye length λ_D, $|\lambda_D k|\ll 1$, the Maxwell equations are simplified under the charge neutrality condition

$$\sum_{s=i} e_s\int \mathrm{d}\boldsymbol{v}\, f_s=0. \tag{15.15}$$

There are three (and only three) independent transformations that leave these equations (15.14) and (15.15) invariant:

$$\mathcal{T}_1:\ f_j\to a_1 f_j, \tag{15.16a}$$

$$\mathcal{T}_2:\ \boldsymbol{v}\to a_2\boldsymbol{v},\quad \boldsymbol{B}\to a_2\boldsymbol{B},\quad t\to a_2^{-1}t,\quad \boldsymbol{E}\to a_2^2\boldsymbol{E}, \tag{15.16b}$$

$$\mathcal{T}_3:\ \boldsymbol{x}\to a_3\boldsymbol{x},\quad \boldsymbol{B}\to a_3^{-1}\boldsymbol{B}\quad t\to a_3 t,\quad \boldsymbol{E}\to a_3^{-1}\boldsymbol{E}. \tag{15.16c}$$

(Note that the temperature scales as $T\to a_2^2T$ by $\mathcal{T}_2$.) One finds that quantities tB and $Ta^{-2}B^{-2}$ (where a is the plasma size) are invariant under the transformations $\mathcal{T}_1$, $\mathcal{T}_2$, and $\mathcal{T}_3$. That is, the confinement time τ_E must satisfy a relation

$$\tau_E B=F_{\mathrm{a}}\left(\frac{T}{a^2B^2}\right) \tag{15.17}$$

where F_{a} is an arbitrary function. In the history of plasma confinement research, the importance of the Bohm diffusion, for which the diffusivity has a dependence like $D_{\mathrm{B}}\propto T/eB$, and the gyro-Bohm diffusion $D\propto(\rho_i a^{-1})D_{\mathrm{B}}$ has been recognized empirically (see, e.g., [2.39, 15.5]). If the function F_{a} has the form $F_{\mathrm{a}}(z)\approx z^{-1}$, one finds so-called Bohm diffusion, i.e.,

$$\tau_E\propto\frac{a^2B}{T}. \tag{15.18}$$

For another power form of $F_a(z) \approx z^{-3/2}$, one obtains

$$\tau_E \propto \frac{a^3 B^2}{T^{3/2}}, \tag{15.19}$$

which is called the gyro-reduced Bohm diffusion.

The energy confinement time (the time scale of the transport across a global scale length) could be different from the autocorrelation time of the microscopic fluctuations in general. In such a case, F_a takes a different form depending on the choice of characteristic times.

A different choice of dynamical equations leads to a different invariant transformation. For electromagnetic MHD fluctuations, the invariant relation is given by

$$\tau_E B = (n^{1/2} a) F_a\left(\frac{nT}{B^2}, Ta^{1/2}\right). \tag{15.20}$$

In specific circumstances, more explicit forms of turbulent transport can be derived (see the exhaustive references in [15.1]). For instance, one may write the transport coefficient $\chi = a^2 \tau_E$ as

$$\chi = \frac{T^{3/2}}{aB^2}\left(\frac{na}{T^2}\right)^q. \tag{15.21}$$

The index q is summarized for various drift-wave range instabilities in table 15.1.

It should be noted that the scale invariance method gives the same result as the mixing length estimate for supercritical instabilities [15.16].

Table 15.1 Summary of scale-invariance relations for various specific turbulence models. (Quoted from [15.1] with supplement.) References 1–11 correspond to [15.6–15.8, 13.4, 15.9–15.15], respectively.

Instability modes	Index q	Reference
Collisional drift	1	1, 2, 3
	0	2, 3
	2/3	4
	1/3	2
Dissipative trapped electrons	−1	2, 3, 5, 6
	−4/3	7
	−2	8
Collisionless trapped electron	0	2, 3, 5
Dissipative trapped ion	−1	2
Ion-temperature gradient	0	3, 5, 6, 9, 10
Current-diffusive turbulence	0	11

15.2.2 Spectrum

A discussion of the fluctuation spectrum is also feasible. For the current diffusive interchange mode (CDIM) turbulence, a basic model equation is

$$\frac{\mathrm{d}}{\mathrm{d}t}\begin{pmatrix}\nabla_\perp^2\phi \\ j \\ p\end{pmatrix} + \begin{pmatrix}0 & -\mathrm{i}k_y sx & \mathrm{i}k_y G_0 \\ \mathrm{i}k_y sx & 0 & 0 \\ \mathrm{i}k_y & 0 & 0\end{pmatrix}\begin{pmatrix}\phi \\ j \\ p\end{pmatrix} = 0, \qquad (15.22)$$

where $\mathrm{d}/\mathrm{d}t = \partial/\partial t + [\phi, \ ]$. The normalization follows MHD models [15.17]. The driving source of fluctuations G_0 ($G_0 = \nabla \ln \bar{p} \cdot \nabla \ln \bar{B}$) is also transformed. The system described by equation (15.22) is found to be invariant under the transformation [15.15]

$$x \to a_1 x, \qquad t \to a_1^{-1} t \qquad (15.23)$$

with

$$(\phi, j, p) \to (a_1^3\phi, a_1^2 j, a_1 p) \quad \text{and} \quad G_0 \to a_1^2 G_0. \qquad (15.24)$$

The turbulent transport coefficient, D, is transformed as $D \to a_1^3 D$, and $D/G_0^{3/2}$ is found to be invariant. The dependence of the diffusivity

$$D \propto G_0^{3/2} \qquad (15.25)$$

can be satisfied.

From the scaling properties of ϕ and k, $\phi^2 G_0^{-3}$ and $k\sqrt{G_0}$ are invariant. As a result, $\phi^2 G_0^{-3}$ is expressed as a function of $k\sqrt{G_0}$ as $\tilde{\phi}^2 G_0^{-3} \approx F_a(\sqrt{G_0}k)$ where F_a is an arbitrary function. For a quasi-two-dimensional turbulence in magnetized plasmas, ($d = 2$), the energy spectrum, $E(k) = \{k_\perp^2 |\phi(k)|^2\} k^{d-1}$ is given by

$$E(k_\perp) \approx G_0^3 F_a(\sqrt{G_0}k_\perp) k_\perp. \qquad (15.26)$$

If the energy spectrum is expressed by a power law by $E(k) \propto k^{-\nu}$, F_a has a dependence $F_a(z) \propto z^{-1-\nu}$. By use of this form, equation (15.26) is rewritten as

$$E(k_\perp) \approx G_0^{5/2-\nu/2} k_\perp^{-\nu}. \qquad (15.27)$$

The index $\nu = 3$ has been obtained for a nonlinear stationary state, in the wavelength regime where turbulence is excited [2.28].

References

[15.1] Connor J W 1988 *Plasma Phys. Contr. Fusion* **30** 619

[15.2] Lamb H 1932 *Hydrodynamics* 6th edition (Cambridge: Cambridge University Press)

[15.3] Landau L D and Lifshitz E M 1959 *Fluid Mechaniccs* (Oxford: Pergamon Press)

[15.4] Kolmogorov A N 1941 *Dokl. Akad. Nauk SSSR* [*Sov. Phys. Dokl.*] **30** 299
[15.5] Yoshikawa S 1973 *Nucl. Fusion* **13** 433
[15.6] Yoshikawa S 1970 *Phys. Rev. Lett.* **25** 353
[15.7] Düchs D F, Post D E and Rutherford P H 1977 *Nucl. Fusion* **17** 565
[15.8] Waltz R E, Dominguez R R, Wong S K, Diamond P H, Lee G S, Hahm T S and Mattor N 1986 *Plasma Phys. Contr. Nuclear Fusion Research* (Proceedings of 11th Conference, Kyoto 1986) vol 1, p 345 (Vienna: IAEA)
[15.9] Romaneli F, Tang W M and White R B 1986 *Nucl. Fusion* **26** 1515
[15.10] Kadomtsev B B and Pogutse O P 1970 in *Review of Plasma Physics* ed M A Leontovich (New York: Consultants Bureau) vol 5, p 249
[15.11] Similon P L and Diamond P H 1984 *Phys. Fluids* **28** 1419
[15.12] Chen L, Berger R L, Lominadze J G, Rosenbluth M N and Rutherford P H 1977 *Phys. Rev. Lett.* **39** 754
[15.13] Horton W, Choi D-I and Tang W M 1981 *Phys. Fluids* **24** 1077
[15.14] Lee G S and Diamond P H 1986 *Phys. Fluids* **29** 3291
[15.15] Connor J W 1993 *Plasma Phys. Contr. Fusion* **35** 757
[15.16] Yagi M, Wakatani M and Shaing K C 1998 *J. Phys. Soc. Jpn.* **57** 117
[15.17] Yagi M 1995 *J. Plasma Fusion Res.* **71** 1123

Chapter 16

Methods for Strong Turbulence III. Model Based on Reduced Variables

By a reduction of variables, specific processes in inhomogeneous turbulence may be studied. Global perturbations that are selectively chosen by boundary conditions are sometimes excited. In such a case, a truncated model is often used, and a few degrees of freedom for the global perturbations are kept. A typical example is the Lorenz model in fluid dynamics [16.1]. When a cascade plays a dominant role, the shell model is a useful approach [3.19–3.22]. The shell model for a system with linear instabilities will be discussed here. The third is the K-ε model [16.2]. The influence of the boundary appears through the spatial transport of turbulence. The K-ε model based on the TSDIA is used to study such situations.

16.1 Lorenz Model

Perturbations in a bounded region are analyzed by keeping only a small number of Fourier modes. The Lorenz model [16.1] employs a closed set of equations for three scalar quantities. An extension to a model of five-mode truncation was made to study the role of standing waves [16.3]. A model of eight-mode truncation was proposed to introduce the effect of travelling waves [16.4].

The extended Lorenz model has been applied to the study of plasma instabilities in the scrape-off layer (thin plasma region that surrounds the plasma surface) [16.5–16.9]. Five variables are often taken into account: the fundamental mode (X for potential and Y for pressure perturbation), the background profile modification for the plasma pressure (Z), the zonal flow component (V) and the second harmonics of electrostatic potential (W). The equations for truncated components are:

$$\frac{\mathrm{d}}{\mathrm{d}t}X = P_r(-X + Y) + VW, \tag{16.1a}$$

$$\frac{d}{dt}Y = -XZ + r_a X - Y, \tag{16.1b}$$

$$\frac{d}{dt}Z = XY - bZ, \tag{16.1c}$$

$$\frac{d}{dt}V = -XW - \gamma_V V, \tag{16.1d}$$

$$\frac{d}{dt}W = -cXV - \gamma_W W, \tag{16.1e}$$

where P_r is the Prandtl number,

$$r_a = R_a \pi^{-4} k_a^2 (1 + k_a^2)^{-3}, \qquad b = 4(1 + k_a^2)^{-1}, \qquad \gamma_V = P_r (1 + k_a^2)^{-1},$$

$$\gamma_W = P_r (4 + k_a^2)(1 + k_a^2)^{-1}, \qquad c = (3/4) k_a^2 (4 + k_a^2)^{-1},$$

and k_a is the inverse of wavelength y in the direction normalized to the thickness of the layer. R_a is the Rayleigh number. The influence of the VW term in equation (16.1a) shows the stabilizing influence of sheared flow. (If one truncates equation (16.1) with $V = W = 0$, equations (16.1a–c) reduce to the Lorenz model.)

This set of dynamical equations is known to show a chaotic nature. Only one positive Lyapunov number exists in a certain range of the Rayleigh number, even though the number of variables increases from three to five [16.8, 16.10]. (Effective Rayleigh number for the magnetized plasmas is defined by using the magnetic gradient instead of gravitational acceleration, as is explained in §12.2.3.) This approach can capture a chaotic behavior, but not a turbulent feature of excited fluctuations.

16.2 Shell Model

16.2.1 One-Dimensional Model

The energy cascade in the Navier–Stokes equation has been studied by shell models. In particular, the Gledzer–Ohkitani–Yamada (GOY) model is often employed [3.21, 3.22]. A shell model has also been discussed for passive advection [16.11]. With these shell models, a system involving a pressure-driven instability (like Rayleigh–Benard instability or flute instability) may be described [16.12]. A set of equations for the fluid velocity and temperature is:

$$\frac{du_n}{dt} = \mathrm{i}(a_n u_{n+1}^* u_{n+2}^* + b_n u_{n-1}^* u_{n+1}^* + c_n u_{n-1}^* u_{n-2}^*) + P_r \theta_n - P_r k_n^2 u_n \tag{16.2a}$$

$$\frac{d\theta_n}{dt} = \mathrm{i}\{e_n(u_{n-1}^* \theta_{n+1}^* - u_{n+1}^* \theta_{n-1}^*) + g_n(u_{n-2}^* \theta_{n-1}^* + u_{n-1}^* \theta_{n-2}^*) + h_n(u_{n+1}^* \theta_{n+2}^* - u_{n+2}^* \theta_{n+1}^*)\} + R_a u_n - k_n^2 \theta_n, \tag{16.2b}$$

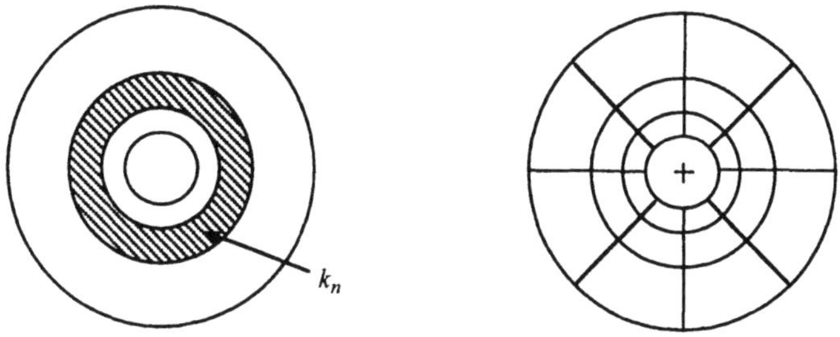

Figure 16.1. Shell in the k-space for a one-dimensional model (left) and for a multiple bin model (right).

where * represents the complex conjugate, $n = 1, \ldots, N$ is the number of each shell, $k_n = C2^{n-1}$, u_n is the fluctuating velocity, θ_n is the fluctuating temperature, R_a is the Rayleigh number,

$$R_a = \frac{L^3 g\alpha_{\text{th}}(\theta(L) - \theta(0))}{\chi\mu_f}, \tag{16.3}$$

μ_f is the kinematic viscosity, and P_r is the Prandtl number (figure 16.1(left)). One takes $a_n = k_n$, $b_n = -k_{n-1}/2$, $c_n = -k_{n-2}/2$, $e_n = k_n/2$, $g_n = -k_{n-1}/2$, and $h_n = k_{n+1}/2$.

One may use this system to examine the consequence of linear insatiability with growth rate:

$$\gamma_n = \frac{-k_n^2(1 + P_r) + \sqrt{k_n^4(1 - P_r)^2 + 4P_r R_a}}{2}. \tag{16.4}$$

The conservation relations for the kinetic and internal energy of fluctuations,

$$\frac{\mathrm{d}}{\mathrm{d}t}\left(\frac{1}{2}\sum_{n=1}^{N}|u_n|^2\right) = -P_r\sum_{n=1}^{N}k_n^2|u_n|^2 + P_r\sum_{n=1}^{N}\theta_n u_n^* \tag{16.5a}$$

and

$$\frac{\mathrm{d}}{\mathrm{d}t}\left(\frac{1}{2}\sum_{n=1}^{N}|\theta_n|^2\right) = -\sum_{n=1}^{N}k_n^2|\theta_n|^2 + R_a\sum_{n=1}^{N}u_n\theta_n^*, \tag{16.5b}$$

show that a competition takes place between the excitation manifested by the heat transport and viscosity damping. A power law solution of the form

$$u_n = Ak_n^{-1/3} \quad \text{and} \quad \theta_n = \sqrt{R_a}Ak_n^{-1/3} \tag{16.6}$$

is obtained for the steady state.

This system has a large number of degrees of freedom with positive Lyapunov exponents and also describes the chaos. In a nonlinear steady state, quantities fluctuate around the levels of the long-time average resulting

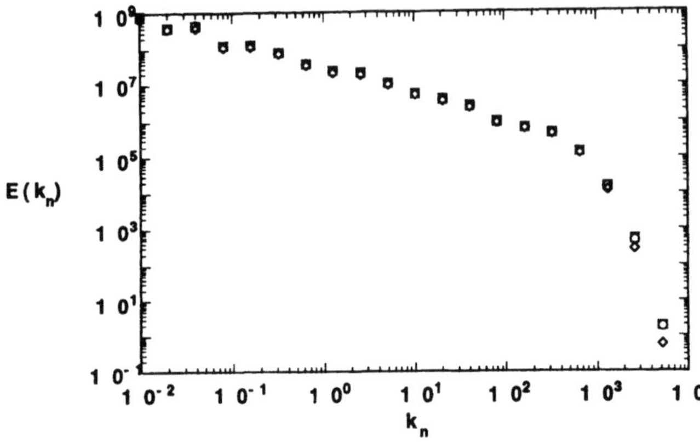

Figure 16.2. Time-average spectrum of the shell model. (Quoted from [16.12].)

in a power spectrum

$$\langle |u_n|^2 + |\theta_n|^2 \rangle \propto k_n^{-2/3} \tag{16.7}$$

and the dependencies on the Rayleigh number as

$$\left\langle \sum_{n=1}^{N} |u_n|^2 \right\rangle \propto R_a \quad \text{and} \quad \left\langle \sum_{n=1}^{N} |\theta_n|^2 \right\rangle \propto R_a^2. \tag{16.8}$$

Figure 16.2 shows an example of the fluctuation spectrum; the maximum Lyapunov number λ_1 scales as

$$\lambda_1 \propto R_a^{0.7}. \tag{16.9}$$

This dependence is stronger than that of the linear growth rate $\gamma_n \propto R_a^{0.5}$. In a large Rayleigh number limit, the nonlinear decorrelation rate dominates the linear instability growth rate.

16.2.2 Multiple-Bin Model

In the presence of anisotropy, the shell model may be extended to a two-dimensional multiple-bin model. Figure 16.1(right) illustrates a possible division of segments in the space comprising the absolute value of the wave-number and the direction orthogonal to the global density gradient.

The Hasegawa–Mima equation (11.21), with a linearly unstable term added,

$$(1 - \rho_i^2 \nabla_\perp^2) \frac{\partial}{\partial t} \phi + V_{\text{de}} \frac{\partial}{\partial y} \phi - \frac{T}{eB} [\phi, \rho_i^2 \nabla_\perp^2 \phi] = +\gamma_{\text{L}} \phi, \tag{16.10}$$

has been analyzed by use of the multiple-bin model. The right-hand side indicates the additional term. Following [16.13], the numerical results for fluctuation energy were compared under a properly chosen linear growth rate

γ_L in [14.22]. Further reading about the derivation of the bin-averaged mode-coupling coefficients and the result is given in [14.22] and references therein.

16.3 *K*-ε Model

The K-ε model has also been applied to plasma turbulence. For an interchange mode, the fluctuation energy K and the dissipation rate ε obey [16.14]

$$\frac{\partial K}{\partial t} = C_p g_0 \frac{K^2}{\varepsilon} - \varepsilon - C_J \frac{1}{\rho_i^2} \frac{K^3}{\varepsilon} + \nabla \cdot \left(C_K \frac{K^2}{\varepsilon} \nabla K \right), \tag{16.11a}$$

$$\frac{\partial \varepsilon}{\partial t} = C_{\varepsilon 1} C_p g_0 K - C_{\varepsilon 2} \frac{\varepsilon^2}{K} - C_{\varepsilon J} C_J \frac{1}{\rho_i^2} K^2 + \nabla \cdot \left(C_\varepsilon \frac{K^2}{\varepsilon} \nabla \varepsilon \right), \tag{16.11b}$$

where specific properties of the plasma turbulence are included in geometrical coefficients of the production terms (the first terms in the right-hand side), that is, $g_0 = (\nabla \bar{p} \nabla \ln \bar{B}) \rho^{-1}$, and the Ohmic dissipation terms (the third terms). The dissipation effect in the vorticity equation and in Ohm's law (e.g., equation (11.24b)) are treated independently and modelled in different forms. C_p, C_K, C_J, C_ε, $C_{\varepsilon 1}$, $C_{\varepsilon 2}$, and $C_{\varepsilon J}$ are numerical constants [16.2]. The numerical solution of these transport equations is discussed in [16.14].

When the transport terms in equation (16.11) are neglected, the stationary local solution is obtained as

$$K \propto g_0 \rho_i^2 \quad \text{and} \quad \varepsilon \propto g_0^{1/2} K. \tag{16.12}$$

The diffusivity is given by

$$D = K^2 \varepsilon^{-1} = C \sqrt{g_0} \rho_i^2, \tag{16.13}$$

where $C = (1 - C_{\varepsilon 1}) / C_J \sqrt{C_p} \sqrt{(C_{\varepsilon 1} - C_{\varepsilon J})(1 - C_{\varepsilon J})}$.

This local result belongs to a family of solutions possessing gyro-reduced Bohm diffusion. When the model is extended to include the transport effects of fluctuation energy and dissipation, the turbulent transport D deviates from the local balance value of $D = C \sqrt{g_0} \rho_i^2$. It should be noticed that the local terms (i.e., the first, second, and third terms in the right-hand side) are of the order $(D \rho_i^{-2}) K$ while the diffusion term (the fourth term) is of the order of $(D L^{-2}) K$, where L is the characteristic scale length. When the inhomogeneity is weak, $L^2 \gg \rho_i^2$, the turbulence level is predominantly determined by the local balance between the production and dissipation.

16.4 Mapping Models

Statistical properties of nonlinear dynamics are also investigated by use of the mapping models.

16.4.1 Standard Map

Standard mapping is used to study the magnetic stochasticity, plasma heating problems, and so on. It is formulated as [13.11]

$$q_{n+1} = q_n + p_{n+1} \tag{16.14a}$$

$$p_{n+1} = p_n + A \sin 2\pi q_n. \tag{16.14b}$$

If the nonlinear parameter (normalized amplitude) exceeds the threshold value

$$A > A_c = 0.1546\ldots, \tag{16.15}$$

stochasticity sets in. In the stochastic part of the phase space, the diffusivity is given as

$$D = \frac{A^2}{4} \frac{1 - 2J_1^2(2\pi A) - J_2^2(2\pi A) + 2J_3^2(2\pi A)}{(1 + J_2(2\pi A))^2} \tag{16.16}$$

where J_n is the nth-order Bessel function of the first kind. In the large-A limit, one has

$$D = \frac{A^2}{4} \tag{16.17}$$

[16.15]. For an intermediate value of A, accelerator modes appear for some special values of A [16.16, 16.17]. The presence of accelerator modes suggests that the transport is not a purely diffusive process in the braided magnetic structure. Nondiffusive transport has been discussed in, e.g., [16.18]. The dispersion increases faster than time, i.e., 'super-diffusion' appears.

Below the threshold value,

$$A < A_c, \tag{16.18}$$

the memory effect of the orbits has been investigated statistically. A recent survey of the statistical theory of the standard map is given in [16.19].

16.4.2 Other Maps

In addition to the standard map, a tokamap has been proposed to describe the magnetic field lines in a tokamak [16.20]. The selective interaction of particles with the toroidal Alfvén wave eigenmode (TAE mode) is captured by the TAE mode Mapping [16.21].

This method has also been applied to the transport by drift wave. In the vicinity of the magnetic surface with vanishing magnetic shear ($q' = 0$), the standard nontwist map (SNM) is applicable [16.22]. The importance of such a magnetic surface for turbulent transport is explained in §17.6. The wave phase and radial position (X, Y), are chosen as variables and a set of mapping equations is given as follows [16.23]:

$$X_{N+1} = X_N + \alpha_{\rm snm}(1 - Y_{N+1}^2) \tag{16.19a}$$

$$Y_{N+1} = Y_N - \beta_{\rm snm} \sin 2\pi X_N \tag{16.19b}$$

where

$$\alpha_{\rm snm} = v_{\parallel}(m - nq_*)/\omega_0 R q_*,$$

$$\beta_{\rm snm} = (-2\pi m\tilde{\phi}/a^2 B\omega_0)(2mq''/q_*(m - nq_*))^{1/2},$$

m and n are the poloidal and toroidal mode numbers, ω_0 is the wave frequency, q_* is the safety factor on the surface where $q' = 0$ and $\tilde{\phi}$ is the electrostatic potential. This mapping is used to study the transport near the surface with $q' = 0$.

References

[16.1] Lorenz E N 1993 *The Essence of Chaos* (Seattle: University of Washington Press) chapter 4

[16.2] Bradshaw P, Cebeci T and Whitelaw J H 1981 *Engineering Calculation Method for Turbulent Flow* (London: Academic Press) p 37

[16.3] Moore D R, Toomre J, Knobroch E and Weiss N O 1983 *Nature* **303** 663

[16.4] Cross M C *Phys. Letters A* 1986 **119** 21

[16.5] Bazdenkov S V and Pogutse O P 1990 *JETP Letters* **57** 410

[16.6] Pogutse O P, Kerner W, Gribkov V, Dazdenkov S and Osipenko M 1994 *Plasma Phys. Contr. Fusion* **36** 1963

[16.7] Sugama H and Horton W 1995 *Plasma Phys. Contr. Fusion* **37** 345

[16.8] Aoyagi T, Yagi M and Itoh S-I 1997 *J. Phys. Soc. Jpn.* **66** 2689

[16.9] Thyagaraya A, Haas F A and Harvey D J 1999 *Phys. Plasmas* **6** 2380

[16.10] In some models, the second positive Lyapunov exponent has been observed (Bekki N 2001, private communications)

[16.11] Jensen M H, Paladin G and Vulpiani A 1992 *Phys. Rev. A* **45** 7214

[16.12] Yagi M, Itoh S-I, Itoh K and Fukuyama A 1999 *Chaos* **9** 393

[16.13] Waltz R E 1983 *Phys. Fluids* **26** 169

[16.14] Sugama H, Okamoto M and Wakatani M 1992 *Plasma Phys. Contr. Nuclear Fusion Research* (Proceedings of the 14th Conference, Würzburg 1992) vol 2, p 353

[16.15] Meiss J D, Cary J R, Grebogi C, Crawfield J D and Abarbanel H D I 1983 *Physica D* **6** 375

[16.16] Karney C F F 1983 *Physica D* **8** 360

[16.17] Ichikawa Y H, Kamimura T and Hatori T 1987 *Physica D* **29** 247

[16.18] Zaslavsky G M, Stevens D and Weitzner H 1993 *Phys. Rev. E* **48** 1683

[16.19] Balescu R 2000 *J. Statist. Phys.* **98** 1169

[16.20] Balescu R, Vlad M and Spineau F 1998 *Phys. Rev. E* **58** 951

[16.21] Berk H L, Breizman B N and Ye H C 1993 *Phys. Fluids B* **5** 1506

[16.22] del-Castillo-Negrete D and Morrison P J 1993 *Phys. Fluids A* **5** 948

[16.23] Horton W, Park H-B, Kwon J-M, Strozzi D, Morrison P J and Choi D-I 1998 *Phys. Plasmas* **5** 3910

Chapter 17

Inhomogeneity-Driven Turbulence

The coexistence of various inhomogeneities and a background magnetic field induces varieties of plasma turbulence. By analyzing the nonlinear dispersion relation, equation (14.50), various properties of turbulence have been analyzed as illustrated in this chapter. Fluctuations with a scale length of the ion gyroradius [11.18, 12.4, 12.5], those with a scale length of collisionless skin depth (or longer) [17.1–17.5] and those with a scale length equal to the electron gyroradius [17.6] have been considered in conjunction with anomalous transport. The relationship of modes with various length scales is given in figure 17.1. This chapter shows several examples of modes. The flux which is driven by turbulence is surveyed in chapter 18. Generation of fluid structure is discussed in chapter 19. The influence of the inhomogeneous flow (electric field in plasmas) is discussed in chapter 20.

17.1 Typical Examples

17.1.1 Dissipative Interchange Mode

The instability of inhomogeneous plasmas known as the interchange mode is an analog of the Rayleigh–Benard instability in neutral fluid dynamics. Free energy source is released when the electron free motion along the field line is impeded by the dissipative process (i.e., the frozen-in condition is relaxed).

17.1.1.1 Resistive interchange mode

Ranging from the regimes of MHD modes and ITG modes in figure 17.1, the resistive interchange mode turbulence can develop in a system of magnetic hill. When the classical resistivity is high, $\eta_{\parallel} > (c/a\omega_p)^2 G_0^{1/2}$, the resistance in the Ohm's law equation (11.24b) is given by the electric resistivity. In this case the renormalized dispersion equation (without the noise source

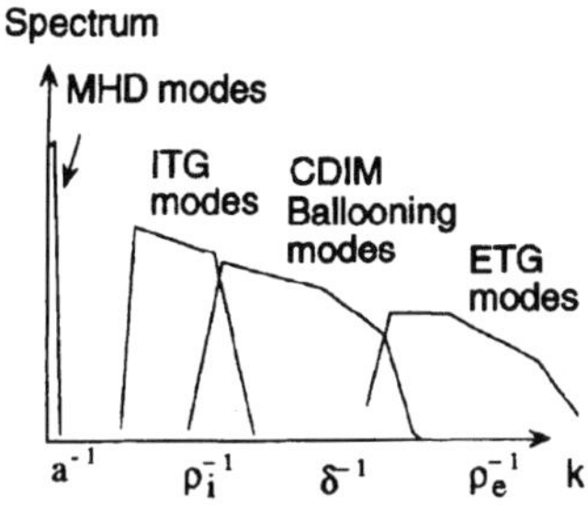

Figure 17.1. Characteristic scale lengths and various typical modes of plasma turbulence.

term) equation (14.51) takes the form

$$\left\{ \frac{\partial}{\partial k_x} \frac{s^2 k_y^2}{\eta_\parallel} \frac{\partial}{\partial k_x} + \frac{k_y^2 G_0}{\chi k_\perp^2} - \mu k_\perp^4 \right\} \phi(k_x; k_y) = 0, \tag{17.1}$$

where χ and μ are renormalized transport coefficients (sums of the collisional and turbulent ones, i.e., $\chi = \chi_c + \chi_N$ and $\mu = \mu_c + \mu_N$), $\eta_\parallel$ is the resistivity, and G_0 is the driving source due to the inhomogeneity equation (12.24c) [17.7]. The critical condition, equation (17.1), can be written as

$$\frac{G_0 \eta_\parallel}{s^2 \chi} = C_r \equiv \left[\frac{\pi}{4 \ln\{2(G_0 \mu^{-1} \chi^{-1} k_y^{-4})^{1/6}\}} \right], \tag{17.2}$$

where C_r is a weakly varying function of G_0, and can be approximated as a constant. From equation (17.2), the turbulent transport coefficient in a stationary state is derived as

$$\chi_N = \frac{\eta_\parallel}{C_r s^2} G_0 \tag{17.3}$$

in the limit of strongly turbulent transport, $\chi_N \gg \chi_c$. The correlation time and correlation length are given as

$$\tau_{\rm cor} \approx G_0^{-1/2}, \tag{17.4}$$

and

$$l_{\rm cor} \approx \eta_\parallel^{1/2} s^{-1} G_0^{1/4}. \tag{17.5}$$

These forms are normalized by use of the MHD normalization.

17.1.1.2 Resistive ballooning mode

In toroidal geometry, the resistive ballooning mode appears [17.8]. The condition for nonlinear marginal stability, equation (14.51), provides the relation

$$\chi_N = 2\alpha \eta_\parallel, \tag{17.6}$$

where

$$\alpha = -Rq^2 \frac{d\beta}{dr}. \tag{17.7}$$

The parameter α for the normalized pressure gradient includes the geometric effect of toroidal plasmas [12.16]. Comparing equations (17.3) and (17.6), one sees the usual role of resistive dissipation ($\eta_\parallel$), in combination with the pressure gradient (G_0 or α). The correlation time and correlation length are given as

$$\tau_{cor} \approx \alpha^{-1/2}, \tag{17.8}$$

and

$$l_{cor} \approx \eta_\parallel^{1/2} \alpha^{1/4} \tag{17.9}$$

with a weak dependence on the magnetic shear. In contrast to the case of interchange mode turbulence, the turbulent transport coefficient and correlation length do not diverge for the case of ballooning mode turbulence even in the limit of zero magnetic shear $s \simeq 0$. This is because the magnetic well works to suppress turbulence in addition to the magnetic shear.

In addition, a variety of resistive turbulence has been investigated, e.g., rippling mode [13.21], resistive drift wave [17.9] and others.

17.1.1.3 Current diffusive interchange/ballooning mode

When the plasma temperature becomes high so that the relation $\eta_\parallel < (c/a\omega_p)^2 G_0^{1/2}$ holds, the dissipation associated with the parallel current is controlled not by the electric resistivity but by the current diffusivity [17.1, 17.5, 17.10–17.19]. The current–diffusive interchange mode (CDIM) turbulence is related to a submarginal instability. (Subcritical excitation is discussed in chapter 21.) The nonlinear decorrelation rate and coherence length are given as

$$\tau_{cor} = \lambda_1^{-1} \simeq \frac{2(1 - C_0)}{C_0} G_0^{-1/2} \tag{17.10}$$

and

$$l_{CDIM} \simeq s^{-1} G_0^{1/2} c\omega_p^{-1} a^{-1}, \tag{17.11}$$

where the numerical constant C_0 originates from the contribution of random noise source, equation (14.74). The spectrum of fluctuation energy is given by

$$E_1(k_\perp) = \frac{2}{(1 - C_0)} G_0 k_\perp^{-3}, \tag{17.12}$$

for $k_\perp > k_* \simeq s G_0^{-1/2} c^{-1} a\omega_p$ [17.20, 17.21]. It is noted that the subcritical excitation of the current–diffusive mode turbulence occurs above a threshold

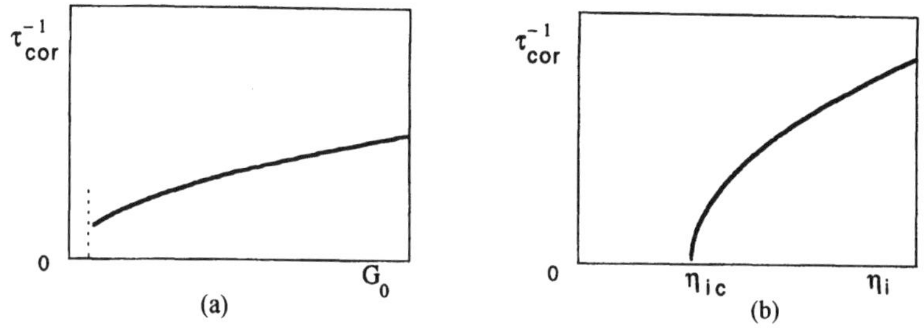

Figure 17.2. Turbulence decorrelation rate as a function of gradient. The case of current diffusive mode turbulence (a) and ion temperature gradient mode (b). In (a) the dotted line indicates a critical gradient for subcritical excitation. (See chapter 21 for explanation of subcritical excitation.)

pressure gradient as is discussed in chapter 21. Expressions like equations (17.10) and (17.12) hold in the limit of large pressure gradient.

In toroidal geometry, one has a current–diffusive ballooning mode (CDBM). The turbulence decorrelation time is given as

$$\tau_{\mathrm{cor}} \propto \alpha^{-1/2}, \tag{17.13}$$

where α is given by equation (17.7). The role of the pressure gradient in determining the turbulence decorrelation time is seen in equations (17.10) or (17.13). Figure 17.2(a) illustrates the influence of the gradient on the decorrelation rate. Notice that the wavelength of an unstable CDBM could be much longer than the collisionless skin depth, as is shown in [17.15]. The coupling with the drift wave has been studied [17.22]. This coupling reduces the growth rate, but strong nonlinear instability still remains.

17.1.2 Ion Temperature Gradient (ITG) Mode

The ITG mode becomes linearly unstable if the temperature gradient exceeds a threshold value [17.6, 17.23–17.25]. One estimate of the stability boundary is given by

$$\eta_i \equiv \frac{L_n}{L_{Ti}} > \eta_{ic} = 0.5 + 2.5\max\left[\frac{L_n}{R}, 0.2\right] \tag{17.14}$$

in toroidal plasmas.

The mode coupling theory has provided a nonlinear dispersion relation [15.13]:

$$\omega\left\{\omega\frac{L_c}{c_s} - \left[1 - k_\perp^2\rho^2\left(1 + \frac{L_n}{L_\mathrm{T}}\right) - 2\frac{L_n}{R}\right]\right\}\frac{L_n}{c_s} + 2k_\perp^2\rho^2\frac{L_n}{R}\left(1 + \frac{L_n}{L_\mathrm{T}}\right)$$
$$- k_x^2\rho^4\int_k^\infty \frac{\mathrm{d}k'}{k'}(k' - k)(k'^2 + k^2)E(k') = 0. \tag{17.15}$$

The last term is the nonlinear interaction term, in which $E(k')$ is the fluctuation spectrum. The balance between linear growth and nonlinear damping yields an evaluation of the stationary fluctuation level:

$$\left\langle \left(\frac{e\tilde{\phi}}{T_e} \right)^2 \right\rangle \simeq \frac{1}{k_x^2 L_n^2} \frac{L_n}{R} \left(1 + \frac{L_n}{L_{\mathrm{T}}} \right) \tag{17.16}$$

with

$$E(k) \propto k_\perp^{-3}. \tag{17.17}$$

For the range $k_\perp \rho_s \approx 1$, the nonlinear decorrelation rate is estimated by $\Gamma_{\mathrm{dec}} \simeq \gamma_{\mathrm{L}}$:

$$\gamma_{\mathrm{L}} \simeq \frac{c_{\mathrm{s}}}{\sqrt{RL_n}} \sqrt{\eta_i - \eta_{ic}} \tag{17.18}$$

[17.24] and is given as

$$\tau_{\mathrm{cor}}^{-1} = \Gamma_{\mathrm{dec}} \simeq \gamma_{\mathrm{L}} \simeq \frac{c_{\mathrm{s}}}{\sqrt{RL_n}} \sqrt{\eta_i - \eta_{ic}}, \tag{17.19}$$

where $\eta_i = L_n / L_{\mathrm{T}}$. Its further elaboration is discussed in [12.7]. Figure 17.2(b) illustrates the decorrelation rate as a function of the temperature gradient.

The origin of the ITG mode instability is an effective negative compressibility owing to the ion temperature gradient, as shown by equation (12.25). (The phase relation between density and pressure perturbations for ITG mode is given by the relation $\tilde{p}_i / \bar{n}_i \bar{T}_i \propto (1 - \omega_{*T_i}/\omega)(\tilde{n}_i / \bar{n}_i)$. A negative compressibility means that the pressure perturbation can be negative at a phase where the density perturbation is positive.)

It should be noted that the linear eigenmode in toroidal geometry has a wavelength in the radial direction which is much longer than the ion gyroradius [17.26, 17.27]. However, this does not mean that the correlation length of the motion of a fluid element is much longer than the ion gyroradius. (See discussion in §18.2.1.)

17.1.3 Electron Temperature Gradient (ETG) Mode

A similar mechanism works for electrons in the presence of their temperature gradient. Modes in the range

$$k_\perp \rho_e \approx 1 \tag{17.20}$$

become unstable if the electron temperature gradient exceeds a threshold value. This instability is called electron temperature gradient (ETG) mode [17.6]. Stability analyses have been made for slab plasmas [17.6, 17.28, 17.29] and toroidal plasmas [17.30]. The stability boundary for the toroidal

ETG mode is given approximately as

$$\eta_e \equiv \frac{L_n}{L_{Te}} > \eta_{ec} = \frac{2}{3} + 2\max\left[\frac{L_n}{R}, \frac{1}{3}\right] \tag{17.21}$$

from figure 1 of [17.30].

The mode-coupling theory has been developed for this mode. In a saturated state, the nonlinear decorrelation rate Γ_{dec} is considered to balance with the linear growth rate γ_{L}, and is estimated as

$$\tau_{\text{cor}}^{-1} = \Gamma_{\text{dec}} \approx k_\perp \rho_e \sqrt{2(\eta_e - \tfrac{2}{3})} \frac{v_{\text{th},e}}{\sqrt{L_n R}}. \tag{17.22}$$

The fluctuation level is also evaluated as

$$\tilde{p}_e = \frac{1}{|k_x|} \left| \frac{\mathrm{d}\bar{p}_e}{\mathrm{d}x} \right|. \tag{17.23}$$

17.1.4 Kinetic Instabilities

A nonlinear dispersion relation has been derived for kinetic instabilities by employing a renormalized kinetic dielectric tensor. For an electrostatic perturbation with the wavelength longer than the Debye length, the dispersion relation is approximated by the charge neutrality condition

$$\boldsymbol{k} \cdot \sum_{s=i,e} \boldsymbol{\sigma}_{s,k\omega} \cdot \boldsymbol{k} = 0 \tag{17.24}$$

where the conductivity tensor $\boldsymbol{\sigma}_{s,k\omega}$ is calculated by use of the renormalized propagator. One example is quoted from the work of [17.31].

In the presence of magnetic shear, the path integral along the field line which appears in, e.g., equation (12.10), takes the form

$$\int_{-\infty}^{t} \mathrm{d}t' \exp\{\mathrm{i}\boldsymbol{k} \cdot [\boldsymbol{x}'(t') - \boldsymbol{x}] - \mathrm{i}\omega(t' - t)\}$$

$$\simeq \int_0^\infty \mathrm{d}\tau \exp[\mathrm{i}(\omega - k_\parallel v_\parallel)\tau - \Gamma_\perp \tau - (\Gamma_{s\parallel}\tau)^3], \tag{17.25}$$

where $\Gamma_{s\parallel}$ is the decorrelation rate (enhanced in the sheared magnetic field)

$$\Gamma_{s\parallel} = [(k'_\parallel v_\parallel)^2 D_\perp / 3]^{1/3} \tag{17.26}$$

and $\Gamma_\perp = D_\perp k_\perp^2$ is the conventional $E \times B$ decorrelation rate, $k'_\parallel = \mathrm{d}k_\parallel/\mathrm{d}r$. For rapidly moving electrons, the decorrelation rate $\Gamma_{s\parallel}$ is much larger than $\Gamma_\perp$. The argument of the plasma dispersion function for electrons is modified as

$$Z\left(\frac{\omega}{\sqrt{2}|k_\parallel v_{\text{th}}|}\right) \to Z\left(\frac{\omega + \mathrm{i}\Gamma_{s\parallel}}{\sqrt{2}|k_\parallel v_{\text{th}}|}\right). \tag{17.27}$$

A nonlinear dispersion relation, with the random noise source neglected, is given for the case of $T_e = T_i$ as

$$\left[\frac{\mathrm{d}^2}{\mathrm{d}x^2} - \left\{\frac{1}{d}\left(2 - \frac{\Lambda_0(\omega - \omega_*)}{\omega}\right) - \mu^2 x^2 - \frac{\alpha_s(\omega - \omega_*)}{\omega x} Z\left(\frac{\omega + \mathrm{i}\Gamma_{s\parallel}}{\sqrt{2}|k_\parallel v_{\mathrm{th},e}|}\right)\right\}\right]\phi(x)$$
$$= 0 \qquad (17.28)$$

where

$$x = (r - r_s)/\rho_s, \qquad \Lambda_n = I_n(b)\exp(-b),$$
$$b = k_\theta^2 \rho_i^2, \qquad \alpha_s = (m_e/2m_i)^{1/2} L_s/L_n,$$
$$d = (\Lambda_0 - \Lambda_1)\left(\frac{\omega + \omega_*}{\omega}\right) - 3\mathrm{i}\left(\frac{\omega - \omega_*}{\omega_*}\right)\frac{\alpha_s^2 \Gamma_{s\parallel}}{\omega_*},$$
$$\mu = \frac{L_n}{L_s}\frac{\omega_*}{\omega}\left(\Lambda_0\left(\frac{\omega + \omega_*}{\omega}\right)d^{-1}\right)^{1/2}$$

and $L_s = Rqs^{-1}$ is the magnetic shear length. This nonlinear dispersion relation provides the stationary state

$$D_\perp \approx 15\left(\frac{m_e}{m_i}\right)^{3/2}\left(\frac{L_s}{L_n}\right)^{7/2}\frac{c_s \rho_s^2}{L_n}. \qquad (17.29)$$

17.2 Influence of Magnetic Field Structure

17.2.1 Drift Due to the Magnetic Field Gradient

The structure of the magnetic field is very influential on plasma turbulence. Particular emphasis has been put on the role of magnetic shear. In the presence of strong magnetic shear, the fluctuations tend to be localized near the mode rational surface and the level is suppressed. Its sign also affects the turbulence.

Figure 17.3 illustrates the effect of the magnetic shear in toroidal plasmas. The drift orbit of trapped particles, which forms a banana orbit when projected onto the poloidal cross-section, drifts in the toroidal direction. If the magnetic shear is positive, e.g.,

$$\frac{\mathrm{d}q}{\mathrm{d}r} > 0, \qquad (17.30)$$

the toroidal drift direction coincides, i.e., the toroidal drift motion is larger when staying in the outer region. In the diamagnetic drift direction, the magnetic drift in bad magnetic curvature increases. In contrast, if the

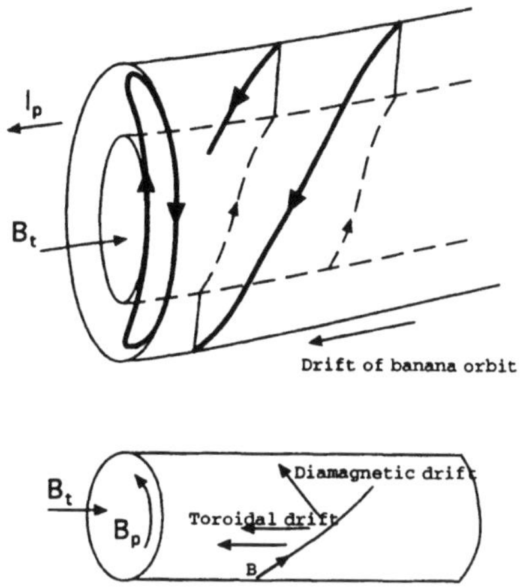

Figure 17.3. Trajectory of an ion in toroidal plasma. The orbit of a banana particle is illustrated together with its projection on the poloidal cross-section. When the magnetic field shear is positive, the toroidal drift is in the same direction as the diamagnetic drift.

magnetic shear is negative,

$$\frac{\mathrm{d}q}{\mathrm{d}r} < 0, \tag{17.31}$$

the toroidal drift is reduced [15.10].

The other issue is the outward toroidal shift of the magnetic axis. When the shift in the inner magnetic surfaces becomes larger, e.g., owing to the higher plasma pressure, the magnetic field line, in the bad magnetic curvature region, becomes shorter. The gradient-B drift in the destabilizing direction on average becomes smaller (see [2.23, 2.28]).

The turbulence level and turbulent transport coefficient reduce in the case of negative magnetic shear and a large Shafranov shift.

17.2.2 Trapped Particle Instability

The growth rate of the trapped ion instability is estimated as

$$\gamma_{\mathrm{L}} \simeq \left(\frac{r}{R}\right)^{1/4} \sqrt{\omega_* \omega_{\mathrm{M}}} \tag{17.32}$$

where $\omega_{\mathrm{M}} = k_\zeta V_{\mathrm{D}}$, k_ζ is the mode number in toroidal direction, and V_{D} is the drift due to the gradient of magnetic field (see figure 17.3). As a result of the reduction of the gradient-B drift velocity V_{D}, the growth rate is reduced in the case of negative magnetic shear. Controlling the cross-section shape leads to a similar result [17.32, 17.33].

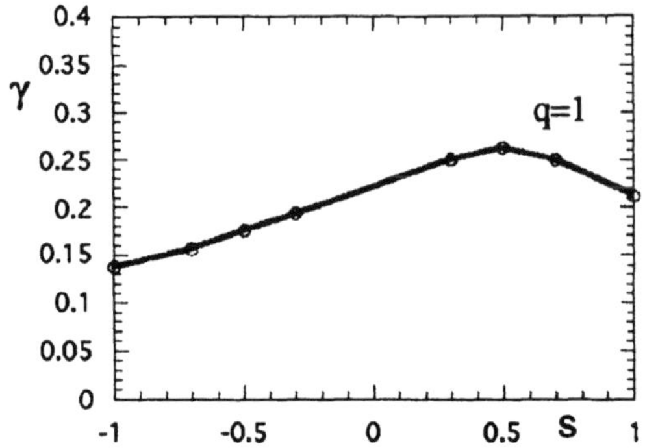

Figure 17.4. Growth rate of toroidal ion-temperature-gradient-driven instability. In the negative shear region, the growth rate becomes smaller. (Reproduced from [17.34].)

17.2.3 Toroidal Ion Temperature Gradient (ITG) Mode

A simple analytic limit of the dispersion relation of the ion temperature gradient (ITG) mode in toroidal plasma is given by

$$\frac{k_{\parallel}^2 v_{\mathrm{th},i}^2}{\omega^2} + k_{\perp}^2 \rho_i^2 + \frac{\omega - \omega_{*i}}{\omega - \omega_{*i}(1 + L_n/L_{\mathrm{T}})} - \frac{\omega_{\mathrm{M}}}{\omega}(\cos\theta + s\theta\sin\theta) = 0, \quad (17.33)$$

where θ is the poloidal angle. When the magnetic field shear is negative, this causes a reduction in ω_{M}, decreasing the growth rate [17.34–17.36]. The growth rate is illustrated in figure 17.4 [17.34]. The impact of weak magnetic shear has been pointed out. In the case of weak magnetic shear, the distance between adjacent rational magnetic surfaces becomes longer. This effect further decreases the particle drift, which is averaged over the eigenmode structure [17.37].

The influence of a helical plasma shape has also been studied, including the effect of helical ripples [17.38, 17.39].

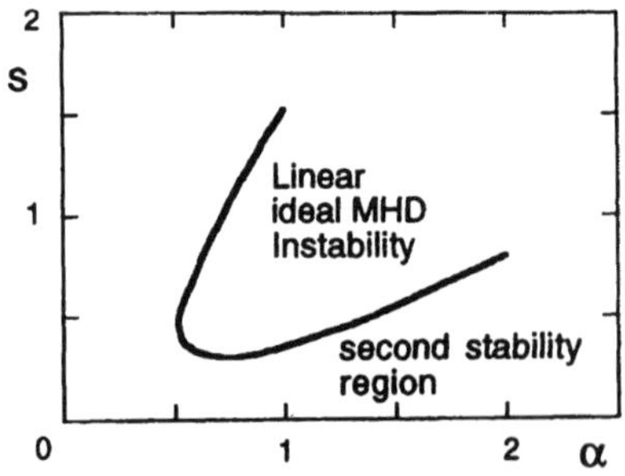

Figure 17.5. The region of ideal MHD instability is shown on the plane of pressure gradient and magnetic shear. The regime of second stability is denoted.

17.2.4 Current-Diffusive Ballooning Mode (CDBM) Turbulence

The ideal MHD ballooning instability appears at a sufficiently high-pressure gradient. The critical pressure gradient for instability becomes higher when the magnetic shear becomes negative. (It is often called the *second stability regime* [12.16–12.18]. See figure 17.5.) CDBM turbulence becomes weaker if the magnetic shear is negative [17.15, 17.40, 17.41].

References

[17.1] Schmidt J and Yoshikawa S 1971 *Phys. Rev. Lett.* **26** 753
[17.2] Ohkawa T 1978 *Phys. Lett.* **67A** 35
[17.3] Parail V V and Pogutse O P 1980 *JETP Letters* **32** 384
[17.4] Kadomtsev B B and Pogutse O P 1985 in *Plasma Physics and Controlled Nuclear Fusion Research 1984* (Vienna: IAEA) vol 2, p 69
[17.5] Itoh K, Itoh S-I and Fukuyama A 1992 *Phys. Rev. Lett.* **69** 1050
[17.6] Kadomtsev B B and Pogutse O P 1970 in *Review of Plasma Physics* ed M A Leontovich (New York: Consultants Bureau) vol 5, p 312
[17.7] Carreras B A *et al* 1983 *Phys. Rev. Lett.* **50** 503
[17.8] Guzdar P N, Drake J F, McCarthy D, Hassma A B and Liu C S 1993 *Phys. Fluids B* **5** 3712
[17.9] Wakatani M and Hasegawa A 1984 *Phys. Fluids* **27** 611
[17.10] Speiser T W 1970 *Planet. Space. Sci.* **18** 613
[17.11] Kaw P J, Valeo E J and Rutherford P H 1979 *Phys. Rev. Lett.* **43** 1398
[17.12] Aydemir A Y 1990 *Phys. Fluids B* **2** 2135
[17.13] Wesson J A 1991 *Plasma Physics and Controlled Nuclear Fusion Research 1990* (Vienna: IAEA) vol 2, p 79
[17.14] Lichtenberg A J, Itoh K, Itoh S-I and Fukuyama A 1992 *Nucl. Fusion* **32** 495
[17.15] Yagi M, Itoh K, Itoh S-I, Fukuyama A and Azumi M 1993 *Phys. Fluids B* **5** 3702
[17.16] Biskamp D and Drake J F 1994 *Phys. Rev. Lett.* **73** 971
[17.17] Ottaviani M and Polcelli F 1993 *Phys. Rev. Lett.* **71** 3802
[17.18] Naitou H, Kobayashi T and Tokuda S 1999 *J. Plasma Phys.* **61** 543
[17.19] Naitou H, Kuramoto T, Kobayashi T, Yagi M, Tokuda S and Matsumoto T 2000 *J. Plasma Fusion Res.* **76** 778
[17.20] Itoh S-I and Itoh K 1996 *Plasma Phys. Control. Fusion* **40** 1729
[17.21] Itoh S-I and Itoh K 1997 *J. Phys. Soc. Jpn.* **66** 1571
[17.22] Uchida M, Fukuyama A, Itoh K, Itoh S-I and Yagi M 1999 *J. Plasma Fusion Res.* **2** 117
[17.23] Romanelli F 1989 *Phys. Fluids B* **1** 1018
[17.24] Hong B-G and Horton W 1990 *Phys. Fluids B* **2** 978
[17.25] Hong B-G, Choi D-I and Horton W 1986 *Phys. Fluids* **29** 1872
[17.26] Connor J W, Taylor J B and Wilson H R 1993 *Phys. Rev. Lett.* **70** 1803
[17.27] Kim J Y and Wakatani M 1994 *Phys. Rev. Lett.* **73** 2200
[17.28] Rozhanskii V A 1981 *JETP Letters* **34** 56 [1981 *Pis'ma Zh. Eksp. Teor. Fiz.* **34** 60]
[17.29] Lee Y C, Dong J Q, Guzdar P N and Liu C S 1987 *Phys. Fluids* **30** 1331
[17.30] Horton W, Hong B-G and Tang W M 1988 *Phys. Fluids* **31** 2971

[17.31] Hirshman S P and Molvig K 1979 *Phys. Rev. Lett.* **42** 648
[17.32] Glasser A, Frieman E A and Yoshikawa S 1974 *Phys. Fluids* **17** 181
[17.33] Roach C M *et al* 1995 *Plasma Phys. Contr. Fusion* **37** 679
[17.34] Kim J Y and Wakatani M 1995 *Phys. Plasmas* **2** 1012
[17.35] Kishimoto Y *et al* 1996 in Proceedings of the 16th International IAEA Fusion Energy Conference (Montreal 1996) vol 2, p 581
[17.36] Romaneli F 1993 in *AIP Conference Proceedings 284* (New York: AIP) p 391
[17.37] Kishimoto Y, Kim Y-J, Horton W, Tajima T, LeBrun M J and Shirai H 1999 *Plasma Phys. Contr. Fusion* **41** A663
[17.38] Lewandowski J L V 1998 *Plasma Phys. Contr. Fusion* **40** 283
[17.39] Kuroda T, Sugama H, Kanno R and Okamoto M 2000 *J. Phys. Soc. Jpn.* **69** 2485
[17.40] Yagi M *et al* 1994 *J. Phys. Soc. Jpn.* **63** 10
[17.41] Fukuyama A, Itoh K, Itoh S-I, Yagi M and Azumi M 1994 *Plasma Phys. Contr. Fusion* **36** 1385

Chapter 18

Global Flow Driven by Turbulence

18.1 $E \times B$ Transport and Magnetic Transport

18.1.1 $E \times B$ Transport

In magnetized plasmas, the cross-field velocity is governed by the $\boldsymbol{E} \times \boldsymbol{B}$ drift velocity ($\boldsymbol{E} = -\nabla\phi$) in the simple case of electrostatic perturbations. The flux of a quantity X across the magnetic surface, F_X, is given as

$$F_X = \frac{1}{\bar{B}}\langle \mathrm{i}k_y \tilde{X}\tilde{\phi}^* \rangle \tag{18.1}$$

in terms of the cross-correlation function between $\tilde{X}$ and $\tilde{\phi}$. The phase relation between two quantities is related to the imaginary part of the linear propagator, equation (14.1). A goal of the analysis is to express the flux in terms of global parameters:

$$F_X = -\mu_X \nabla \bar{X} + F_{\mathrm{rest}}, \tag{18.2}$$

where F_{rest} does not include the gradient $\nabla \bar{X}$. It includes the term which is called the 'pinch' term [2.30]. The pinch flux which is induced by the gradient of magnetic field in toroidal plasmas [12.30] is discussed in Appendix 12A.

In the quasi-linear approximation, the flux is expressed as a quadratic function of the fluctuation amplitude. For the particle flux in the presence of the density gradient, one has

$$\langle \mathrm{i}\tilde{n}^*_{s,k,\omega}\tilde{\phi}_{k,\omega}\rangle = -\Im \int \left(\frac{(\omega - \omega_{*s})}{\omega - k_\parallel v_\parallel} J_0^2\left(\frac{k'_\perp v_\perp}{\omega_{cs}}\right) \bar{f}_s \right) \mathrm{d}v \, \frac{e_s}{T_s} \langle \tilde{\phi}^*_{k,\omega}\tilde{\phi}_{k,\omega}\rangle,$$
$$(s = e, i). \tag{18.3}$$

Formula (18.3) is valid within a framework of weak turbulence theory and holds when equation (12.36) is satisfied. In usual circumstances, the nonlinear modification of the propagator is important to the evaluation of global flux. The simplest evaluation is to balance between the linear growth rate γ_{L} and the turbulent decorrelation rate $Dk_{\perp}^2$, equation (14.22), giving

$$D \simeq \frac{\gamma_{\mathrm{L}}}{k_{\perp}^2}. \tag{18.4}$$

This is a mixing length estimate. A non-Markovian effect is discussed in [14.33], and a correction is proposed as

$$D \simeq \frac{\gamma_{\mathrm{L}}^2}{\omega_k^2 + \gamma_{\mathrm{L}}^2} \frac{\gamma_{\mathrm{L}}}{k_{\perp}^2}, \tag{18.5}$$

where ω_k is a real frequency of a linear mode.

However, the use of the linear growth rate in the mixing length estimate (18.4), which is called the Kadomtsev formula, might not be relevant in turbulent plasmas. Examples of nonlinear analysis are illustrated in this subsection.

18.1.2 Magnetic Braiding and Transport

When magnetic perturbations exist and equation (13.19) is satisfied, braiding of magnetic surfaces occurs, and the trajectory of the magnetic field line is subject to diffusion as

$$\langle x(l)x(0) \rangle = D_{\mathrm{M}} l, \tag{18.6}$$

where l is the distance along the field line and D_{M} is called the diffusion coefficient of magnetic field line. Magnetic-field fluctuations are characterized by two correlation lengths, one in the direction of the unperturbed field line, $l_{c,\parallel}$, and the other across the magnetic field line $l_{c,\perp}$. ($l_{c,\parallel}$ is the Eulerian view of correlation length.) The correlation length along the *perturbed* magnetic field line L_{ac}, that determines D_{M} as

$$D_{\mathrm{M}} = (\tilde{B}_r/B)^2 L_{\mathrm{ac}}, \tag{18.7}$$

depends on the relative magnitude of perturbation $\tilde{B}_r/B$. In a weak turbulence, one has

$$L_{\mathrm{ac}} = l_{c,\parallel}. \tag{18.8a}$$

In the strong-turbulence case, the autocorrelation length L_{ac} is estimated by

$$L_{\mathrm{ac}} \simeq l_{c,\perp} \left(\frac{\tilde{B}_r}{B} \right)^{-1}. \tag{18.8b}$$

The diffusion coefficient of the magnetic field line is given from equations (18.7) and (18.8) as

$$D_{\mathrm{M}} = \begin{cases} l_{c,\parallel}(\tilde{B}_r/B)^2 & (\tilde{B}_r/B < l_{c,\perp}/l_{c,\parallel}), \\ l_{c,\perp}(\tilde{B}_r/B) & (\tilde{B}_r/B > l_{c,\perp}/l_{c,\parallel}). \end{cases} \tag{18.9}$$

The small and large amplitude limits correspond to the quasi-linear and the strong turbulence cases, respectively.

The plasma transport in the presence of stochastic magnetic field has been discussed [2.26, 18.1–18.6]. The case of low collisionality is surveyed, i.e.,

$$\nu^{-1} > \tau_{\mathrm{t}} \quad \text{and} \quad \nu^{-1} \gg \tau_{\mathrm{dec}}, \tag{18.10}$$

where ν is the collision frequency, τ_{t} is the transit time of particles,

$$\tau_{\mathrm{t}} = L_{\mathrm{ac}}/v_{\mathrm{th}}, \tag{18.11}$$

and τ_{dec} is the decorrelation time of plasma particles due to the cross-field diffusion. (τ_{dec} could be determined by various processes in addition to this magnetic braiding.) In this simplified situation, the transport coefficient due to the magnetic stochasticity is given by [18.3]

$$\chi \simeq \frac{\tau_{\mathrm{dec}}}{\tau_{\mathrm{t}}}(v_{\mathrm{th}} D_{\mathrm{M}}) \quad \text{if } \frac{\tau_{\mathrm{dec}}}{\tau_{\mathrm{t}}} \le 1 \tag{18.12a}$$

$$\chi \simeq v_{\mathrm{th}} D_{\mathrm{M}} \qquad \text{if } \frac{\tau_{\mathrm{dec}}}{\tau_{\mathrm{t}}} > 1 \tag{18.12b}$$

For given correlation lengths, $l_{c,\perp|}$, $l_{c,\|}$ and L_{ac}, and amplitude of magnetic fluctuations, the diffusion coefficient for each species depends on the ratio $\tau_{\mathrm{dec}}/\tau_{\mathrm{t}}$ as well. When the cross-field decorrelation time τ_{dec} is determined by the magnetic perturbation (e.g., high-temperature electrons in a braided magnetic region), the relation

$$\tau_{\mathrm{dec}} = \tau_{\mathrm{t}} \tag{18.13}$$

holds. More detailed formulae in general circumstances are summarized in [11.16, 18.3, 18.4]. It is possible for superdiffusion to occur because of the accelerator modes discussed by using a mapping model (§16.4). In addition, a subdiffusive behavior has also been found for a stochastic magnetic field [18.7].

The turbulence amplitude usually depends on the renormalized transport coefficients of plasma. It is necessary to determine the correlation length and time, L_{ac} and τ_{dec}, and amplitude $\tilde{B}_r/B$ simultaneously.

18.2 Heat Flux

The energy flux is given by the formula (in the electrostatic limit)

$$q_x = \frac{-1}{\overline{B}}\langle \tilde{p}\nabla_y \tilde{\phi}\rangle. \tag{18.14}$$

An application of the statistical approach described in §14.4 is illustrated here. In the case of equation (14.73), there is no velocity gradient. In this case we have

$$\bar{q}_x = -\left\{ \sum_k \frac{k_y^2}{\gamma_p - \lambda_1} \langle f_{1,k}^{(1)^*} f_{1,k}^{(1)} \rangle \right\} \bar{p}'. \tag{18.15}$$

It is noted that the turbulence decorrelation rate λ_m and eddy damping rate γ_j are given by the nonlinear balance. Once the autocorrelation function $\langle f_1 f_1^* \rangle$ is given, a decomposition of the fluctuating fields, e.g., equation (14.72), allows one to relate the cross-correlation functions $\langle f_i f_j^* \rangle$ and autocorrelation function $\langle f_1 f_1^* \rangle$. One has an expression for the turbulent thermal conductivity,

$$\bar{q}_x = -\chi_{\text{turb}} \bar{p}'. \tag{18.16}$$

Under many circumstances, the turbulent heat flux which is caused by the pressure gradient dominates the dissipation $\sum_X F_X \nabla(\bar{X}^{-1})$ (e.g., the left-hand side of equation (11.27)). The dissipation associated with the 'pinch terms' can be negative, but the total dissipation coming from all components of fluxes remains positive. If one takes an example of drift waves, the flux of perpendicular electron energy is much larger than that of parallel electron energy [18.8]. For this example, the inward pinch of the particle induced by the temperature gradient is obtained.

Three typical examples of ITG mode, CDIM and ETG mode are shown below. The influence of the shear flow due to the inhomogeneous radial electric field is discussed in §20.2.

18.2.1 ITG Mode Turbulence

In the case of ITG mode turbulence, the transport coefficient has the dimensional form of $\chi \simeq c_s \rho_s^2 / R$. ($\lambda_1$ scales with c_s/R, and l_{cor} is of the order of ρ_s.) It is given by

$$\chi_{\text{ITG}} = \frac{c_s \rho_s^2}{R} F\left(q, s, \frac{r}{R}, \frac{L_n}{L_T}, \frac{T_e}{T_i}, \ldots \right), \tag{18.17}$$

where the factor $F(q, s, r/R, L_n/L_T, T_e/T_i, \ldots)$ denotes a geometrical factor. One model has been proposed taking this fact into account [17.23]:

$$\chi_{\text{ITG}} = 14 \left(\frac{T_i}{T_e} \right)^{3/2} \sqrt{\frac{R(1 - \eta_{ic} L_T / L_n)}{L_T}} \frac{c_s \rho_s^2}{R}. \tag{18.18}$$

Other models are listed in, e.g., [12.5]. The inward pinch of particles is discussed in relation with the ITG mode in [18.9].

The result in [17.26] and those in numerical simulation [17.37, 18.10] suggest that the radial wavelength is of the order of $\sqrt{a \rho_i}$. For the turbulence level of equation (13.3), the $E \times B$ velocity is given as $c_s \rho_i L_n^{-1}$ and the eddy turnover time τ_{et} is estimated as

$$\tau_{\text{et}} \approx \sqrt{a / \rho_i} L_n c_s^{-1}. \tag{18.19}$$

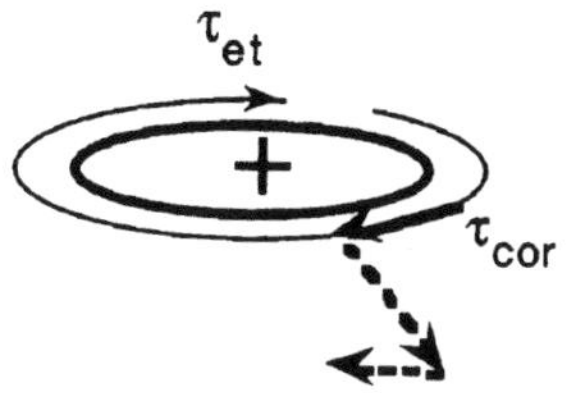

Figure 18.1. Fluctuation with small Kubo number $\mathcal{K} \ll 1$. If the correlation time is much shorter than the eddy turnover time, the fluctuating motion of plasma is decorrelated before it circumnavigates the equipotential contour. The coherence length of the fluctuating plasma motion (shown by thick arrows) is much shorter than the one-time correlation length of fluctuating fields.

This value of the eddy turnover time is much longer than the characteristic coherence time of turbulence, $L_n c_s^{-1}$. Even in this case, the radial coherence length of random motion remains of the order of the ion gyroradius, and the thermal conductivity has a scaling property like equation (18.18). The Kubo number is given as

$$\mathcal{K} \approx \sqrt{\rho_i / a} \tag{18.20}$$

and is much smaller than unity, if $\mathcal{K}$ is estimated by use of equation (14.13). Figure 18.1 illustrates the case where the Kubo number is much smaller than unity. In this case, the fluctuating motion of plasma is decorrelated before it circumnavigates the equipotential contour. The coherence length of the turbulent field in the Eulerian view is not always relevant for the step size of random motion. For the step size one may choose the correlation length of the fluctuating motion (the Lagrangian view), which can be different from the one-time coherence length. In the case of figure 18.1, the correlation length of the fluctuating plasma motion is much shorter than the one-time correlation length of electric perturbations.

18.2.2 CDIM Turbulence

Fluctuations in the range of the collisionless skin depth are also important for plasma transport [17.1–17.5]. Such turbulence belongs to a family of electromagnetic turbulence, and a typical time scale is given by the poloidal-Alfvén transit time τ_{Ap}. The CDIM leads to the characteristic coherence length and nonlinear decorrelation rate, equations (17.10) and (17.11) [2.28, 14.23]:

$$\tau_{cor}^{-1} = \lambda_1 \simeq \frac{C_0}{2(1 - C_0)} G_0^{1/2} \tau_{Ap}^{-1}.$$

The turbulent transport coefficient is estimated as

$$\chi_{turb} \approx \frac{1}{(2 - C_0)} \frac{1}{s^2} G_0^{3/2} \frac{\delta^2}{\tau_{Ap}}. \tag{18.21}$$

(δ: collisionless skin depth.) In the limit of small magnetic shear $s \to 0$, equation (18.21) converges to a large but finite value [18.11].

18.2.3 ETG Mode Turbulence

The electron temperature gradient (ETG) mode could be important for energy transport [17.6]. The characteristic scale length and time rate are given by the electron gyroradius ρ_e and the electron transit time $R/v_{\text{th},e}$, respectively. The transport coefficient is given as

$$\chi_{\text{ETG}} = \frac{v_{\text{th},e}\rho_e^2}{R} F\left(q, s, \frac{r}{R}, \frac{L_n}{L_T}, \frac{T_e}{T_i}, \ldots\right). \tag{18.22}$$

Upon comparing equations (18.18) and (18.22), one sees that the typical form for the ETG mode, $v_{\text{th},e}\rho_e^2/R$, is about $\sqrt{m_e/m_i}$ times smaller than $\chi_{\text{ITG}} \simeq c_s\rho_s^2/R$ for the ITG mode.

A possible larger transport coefficient has been studied in [17.28, 17.29, 17.30]. For example, [17.30] gives

$$\chi_{\text{ETG}} = \frac{c\rho_e}{\omega_{pe}} \frac{v_{\text{th},e}}{L_{Te}} \tag{18.23a}$$

from random $E \times B$ motion, and

$$\chi_{\text{ETG}} = \frac{r}{qR}\left(\frac{c}{\omega_{pe}}\right)^2 v_{\text{th},e} \tag{18.23b}$$

owing to the stochastic motion induced by the magnetic perturbations. The latter is independent of the gradient length of the electron temperature, and has a dependence on the collisionless skin depth similar to that of the CDBM turbulence. This is because the collisionless skin depth is chosen as a relevant length.

In addition, the ETG mode induces a streamer [18 12], which is able to enhance the geometrical factor $F(q, s, r/R, L_n/L_T, T_e/T_i, \ldots)$. The direct numerical simulation provides [18.13]

$$\chi_{\text{ETG}} \approx 60 \frac{v_{\text{th},e}\rho_e^2}{L_n}, \tag{18.24}$$

which is not far away from equation (18.18) in magnitude.

18.2.4 Low or Negative Magnetic Shear

Associated with the reduction of the free energy source, the turbulent transport coefficient is also reduced in the low or negative magnetic shear region of tokamak plasma,

$$\chi_{\text{CDBM}} = F(\alpha, s)\alpha^{3/2} \frac{\delta^2}{\tau_{\text{Ap}}}. \tag{18.25}$$

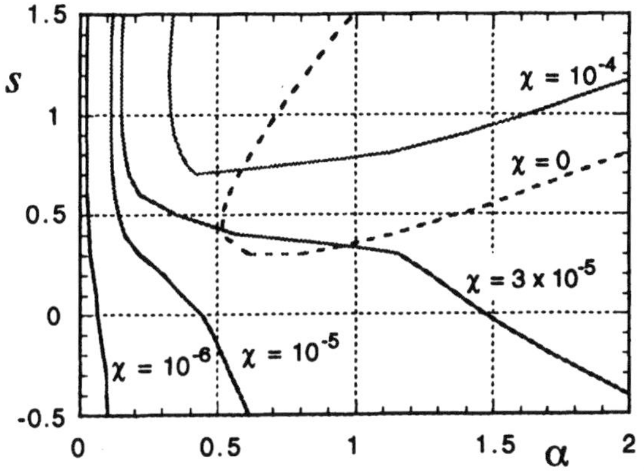

Figure 18.2. The contour of the transport coefficient on the s–α plane, which is obtained from the numerical solution of the nonlinear current–diffusive ballooning mode. The normalized values for $\hat{\chi} = \tau_{\mathrm{Ap}} a^{-2} \chi$ are shown. For a fixed pressure gradient, the thermal diffusivity becomes smaller for the weaker magnetic shear. The broken line indicates an instability boundary for the linear ideal MHD ballooning modes [17.41].

The contour of thermal conductivity is plotted on a plane of pressure gradient and magnetic shear in figure 18.2.

The connection length between the bad and good curvature regions for resistive turbulence becomes shorter in the case of negative magnetic shear. This effect is favourable for reducing the turbulence level [18.14].

The influence of the negative magnetic shear suggests that the current profile has an impact modifying the transport coefficients. The pressure and current profiles are two keys for the transport and structural formation in confined plasmas.

18.3 Momentum Flux and Reynolds Stress

The momentum transport is induced by pressure-gradient-driven turbulence, even in the limit of $\nabla \bar{V} \to 0$. The off-diagonal element of the transport matrix is also important. The turbulent viscosity and spontaneous torque are discussed below. Then the convective cell formation through a parametric decay instability [18.15] is reviewed.

18.3.1 Anomalous Viscosity and Spontaneous Torque

In the fluid picture of turbulence, all plasma particles respond in a similar way. In this case, the energy and momentum transport coefficients are close to each other. That is, the turbulent Prandtl number is of the order unity [18.16, 18.17]:

$$P_r^{\mathrm{turb}} \simeq 1, \tag{18.26}$$

upon assuming the isotropy of fluctuation spectrum. (This is in contrast to the case of kinetic instabilities, in which only resonant particles contribute to turbulent fluxes.) By employing a reduced set of equations, the flux of the perpendicular momentum in the radial direction, $\Pi^{\mathrm{d}}_{\perp r}$, has been obtained for CDBM turbulence as [18.18]

$$\frac{\Pi^{\mathrm{d}}_{\perp r}}{m_i n_i} = M_{11}\nabla\left(\frac{\nabla\phi}{B}\right) - M_{12}v_{\mathrm{A}}\nabla(\beta) \tag{18.27}$$

where M_{ij} is the (i, j) element of the transport matrix, and M_{11} is the shear viscosity $\mu_\perp$. The second term on the right-hand side is the off-diagonal element driven by the asymmetry of the propagation direction of the perturbations. In this case, a rotation torque is induced by the pressure gradient. It is estimated to be

$$\frac{M_{12}}{M_{11}} \approx \frac{1}{4q}\sqrt{\frac{m_i}{m_e}}\frac{c}{a\omega_{\mathrm{p}}} \tag{18.28}$$

based on a mechanism of $m = 1$ convection (m being the poloidal mode number) or loss of wave torque near the plasma periphery.

18.3.2 Excitation of Convective Cell (Zonal Flow and Streamer)

When turbulence is highly anisotropic, the coefficient μ_X in the gradient–flux $F_X = -\mu_X\nabla\bar{X} + F_{\mathrm{rest}}$ relation could be negative. That is, a certain turbulent-driven flux could enhance $\nabla\bar{X}$.

The generation of a convective cell by the drift waves has been observed in numerical simulation [12.14, 18.19] and discussed in [11.20]. As is shown by the dispersion relation $\omega \simeq -k_\parallel^2 c_{\mathrm{s}}^2 \omega_*^{-1}$, there is a damped oscillation with $k_\parallel \simeq 0$. The dispersion relation of drift waves, with finite-ion-gyroradius correction, is illustrated in figure 18.3. A decay instability [18.15] is possible for drift waves. Consider the case of a drift wave with wavenumber and frequency $(k_{\mathrm{d},0}, \omega_{\mathrm{d},0})$. This drift wave can decay into the convective cell with $(k_{\mathrm{c}}, \omega_{\mathrm{conv}})$ and another drift wave with $(k_{\mathrm{d},1}, \omega_{\mathrm{d},1})$, if the conditions

$$\boldsymbol{k}_{\mathrm{d},0} = \boldsymbol{k}_{\mathrm{c}} + \boldsymbol{k}_{\mathrm{d},1} \tag{18.29a}$$

$$\omega_{\mathrm{d},0} = \omega_{\mathrm{conv}} + \omega_{\mathrm{d},1} \tag{18.29b}$$

are satisfied. (The suffices d and c (or conv) indicate drift wave and convective cell, respectively.) As can be seen from figure 18.3, the decay instability occurs in the vicinity of the peak of the dispersion curve of the drift wave.

The growth rate of the convective cell is expressed in terms of the amplitude of drift-wave fluctuations. Density and potential perturbations for drift

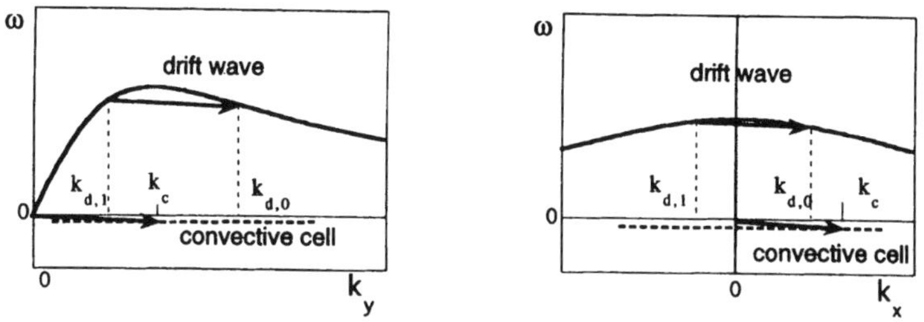

Figure 18.3. The decay process of the drift wave ($k_{\rm d,0}$) into the convective cell ($k_{\rm c}$) and the drift wave ($k_{\rm d,1}$). The dispersion relations for the drift wave (full curve) and the convective cell (broken line) are illustrated. The absolute value of the frequency of the convective cell $\omega_{\rm conv}$ is exaggerated. (Left) a streamer and (right) zonal flow.

and convective cell components are considered. They satisfy

$$\frac{\tilde{n}_{\rm d}}{\bar{n}} \simeq \frac{e\tilde{\phi}_{\rm d}}{T_e}, \qquad \frac{\tilde{n}_{\rm c}}{\bar{n}} \gg \frac{e\tilde{\phi}_{\rm c}}{T_e}. \tag{18.30}$$

The parametric decay instability is analyzed below. Quasi-linear equations are deduced; a bi-linear correction due to convective cell is kept, and the quasi-linear excitation of a convective cell by drift waves is considered. A model set of equations is

$$\left\{(1-\rho_{\rm s}^2\nabla_\perp^2)\frac{\partial}{\partial t} + V_{\rm de}\frac{\partial}{\partial y}\right\}\tilde{n}_{\rm d} = -\frac{T_e}{eB\bar{n}}[\tilde{n}_{\rm d}, \tilde{n}_{\rm c}], \tag{18.31a}$$

$$\left\{\frac{\partial}{\partial t} - \mu_c\nabla_\perp^2\right\}\tilde{n}_{\rm c} = \frac{T_e}{eB\bar{n}}[\tilde{n}_{\rm d}, \nabla_\perp^2\tilde{n}_{\rm d}]. \tag{18.31b}$$

The growth rate of the convective cell amplitude is given by [11.18]

$$\gamma_{\rm conv} = \gamma_{\rm pi} - \mu_c k_{\perp,{\rm c}}^2 \tag{18.32}$$

(the suffix 'pi' indicates the parametric instability process) with

$$\gamma_{\rm pi} = \gamma_{\rm sd} \equiv \frac{T_e}{eB}|\boldsymbol{k}_{\rm d,0} \times \boldsymbol{k}_{\rm d,1}|\sqrt{\frac{(k_{\rm d,0}^2 - k_{\rm d,1}^2)}{k_{\rm c}^2}}\left|\frac{\tilde{n}_{\rm d,0}}{\bar{n}}\right| \tag{18.33}$$

for the case of strong decay instability. If the decay instability is weak, one has

$$\gamma_{\rm pi} = \frac{\gamma_{\rm sd}^2}{\mu_\perp k_{\perp{\rm c}}^2}. \tag{18.34}$$

When the wavenumber is directed into the radial direction, $\boldsymbol{k}_{\rm c} \simeq (k_r, 0, 0)$, one has a zonal flow. In the case of $\boldsymbol{k}_{\rm c} \simeq (0, k_\theta, 0)$, one has a streamer.

18.4 Resistivity and Current Diffusivity

Turbulent resistivity is an important element that dictates the plasma dynamics, in particular, for the evolution of magnetic field. Taking the diffusion of magnetic flux and current into account, Ohm's law takes the form

$$\bar{\boldsymbol{E}} + \bar{\boldsymbol{V}} \times \bar{\boldsymbol{B}} = (\eta_c + \eta_{\mathrm{turb}})\bar{\boldsymbol{J}} - (\lambda_c + \lambda_{\mathrm{turb}})\,\Delta_\perp \bar{\boldsymbol{J}}. \tag{18.35}$$

In the presence of $\boldsymbol{E} \times \boldsymbol{B}$ convective nonlinearity, the turbulent resistivity η_{turb} and the turbulent current diffusivity λ_{turb} are estimated as (see, e.g., [2.28, 17.1])

$$\eta_{\mathrm{turb}} \simeq \mu_0 \chi_{\mathrm{turb}} \tag{18.36}$$

and

$$\lambda_{\mathrm{turb}} \simeq \mu_0 \left(\frac{c}{\omega_{\mathrm{p}}}\right)^2 \chi_{\mathrm{turb}}. \tag{18.37}$$

In the case of magnetic turbulence, one has

$$\lambda_{\mathrm{turb}} \simeq \frac{\sqrt{\pi}}{2}\mu_0 \left(\frac{c}{\omega_{\mathrm{p}}}\right)^2 v_{\mathrm{th},e} D_{\mathrm{M}} \sim \mu_0 \left(\frac{c}{\omega_{\mathrm{p}}}\right)^2 \chi_{\mathrm{turb},e} \tag{18.38}$$

where D_{M} is given by equation (18.9) [18.20–18.24]. The resistivity is less affected by the magnetic perturbation and remains as

$$\eta_{\mathrm{turb}} \simeq \mu_0 \chi_{E \times B}, \tag{18.39}$$

where $\chi_{E \times B}$ is the diffusivity of the turbulent motion [18.20].

Off-diagonal elements in the transport matrix could also be important for the electromotive force. The global electromotive force that drives global current is influenced by the pressure gradient, but its influence is found to be small in tokamaks [18.25]. The electromotive force caused by the velocity inhomogeneity has been investigated with the name of cross-helicity or γ-dynamo.

References

[18.1] Rechester A B and Rosenbluth M N 1978 *Phys. Rev. Lett.* **40** 38

[18.2] Galeev A A and Zelenyi L M 1979 *JETP Lett.* **29** 614 [1979 *Pis'ma Zh. Eksp. Teor. Fiz.* **29** 669]

[18.3] Krommes J A *et al* 1983 *J. Plasma Phys.* **30** 11

[18.4] Galeev A A 1984 in *Basic Plasma Physics II* ed A A Galeev and R N Sudan (Amsterdam: North-Holland) chapter 6.2

[18.5] Mel'nikov Yu P 1996 *JETP* **82** 860 [1996 *Zh. Eksp. Teor. Fiz.* **109** 1599]

[18.6] Eijnden E V and Balescu R 1997 *Phys. Plasmas* **4** 270

[18.7] Balescu R, Wang H-D and Misguich 1994 *Phys. Plasmas* **1** 3826

[18.8] Tange T, Inoue S, Itoh K and Nishikawa K 1979 *J. Phys. Soc. Jpn.* **46** 266

[18.9] Coppi B and Spight C 1978 *Phys. Rev. Lett.* **41** 551
[18.10] Waltz R E, Kerbel G D and Milovich J 1994 *Phys. Plasmas* **1** 2229
[18.11] Itoh K, Yagi M, Itoh S-I and Fukuyama A 1997 *Plasma Phys. Control. Fusion* **39** 1561
[18.12] Drake J F, Guzdar P N and Hassam A B 1988 *Phys. Rev. Lett.* **61** 2205
[18.13] Jenko F, Dorland W, Kotschenreuther M and Rogers B N 2000 *Phys. Plasmas* **7** 1904
[18.14] Antonsen T *et al* 1996 *Phys. Plasmas* **3** 2221
[18.15] Mima K and Nishikawa K 1984 in *Basic Plasma Physics II* ed A A Galeev and R N Sudan (Amsterdam: North-Holland) chapter 6.5
[18.16] Itoh S-I *et al* 1994 *Phys. Plasmas* **1** 1154
[18.17] Itoh K, Itoh S-I, Fukuyama A, Yagi M and Azumi M 1993 *J. Phys. Soc. Jpn.* **62** 4269
[18.18] Itoh K, Itoh S-I, Fukuyama A and Yagi M 1996 *J. Phys. Soc. Jpn.* **65** 760
[18.19] Cheng C Z and Okuda H 1977 *Phys. Rev. Lett.* **38** 708
[18.20] Diamond P H *et al* 1984 *Phys. Fluids* **27** 1449
[18.21] Strauss H R 1986 *Phys. Fluids* **29** 3668
[18.22] Bahattacharjee A and Hameiri E 1986 *Phys. Rev. Lett.* **57** 206
[18.23] Boozer A A 1986 *J. Plasma Phys.* **35** 133
[18.24] Yagi M, Itoh S-I, Itoh K and Fukuyama A 1997 *J. Korean Phys. Soc.* (*Proc. Suppl.*) **31** S189
[18.25] Itoh S-I and Itoh K 1988 *Phys. Letters A* **127** 267

Chapter 19

Generation of Structure in Flow

In inhomogeneous plasmas a spontaneous radial electric field is generated and induces transitions [19.1]. There are many processes that induce the radial electric field, and much work has been done on these processes, e.g., [19.1–19.9].

19.1 Breakdown of Ambipolarity of Turbulent Flow

It is often argued that the turbulent-driven particle flux is ambipolar in a steady state; in other words, the turbulent transport does not continue to induce a radial electric field. This is not correct. Turbulent-driven flux does generate a radial electric field, causing a variety of nonlinear dynamics of plasma turbulence [2.28].

The rate of change of the wave momentum is

$$\frac{\partial}{\partial t} P_{\mathrm{wave},y} = -(\boldsymbol{J} \times \boldsymbol{B})_y \tag{19.1}$$

with an assumption of no net wave momentum flux across a surface. (The injection of a torque by radio frequency (rf) waves is discussed in [19.10–19.13].) In a stationary state, $\partial P_{\mathrm{wave},y}/\partial t = 0$ holds and the momentum balance equation leads to the ambipolar condition

$$J_x = \sum_s e_s \Gamma_{s,x} = 0 \tag{19.2}$$

so long as an *average in a certain region* is treated. The basis and validity of ambipolarity are discussed in [19.14, 19.15]. In general, the wave can propagate in the direction of the gradient ($\hat{x}$ direction). Momentum exchange between the different magnetic surfaces can take place. The local momentum balance does not hold in every region. The ambipolarity of the turbulent flux, in a stationary condition, holds in a scale length which is longer than the autocorrelation length of the turbulence.

19.2 Generation of Zonal Flow by Drift Wave Turbulence

The perturbation amplitude of the convective cell (including zonal flow and streamer) is constant along the field line. If one calculates the average density (or potential) over some flux tube, the contribution from drift waves (and other waves) is averaged out, but that from a convective cell remains finite.

The excitation of a convective cell, zonal flow or streamer is interpreted as the occurrence of a structure with a mesoscale. *Mesoscale* [19.16] means that the characteristic scale length is longer than that of the turbulent fluctuations, but is shorter than the system size. The flow generation mechanism has an analogy to the magnetic dynamo. The generation of small scale dc radial electric field, i.e., a convective cell, has been confirmed by numerical simulations [18.19]. The generation process on a global scale has also been studied by numerical simulations [19.17–19.22]; see also discussions in [19.23, 19.24]. The generation of the flow by electrostatic turbulence is briefly discussed here.

One can extend the analysis of equation (18.31) to obtain a nonlinear evolution equation with respect to the amplitude of a convective cell [19.25]. If one employs the simple assumption that the drift wave fluctuations are in a stationary state, the action

$$N_k \equiv \frac{\mathcal{E}_k}{\omega_k} \propto (1 + k_\perp^2)^2 |\tilde{\phi}_k|^2 \tag{19.3}$$

is conserved. For a conserved quantity the ray-tracing equation holds. A detailed description of the ray tracing equation is given in, e.g., [19.26]. It is applied to drift waves as

$$\frac{\partial}{\partial t} N_k + \frac{\partial \omega_k}{\partial \boldsymbol{k}} \frac{\partial}{\partial \boldsymbol{x}} N_k - \frac{\partial \omega_k}{\partial \boldsymbol{x}} \frac{\partial}{\partial \boldsymbol{k}} N_k = 0 \tag{19.4}$$

(see also [19.27]). In the case of zonal flow generation $\boldsymbol{k}_{\mathrm{c}} \simeq (k_r, 0, 0)$, one has

$$\frac{\partial}{\partial t} V_{\mathrm{c}} - u \frac{\partial}{\partial r} V_{\mathrm{c}} + b \frac{\partial}{\partial r} V_{\mathrm{c}}^2 + D_{rr} \frac{\partial^2}{\partial r^2} V_{\mathrm{c}} = 0 \tag{19.5}$$

which describes an evolution of the zonal flow (being represented by V_{c}) in the presence of the stationary drift wave turbulence. The coefficients for the convective and nonlinear terms are given by

$$u = \frac{1}{B^2} \int \mathrm{d}^2 k \frac{k_\theta^2 k_r}{(1 + k_\perp^2)^2} \left(\frac{\partial \omega_k}{\partial k} \right)^{-1} \frac{\partial N_k}{\partial k_r}, \tag{19.6a}$$

$$b = \frac{-1}{2B^2} \int \mathrm{d}^2 k \frac{k_\theta^3 k_r}{(1 + k_\perp^2)^2} \left(\frac{\partial \omega_k}{\partial k} \right)^{-1} \frac{\partial}{\partial k_r} \left\{ \left(\frac{\partial \omega_k}{\partial k} \right)^{-1} \frac{\partial N_k}{\partial k_r} \right\}. \tag{19.6b}$$

The last term corresponds to the growth rate of the zonal flow

$$-D_{rr}\frac{\partial^2}{\partial r^2}V_{\mathrm{c}} = \gamma_{\mathrm{conv}}V_{\mathrm{c}} \tag{19.7}$$

and γ_{conv} is given by equation (18.33). When the spectrum of the drift wave has anisotropy, the growth rate of the convective cell γ_{conv} can be positive for short wave-length zonal flows.

19.3 Generation of Poloidal Flow by Collisional Processes

Collisional processes can induce the poloidal rotation. The ambipolarity breaks down in neoclassical transport theory at the order of ρ_i^2/L_n^2 [13.3]. In the presence of a steep radial gradient, collisional processes may drive the radial electric field and plasma rotation [19.1–19.4, 19.8]. (See reviews [2.41–2.51].) The spontaneous onset of poloidal rotation of tokamak plasmas in the framework of neoclassical theory is given in, e.g., [19.28], and related work is also found in [19.29, 19.30]. In plasmas without toroidal symmetry (like stellarators or helical systems), neoclassical transport drives a strong radial electric field, inducing transitions [19.31, 19.32].

Mechanisms that cause the plasma flow and radial electric field are shown in [2.28]. In order to study the mechanisms of structure formation of global flow, it is convenient to employ the Poisson equation to describe the radial electric field structure. The evolution of the global radial electric field E_{r}, being averaged over the magnetic surface, is governed by the charge conservation relation combined with the Poisson's equation as

$$\frac{\partial}{\partial t}E_{\mathrm{r}} = -\frac{1}{\varepsilon_0\varepsilon_\perp}\left(J_{\mathrm{r}}^{\mathrm{net}} - J_{\mathrm{ext}}\right), \tag{19.8}$$

where $J_{\mathrm{r}}^{\mathrm{net}}$ is the net radial current density in the plasma which flows across the magnetic surface, and ε_0 is the vacuum susceptibility. $\varepsilon_\perp$ is a dielectric constant of the magnetized plasma, and it takes the form

$$\varepsilon_\perp = (1 + 2q^2)c^2 v_{\mathrm{A}}^{-2}. \tag{19.9}$$

In comparison with the case of slab plasma, $\varepsilon_\perp$ is enhanced by a factor $1 + 2q^2$ in toroidal plasmas. This enhancement factor holds when the poloidal flow is induced by the radial electric field. The radial current density which is driven into the electrode by the external circuit is denoted by J_{ext}. The radial current is composed of two components,

$$J_{\mathrm{r}}^{\mathrm{net}} = J_{\mathrm{r}} - \varepsilon_0\varepsilon_\perp\nabla\cdot\mu_i\nabla E_{\mathrm{r}}. \tag{19.10}$$

The first term J_{r} is the 'local current', which is determined by the radial electric field at the same radial location. The second term is caused by the shear viscosity of ions, μ_i, and includes the diffusion operator. The shear viscosity, which is caused either by the isotropic part of fluctuations or by the ion collisions, tends to make the $E \times B$ velocity uniform. The drag force that makes the $E \times B$ velocity uniform, $m_i n_i \nabla \cdot \mu_i \nabla (E_r B^{-1})$, is in the direction perpendicular to the magnetic field, and induces the $\boldsymbol{F} \times \boldsymbol{B}$ drift of ions. ($\boldsymbol{F}$ is the drag force.) This $\boldsymbol{F} \times \boldsymbol{B}$ drift is in the radial direction and works for ions selectively. That is, the radial current is induced by this shear viscosity. The equation of E_r is a nonlinear diffusion equation as

$$\frac{\partial}{\partial t} E_r = \nabla \cdot \mu_i \nabla E_r - \frac{1}{\varepsilon_0 \varepsilon_\perp} (J_r - J_{ext}). \tag{19.11}$$

Contributions of the drive by drift wave fluctuations in equation (19.5) are also included in J^{net}. Even in the limit of $\tilde{n}_d / \bar{n} \to 0$, i.e., without the drive by drift-waves, equation (19.11) can show a self-sustaining of the plasma flow. ($\tilde{n}_d$ is the density fluctuation associated with drift wave fluctuations.)

19.4 Electric Field Domain Interface

Equations (19.5) and (19.8) include nonlinear terms with respect to the radial electric field. This nonlinearity gives rise to a self-sustained structure of the radial electric field. It allows solitary structures of radial electric field and flow velocity. A kink-soliton-like structure and a soliton-like structure can be obtained. The electric domain interface is demonstrated in [19.33] and following work (see, e.g., [19.34]).

In such structures, characteristic scale lengths are determined by the balance between the nonlinearity and spatial diffusion, not by a global size of a system. A mesoscopic scale length is self-sustained.

19.4.1 Domain and Domain Interface

The radial current $J_r(E_r)$ is a nonlinear function of the radial electric field. The local balance equation, which means that the space–time derivatives are neglected in equation (19.11),

$$J_r = J_{ext}, \tag{19.12}$$

often has multiple solutions, as is illustrated in figure 19.1. Consider the case where three solutions,

$$E_r = E_A, \qquad E_r = E_B, \qquad E_r = E_C, \tag{19.13}$$

are allowed for equation (19.12) in some radial region of the plasma. In this case, the radial electric field takes one branch of solution in one region of plasma,

$$E_r \simeq E_A \quad \text{for } r < r_t \tag{19.14a}$$

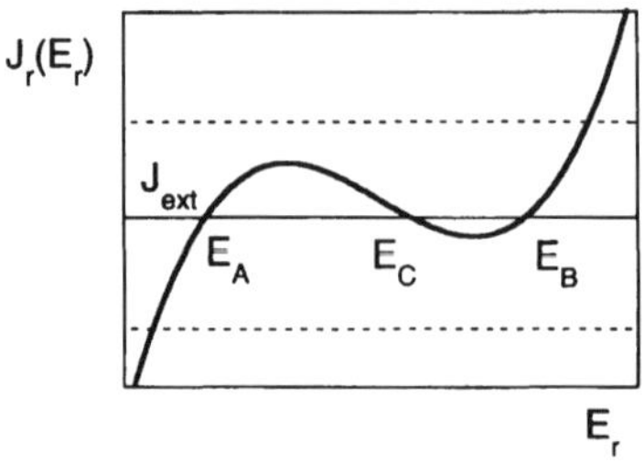

Figure 19.1. Multiple solutions of the local balance equation $J_r = J_{ext}$. A case of multiple solutions (shown by solid line) and cases of only one branch of solution (shown by dashed lines).

and the other branch in other region,

$$E_r \simeq E_B \quad \text{for } r > r_t \tag{19.14b}$$

(see figure 19.2). They are called electric field domains, and the boundary between them is called the domain interface.

The location of the domain interface is expressed in terms of the Maxwell's construction rule. (See Appendix 19A for details.)

A stationary state is studied for equation (19.11). Multiplying ∇E_r to equation (19.11) with $\partial E_r/\partial t = 0$, and integrating it in space across the domain interface $r = r_t$, one has a relation

$$(\nabla E_r)^2 = \frac{2}{\mu_i}\int_{E_A}^{E_r} \frac{1}{\varepsilon_0 \varepsilon_\perp}(J_r - J_{ext})\, dE_r \equiv F(E_r). \tag{19.15}$$

with the help of the boundary condition $dE_r/dr \simeq 0$ at $E_r \simeq E_A$ for $r \ll r_t$. From equation (19.15), an equation is derived as

$$\nabla E_r = \sqrt{F(E_r)} \tag{19.16}$$

which determines the relation $(E_r, dE_r/dr)$. Integrating equation (19.16), a

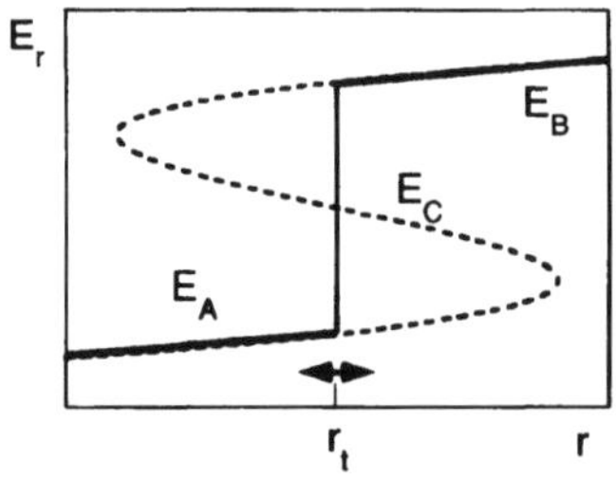

Figure 19.2. Schematic drawing of the domain and domain interface. The local solution is plotted by the dotted line, and thick solid lines show the realized solutions. Domains of branches A and B are separated by the domain interface at $r = r_t$. The region of integral in equation (19.15) is denoted by an arrow.

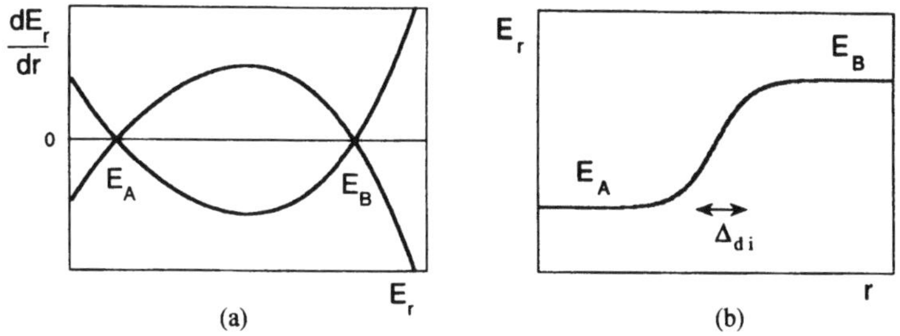

Figure 19.3. A functional relation (E_r, dE_r/dr) and the solution $E_r(r)$, when the Maxwell's construction equation (19.18) is satisfied (a). Solution $E_r(r)$ is shown in (b).

solution $E_r(r)$ is given in an implicit form as

$$\int^{E_r} \frac{dE_r}{\sqrt{F(E_r)}} = r - r_t. \tag{19.17}$$

The steep gradient of the radial electric field is restricted to the domain interface, and the gradient is small away from the interface. Therefore the condition

$$\int_{E_A}^{E_B} (J_r - J_{ext})\, dE_r = 0 \tag{19.18}$$

must be satisfied at the domain interface. Equation (19.18) is called *Maxwell's construction* and determines the location of the domain interface. Figure 19.3(a) illustrates the curve (E_r, dE_r/dr), and the solution of it $E_r(r)$ is shown in figure 19.3(b).

The thickness of the interface is also given from this work function $\int J_r\, dE_r$. The gradient of the radial electric field is maximum when $E_r = E_C$ holds and

$$\tfrac{1}{2}\mu_i(\nabla E_r)^2|_{max} = \int_{E_A}^{E_C} \frac{1}{\varepsilon_0\varepsilon_\perp}(J_r - J_{ext})\, dE_r. \tag{19.19}$$

The thickness of the layer, $\Delta_{di} \simeq |E_B - E_A|\,|\nabla E_r|_{max}^{-1}$, is given as

$$\Delta_{di} \simeq \frac{\sqrt{\mu_i\varepsilon_0\varepsilon_\perp}\,|E_B - E_A|}{\sqrt{2\int_{E_A}^{E_C} (J_r - J_{ext})\, dE_r}}. \tag{19.20}$$

Some examples are shown in the following.

19.4.2 Kink-Soliton-Like Structure in Zonal Flows

Equation (19.5) has stationary solutions as

$$V_c = 0 \tag{19.21a}$$

and

$$V_c = \frac{u}{b}. \tag{19.21b}$$

The nontrivial solution equation (19.21b) is determined by the balance between the convective term and the nonlinear term. A kink-soliton-like solution, which connects two solutions, equations (19.21a) and (19.21b), is also allowed from equation (19.5). Imposing the boundary conditions

$$V_c \to 0 \quad \text{at } r \to -\infty \qquad \text{and} \qquad V_c \to \frac{u}{b} \quad \text{at } r \to \infty, \tag{19.22}$$

one has a stationary solution of equation (19.5) as

$$V_c = \frac{1}{2}\frac{u}{b} + \frac{1}{2}\frac{u}{b}\tanh[l_{zf}^{-1}(r - r_c)], \tag{19.23a}$$

where

$$l_{zf} = \frac{D_{rr}}{u} \tag{19.23b}$$

dictates the thickness of the kink layer, and r_c is the location of the knee of the kink [19.35]. (See figure 9.3 for illustration.) The radial convection velocity is of the order of drift velocity,

$$u \simeq V_{de} = T_e/eBL_n. \tag{19.24}$$

19.4.3 Soliton-Like Structure

Consider a situation where two electrodes are placed on magnetic surfaces at some distance apart, and a current is driven into the electrode by the external circuit. Another example of a soliton-like solution is induced by an external bias current.

In toroidal plasmas, collisional processes have large contributions in generating radial electric field. A dc conductivity has a nonlinear dependence on the radial electric field. When the bulk viscosity damping of ion flow is considered, the radial current J_r in tokamak plasmas is given as

$$J_r = \frac{en_i\rho_{pi}r}{R^2B}\Im Z\left(\frac{e\rho_{pi}E_r}{T_i} + \mathrm{i}\frac{r\nu_{ii}B}{v_{th,i}B_p}\right)(E_r - E_{nc}), \tag{19.25}$$

where E_{nc} is the neoclassical electric field induced by the pressure gradient, and Z stands for the plasma dispersion function [19.8]. (ρ_{pi} is the poloidal ion gyroradius, $\rho_{pi} = m_i v_{th,i}/eB_p$, ν_{ii} is the ion–ion collision frequency.) The radial current is an increasing function of E_r when the magnitude of electric field is weak. If the radial electric field becomes strong,

$$E_r \approx T_i/e\rho_{pi}, \tag{19.26a}$$

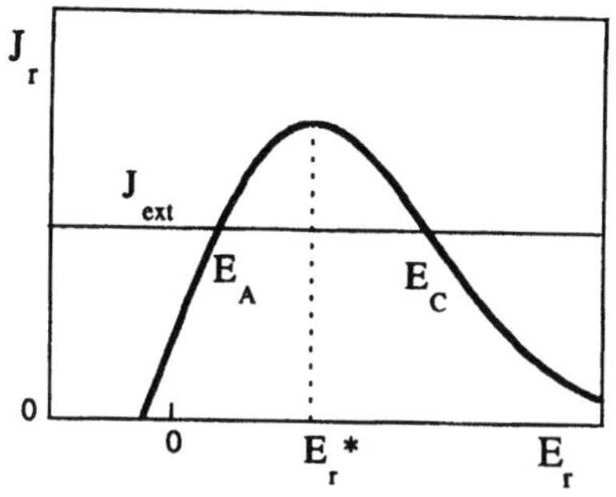

Figure 19.4. Radial current as a function of the radial electric field.

i.e.,

$$\frac{E_r}{B} \approx \frac{r}{Rq} c_s, \tag{19.26b}$$

the radial current J_r becomes a decreasing function of the radial electric field. A negative resistance, $\partial J_r/\partial E_r < 0$, appears. As is illustrated in figure 19.4, multiple solution of E_r is possible for the local equation $J_r(E_r) = J_{ext}$. Solution $E_r = E_A$ is the solution that connects to one in the limit of weak electric field.

Equation (19.11) with equation (19.25) has a solitary solution with the boundary condition $E_r \to E_A$ for $|r| \to \infty$. An example of the solitary solution of the radial electric field is illustrated in figure 19.5(a) [19.36]. The solitary solution (solid line in figure 19.5(a)) and the homogeneous solution (dotted line) are solutions of equation (19.11) with the same external current.

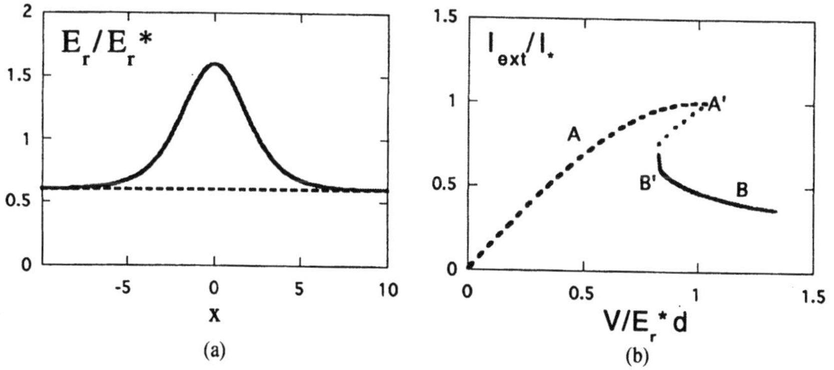

Figure 19.5. The solitary structure of a radial electric field. (a) Electric field as a function of the normalized radius. The spatially homogeneous solution is shown by the dashed line. (b) Applied voltage across the plasma layer and external radial current. The case with a homogeneous radial electric field is shown by dotted line A and the case for the solitary structure is shown by the solid line B. See [19.36] for the details.

The applied voltage across the electrode, which is an integral of E_r over radius, is larger for the solitary solution than the homogeneous solution. The solitary solution and homogeous solution have different voltage–current relationships.

This solitary structure appears when the applied voltage exceeds a threshold value. A bifurcation curve of the applied voltage and radial current is illustrated in figure 19.5(b). In the limit of a small applied voltage, the radial electric field is homogeneous in radius, and the current is an increasing function of the applied voltage. When the threshold is reached (A′ in figure 19.5(b)), the bifurcation to the solitary electric field takes place. When the applied voltage is decreased, a back transition takes place at the point B′ in figure 19.5(b).

A detailed explanation is given in Appendix 19A.

19.4.4 Poloidal Shock

A poloidal shock can occur if the poloidal Mach number is close to unity [19.37–19.40]. In the limit of a weak radial gradient, the poloidal flow V_θ is governed by the source rotating the plasma and by the poloidal derivatives. An example of the V_θ equation is [19.40]

$$\frac{\partial}{\partial t}V_\theta + V_\theta \frac{\partial}{r\partial\theta}V_\theta - \frac{\mu}{2}\frac{\partial^2}{r^2\partial\theta^2}V_\theta = -\frac{F_p}{2}\sin\theta \qquad (19.27)$$

where $S_c = -(F_\theta/2)\sin\theta$ models the drive by $m = 1$ component.

19.5 Streamer Formation

When a streamer is excited by drift-wave fluctuations, $\boldsymbol{k}_c \simeq (0, k_\theta, 0)$, it induces a modification of the density profile, which is estimated from the balance

$$D\nabla_\perp^2 N_{\text{streamer}} \simeq V_{\text{streamer}} \nabla_\perp \bar{N}. \qquad (19.28)$$

The deformed density contour in the presence of a streamer is illustrated in figure 19.6. Figures 19.6 (left) and (center) show schematically the potential perturbation and deformed density contour in toroidal plasmas, respectively. A detailed profile of the density deformation is given in figure 19.6 (right).

The conservation form of the wave kinetic equation for the drift-wave fluctuations equation (19.4) is an idealized simplification. Back interaction allows the possibility of nonlinear instability. This process is discussed in §21.2.

The analysis shows that both the zonal flow and streamer are excited by drift wave fluctuations. Depending on the waveform of drift waves,

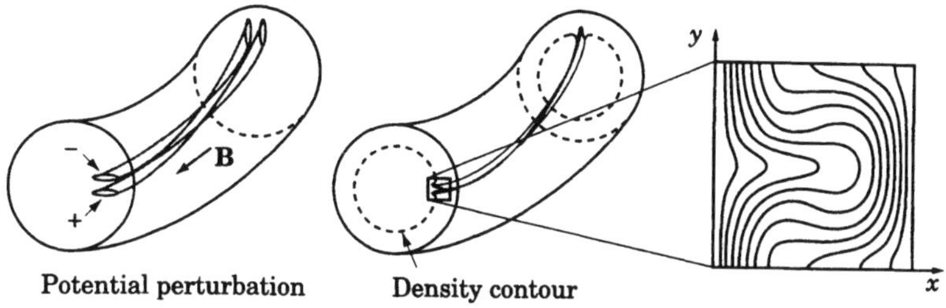

Figure 19.6. Illustration of density contour in the presence of a streamer. Potential perturbation (left) and density contour (center). The contour of the unperturbed state is denoted by the dashed curve and the deformed state by the solid line. Expanded view of density contour (right), based on [2.49].

zonal flow and streamer are excited. The conditions for their excitation have been discussed. A plane drift wave is known to induce them through parametric instabilities. Modulational instability of the plane drift wave has been analyzed. The condition for the generation of zonal flow and streamers from plane drift wave was discussed in [19.41]. The excited zonal flow and streamer might be unstable and could be torn into convective cells. A plausible shape of the sustained convective cell has been analyzed, and the shape of the cell has been analyzed in [19.42]. It has been shown that the sustained convective cell is stretched in the poloidal direction.

Appendix 19A

Maxwell's Construction and Domain Interface

In plasmas, nonlinearlity is found in various gradient–flux relationships. One example is the conductivity of radial current, which is given by equation (19.25). When the electric field is weak, the radial current J_r is an increasing function of the radial electric field E_r. (Both E_r and J_r are global components, not fluctuation components.) Under this circumstance, multiple values of E_r are possible for given value of J_r. Another example is the nonlinearlity between the pressure gradient and energy flux, as is illustrated in figure 2.18. Multiple values of pressure gradient are allowed for a given value of flux. In these cases, one branch of solution is chosen among possible branches of solutions. The selection rule for branches is known as *Maxwell's construction*, and is constructed for the structure of flows in plasmas.

19A.1 Nonlinear Diffusion Equation of Radial Electric Field

The evolution of the global radial electric field E_r, being averaged over the magnetic surface, is governed by the charge conservation relation combined with the Poisson's relation as equation (19.11). The equation of E_r is a nonlinear diffusion equation as

$$\frac{\partial}{\partial t}E_r = \nabla \cdot \mu_i \nabla E_r - \frac{1}{\varepsilon_0 \varepsilon_\perp}(J_r - J_{ext}). \tag{19A.1}$$

We are interested in the very steep gradient of E_r. Compared with the structure of E_r, the other plasma parameters are slowly varying in space, so that the others and the ion viscosity μ_i are treated as constant for simplicity in equation (19A.1). (The effect of the electric field shear on μ_i is discussed in the next section.)

It might be convenient to introduce a normalized form. A conductivity in the linear regime

$$\sigma(0) \equiv \partial J_r / \partial E_r|_{E_r \to 0} \tag{19A.2}$$

is introduced and the length, time, E_r and current density are normalized as

$$x = \frac{(r - r_0)}{l} \quad \text{with } l = \sqrt{\frac{\mu_i \varepsilon_0 \varepsilon_\perp}{\sigma(0)}}, \qquad \tau = \frac{t}{t_n} \quad \text{with } t_n = \frac{\varepsilon_0 \varepsilon_\perp}{\sigma(0)}$$
$$X = \frac{e\rho_{pi} E_r}{T}, \qquad J = \frac{e\rho_{pi}}{T\sigma(0)} J_r, \qquad I = \frac{e\rho_{pi}}{T\sigma(0)} J_{ext}, \tag{19A.3}$$

(ρ_{pi} is the ion poloidal gyroradius, and T is the ion temperature), respectively. The radius r_0 is chosen at some reference position. Then the basic equation for E_r is rewritten as

$$\frac{\partial}{\partial t}X = \frac{\partial^2}{\partial x^2}X - J(X) + I. \tag{19A.4}$$

A steady-state solution is given by the equation

$$\frac{\partial^2}{\partial x^2}X - J(X) + I = 0. \tag{19A.5}$$

19A.2 Local Solution

A local stationary solution is given by

$$J(X) = I. \tag{19A.6}$$

This equation could have multiple solutions as is illustrated in figure 19.1. In an appropriate range of I (shown by the solid line), the local solution of the radial electric field has multiple branches, X_A, X_B, and X_C. (Solutions X_A and

X_B are stable, and X_C is unstable.) In the cases which are shown by dashed lines in figure 19.1, only one branch of solution exists.

19A.3 Electric Field Domain and Domain Interface

When multiple branches of local solutions are allowed, one branch might be selected in one region of plasma radius, and another branch might be selected in the other region of plasma. Domains with different branches are separated by a domain interface. A schematic drawing is given in figure 19.2.

The location of the domain interface $x = x_t$ and the structure of the electric field are determined by the nonlinear diffusion equation (19A.5). Consider a case that the electric field is given by

$$X_A \quad \text{for } x \ll x_t \tag{19A.7}$$

as is illustrated in figure 19A.1.

Equation (19A.5) is multiplied by $\partial X/\partial x$ and is integrated as

$$\left(\frac{dX}{dx}\right)^2 = 2\int_{X_A}^{X} \{J(X) - I\}\, dX + \text{const} \equiv F(X). \tag{19A.8}$$

In the present application, where the gradient of parameter is weak, the integral constant is approximately treated as const $\simeq 0$, because dX/dx is small in the region of $X = X_A$ at $x \ll x_t$. $F(X)$ takes the minimum at $X = X_A$ and $X = X_B$, and the maximum at $X = X_C$.

The domain interface requires that the solution smoothly continues to the branch X_B. In the domain of X_B, dX/dx is also small at $x \gg x_t$. The boundary condition is written as

$$X = X_B \quad \text{and} \quad dX/dx = 0 \quad \text{at } x \gg x_t. \tag{19A.9}$$

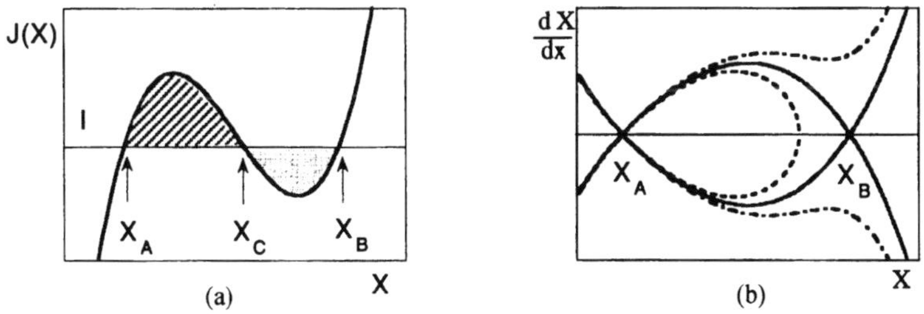

Figure 19A.1. Condition that determines the domain interface (a). When areas of shaded and gray regions are equal, equation (19A.10) is satisfied. The curve $(X, dX/dx)$ is shown in (b). The solid line indicates the case where the condition equation (19A.10) is satisfied. Dashed and chained lines indicate the cases of $F(X_B) < 0$ and $F(X_B) > 0$, respectively.

Substituting equation (19A.9) in to equation (19A.8), one has

$$F(X_B) = 2\int_{X_A}^{X_B} \{J(X) - I\}\,dX = 0 \quad \text{at } x = x_t. \tag{19A.10}$$

This condition implies that the areas between lines $J(X)$ and $J = I$ are equal (see figure 19A.1). The condition equation (19A.10) determines the location of the domain interface. This is an extension of Maxwell's construction rule in thermodynamics to that in nonlinear nonequilibrium systems.

A portrait (X, dX/dx) is illustrated in figure 19A.1(b). The solid line shows the case where the Maxwell's construction rule holds. The solid line is the solution of equation (19A.5) with the boundary conditions equations (19A.7) and (19A.9).

19A.4 Structure of the Domain Interface

The domain interface is characterized by a steep gradient. The characteristic thickness of the interface is obtained from equation (19A.10). $F(X)$ takes the maximum value at $X = X_C$. That is, the maximum value of the gradient dX/dx is given as

$$dX/dx|_{max} = \sqrt{F(X_C)}. \tag{19A.11}$$

The thickness of the domain interface Δ_{di} is approximately given by $(X_B - X_A)(dX/dx|_{max})^{-1}$, i.e.,

$$\Delta_{di} = \frac{X_B - X_A}{\sqrt{F(X_C)}}. \tag{19A.12}$$

A more explicit form of the structure of the domain interface is obtained by choosing a model form of $J(X)$. We choose a cubic-nonlinear model as

$$J - I = \hat{J}E(E^2 - 1) \tag{19A.13a}$$

with $E = (2X - X_A - X_B)/(X_B - X_A)$, for which $F(X)$ is given as

$$F(E) = \frac{\hat{J}}{2}(1 - E^2)^2. \tag{19A.13b}$$

The solution $X(x)$ is given from equation (19A.8) as

$$x = \int^X F(X)^{-1/2}\,dX. \tag{19A.14}$$

Substituting model equation (19A.13) into equation (19A.14), the solution is given as

$$X = \frac{(X_A + X_B)}{2} + \frac{(X_B - X_A)}{2}\tanh\left(\frac{x - x_t}{\Delta_{di}}\right) \tag{19A.15}$$

and

$$\Delta_{\mathrm{di}} = \sqrt{\frac{X_{\mathrm{B}} - X_{\mathrm{A}}}{\hat{J}}}. \tag{19A.16}$$

19A.5 Relaxation of the Interface

When the interface is imposed, as an initial condition, away from the stationary solution, a relaxation to the stable state is realized. Dynamical processes are explained briefly.

First, a competition between the stable and unstable branches are discussed. Let us consider the case where one stable branch $X = X_{\mathrm{A}}$ is realized in one region and an unstable branch $X = X_{\mathrm{C}}$ is imposed in the other region. The solution smoothly connects $X = X_{\mathrm{A}}$ and $X = X_{\mathrm{C}}$ across the interface (see figure 19A.2).

This state does not satisfy the stationary condition, and the interface moves. The velocity of the interface for the relaxation can be estimated from equation (19A.4). Multiplying $\partial X/\partial x$ on equation (19A.4) and integrating over the interface, one has

$$\int \frac{\partial X}{\partial x}\frac{\partial X}{\partial \tau}\,\mathrm{d}x = \frac{1}{2}\left(\frac{\mathrm{d}X}{\mathrm{d}x}\right)^2\Bigg|_{X=X_{\mathrm{A}}}^{X=X_{\mathrm{C}}} - \int_{X_{\mathrm{A}}}^{X_{\mathrm{C}}} \{J(X) - I\}\,\mathrm{d}X. \tag{19A.17}$$

From the boundary condition that the solution smoothly connects to $X = X_{\mathrm{A}}$ and $X = X_{\mathrm{C}}$ across the interface, one uses the relation that $\mathrm{d}X/\mathrm{d}x \simeq 0$ holds at $X = X_{\mathrm{A}}$ and $X = X_{\mathrm{C}}$ in equation (19A.17). One has

$$\int \frac{\partial X}{\partial x}\frac{\partial X}{\partial \tau}\,\mathrm{d}x = -\int_{X_{\mathrm{A}}}^{X_{\mathrm{C}}} \{J(X) - I\}\,\mathrm{d}X. \tag{19A.18}$$

The integrand of the left-hand side of equation (19A.18) has a dominant contribution at the interface, where the field X has a large gradient. When an interface is moving in space with the velocity U, the time derivative is

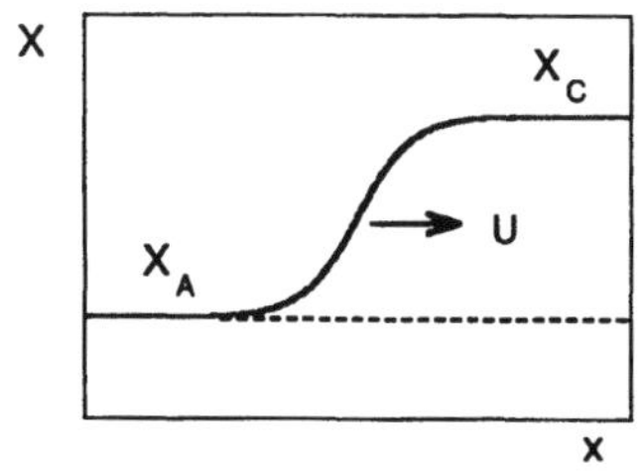

Figure 19A.2. Domains of stable branch $X = X_{\mathrm{A}}$ and unstable branch $X = X_{\mathrm{C}}$ are separated by an interface.

estimated as

$$\frac{\partial X}{\partial \tau} \simeq -U\frac{\partial X}{\partial x}. \tag{19A.19}$$

This gives

$$\int \frac{\partial X}{\partial x}\frac{\partial X}{\partial \tau}\,\mathrm{d}x \simeq -U\int\left(\frac{\partial X}{\partial x}\right)^2 \mathrm{d}x \simeq -U\,\Delta_{\mathrm{di}}\left(\left.\frac{\partial X}{\partial x}\right|_{\max}\right)^2. \tag{19A.20}$$

This estimate is derived by noting the fact that the steep gradient of X is localized in the vicinity of the interface. Combining equations (19A.17) and (19A.20), the velocity of the interface is estimated as

$$U \simeq \Delta_{\mathrm{di}}^{-1}\left(\left.\frac{\partial X}{\partial x}\right|_{\max}\right)^{-2}\int_{X_{\mathrm{A}}}^{X_{\mathrm{C}}}\{J(X)-I\}\,\mathrm{d}X, \tag{19A.21a}$$

i.e.,

$$U \simeq \frac{\sqrt{F(X_{\mathrm{C}})}}{2(X_{\mathrm{B}}-X_{\mathrm{A}})}, \tag{19A.21b}$$

where equations (19A.11) and (19A.12) are used. For the cubic model of equation (19A.13), one has

$$U \simeq \frac{\sqrt{\hat{J}}}{4\sqrt{X_{\mathrm{B}}-X_{\mathrm{A}}}}. \tag{19A.22}$$

With this velocity, the interface moves so as to reduce the area of the domain with unstable solution $X = X_{\mathrm{C}}$. After the passage of the domain interface, the solution is dominated by the stable branch as is shown by the dotted line in figure 19A.2. Noting the fact that the length and time are normalized, a typical value of the velocity is of the order of l/t_{n}. Equation (19A.22) gives a velocity in a dimensional form as

$$\frac{\sqrt{\hat{J}}}{4\sqrt{X_{\mathrm{B}}-X_{\mathrm{A}}}}\sqrt{\frac{\mu_i\sigma(0)}{\varepsilon_0\varepsilon_\perp}}. \tag{19A.23}$$

The velocity is in proportion to $\sqrt{\hat{J}}/(X_{\mathrm{B}}-X_{\mathrm{A}})$, and scales with a hybrid mean of viscosity and conductivity. The region in which the unstable branch of solution is imposed as an initial condition disappears in a time scale of $t_N L/l$ where L is a size of this domain. The change of domain appears as a propagation of the steep interface, not as a diffusion. The change of the electric field propagates much faster than the diffusive change. Discussion based on an alternative picture (equivalent particle dynamics) is made in [19.35].

Next, an approach to Maxwell's construction rule is explained. When the interface between two stable branches $X = X_{\mathrm{A}}$ and $X = X_{\mathrm{B}}$ is placed away from the equilibrium position, the interface moves to the location that satisfies equation (19A.10) (see figure 19A.3). The velocity of the interface

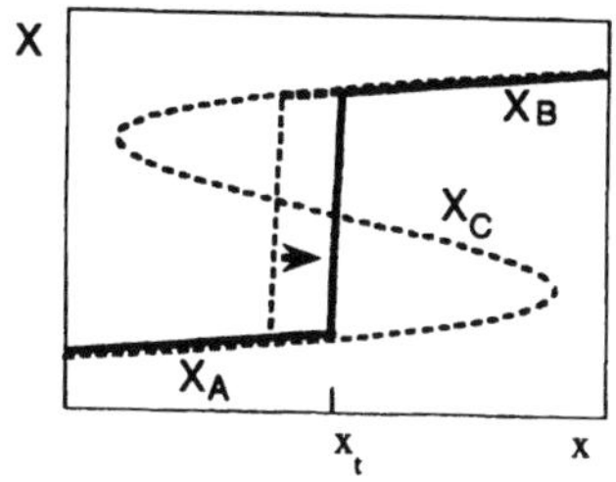

Figure 19A.3. When an interface is imposed away from the equilibrium position $x = x_t$, it moves to the stationary solution.

for the relaxation can be estimated from equation (19A.4). Multiplying $\partial X/\partial x$ on equation (19A.4) and integrating over the space, one has

$$\int \frac{\partial X}{\partial x}\frac{\partial X}{\partial \tau}\,dx = \frac{1}{2}\left(\frac{dX}{dx}\right)^2\Bigg|_{X=X_A}^{X=X_B} - \int_{X_A}^{X_B}\{J(X)-I\}\,dX. \qquad (19A.24)$$

With the help of the boundary conditions equations (19A.7) and (19A.9), equation (19A.17) reduces to

$$\int \frac{\partial X}{\partial x}\frac{\partial X}{\partial \tau}\,dx = -\int_{X_A}^{X_B}\{J(X)-I\}\,dX. \qquad (19A.25)$$

Following the same argument that leads to equation (19A.21), the velocity U of the interface between two stable branches is given as

$$U \simeq \frac{F(X_B)}{2(X_B - X_A)\sqrt{F(X_C)}}. \qquad (19A.26)$$

The interface moves until the Maxwell's construction, equation (19A.10), is satisfied. In the vicinity of an equilibrium position $x = x_t$, where equation (19A.10) holds, the velocity is Taylor-expanded as

$$U \simeq \frac{1}{2(X_B - X_A)\sqrt{F(X_C)}}\frac{dF(X_B)}{dx}(x - x_t) \equiv \frac{-1}{\tau_{\text{interface}}}(x - x_t). \qquad (19A.27)$$

The location of the interface approaches to the equilibrium position as

$$x = x_t + [x(0) - x_t]\exp\left(\frac{-\tau}{\tau_{\text{interface}}}\right) \qquad (19A.28)$$

where $x(0)$ is the initial location of the interface.

In the cubic model of the nonlinear relation, equation (19A.13) is asymmetrized as

$$J - I = \hat{J}(E^2 - 1)(E - E_{as}) \qquad (19A.29)$$

away from the equilibrium position of the interface. The asymmetry parameter E_{as} characterizes the dominant solution. (Here we take $|E_{as}| < 1$.)

When $E_{\rm as} > 0$ holds, the branch $X_{\rm A}$ is dominant, and $X_{\rm B}$ is dominant if $E_{\rm as} < 0$. $E_{\rm as}$ vanishes at $x = x_{\rm t}$ where Maxwell's construction holds, as is shown in figure 19A.3. As the first-order correction with respect to $E_{\rm as}$, equation (19A.26) is expressed as

$$U \simeq \frac{4\sqrt{\hat{J}}}{3\sqrt{X_{\rm B} - X_{\rm A}}} E_{\rm as}. \tag{19A.30}$$

This velocity is smaller than equation (19A.22) by a smallness factor $E_{\rm as}$. The asymmetry parameter is given by Taylor expansion near the equilibrium position $x = x_{\rm t}$ as

$$E_{\rm as} = (x_{\rm t} - x)l_{\rm g}^{-1} + \cdots \tag{19A.31}$$

where $l_{\rm g}$ is a global gradient scale length (normalized to l). The rate of the change of the location of interface is given as

$$\frac{1}{\tau_{\rm interface}} = \frac{4\sqrt{\hat{J}}}{3\sqrt{X_{\rm B} - X_{\rm A}}} \frac{1}{l_{\rm g}}. \tag{19A.32}$$

In a dimensional form, the gradient scale length of $E_{\rm as}$ is denoted by $L_{\rm gl}$ here. ($L_{\rm gl} = l_{\rm g} l$.) In a dimensional form, the characteristics time $\tau_{\rm interface}$, by which the Maxwell's construction equation (19A.10) is reached, has an order of magnitude of $L_{\rm gl}\sqrt{\varepsilon_0 \varepsilon_\perp / \mu_i \sigma(0)}$. This characteristic time is linear in scale length $L_{\rm gl}$ and is inversely proportional to the hybrid mean between diffusion coefficient and conductivity. The motion of the interface is faster than the diffusive change.

19A.6 Solitary Radial Electric Field

In some circumstances, the local solution of equation (19A.6) has only one stable branch and one unstable branch, as is illustrated in figure 19A.4. In this case, the solitary structure of the radial electric field is realized.

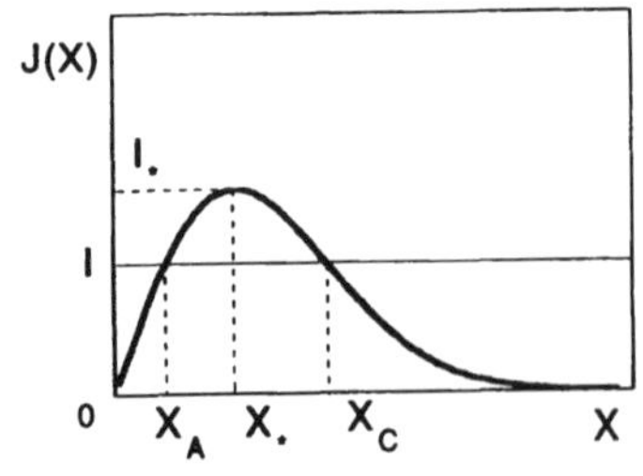

Figure 19A.4. The case of one stable branch and one unstable branch. The peak of $J(X)$ is denoted by I_* at $X = X_*$.

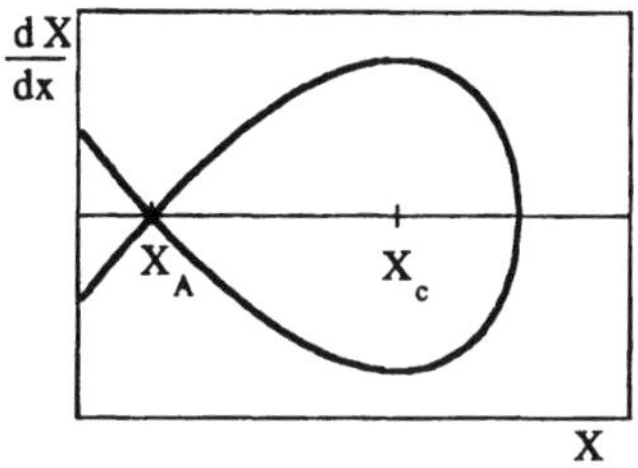

Figure 19A.5. The solution curve (X, $\mathrm{d}X/\mathrm{d}x$) for the case of figure 19A.3.

The boundary condition is chosen as

$$X = X_{\mathrm{A}} \quad \text{and} \quad \mathrm{d}X/\mathrm{d}x \to 0 \quad \text{at } |x| \to \infty. \tag{19A.33}$$

There is one trivial solution,

$$X = X_{\mathrm{A}}. \tag{19A.34}$$

Besides this trivial solution, there is a nontrivial solution with the solitary radial electric field. The integral equation (19A.8) is obtained, and the solution (X, $\mathrm{d}X/\mathrm{d}x$) is illustrated in figure 19A.5.

The solution is studied near the critical current, $I \simeq I_*$, where the local current $J(X)$ takes the maximum with respect to X, as is shown in figure 19A.4. Expanding $F(X)$ in equation (19A.10) in the vicinity of $I \simeq I_*$, and keeping terms up to $(X - X_{\mathrm{A}})^3$ in $F(X)$, we have

$$F(X) = C[(X_* - X_{\mathrm{A}})(X - X_{\mathrm{A}})^2 - \tfrac{1}{3}(X - X_{\mathrm{A}})^3 + \cdots], \tag{19A.35a}$$

with

$$C = -\left.\frac{\partial^2 J(X)}{\partial X^2}\right|_{X=X_*}. \tag{19A.35b}$$

The solution is given by the relation equation (19A.14). Substituting equation (19A.35a) into equation (19A.14), the radial electric field profile is explicitly obtained as

$$X(x) = X_{\mathrm{A}} + (X_* - X_{\mathrm{A}})\frac{12\,\mathrm{e}^{\alpha x}}{(1 + \mathrm{e}^{\alpha x})^2} \tag{19A.36}$$

where

$$\alpha \equiv \sqrt{C(X_* - X_{\mathrm{A}})}. \tag{19A.37}$$

The solution equation (19A.36) represents a solitary solution. Noting the relation

$$(X_* - X_{\mathrm{A}}) \simeq \sqrt{\frac{2}{C}(I_* - I)} \tag{19A.38}$$

in the small $I_* - I$ limit, one finds that the peak height scales as $(I_* - I)^{1/2}$ and the width scales like $(I_* - I)^{-1/4}$. Applications to tokamak plasmas are reported in [19.43, 19.44]. Solutions with multiple peaks are discussed in [19.44]. Stability of solutions with multiple peaks are discussed in [19.45], and a selection rule among various types of solutions is discussed.

References

[19.1] Itoh S-I and Itoh K 1988 *Phys. Rev. Lett.* **60** 2276

[19.2] Shaing K C *et al* 1989 in *Plasma Physics and Controlled Nuclear Fusion Research 1988* (Vienna: IAEA) vol 2, p 13

[19.3] Itoh S-I and Itoh K 1989 *Nucl. Fusion* **29** 1031

[19.4] Shaing K C and Crume E Jr 1989 *Phys. Rev. Lett.* **63** 2369

[19.5] Yoshizawa A 1991 *Phys. Fluids B* **3** 2723

[19.6] Drake J F *et al* 1992 *Phys. Fluids B* **4** 488

[19.7] Rozhanskii V and Tendler M 1992 *Phys. Fluids B* **4** 1877

[19.8] Stringer T E 1993 *Nucl. Fusion* **33** 1249

[19.9] Diamond P H, Liang Y-M, Carreras B A and Terry P W 1994 *Phys. Rev. Lett.* **72** 2565

[19.10] Inoue S and Itoh K 1981 in *Plasma Physics and Controlled Nuclear Fusion Research 1980* (Vienna: IAEA) vol II, p 649

[19.11] Craddock G G and Diamond P H 1991 *Phys. Rev. Lett.* **67** 1535

[19.12] Qiu X M, Bai L and Ran L B 1993 in *Plasma Physics and Controlled Nuclear Fusion Research 1992* (Vienna: IAEA) vol 2, p 269

[19.13] Ono M 1993 *Phys. Fluids B* **5** 241

[19.14] Inoue S *et al* 1979 *Nucl. Fusion* **19** 1252

[19.15] Waltz R E 1982 *Phys. Fluids* **25** 1269

[19.16] Itoh S-I, Itoh K, Fukuyama A and Miura Y 1991 *Phys. Rev. Lett.* **67** 2485

[19.17] Hasegawa A and Wakatani M 1987 *Phys. Rev. Lett.* **59** 1581

[19.18] Carreras B A, Lynch V E and Garcia L 1991 *Phys. Fluids B* **3** 1438

[19.19] Wakatani M, Watanabe K, Sugama H and Hasegawa A 1992 *Phys. Fluids B* **4** 1754

[19.20] Drake J F *et al* 1993 in *Plasma Physics and Controlled Nuclear Fusion Research 1992* (Vienna: IAEA) vol 2, p 115

[19.21] Carreras B A *et al* 1993 *Phys. Fluids B* **5** 1491

[19.22] Hallatscheck K 2000 *Phys. Rev. Lett.* **84** 5145

[19.23] Diamond P H and Kim Y B 1991 *Phys. Fluids B* **3** 1626

[19.24] Takayama A, Wakatani M and Sugama H 1996 *Phys. Plasmas* **3** 3

[19.25] Smolyakov A I and Diamond P H 1999 *Phys. Plasmas* **6** 4410; Smolyakov A I, Diamond P H and Malkov M 2000 *Phys. Rev. Lett.* **84** 491

[19.26] Bernsteain I B and Friedland L 1983 in *Basic Plasma Physics I* ed A A Galeev and R N Sudan (Amsterdam: North Holland) chapter 2.5

[19.27] Krommes J A and Kim C-B 2000 *Phys. Rev. E* **62** 8508

[19.28] Galeev A A, Sagdeev R Z, Liu C S and Novakovskii V 1996 *JETP* **82** 875 [1996 *Zh. Eksp. Teor. Fiz.* **109** 1626]

[19.29] Taguchi M 1991 *Plasma Phys. Contr. Fusion* **33** 859

[19.30] Hsu C T, Shaing K C and Gormly R 1994 *Phys. Plasmas* **1** 132

[19.31] Kovrizhnykh L M 1984 *Nucl. Fusion* **24** 851
[19.32] Hastings D E and Kamimura T 1985 *J. Comp. Phys.* **61** 286
[19.33] Hastings D E, Hazeltine R D and Morrison P J 1986 *Phys. Fluids* **29** 69
[19.34] Yahagi E, Itoh K and Wakatani M 1988 *Plasma Phys. Contr. Fusion* **30** 1009
[19.35] Diamond P H *et al* 1995 *Phys. Plasmas* **2** 3685; 1997 *Phys. Rev. Lett.* **78** 1472
[19.36] Itoh K, Itoh S-I, Yagi M and Fukuyama A 1998 *Phys. Plasmas* **5** 4121
[19.37] Hazeltine R D, Lee E P and Rosenbluth M N 1971 *Phys. Fluids* **14** 361
[19.38] Greene J M, Johnson J L, Weimer K E and Winsor N K 1971 *Phys. Fluids* **14** 1258
[19.39] Shaing K C, Hazeltine R D and Sanuki H 1992 *Phys. Fluids B* **4** 404
[19.40] Taniuti T, Moriguchi H, Ishii Y, Watanabe K and Wakatani M 1992 *J. Phys. Soc. Jpn.* **61** 568
[19.41] Champeaux S and Diamond P H 2001 *Phys. Lett. A* **288** 214
[19.42] Sanuki H and Weiland J 1980 *J. Plasma Phys.* **23** 29
[19.43] Heikkinen J A, Jachamich S, Kiviniemi T P, Kurki-Suonio T and Peeters A G 2001 *Phys. Plasmas* **8** 2824
[19.44] Kasuya N, Itoh K and Takase Y 2002 *J. Phys. Soc. Jpn.* **71** 93
[19.45] Kasuya N, Itoh K and Takase Y 2002 *Plasma Phys. Contr. Fusion* **44** A287

Chapter 20

Flow-Shear Suppression

Flow shear is itself a source of turbulence. However, if the flow shear exists together with a pressure gradient (another source of turbulence) it may suppress the turbulence.

This topic has attracted attention in the context of the improved confinement of plasmas [2.28, 2.41–2.51].

20.1 Effect of Flow Shear on Linear Stability

20.1.1 Linear Stability in Fluid Dynamics

The effects of shear and gravity on flow instability have been studied in fluid dynamics [20.1–20.3]. Gravity in the presence of a heat source can cause a Rayleigh–Benard instability (figure 20.1(a)) and flow shear (figure 20.1(b)) may drive Kelvin–Helmholtz (KH) instability. Their stability diagram is shown in figure 20.2 [20.3]. The regions of Rayleigh–Benard instability and Tollmein–Schlichting instability are dictated by the Richardson number

$$R_i \equiv \frac{R_a}{64 R_e^2 P_r}. \tag{20.1}$$

The transition from roll to wave structures occurs at $R_i = R_{i^*} \simeq 10^{-5}$. When the velocity shear is weak and the Reynolds number is small, the stability boundary for Rayleigh–Benard instability, R_{ac}, becomes higher by the effect of shear flow as

$$R_{ac} = R_{ac,0}(1 + c_1 R_e^2 + \cdots) \tag{20.2}$$

in the small R_e limit, where $c_1 \approx 10^{-1}$ for the case of figure 20.2. If the flow shear becomes too strong, KH-type instability takes place.

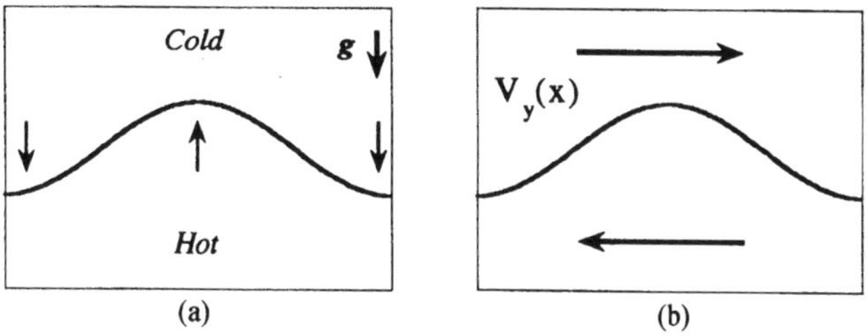

Figure 20.1. Rayleigh–Benard instability in the presence of gravity (a) and KH instability in the presence of sheared flow (b).

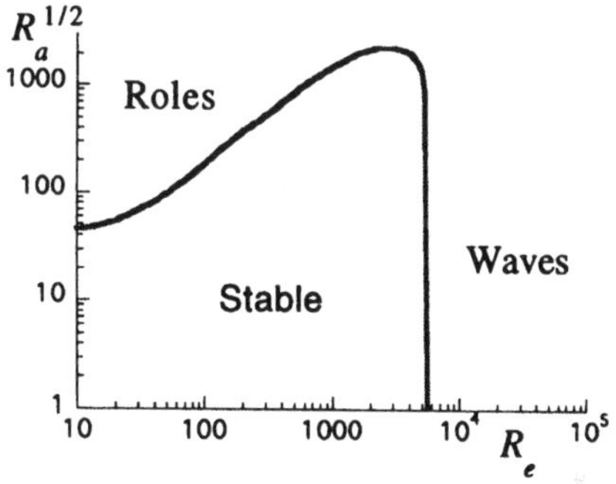

Figure 20.2. Stability diagram of the Poiseuille flow, in the presence of a heat source at the bottom, is given in the Rayleigh number–Reynolds number plane (R_a, R_e). The case in which the wavevector is in the direction of shear flow is reproduced from [20.3].

20.1.2 Linear Stability in Plasma Dynamics

In the presence of a pressure gradient ∇p parallel to a 'gravity' g (either real gravity or a centrifugal force due to magnetic curvature) a fluid-like instability, the flute mode, can occur with a linear growth rate $\gamma_L \approx \sqrt{g/L_p}$ ($L_p = |n/n'|$ for $\nabla T = 0$). When the electric field is radially inhomogeneous, the growth rate is reduced [20.4]. An off-resonant type of stabilization is possible, if a condition

$$|V'_{E\times B}| \approx \gamma_{L0} \tag{20.3}$$

is satisfied, where γ_{L0} is the linear growth rate in the limit of $V'_{E\times B} = 0$ [20.5–20.7]. This order-of-magnitude estimate is widely applicable for *linear-stability* analysis.

Another type of stabilization mechanism is seen in a wave-particle resonance. Landau damping is one of the main mechanisms for instabilities and can also be important as a nonlinear mechanism. The ion orbit is modified by the inhomogeneous electric field, and Landau damping is expected to occur. Strong ion Landau damping takes place if the inhomogeneity is strong enough [20.8]

$$\frac{L_{n2}\rho_i e E_r'}{T_i} \simeq \frac{\omega}{k_\theta v_{\mathrm{th},i}}, \tag{20.4}$$

where $\mathrm{d}^2n/\mathrm{d}r^2 = -L_{n2}^{-2}n$, even for a flute mode with $k_\parallel \simeq 0$.

The drift reversal of trapped particles due to an inhomogeneous electric field also influences stability. The toroidal drift velocity of trapped ions is modified by a factor $(1 + 2u_g)$, where

$$u_g = \rho_{pi} v_{\mathrm{th},i}^{-1} B_p^{-1} (\mathrm{d}E_r/\mathrm{d}r). \tag{20.5}$$

If the condition $u_g < -\frac{1}{2}$ is satisfied, trapped particles drift as if the magnetic curvature were favourable. As a result of this modification, the growth rate turns out to be $\Im(\omega) \simeq (2r/R)^{1/4}\sqrt{(1+u_g)\omega_{Me}\omega_*}$. The trapped-ion mode is stabilized by a drift reversal in the range of

$$u_g < -1. \tag{20.6}$$

This stabilization mechanism has an asymmetry with respect to the sign of E_r' [20.9].

An inhomogeneity of the toroidal flow is also known to stabilize plasma instabilities. The effects of toroidal flow on various instabilities are seen in [20.10–20.13].

When the flow shear becomes too large, a KH-type instability may occur in the plasma. The competition between the flute and KH instabilities is illustrated in figure 20.3 [20.14]. In this model, an inhomogeneous $E \times B$ flow of the form

$$V_{E\times B} = V_0 \tanh(x/L_E)$$

in the poloidal direction is imposed on the plasma. The influence of the shear flow on the linear growth rate is plotted in figure 20.3 for a given value of Δ_x/L_E. (Δ_x and γ_g are the radial width and growth rate, respectively, of

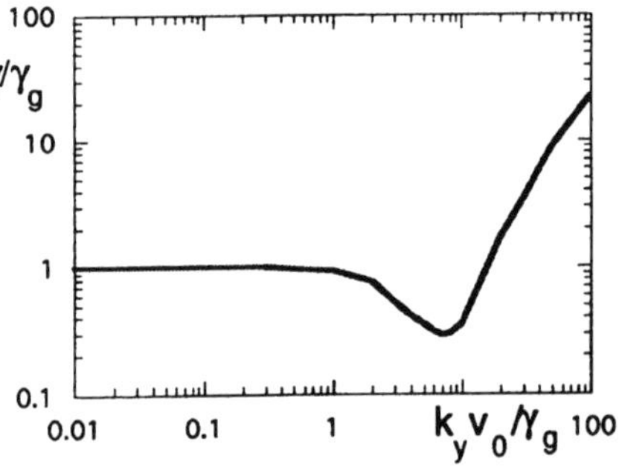

Figure 20.3. Linear growth rate of pressure-gradient-driven instability as a function of the flow shear for a case of $\Delta_x/L_E = 0.5$. (Quoted from [20.14].)

the resistive interchange mode in the absence of an inhomogeneous radial electric field.) For a fixed pressure gradient, the flow shear first stabilizes the mode when $k_y V_0 \Delta_x / L_E \approx \gamma_g$. In a strong flow shear limit, KH instability dominates and a strong instability may appear.

20.2 Suppression of Turbulence

20.2.1 Decorrelation Rate

In the model equation for a passive scalar advected by background fluctuations, the effect of rapidly changing fluctuations are included in the turbulent transport coefficient, and equation (14.40) has the form

$$\frac{\partial}{\partial t}\tilde{X} + \bar{V}_y(x)\frac{\partial}{\partial y}\tilde{X} - D\nabla^2\tilde{X} = \tilde{S}^{\text{ext}}. \tag{20.7}$$

The stretching of the test perturbation takes place, and the turbulence level is suppressed by $\partial \bar{V}_y / \partial x$ (i.e., radial electric field shear E_r' in magnetized plasmas). Figure 20.4 illustrates a sheared flow and a stretching of fluid element. Its mean velocity in the y direction (poloidal direction), which has a shear in the x direction (radial direction), is expressed as

$$\bar{V}_y = S_v x \tag{20.8}$$

in local coordinates. (The flow shear is interpreted as

$$S_v = r\frac{\mathrm{d}}{\mathrm{d}r}\left(\frac{E_r}{Br}\right) \tag{20.9}$$

in a cylindrical geometry.) After time t, a circular element is stretched to an ellipse with a major axis of the length $L_l \approx \sqrt{L^2 + (LS_v t)^2}$. Since an area is preserved by this stretching, the length of the minor axis is given by

$$L_\perp = \frac{L}{\sqrt{1 + S_v^2 t^2}}. \tag{20.10}$$

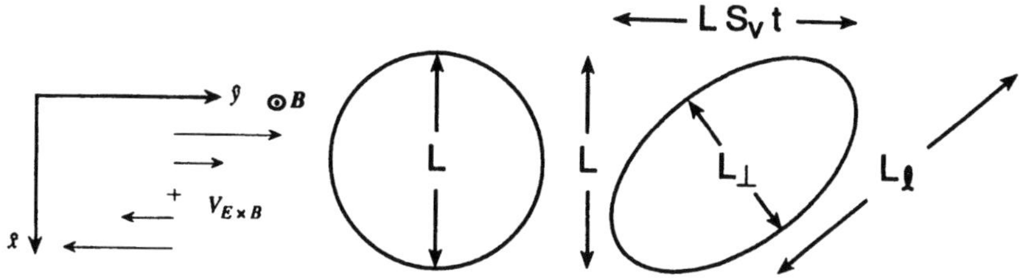

Figure 20.4. Flow in the y direction, which is inhomogeneous in the x direction, and a stretching of fluid element.

A perpendicular wavelength $k_\perp^{-1}$, of any mode is compressed due to this shear flow. The characteristic perpendicular wavenumber for the test field $\tilde{X}$ is effectively enhanced by a factor $(1 + S_v^2 t^2)$ [19.2, 20.15–20.17],

$$l_{\perp \text{eff}}^{-2} = l_c^{-2}(1 + S_v^2 t^2).$$

where l_c is an initial value of the characteristic scale length of test perturbation. As a consequence, the decorrelation rate, τ_D^{-1}, which is defined by the relation (14.22)

$$1/\tau_D \approx D l_{\perp \text{eff}}^{-2}, \tag{20.11}$$

becomes larger (for a given value of D).

The stretching of the turbulent vortex continues until the time reaches τ_D. Substituting the relation $t = \tau_D$ into equation (20.10), an effective mode-number $k_{\perp \text{eff}}$ (i.e., an inverse of the correlation length) is estimated as

$$k_{\perp \text{eff}}^2 \simeq l_{\perp \text{eff}}^{-2} = l_c^{-2}(1 + S_v^2 \tau_D^2). \tag{20.12}$$

The equation that determines a consistent decorrelation rate $1/\tau_D$ is derived by substituting it into equation (20.11) as,

$$\frac{1}{\tau_D} = D l_c^{-2}(1 + S_v^2 \tau_D^2). \tag{20.13}$$

Depending on the magnitude of the flow shear, $|S_v|$, two limiting cases are derived from equation (20.13):

$$\frac{1}{\tau_D} = \begin{cases} D l_c^{-2} + S_v^2 D^{-1} l_c^2 + \cdots & |S_v| \ll D l_c^{-2} \\ (D l_c^{-2})^{1/3} S_v^{2/3} & |S_v| \gg D l_c^{-2} \end{cases} \tag{20.14}$$

In equation (20.14), l_c stands for the correlation length in the absence of the flow shear S_v. The limiting form of $\tau_D \approx (D l_c^{-2})^{-1/3} S_v^{-2/3}$ is similar to the one derived from the renormalization of the turbulent electron parallel motion in a sheared magnetic field, equation (17.26) [17.31].

The decorrelation due to the shear flow is effective if S_v reaches a level $D k_{\perp 0}^2$. For a constant D, the relation

$$S_v \geq D k_{\perp 0}^2 \tag{20.15}$$

shows that stabilization by shear flow is more effective for longer wavelength modes.

The reduction of the correlation length leads to suppression of the fluctuation amplitude of the test field $\tilde{X}$, as

$$\frac{\langle \tilde{X}^2 \rangle}{\langle \tilde{X}^2 \rangle_{\text{ref}}} \approx \frac{1}{1 + S_v^2 \tau_D^2}, \tag{20.16}$$

assuming that the magnitude of the source term $\lim_{l \to 0} S^{\text{ext}}(l)$ is unaffected, where the suffix 'ref' indicates a reference case $S_v = 0$. A Lorentzian correction appears [19.2, 20.16, 20.17]. In the large shear limit, $\tau_{cor} \approx (Dk_{\perp 0}^2)^{-1/3} S_v^{-2/3}$, reduced fluctuations were discussed.

A stochastic Doppler shift is also effective in reducing the turbulence level, as shown in equation (14.34). This process has also attracted attention recently [14.16, 20.18].

20.2.2 Turbulence Level and Turbulent Transport

As was discussed in [19.2], these analyses have been developed assuming that the basic properties of the fluctuations (such as the wave number $k_{\perp 0}$ of a relevant mode and so on) are unchanged. These terms can also be functions of E_r and dE_r/dr. Effects of E_r and dE_r/dr, as a whole, are determined simultaneously after the turbulence structure is properly solved.

Analyses have been performed for ITG modes, e.g., [12.7, 20.19–20.21]. A formula for the turbulent transport coefficient has been proposed [20.21]:

$$\chi_{\rm turb} \simeq \frac{(\gamma_{\rm L} - \omega_{E1} - \gamma_{*1})^{1/2} \gamma_{\rm d}^{1/2}}{k_y^2} \tag{20.17}$$

where $\gamma_{\rm L}$ is the linear growth rate in the absence of flow shear, ω_{E1} is the $E \times B$ flow shear

$$\omega_{E1} = \frac{r}{q} \frac{\rm d}{{\rm d}r} \left(\frac{qE_{\rm r}}{rB} \right), \tag{20.18}$$

[20.22], γ_{*1} is the shear of diamagnetic flow, $\gamma_{\rm d}$ is the damping rate of a representative beat mode, being approximated as

$$\gamma_{\rm d} \simeq 0.3(T_i/T_e)\omega_{\rm M} \tag{20.19}$$

($\omega_{\rm M}$ is the toroidal gradient-B drift frequency), and k_y is the poloidal wavenumber of the most unstable mode. (The dependence of $\chi_{\rm turb}$ on ω_{E1} is adjusted to the observation of nonlinear simulation.)

In the case of self-sustaining turbulence, the thermal diffusivity has been derived for CDIM turbulence as follows [20.23]:

$$\chi_{\rm turb} \approx \frac{1}{(1 + 0.5G_0^{-1}\omega_{E1}^2)} \frac{G_0^{3/2}}{s^2} \left(\frac{c}{\omega_{\rm p}} \right)^2 \frac{v_{\rm Ap}}{a}, \tag{20.20}$$

where $\omega_{E1} = k_\theta \tau_{\rm Ap} E_{\rm r}'/B$ and $\langle k_\perp^2 \rangle \propto (1 + 0.5G_0^{-1}\omega_{E1}^2)G_0^{-1}$. As the gradient of the radial electric field becomes larger, the correlation length becomes shorter. In a toroidal geometry (i.e., CDBM turbulence), the normalized parameter

$$\omega_{E1} = \tau_{\rm Ap} \frac{1}{srB} \frac{{\rm d}E_{\rm r}}{{\rm d}r} \tag{20.21}$$

controls the turbulence level and turbulent transport [20.17]. The $E \times B$ flow shear effect is more effective for the lower-magnetic shear case. This shear dependence is also found for the ITG mode [12.7].

The electron temperature gradient (ETG) mode has a shorter characteristic wave length. This fact suggests that the $E \times B$ flow shear has a weaker effect. Streamers could be affected by the $E \times B$ shear and so transport by ETG modes could also be affected. Further study is required for the transport linkage between $E \times B$ shear and shorter wave length turbulence.

In addition to the inhomogeneity of flow across the magnetic surfaces, the inhomogeneity within the magnetic surface is also effective in suppressing turbulence. The toroidal flow in tokamaks varies in the poloidal direction if a hot ion component exists. This poloidal dependence suppresses nonlinear turbulence [20.24].

The dependence of χ_{turb} on ω_{E1} has also been discussed. The expression

$$\chi_{\text{turb}} \propto \frac{1}{1 + (\omega_{E1}/\gamma)^h} \tag{20.22}$$

has been derived analytically with an index h (γ is the decorrelation rate or instability growth rate in the absence of $E \times B$ shear). The index is given as $h = 2$ in the models described [19.2, 20.16, 20.17, 20.23] and as $h = \frac{2}{3}$ in [20.15]. A nonlinear simulation has suggested a dependence such as that in equation (20.17) for ITG mode turbulence. Further elaboration of theory might be required to derive the formula which is relevant in a wide parameter region. A comparison of the index h with experimental observations has been reported [20.25] when the electric bifurcation is controlled by an external bias current [20.26]. Other examples are summarized in the review article [2.50].

The turbulence suppression by radial electric field shear is essential in the structural formation of plasmas. Analyses along the theories which are introduced in this chapter have been developed in understanding transport barriers, e.g., the edge barrier for H-mode and those for internal transport barriers (ITB).

In this monograph, influences of global flows on turbulence are illustrated from various aspects. In particular, mechanisms for suppression of inhomogeneity-driven turbulence are important. They are summarized in table 20.1. Impacts of inhomogeneous d.c. flow are explained in this chapter.

Table 20.1 Suppression of micro fluctuations by flows

	Static but inhomogeneous global flow	Rapidly-oscillating homogeneous flow
Linear theory	Deformed eigenmode, drift reversal, etc. Refs. [20.1–20.13]	Dynamic stabilization. Ref. [20.27]
Nonlinear theory	Stretching of eddy and enhanced decay by turbulence. This monograph §20.2	Random Doppler-shift. This monograph §14.2.2

Those of homogeneous but rapidly changing flow are described in chapter 14. Linear stabilization mechanisms as well as nonlinear mechanisms are explained in these chapters. Both types of flows are possible and have nonlinear interactions with turbulence.

Appendix 20

Effect of Radial Electric Field Inhomogeneity on Domain Interface

As is discussed in this chapter, the electric field shear can influence the ion shear viscosity. When this effect is taken into account, the term μ_i in equation (19.11) is no longer a constant but depends explicitly on $\mathrm{d}E_\mathrm{r}/\mathrm{d}r$.

The normalized form in Appendix 19A is used here, and μ_i has a dependence on $|\mathrm{d}X/\mathrm{d}x|$. Considering this, one writes as

$$\mu_i = \mu_{i,0} f\left(\frac{\mathrm{d}X}{\mathrm{d}x}\right) \tag{20A.1}$$

where $\mu_{i,0}$ is the value of μ_i at $\mathrm{d}X/\mathrm{d}x = 0$. (The function takes $f(0) = 1$ and satisfies $f < 1$ if $|\mathrm{d}X/\mathrm{d}x|$ becomes larger.) The normalization length is redefined as $l = \sqrt{\mu_{i0}\varepsilon_0\varepsilon_\perp/\sigma(0)}$. The coefficient μ_{i0} depends on many plasma parameters that might have spatial dependencies. The inhomogeneity of μ_i due to such dependencies is considered to be weaker than the influence of $|\mathrm{d}X/\mathrm{d}x|$ at the interface. Focusing on the effect of $|\mathrm{d}X/\mathrm{d}x|$ at the interface, a model of equation (20A.1) is used.

With this modification (20A.1), the basic equation (19A.5) is deformed as

$$\frac{\mathrm{d}}{\mathrm{d}x} f\left(\frac{\mathrm{d}X}{\mathrm{d}x}\right)\frac{\mathrm{d}}{\mathrm{d}x}X - J(X) + I = 0. \tag{20A.2}$$

Solving equation (20A.2), equation (19A.8) is rewritten as

$$f(X')(X')^2 - \int_0^{X'} f(X')X'\,\mathrm{d}X' = \tfrac{1}{2}F(X). \tag{20A.3}$$

where $X' = \mathrm{d}X/\mathrm{d}x$. Equation (20A.3) gives $\mathrm{d}X/\mathrm{d}x$ as a function of X.

From equation (20A.3), one sees the following. First, the equilibrium condition equation (19A.10) is not modified. Substituting boundary condition $\mathrm{d}X/\mathrm{d}x = 0$ at $X = X_\mathrm{B}$ into equation (20A.3), one obtains equation (19A.10). The Maxwell's construction rule is not altered, and the dependence of μ_i on $\mathrm{d}E_\mathrm{r}/\mathrm{d}r$ does not directly influence the location of the interface. Next,

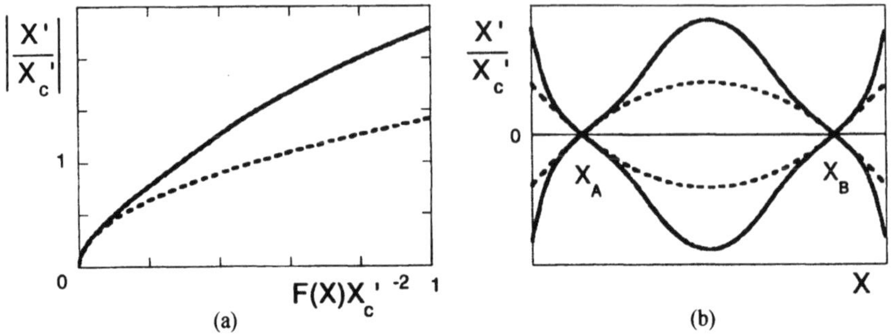

Figure 20A.1. Normalized gradient of the radial electric field dX/dx as a function of the normalized work function. Solid line for the case of equation (20A.4) and dashed line for $f = 1$. ($C_{\rm NC} = 0.4$.) (a) A schematic portrait (X, /dx) is given in (b). Cases with turbulence suppression (solid line) and without (dashed line) are shown.

a phase relation (X, dX/dx) is similar to that in figure 19A.1(b) and a quantitative modification is made near $X \simeq X_{\rm C}$. The maximum value of $|{\rm d}X/{\rm d}x|$ is increased by the reduction of the ion viscosity. The interface becomes thinner.

One may choose a simple model as

$$f(X') = \frac{1 - C_{\rm NC}}{1 + (X')^2 (X'_{\rm c})^{-2}} + C_{\rm NC} \tag{20A.4}$$

where $X'_{\rm c}$ is the critical level of the electric field shear above which the suppression of turbulence becomes substantial, and $C_{\rm NC}$ is a residual part of ion shear viscosity (such as the one by the neoclassical transport). ($C_{\rm NC} < 1$.) Equation (20A.3) leads to

$$(1 - C_{\rm NC})\left(\frac{2X'^2}{1 + X'^2 X_{\rm c}'^{-2}} - X_{\rm c}'^2 \ln\left(1 + \frac{X'^2}{X_{\rm c}'^2}\right)\right) + C_{\rm NC} X'^2 = F(X). \tag{20A.5}$$

The normalized gradient of the radial electric field dX/dx is illustrated in figure 20A.1 as a function of the normalized work function $F(X)$. The solid line is for the case of equation (20A.4) in which the effect of turbulence suppression is introduced, and the dashed line for $f = 1$. When the work function $F(X)$ approaches the value of $(X'_{\rm c})^2$, the effect of the electric field inhomogeneity becomes appreciable in the structure of the domain interface. A schematic drawing of the portrait (X, dX/dx) is shown in figure 20A.1(b). Cases with and without suppression factor $f({\rm d}X/{\rm d}x)$ are shown. The gradient of the electric field becomes stronger in the interface.

Quantitative analysis of the electric field domain interface in toroidal plasmas has been explained in detail in [20.27].

References

[20.1] Kuo H L 1963 *Phys. Fluids* **6** 195
[20.2] Deardorff J W 1965 *Phys. Fluids* **8** 1027
[20.3] Gage K S and Reid W H 1968 *J. Fluid Mech.* **33** 21
[20.4] Lehnert B 1966 *Phys. Fluids* **9** 1367
[20.5] Shaing K C 1991 *Comments Plasma Phys. Contr. Fusion* **14** 41
[20.6] Hassam A B 1991 *Comments Plasma Phys. Contr. Fusion* **14** 275
[20.7] Waltz R E, Kerbel G D, Milovich J and Hammett G W 1994 *Phys. Plasmas* **1** 2229; 1995 *Phys. Plasmas* **2** 2408
[20.8] Sanuki H 1984 *Phys. Fluids* **27** 2500
[20.9] Itoh S-I, Itoh K, Ohkawa T and Ueda N 1989 in *Plasma Physics and Controlled Nuclear Fusion Research 1988* (Vienna: IAEA) vol 2, p 23
[20.10] Inoue S, Itoh K and Yoshikawa S 1980 *J. Phys. Soc. Jpn.* **49** 367
[20.11] Waelbroeck F L *et al* 1992 *Phys. Fluids B* **4** 1441
[20.12] Chu M S *et al* 1995 *Phys. Plasmas* **2** 2236
[20.13] Bai L, Fukuyama A and Uchida M 1998 *Plasma Phys. Contr. Fusion* **40** 785
[20.14] Sugama H and Wakatani M 1991 *Phys. Fluids B* **3** 1110
[20.15] Biglari H, Diamond P H and Terry P W 1990 *Phys. Fluids B* **2** 1
[20.16] Zhang Y Z and Mahajan S M 1992 *Phys. Fluids B* **4** 1385
[20.17] Itoh S-I, Itoh K, Fukuyama A and Yagi M 1994 *Phys. Rev. Lett.* **72** 1200
[20.18] Hahm T S, Burrell K H, Lin Z, Nazikian R and Synakowski E J 2000 *Plasma Phys. Contr. Fusion* **42** A205
[20.19] Hamaguchi S and Horton W 1992 *Phys. Fluids B* **4** 319
[20.20] Kim Y-J, Kishimoto Y, Wakatani M and Tajima T 1996 *Phys. Plasmas* **3** 3689
[20.21] Waltz R E, Staebler G M, Dorland W, Hammett G W, Kotschenreuther M and Konings J A 1997 *Phys. Plasmas* **4** 2482
[20.22] Hahm T S and Burrell K H 1995 *Phys. Plasmas* **2** 1648
[20.23] Itoh K, Itoh S-I, Fukuyama A, Sanuki H and Yagi M 1994 *Plasma Phys. Contr. Fusion* **36** 123
[20.24] Itoh K, Ohkawa T, Itoh S-I, Yagi M and Fukuyama A 1998 *Plasma Phys. Contr. Fusion* **40** 661
[20.25] Weynants R R, Jachmich S and Van Oost G 1998 *Plasma Phys. Contr. Fusion* **40** 635
[20.26] Cornelis J, Sporken R, Van Oost G and Weynants R R 1994 *Nucl. Fusion* **34** 171
[20.27] Wolf G H 1970 *Phys. Rev. Lett.* **24** 444
[20.28] Toda S and Itoh K 2001 *Plasma Phys. Contr. Fusion* **43** 629; 2002 *Plasma Phys. Contr. Fusion* **44** 325

Chapter 21

Subcritical Excitation

A nonlinear theory has been developed for the study of turbulence called submarginal turbulence or subcritical excitation. Such a turbulence has a larger amplitude at a higher gradient and is sometimes sustained under the linearly-stable condition. Figure 21.1 illustrates a schematic relation between the fluctuation amplitude and a driving parameter.

If a dynamical equation is expanded in a series of the perturbation amplitude $\tilde{A}$, the time-evolution equation of $\tilde{A}$ may be written as

$$\frac{\partial}{\partial t}\tilde{A} = \gamma_{\mathrm{L}}\tilde{A} + N_1\tilde{A}^3 + N_2\tilde{A}^5 + \cdots. \tag{21.1}$$

From a symmetry point of view, one sees that only odd-order terms remain for symmetry-breaking perturbations. In the case of supercritical excitation, the mode is suppressed by the low-order nonlinear effects $N_1\tilde{A}^3$ with $N_1 < 0$. However, in the case of subcritical excitation, the nonlinear terms have destabilizing influence, e.g.,

$$N_1 > 0 \quad \text{or} \quad N_2 > 0, \tag{21.2}$$

so that a finite-amplitude solution may exist even in the regime of $\gamma_{\mathrm{L}} < 0$.

21.1 Subcritical Excitation in Neutral Fluid

There are several examples of submarginal turbulence in fluids: planar Couette flow, planar Poiseuille flow, and pipe flow. Linear stability analysis has shown that a pipe flow is linearly stable. However, in reality, turbulence develops in the pipe flow if $R_e > 2200$ holds. Nonlinear analyses are inevitable.

21.1.1 Nonlinear Marginal Stability Condition

Planar Poiseuille flow has been analyzed, by truncating the series in equation (21.1) [21.1]. The nonlinear marginal stability boundary is illustrated in the

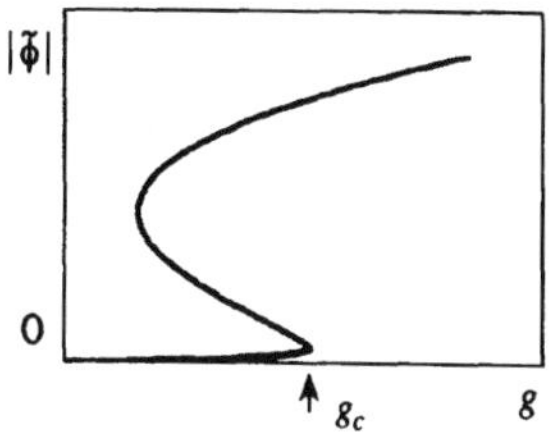

Figure 21.1. Excitation of submarginal turbulence. The steady-state amplitude of turbulence is illustrated as a function of the driving parameter g (e.g., pressure gradient, Reynolds number, etc.). g_c denotes the linear stability boundary. The lower branch is the result of thermodynamical excitations.

space of the Reynolds number R_e and perturbation amplitude $\tilde{A}$ in figure 21.2. A backward bifurcation is demonstrated. The growth rate of fluctuation becomes positive if amplitude exceeds the nonlinear marginal condition. One must note that truncation at a finite order of $\tilde{A}$ prevents the nonlinear marginal stability condition converging to an asymptotic limit.

21.1.2 Self-Sustaining Mechanism

Subcritical turbulence has been studied by taking into account secondary flow (roll) generation. When the planar flow is slightly deformed by the roll, this deformed two-dimensional flow has been known to be unstable for three-dimensional perturbations [21.2]. If the roll is driven by three-dimensional

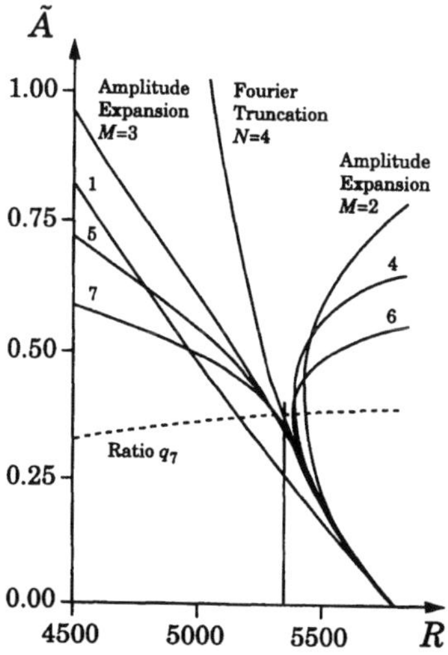

Figure 21.2. Nonlinear stability boundary for Poiseuille flow on the $(R, \tilde{A})$ plane (R: Reynolds number, reproduced from [21.1]): stable in left of curves; unstable in right of curves.

perturbation, then self-sustaining instability is possible. The mechanism of subcritical turbulence has been discussed by taking into account this secondary flow generation.

Waleffe has proposed a model to study self-sustaining turbulence in shear flows ([21.3] and references therein). Planar shear flow,

$$\boldsymbol{V}_0 = [0, 0, U(x)], \tag{21.3}$$

is considered. The x axis is in the direction of inhomogeneity (the 'radial' direction), the y axis is in the span-wise direction (the 'poloidal' direction), and the z axis is in the direction of flow. (Coordinates are reoriented for the convention of plasma physics.) To study the effects of streaks, the roll-type flow

$$\boldsymbol{V}_{\mathrm{roll}} = [V(x, y), W(x, y), 0] \tag{21.4}$$

is considered (an analogue of the convective cell in plasma physics). The streak,

$$\bar{\boldsymbol{V}}_{\mathrm{streak}} = [0, 0, \delta U(x, y)], \tag{21.5}$$

is induced by convection. The simple shear flow $\boldsymbol{V}_0$ is linearly stable. However, the deformed shear flow,

$$\boldsymbol{V}_s = \boldsymbol{V}_0 + \boldsymbol{V}_{\mathrm{streak}} = [0, 0, U(x) + \delta U(x, y)] \tag{21.6}$$

can become unstable against waves of the form

$$\tilde{\boldsymbol{V}} = [v(x) \sin k_y y, w(x) \cos k_y y, 0] \exp(\mathrm{i} k_z z). \tag{21.7}$$

if the amplitude of the streak is large enough. By nonlinear interactions, roll-type flow is induced by wave-like perturbations.

This nonlinear chain is summarized in figure 21.3. The self-sustaining mechanism consists of three processes:

(i) advection of mean flow shear by rolls (vortices) inducing a streak-like structure;
(ii) if the streak is strong enough, an instability occurs; and
(iii) this instability enhances the roll, closing the nonlinear chain.

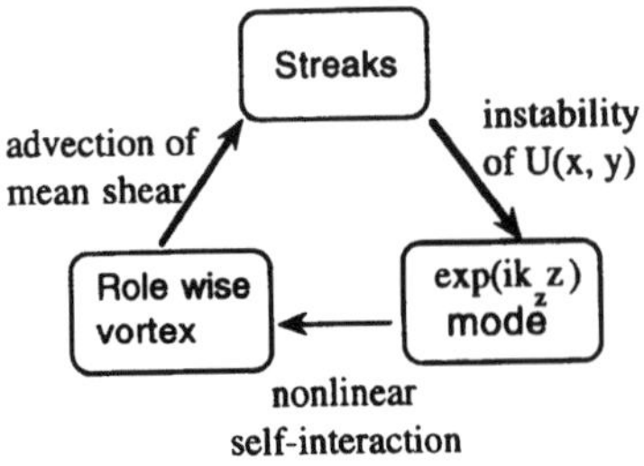

Figure 21.3. Subcritical excitation mechanism for shear flow turbulence (based on [21.3].)

In [21.3], Lorenz-model-like equations with eight variables are proposed, and a model with four variables is also discussed. The similarity of this nonlinear instability mechanism to those in plasma turbulence has been discussed in [2.49].

21.2 Subcritical Excitation in Plasma Turbulence

Nonlinear excitation of instabilities is important in understanding dynamic events in plasmas. There are abundant observations on the abrupt excitation of strong perturbations in plasmas, which may not be explained by the growth of linear instabilities. (For a review, see [21.4] and articles in [2.53].) Self-sustained fluctuations and perturbations have been obtained both theoretically and by numerical simulations [17.5, 17.14, 17.31, 21.5–21.21].

Subcritical excitation in the presence of dissipation is an important process for plasma turbulence. In plasmas, dissipation can induce instabilities. This is in contrast to neutral fluids, where dissipation-like viscosity and thermal conductivity usually stabilize the system. When the electron motion is impeded by a dissipation mechanism, the electron response deviates from the Boltzmann response ($\tilde{n}_e/\bar{n} = e\tilde{\phi}/T_e$) and the perturbed electric field appears on the magnetic surface. This mechanism drives instabilities. The stability property can be different in the presence of background turbulence, because the electron response is easily modified by turbulence. Examples include interchange mode turbulence, drift-wave turbulence, tearing-mode excitation, and MHD turbulence.

21.2.1 Current-Diffusive Interchange Mode Turbulence

The interchange mode is stabilized if the free electron motion along the magnetic field line neutralizes the charge separation. However, if this motion is impeded, the neutralization is not perfect for the instability to occur. The marginal stability boundary determined by the balance between the excitation by current diffusivity (i.e., electron viscosity μ_e) and the damping by viscosity μ_e and thermal conductivity χ was derived as [21.17]

$$\mathscr{I}_t \equiv \frac{G_0}{s^{4/3}}\left(\frac{\delta}{a}\right)^{4/3} \frac{(\mu_{e,N} + \mu_{e,c})^{2/3}}{(\chi_N + \chi_c)(\mu_N + \mu_c)^{1/3}} = \mathscr{I}_c \tag{21.8}$$

where $\mathscr{I}_c$ is the critical Itoh number (of the order of unity). (Suffices N and c indicate the turbulent and collisional contributions, respectively.)

As has been explained in chapter 18, the turbulence can enhance all the transport coefficient $\mu_{e,N}$, χ_N, and μ_N. Equation (21.8) shows that the turbulent electron viscosity destabilizes the mode. In the limit of small-amplitude fluctuations, the destabilizing effect dominates. The turbulent viscosity and

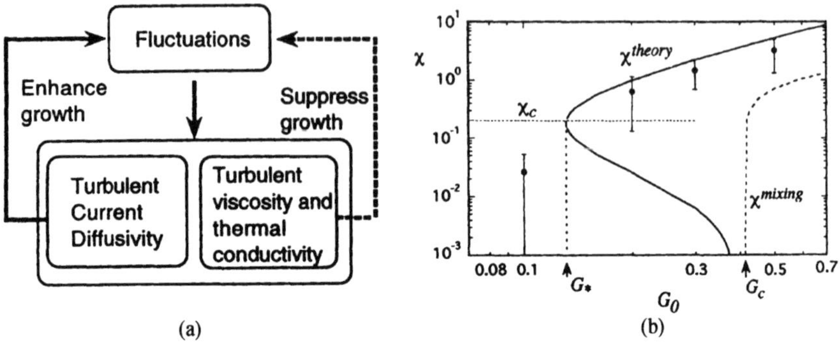

Figure 21.4. (a) The mechanism of nonlinear instability for CDIM turbulence. (b) The turbulence fluctuation level as a function of the pressure gradient. The analytic theory (full curve) predicts subcritical excitation, and direct nonlinear simulation (dots) demonstrates the self-sustained turbulent state below the critical pressure gradient for linear instability G_c [21.20].

thermal conductivity suppress the turbulence. The nonlinear chain is illustrated in figure 21.4(a). There are both a similarity and a dissimilarity to the neutral fluid. Rayleigh–Benard convection is excited if the Rayleigh number of equation (16.3),

$$R_a = \frac{L^3 g\alpha_{\text{th}}[\theta(L) - \theta(0)]}{\chi\mu_f},$$

exceeds a threshold value. ($\theta(L) - \theta(0)$ is the temperature difference.) Both $\Im_t$ and R_a increase with the gradient of pressure or temperature, and decrease with increasing viscosity and thermal conductivity. In this plasma turbulence, an increment of electron viscosity makes $\mathscr{I}_t$ larger and destabilizes the mode.

By expanding equation (21.8) near $G_0 \simeq G_c$, which is the linear stability boundary, the amplitude $\tilde{\phi}$ near the marginal condition is given by

$$\tilde{\phi}^2 \simeq -\frac{3\mu_{e,c}^2}{2}\left(\frac{G_0}{G_c} - 1\right), \tag{21.9}$$

where $\mu_{e,c}$ is the collisional electron viscosity. This result shows the existence of a backward bifurcation. The fluctuation level as a function of the gradient is illustrated in figure 21.4(b).

21.2.2 Nonlinear Drift Instabilities

Linear theories in slab geometry show that the drift wave in a sheared magnetic field is stable (marginal at most) [21.22–21.26]. However, nonlinear theory shows that the drift wave is unstable, if the fluctuation amplitude is

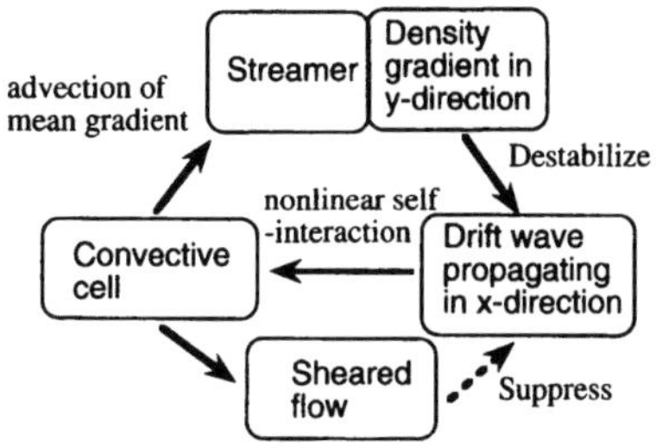

Figure 21.5. A nonlinear self-sustaining mechanism for the streamer and drift wave. The loop of full lines should be compared with figure 21.3.

high enough. An analysis based on equation (17.28) indicates that the mode is stable if $\Gamma_{s\|} = 0$ (linear theory), but that it is unstable if $\Gamma_{s\|}$ is large enough. The electron motion is impeded by the scattering by turbulence.

Another mechanism has been recently proposed [21.15]. This is a nonlinear linkage that includes the excitation of a streamer by drift-wave turbulence, which is discussed in §19.5. The nonlinear chain consists of:

(i) streamers ($k_\| \approx 0$, a type of convective cell) are induced by the drift waves ($k_\| \neq 0$), associated with the strongly-sheared radial flow (figure 19.6);
(ii) this system with a radial drift flow is unstable for the drift waves ($k_\| \neq 0$); and
(iii) the drift waves regenerate streamers, closing the nonlinear link.

The mechanism is illustrated in figure 21.5.

A set of equations for the streamer and drift wave has been proposed [21.15]. A potential perturbation of the form

$$\tilde{\phi} = \phi_c \cos \pi y + (\phi_{d1} \cos \pi y + \phi_{d2} \sin 2\pi y) \sin k_z z \exp(ik_x x) \tag{21.10}$$

is considered, where ϕ_c is the streamer amplitude, ϕ_{d1} is the least stable drift wave, and ϕ_{d2} denotes the damped drift wave. The original density gradient is in the x direction, but a driven gradient of density in the y direction appears owing to the streamer formation. The drift wave in equation (21.10) is propagating in the x direction. The potential and density evolution equations for the streamer are

$$\frac{\partial}{\partial t}\phi_c = A_0 |\phi_{d1}|^2 \phi_c \tag{21.11}$$

and

$$\frac{\partial}{\partial t}\left(\frac{\partial \bar{n}}{\partial y}\right) = 2\pi\phi_c \tag{21.12}$$

where A_0 is a coefficient in the growth rate of a convective cell in the limit of weak dissipative growth, equation (18.35). The complex growth rate of the drift wave is calculated by taking the lowest-order correction of the streamer as

$$\gamma\left(1 + k_\perp^2\rho_s^2 + \frac{k_{\perp 1}^2}{k_z^2}\gamma\right) = -ik_x\frac{\partial\bar{n}}{\partial y}. \tag{21.13}$$

From this set of equations, one sees a nonlinear mechanism for self-sustainment: the drift wave causes a convective cell to form, the cell leads to a density streamer; the density gradient in the y direction destabilizes the drift wave. (Normalization in this model is with $L_s = qR/s$ as the parallel scale length, the resistive scale length $L_\perp = (\nu_{ei}L_s^2\rho_s^2/\omega_{ce}L_n)^{1/3}$ as the perpendicular scale length, and $L_nL_\perp/c_s\rho_s$ as the time scale.)

The second-order correction by the streamer is the stabilizing influence of sheared flow. A quasi-stationary state is considered, in which the destabilizing effect ($\partial\bar{n}/\partial y$ term) balances the stabilizing effect (ϕ_c^2 terms) of the streamer on drift waves. This requires

$$\phi_c^4 = \gamma_1\frac{\partial\bar{n}}{\partial y} \tag{21.14}$$

where the coefficient γ_1 is given by the nonlinear interaction terms; its explicit formula is given in [21.15]. Combining equations (21.12) and (21.14), the algebraic growth of the streamer amplitude is given by

$$\frac{\partial\bar{n}}{\partial y} \approx t^{4/3}. \tag{21.15}$$

This result shows that the submarginal drift wave turbulence self-sustains in the presence of the density gradient.

21.2.3 Tearing Mode at High Pressure Gradient

Nonlinear destabilization is possible for large-scale perturbations in magnetized plasmas. When a current perturbation $\tilde{J}_\parallel \propto \exp(im\theta - in\zeta)$ exists on the mode rational surface r_s, $q(r_s) = m/n$, the topology changes and magnetic islands appear. Helical magnetic islands have self-sustaining mechanism, and subcritical excitation is possible.

Their dynamics are conveniently described by introducing a helical flux function A_h

$$A_h = A_\zeta + (nr^2/2Rm)\bar{B}_\zeta, \tag{21.16a}$$

where the suffix h denotes the helical coordinate and $l = \zeta/q - \theta$. Magnetic perturbations are given as $\tilde{\boldsymbol{B}}_\perp = \hat{\boldsymbol{b}} \times \nabla\tilde{A}_\zeta$ and the width of island w is given as

$$w = \sqrt{\frac{R}{r}\frac{q^2}{q'}\frac{1}{\bar{B}}\tilde{A}_h(r_s)}. \tag{21.16b}$$

Induction equation

$$\frac{\partial}{\partial t}\tilde{A}_{\mathrm{h}} = \eta_{\parallel}\tilde{J}_{\mathrm{h}} \tag{21.17}$$

governs the evolution of the perturbation fields. The relation between the perturbed current and the flux function is given by

$$\tilde{J}_{\mathrm{h}}(r_{\mathrm{s}}) = F_J[\tilde{A}_{\mathrm{h}}(r_{\mathrm{s}})], \tag{21.18}$$

and closes the dynamical evolution.

In the presence of a current gradient $\mathrm{d}\bar{J}_{\zeta}/\mathrm{d}r$, an explicit relation can be deduced [21.27]. Away from the mode rational surface $\boldsymbol{J} \times \boldsymbol{B} = 0$ holds (perturbation approximately satisfies the MHD equilibrium equation, because a time for the evolution is much slower than Alfvén time), and one has

$$\bar{B}_{\mathrm{h}}\Delta_{\perp}\tilde{A}_{\mathrm{h}} + \left(\frac{-\mathrm{d}\bar{J}_{\zeta}}{\mathrm{d}r}\right)\tilde{A}_{\mathrm{h}} = 0 \tag{21.19}$$

where $\bar{B}_{\mathrm{h}} = \bar{B}_{\theta} - \bar{B}_{\zeta}/qR$. The logarithmic derivative of this perturbation across the magnetic island

$$\Delta'(w) = \left(\frac{\mathrm{d}}{\mathrm{d}r}\ln(\tilde{A}_{\mathrm{h}})\right)_{r_{\mathrm{s}}+w} - \left(\frac{\mathrm{d}}{\mathrm{d}r}\ln(\tilde{A}_{\mathrm{h}})\right)_{r_{\mathrm{s}}-w} \tag{21.20}$$

is related to the perturbed current averaged over the magnetic island by (see, e.g., [2.26])

$$\tilde{J}_{\mathrm{h}}(r_{\mathrm{s}}) \simeq \frac{1}{\mu_0}\frac{\Delta'}{w}\tilde{A}_{\mathrm{h}}(r_{\mathrm{s}}). \tag{21.21}$$

Combination of equation (21.17) together with the relation to Δ' and $w \propto \sqrt{\tilde{A}_{\mathrm{h}}(r_{\mathrm{s}})}$ provides the temporal evolution equation as

$$\frac{\partial}{\partial t}\frac{w}{a} = \gamma_{\mathrm{R}} \cdot \Delta' a \tag{21.22}$$

where the resistive diffusion rate $\gamma_{\mathrm{R}} = \eta_{\parallel}\mu_0^{-1}r_{\mathrm{s}}^{-2}$ is characterized by the electric resistivity [21.27]. Δ' vanishes if an island width reaches the equilibrium value $w = w_{\mathrm{eq}}$. The growth of a magnetic island is slow, if it is regulated by equation (21.22). Acceleration mechanism of the growth is discussed later in §21.3.2. Note that the parameter Δ' at zero island width, $\Delta'(0)$, dictates the linear stability:

$$\Delta'(0) > 0 \tag{21.23}$$

is the condition for destabilization of the linear tearing mode.

An additional perturbed current $\tilde{J}_{\mathrm{h}}$ can induce a nonlinear destabilization. Reference [21.5] has suggested that local plasma heating or radiation loss drives a perturbation current as $\tilde{J}_{\rm h} \propto w^3$, giving the relation

$$\tilde{J}_{\rm h}(r_{\rm s}) \simeq \frac{1}{\mu_0} \frac{\Delta'}{w} \tilde{A}_{\rm h}(r_{\rm s}) + N_{\rm rad} w^3, \tag{21.24}$$

where $N_{\rm rad}$ is a positive coefficient. The right-hand side can become positive if w becomes large enough even if $\Delta' < 0$. This result allows a self-sustaining state of the magnetic island.

Another origin of perturbed current is the bootstrap current. When a magnetic island appears, the pressure gradient vanishes on it, causing the loss of bootstrap current on the magnetic island [21.28, 21.29]. The collisional diffusion process associated with the radial pressure gradient induces the toroidal current, which is known as the bootstrap current. The perturbed current associated with this process has the form

$$\tilde{J}_{\rm h} \propto \frac{w^2}{w_0^2 + w^2} \beta'_{\rm p}, \tag{21.25}$$

where $\beta'_{\rm p}$ is the pressure gradient normalized using the poloidal magnetic field pressure. (A finite width w_0 is introduced because the pressure flattening is not expected for $w < w_0$, as the perpendicular diffusion tends to keep the pressure gradient finite [21.30].) This process may also induce the self-sustaining magnetic island. The equilibrium island width $w_{\rm eq}$ is determined by the relation

$$\Delta'(w_{\rm eq}) - N_{\rm c,NC} \frac{w_{\rm eq}}{w_0^2 + w_{\rm eq}^2} \beta'_{\rm p} = 0. \tag{21.26}$$

This mechanism is called the neoclassical tearing mode and detailed analyses have been performed [21.28–21.35]. Figure 21.6 shows the saturated island width as a function of the driving parameter. Below the linear stability

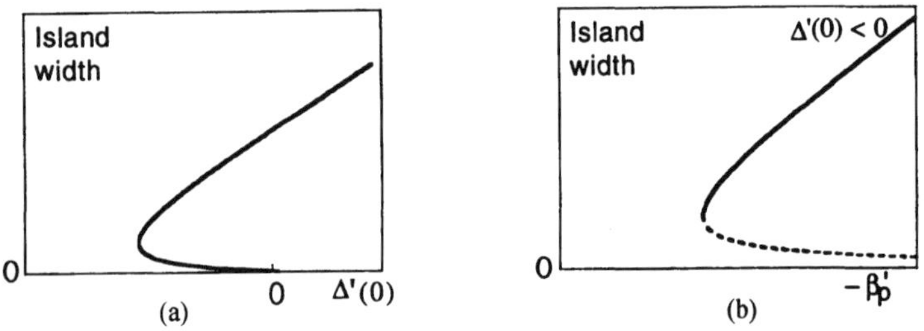

Figure 21.6. Saturation level of the tearing-mode perturbation, as a function of the MHD driving parameter $\Delta'(0)$ for a fixed pressure gradient (a) and as a function of the pressure gradient for a fixed value of $\Delta'(0)$ (b). The tearing mode is linearly stable if $\Delta'(0) < 0$. In the linearly stable case ($\Delta'(0) < 0$), a finite-amplitude island is self-sustaining if the pressure gradient becomes large enough. In (b), the unstable branch of the nonlinear marginal condition is denoted by the dotted line, which is a threshold for subcritical excitation.

boundary, a self-sustaining of perturbation is possible. Applications to tokamak plasmas are discussed in [21.36–21.39].

21.2.4 Turbulence–Turbulence Transition (M-Mode Transition)

Subcritical excitation of turbulence associated with magnetic braiding is predicted. As is discussed in §18.1.2, the transport by magnetic braiding is more influential on electrons than ions. Hence, the electron viscosity is more pronouncedly enhanced than ion viscosity once threshold condition equation (13.19) is satisfied. The electron viscosity is selectively enhanced in (18.12) so that new nonlinear instability occurs.

The first example is an onset of global stochasticity associated with an $m = 1$ magnetic island. When the amplitude of an $m = 1$ magnetic island exceeds a critical size in toroidal plasmas, the system loses symmetry and global stochasticity sets in. The criterion is given by [21.40, 21.41]

$$\sqrt{2\left(\frac{1}{q(0)} - 1\right)\frac{\tilde{B}_r R}{r\bar{B}}} > \frac{1}{7}. \tag{21.27}$$

The magnetic braiding and its associated enhancement of the electron viscosity cause nonlinear instability destabilization [17.14, 21.11].

The transition to turbulence with magnetic braiding occurs for the case of micro instabilities [21.42]. For the self-sustained CDBM turbulence in the low pressure gradient limit, the associated magnetic island width $w_{\rm is}$ scales like $w_{\rm is} \propto \alpha^{3/2}$, while the island separation $d_{\rm is}$ scales like $d_{\rm is} \propto \alpha^{1/2}$. ($\alpha$ is a normalized pressure gradient, equation (17.7), $\alpha = -q^2 R({\rm d}\beta/{\rm d}r)$.) Magnetic stochasticity sets in if $w_{\rm is}/d_{\rm is} > 1$ holds, where the Chirikov parameter, $w_{\rm is}/d_{\rm is}$, has a dependence on the pressure gradient as

$$w_{\rm is}/d_{\rm is} \propto \alpha. \tag{21.28}$$

As a result turbulence is enhanced which induces magnetic braiding. The turbulent transport coefficient is estimated as

$$\begin{aligned} \chi_i^{\rm M} &\simeq \sqrt{\frac{m_i}{m_e}}\,\beta_i \alpha F(\alpha, s)\alpha^{3/2}\left(\frac{\delta^2}{\tau_{\rm Ap}}\right) \\ \chi_e^{\rm M} &\simeq \frac{m_i}{m_e}\,\beta_i \alpha F(\alpha, s)\alpha^{3/2}\left(\frac{\delta^2}{\tau_{\rm Ap}}\right) \end{aligned} \tag{21.29}$$

which is larger than equation (18.25). (The superscript M stands for the magnetic braiding. This state is called the M-mode. s is the magnetic shear parameter, $s = rq^{-1}({\rm d}q/{\rm d}r)$ and $F(\alpha, s)$ is the one in equation (18.25).) The condition for the onset of the M-mode is illustrated in figure 21.7.

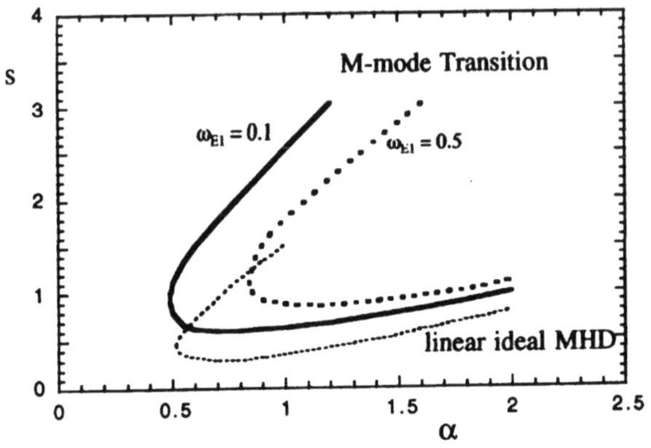

Figure 21.7. M-mode turbulence region in tokamak plasmas [21.42].

21.3 Abrupt Transition

21.3.1 Microscopic Turbulence and Transport Coefficient

The time scale for the onset of H- to M-bifurcation is given by the growth time of nonlinear instability $\tau_{\rm ng}$. It is of the order of the poloidal Alfvén time. During the transition, the magnetic island overlaps associated with a nonlinear growth in the turbulence. The time scale, $\tau_{\rm ng} = \gamma_{\rm ng}^{-1}$, was calculated from [21.42]

$$\gamma_{\rm ng}\tau_{\rm Ap} \approx s^{-1/3}F(s,\alpha)^{-1/6}[1+s^2F(s,\alpha)]^{1/3}\alpha^{1/2} \tag{21.30}$$

for the M-mode transition in a strong-shear limit. ($\gamma_{\rm ng}$ is the nonlinear growth rate.) The typical time scale is given by $\tau_{\rm Ap}/\sqrt{\alpha}$ for the turbulence–turbulence transition which implies a rapid process. An abrupt growth of turbulence is predicted for a steep gradient [21.43]. These subcritical excitations are essential for the dynamics of global structures.

21.3.2 MHD Modes

An acceleration in the MHD growth rate is possible in the presence of electron scattering by turbulence [21.44]. The nonlinear destabilization of global MHD modes, for which the growth rate is increased by a finite amplitude, shows an abrupt explosive growth of the mode. The explosive growth of the mode is more violent than the exponential growth. If the growth rate is expressed as

$$\gamma \propto |\tilde{B}|^{\nu} \tag{21.31}$$

(ν denotes a constant in the power-law index, not a collision frequency), then the mode amplitude shows the time evolution:

$$|\tilde{B}| \propto \frac{1}{(t_0 - t)^{1/\nu}}. \tag{21.32}$$

An explosive growth occurs.

Typical examples are the interactions of the $m/n = 2/1$, $m/n = 3/2$ and $m/n = 5/3$ modes [21.45–21.50]. When $m/n = 2/1$ and $m/n = 3/2$ islands overlap, then a stochastic region appears near the rational surface $q(r_s) \simeq 5/3$. The $m/n = 5/3$ mode is subject to tearing destabilization by the enhanced anomalous resistivity. ($\nu = 3/4$ in equation (21.31).) Nonlinear destabilization through magnetic braiding is also obtained. After slow growth, fast growth is recovered by the simulation. A change of the growth rate from the classical tearing mode to nonlinear instability takes place, since equation (21.22) predicts slow growth of the resistive diffusion time. For the case of the current-diffusive MHD mode, one has $\nu = 2/5$ in equations (21.31) and (21.32).

21.4 Bubble Formation and Suppression by Shear Flow

Bubble formation is a typical phenomenon of subcritical excitation and is strongly influenced by the flow. Two forces work on a bubble in a sheared flow: a shearing force $\bar{\mu} S_v$ tends to stretch a bubble ($\bar{\mu}$ being a coefficient of fluid viscosity). A surface tension $\bar{\sigma}/L$ acts to restore a spherical shape ($\bar{\sigma}$ is a coefficient for surface tension, L the size of bubble) (see figure 21.8(a)). If the shearing force is greater than the surface tension, $\bar{\mu} S_v > \bar{\sigma}/L$, the bubble is destroyed. The critical size above which a bubble collapses,

$$L_{\mathrm{cr}} = \frac{\bar{\sigma}}{\bar{\mu} S_v}, \tag{21.33}$$

can be found.

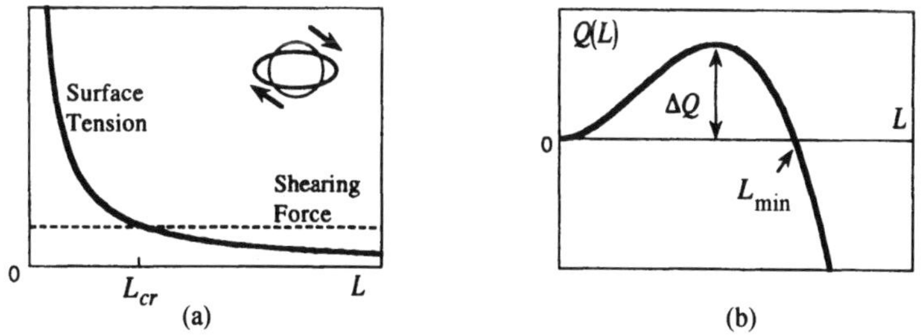

Figure 21.8. (a) Shearing force and surface tension (per unit volume) versus bubble size. (b) Free energy of bubble formation as a function of bubble size. Activation energy ΔQ and minimum size $L_{\min}$ for nucleation are shown.

On the one hand, there is a critical (minimum) size for nucleation, L_{min}, because bubble formation is a subcritical phenomenon [21.51]. Below this size, a bubble will not grow and disappears with a finite lifetime (even in the absence of sheared flow). The latter size is determined by a balance between the surface tension and the difference in the free energy between the inside of the bubble and the outer fluid. The free energy associated with bubble formation is illustrated in figure 21.8(b). If the velocity shear is strong enough so that

$$L_{cr} < L_{min}, \tag{21.34}$$

that is, if

$$S_v > \frac{\bar{\sigma}}{\bar{\mu} L_{min}} \tag{21.35}$$

is satisfied, then the nucleation of the bubble is strongly suppressed.

As the critical size for the onset of nucleation L_{min} becomes larger, the shear-flow suppression appears at the lower level of flow shear rate.

References

[21.1] Herbert T 1980 *AIAA J.* **18** 243

[21.2] Montgomery D 1989 in *Nagoya Lectures in Plasma Physics and Controlled Fusion* ed Y H Ichikawa and T Kamimura (Tokai University Press) p 207

[21.3] Waleffe F 1997 *Phys. Fluids* **9** 883

[21.4] Itoh S-I, Itoh K, Zushi H and Fukuyama A 1998 *Plasma Phys. Contr. Fusion* **40** 879

[21.5] Rebut P H and Hugon M 1985 in *Plasma Physics and Controlled Nuclear Fusion Research 1984* (Vienna: IAEA) vol 2, p 197

[21.6] Waltz R E *Phys. Rev. Lett.* **55** 1098

[21.7] Biskamp D and Walter M 1985 *Phys. Lett.* **109A** 34

[21.8] Sydora R D *et al* 1985 *Phys. Fluids* **28** 528

[21.9] Scott B D 1990 *Phys. Rev. Lett.* **65** 3289; 1992 *Phys. Fluids B* **4** 2468

[21.10] Carreras B A, Sidikman K, Diamond P H, Terry P W and Garcia L 1992 *Phys. Fluids B* **4** 3115

[21.11] Fukuyama A *et al* 1993 in *Plasma Physics and Controlled Nuclear Fusion Research 1992* (Vienna: IAEA) vol 2, p 363

[21.12] Nordman H, Pavlenko V P and Weiland J 1993 *Phys. Plasmas B* **5** 402

[21.13] Itoh K, Itoh S-I, Fukuyama A, Yagi M and Azumi M 1994 *Plasma Phys. Cont. Fusion* **36** 279.

[21.14] Biskamp D and Zeiler A 1995 *Phys. Rev. Lett.* **74** 706

[21.15] Drake J F, Zeiler A and Biskamp D 1995 *Phys. Rev. Lett.* **75** 4222

[21.16] Cowley S, Artun M and Albright B 1996 *Phys. Plasmas* **3** 1848

[21.17] Itoh K *et al* 1996 *J. Phys. Soc. Jpn.* **65** 2749

[21.18] Knobloch E and Weiss N O J 1983 *Physica D* **9** 379

[21.19] Bekki N and Karakisawa T 1995 *Phys. Plasmas* **2** 2945

[21.20] Yagi M, Itoh S-I, Itoh K, Fukuyama A and Azumi M 1995 *Phys. Plasmas* **2** 4140
[21.21] Pfirsch D 1993 *Phys. Rev. E* **48** 1428
[21.22] Pearlstein L D and Berk H L 1969 *Phys. Rev. Lett.* **23** 220
[21.23] Ross D W and Mahajan S M 1978 *Phys. Rev. Lett.* **40** 324; Tsang K T, Catto P J, Whitson J C and Smith J 1978 *Phys. Rev. Lett.* **40** 327
[21.24] Guzdar P N, Chen L, Kaw P K and Oberman C 1978 *Phys. Rev. Lett.* **40** 1566
[21.25] Cordey J G, Jones E M and Start D F H 1979 *Plasma Phys.* **21** 725
[21.26] Itoh K and Inoue S 1980 *Phys. Fluids* **23** 847
[21.27] Rutherford P H 1973 *Phys. Fluids* **6** 1903
[21.28] Callen J D, Qu W X, Siebert K D, Carreras B A, Shaing K C and Spong D 1986 in *Plasma Physics and Controlled Nuclear Fusion Research* (Vienna: IAEA) vol 2, p 157
[21.29] Carrera R, Hazeltine R D and Kotschenreuther M 1986 *Phys. Fluids* **29** 899
[21.30] Waelbroeck F L and Fitzpatrick R 1997 *Phys. Rev. Lett.* **78** 1703
[21.31] Smolyakov A I 1993 *Plasma Phys. Contr. Fusion* **35** 657
[21.32] Zabiego M and Garbet X 1994 *Phys. Plasmas* **1** 1890
[21.33] Wilson H R, Connor J W, Hastie R J and Hegna C C 1996 *Phys. Plasmas* **3** 248
[21.34] Mikhailovskii A B, Pustovitov V D, Tsypin V S and Smolyakov A I 2000 *Phys. Plasmas* **7** 1204
[21.35] Mikhailovskii A B, Pustovitov V D, Smolyakov A I and Tsypin V S 2000 *Phys. Plasmas* **7** 1214
[21.36] Chang Z *et al* 1995 *Phys. Rev. Lett.* **74** 4663
[21.37] Zohm H *et al* 1997 *Plasma Phys. Contr. Fusion* **39** B237
[21.38] Sauter O *et al* 1997 *Phys. Plasmas* **4** 1654
[21.39] See also ITER Physics Basis 1999 *Nucl. Fusion* **39** chapter 3, section 2.3
[21.40] Mercier C 1983 *Sov. J. Plasma Phys.* **9** 82 [1983 *Fiz. Plasmy* **9** 132]
[21.41] Lichtenberg A J 1984 *Nucl. Fusion* **24** 1277
[21.42] Itoh S-I, Itoh K, Fukuyama A and Yagi M 1996 *Phys. Rev. Lett.* **76** 920
[21.43] Fukuyama A, Itoh K, Itoh S-I and Yagi M 2000 *Nucl. Fusion* **40** 685
[21.44] Furuya A, Itoh S-I and Yagi M 2000 *Contrib. Plasma Phys.* **40** 375; 2001 *J. Phys. Soc. Jpn.* **70** 407; 2002 *J. Phys. Soc. Jpn.* **71** 1261
[21.45] Waddel B V, Carreras B, Hicks H R, Holmes J A and Lee D K 1978 *Phys. Rev. Lett.* **41** 1386
[21.46] Carreras B A *et al* 1981 *Phys. Fluids* **24** 66
[21.47] Biskamp D and Welter H 1983 *Plasma Physics and Controlled Nuclear Fusion Research 1982* (Vienna: IAEA) vol 3, p 373
[21.48] Izzo R *et al* 1983 *Phys. Fluids* **26** 2240
[21.49] Kurita G *et al* 1986 *Nucl. Fusion* **26** 449
[21.50] Bondeson A 1986 *Nucl. Fusion* **26** 929
[21.51] Taylor G I 1934 *Proc. R. Soc. London* **A146** 501; for a recent review, see, e.g., Mel'nikov V I 1991 *Phys. Rep.* **209** 1

Chapter 22

Bifurcation

Surveying the properties of plasma turbulence and turbulent-driven transport, the following points about turbulence in inhomogeneous plasma must be made.

(i) Fluctuations do not follow the equipartition law.
(ii) Various turbulent states and plasma profiles exist for common external control parameters.
(iii) Among the different states, differences are seen in the average spectrum and in its dependencies on the global driving parameters (such as gradient), and in the dynamical and spatial structure of the global profile, e.g., the presence of a mesoscale structure (electric field domain interface, zonal flow, streamer, etc.).
(iv) The change between different states happens in a short time of nonlinear decorrelation rate, often being triggered by subcritical excitation.

These features suggest that the dynamical change of turbulence should be understood within the concept of bifurcation. The physics of *phase transition* may be extended for inhomogeneous plasma turbulence. The concept of structural transition in critical phenomena should be used: that is, the change of global structure (gradient, symmetry, shape) is described simultaneously with that of the turbulence spectra. The driving motivation has been the observation of the transition between the L-mode and H-mode [2.35]. (See also monograph [2.53].)

Changes in the turbulence level have been discussed. Self-organized dynamical oscillations have also been investigated. As illustrated in figure 22.1, a transition between states A and B takes place. (A case of subcritical excitation is shown as an example.) The statistical properties of the turbulence are different from those of thermodynamical equilibrium, so the transition probability between different turbulent states is modified. An overview of properties of bifurcations and transitions is given in the following.

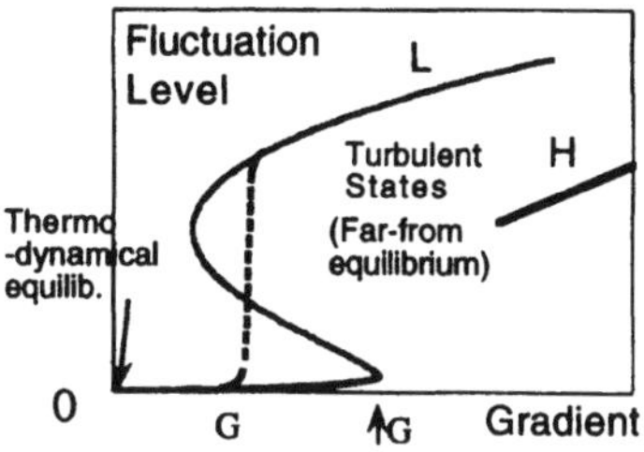

Figure 22.1. Bifurcation of turbulent states and transitions between them. Two turbulent states (H and L) as well as the thermodynamical fluctuation level are schematically shown.

22.1 System with Hysteresis

Under given inhomogeneities in the plasma parameters, turbulence and mesoscale structures develop. Figure 22.2 illustrates a typical concept: figure 22.2(a) illustrates various structures in k-space, and figure 22.2(b) demonstrates the mutual processes of interaction. Microturbulence drives turbulent transport, possibly by nonlinear instabilities, and induces mesoscale structures such as electric field domain interface, zonal flow, solitary radial electric field, streamer, and so on. Some suppress turbulence, but others drive nonlinear destabilization. Microfluctuations with a large number of positive Lyapunov exponents exist in the turbulent state. Mesoscale structures might be described by a system with a small number of degrees of freedom. The combined dynamics of turbulence and chaos are expected to occur in inhomogeneous plasmas.

The mutual interaction of the plasma gradient and turbulent transport is another example. The pressure gradient is the origin of turbulent transport (e.g., figure 17.2), which has a feedback effect on the pressure gradient under the fixed flux condition, figure 22.3(a). However, if the pressure gradient becomes high enough, as discussed in §17.2, it suppresses turbulent transport through the reduction of the magnetic shear or through the

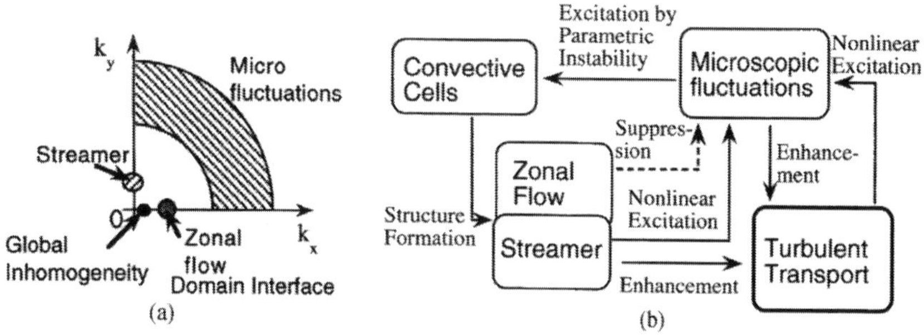

Figure 22.2. Distribution of various kinds of fluctuations (a). Schematic drawing of the self-regulating mechanisms in plasma turbulence and structural formation is given in (b).

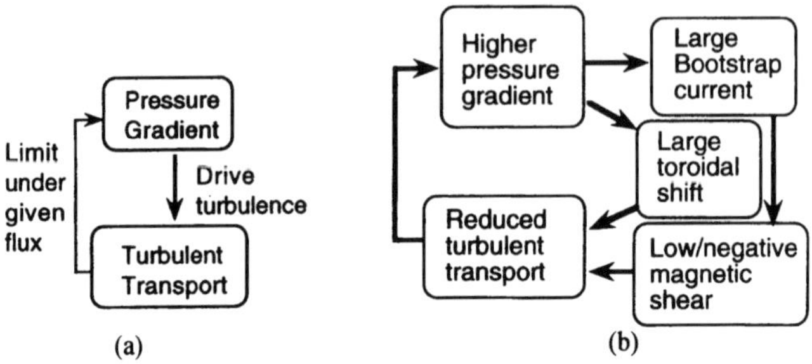

Figure 22.3. Other nonlinear chains. (a) Pressure gradient as an origin of turbulent transport. (b) Structural formation through the modification of the current profile and toroidal shift of magnetic axis.

increment of toroidal shift of the magnetic axis. In this case, a nonlinear linkage that enhances the plasma pressure gradient arises (figure 22.3(b)).

22.1.1 Dynamical Model Equations for Structural Transition

In order to describe these nonlinearly interacting systems, one employs a set of model equations to describe the plasma pressure

$$\frac{\partial \bar{p}}{\partial t} = \nabla \cdot \{\chi_{\text{turb}}[\tilde{\phi}; \nabla \bar{p}; \nabla \bar{E}_{\text{r}}; \ldots] \nabla \bar{p}\} + P_{\text{heat}}, \tag{22.1}$$

as well as the electromagnetic field, e.g., radial electric field (including zonal flow and electric field interface) and poloidal magnetic field

$$\frac{\partial \bar{E}_{\text{r}}}{\partial t} = \nabla \cdot \{\mu_{\text{turb}}[\tilde{\phi}; \nabla \bar{p}; \nabla \bar{E}_{\text{r}}; \ldots] \nabla \bar{E}_{\text{r}}\} + N[\bar{E}_{\text{r}}; \nabla \bar{p}; \ldots], \tag{22.2}$$

$$\frac{\partial \bar{B}_{\theta}}{\partial t} = \nabla \cdot \{\eta_{\parallel} \nabla \bar{B}_{\theta}\} + N_{\text{B}}[\bar{B}_{\theta}; \nabla \bar{p}; \ldots], \tag{22.3}$$

and fluctuating quantities $\tilde{\boldsymbol{f}}$

$$\frac{\partial}{\partial t} \tilde{\boldsymbol{f}} + \mathfrak{L} \tilde{\boldsymbol{f}} = \tilde{\boldsymbol{S}} + \tilde{\boldsymbol{S}}_{\text{th}}. \tag{22.4}$$

The explicit formulae for transport coefficient $\mu_{\text{turb}}[\tilde{\phi}; \nabla \bar{p}; \nabla \bar{E}_{\text{r}}; \ldots]$, the nonlinear source term $N[\bar{E}_{\text{r}}; \nabla \bar{p}; \ldots]$ and the renormalized operator $\mathfrak{L}$ can be discussed for several examples of turbulence models. In equations (22.1)–(22.3), P_{heat} is a heating power, $N[\bar{E}_{\text{r}}]$ is the nonlinear source that generates a radial electric field (i.e., plasma flow), and $N_{\text{B}}[\bar{B}_{\theta}; \nabla \bar{p}; \ldots]$ represents the driving source of magnetic field like a bootstrap current.

22.1.2 Nonlinearity in Gradient–Flux Relation

To analyze the bifurcation and transition phenomena, a time-scale separation is introduced. We consider the case when the evolution time of the turbulence level τ_{fluc} is shorter than that of the plasma inhomogeneities τ_{global}, i.e.,

$$\tau_{\text{fluc}} \ll \tau_{\text{global}}, \tag{22.5}$$

where τ_{fluc} is of the order of nonlinear growth time τ_{ng} or autocorrelation time τ_{cor}. Taking two global parameters (pressure gradient and radial electric field), a system of plasma transport equations

$$\frac{\partial \bar{p}}{\partial t} = \nabla \cdot \{\chi_{\text{turb}} \nabla \bar{p}\} + P_{\text{heat}}, \tag{22.6}$$

$$\frac{\partial \bar{E}_{\text{r}}}{\partial t} = \nabla \cdot \{\mu_{\text{turb}} \nabla \bar{E}_{\text{r}}\} + N[\bar{E}_{\text{r}}], \tag{22.7}$$

has been studied. Examples are discussed in §19.3 as equation (19.11). Coefficients χ_{turb} and μ_{turb} are functionals of $\nabla \bar{p}$ and $\nabla \bar{E}_{\text{r}}(r)$. A relation $\mu_{\text{turb}} \simeq \chi_{\text{turb}}$ ($P_{\text{r}} \simeq 1$) is sometimes employed; see discussions for equation (18.26). This system is known to have a hysteresis [2.28, 2.50, 22.1, 22.2], when equation (22.7) has a faster time scale than equation (22.6). Examples of the dynamics of a radial electric field are explained in the appendix to chapter 19. That is, a stationary solution of equation (22.7), $\bar{E}_{\text{r}}(r) = \bar{E}_{\text{r}}[\nabla \bar{p}; \ldots]$, can be substituted into equation (22.6). On its slow time scale, equation (22.6) is solved using the formulae of thermal transport coefficients. Figure 22.4 illustrates the energy transport coefficient as a function of the pressure gradient and other plasma parameters. It is possible for a cusp-type catastrophe to occur. The thermal conductivity is subject to change at critical pressure gradients. The flow patterns of the upper and lower branches of figure 22.4 are different, as illustrated in figure 22.5(a) and (b). In the state shown in figure 22.5(b), the electric field is weak: damping of poloidal flow due to the transit time magnetic pumping is large. Plasma rotates in the toroidal direction. When the electric field becomes strong enough, it is possible for the plasma to rotate in the poloidal direction (see chapter 19 of [2.28]).

Another transition model is based on the theory of instabilities in the range of the drift waves [22.3–22.10]. The temperature gradient induces

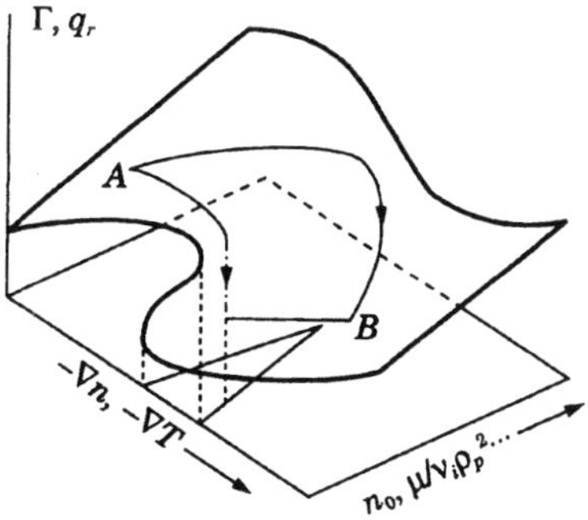

Figure 22.4. Cusp-type catastrophe.

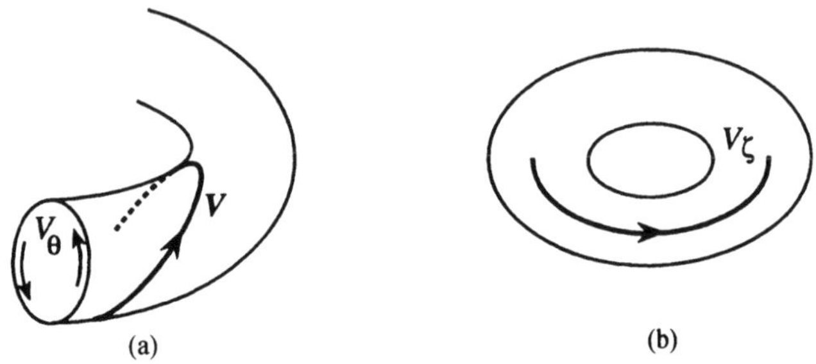

Figure 22.5. Bifurcation in flow pattern. (a) The flow pattern in a reduced transport state and (b) that in an ordinary turbulent state.

drift-wave turbulence, which can be suppressed by the density gradient. According to the balance between them, a phase diagram of drift-wave turbulence has been proposed [22.3]. Figure 22.6 is a diagram for the tokamak edge turbulence for a fixed value of η_i, $\eta_i = 2.5$, in the $(\varepsilon_n, \alpha_d)$ plane, where

$$\varepsilon_n = \frac{2L_n}{R}, \tag{22.8}$$

$$\alpha_d = \frac{\rho_s}{(1 + T_e/T_e)L_\perp}\sqrt{\frac{R}{L_n}}, \tag{22.9}$$

and $L_\perp = (\nu_{ei} L_s^2 \rho_s^2 / \omega_{ce} L_n)^{1/3}$. The parameter ε_n controls the effective compressibility, and α_d measures the strength of the diamagnetic drifts in the plasma edge.

A much simpler model of the bifurcation in energy transport has been proposed, corresponding to one section of figure 22.4, as follows:

$$\chi = \frac{\chi_1}{1 + c_1(a/L_\mathrm{T})^2 + c_2(a/L_\mathrm{T})^4} + \chi_2 \tag{22.10}$$

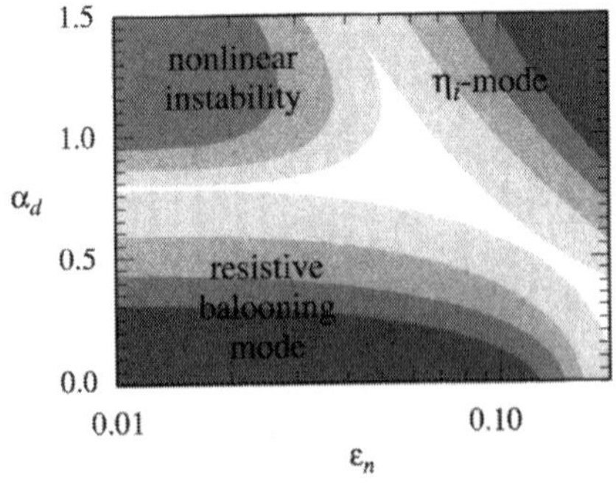

Figure 22.6. Phase diagram of drift-wave turbulence [22.3].

where c_1 and c_2 are numerical coefficients and L_T is the temperature-gradient scale length. χ_1 and χ_2 represent two different mechanisms [22.11, 22.12]. This model predicts a bifurcation of the pressure gradient at certain conditions. In some models, the curvature of temperature, $d^2T_{e,i}/dr^2$, is included in the denominator of equation (22.10). In this case, the temperature profile is not determined uniquely even if boundary conditions are imposed [22.13].

Various turbulence models, combined with the transport picture of equations (22.6) and (22.7), have been used to analyze the transitions in plasma confinement and the establishment of steep gradient in plasmas [22.2, 22.14–22.26]. The objective is to determine whether a turbulence transport model can reproduce the anomalous loss as well as the transition to an improved state [2.32–2.38]. At least a qualitative understanding has been given including the formation of transport barriers and exhaustive reference is found in [2.53, 22.27].

22.1.3 Simultaneous Evolution of Fluctuation, Flow and Gradient

When the system size becomes smaller, the characteristic times for the change in the average profile and of turbulence level become closer to each other.

$$\tau_{\text{fluc}} \approx \tau_{\text{global}}. \tag{22.11}$$

The evolution of the fluctuation level must be simultaneously solved together with plasma gradients and flows. Then mesoscale structures and turbulence regulate each other nonlinearly. A complicated spatio-temporal evolution is expected to occur (see also figure 22.2).

Under the ordering of equation (22.11), a soft-type bifurcation has been examined in a dynamical model with a local approximation. An example is presented here, in the form of a set of equations for the plasma pressure, the electric field inhomogeneity, and the fluctuation level [19.9]:

$$\frac{\partial}{\partial t}A = (\gamma_0 - \alpha_1 A - \alpha_2 U)A, \tag{22.12a}$$

$$\frac{\partial}{\partial t}U = -\mu U + \alpha_3 AU, \tag{22.12b}$$

$$\frac{\partial}{\partial t}G = -\alpha_5 G - \alpha_4 AG + P, \tag{22.12c}$$

where A is the fluctuation level, $A \approx |\tilde{n}/n|^2$, U is the flow shear, $U \approx |V_E'|^2$, G is the pressure gradient, and P is the heating power, respectively. In the first equation, the growth rate of fluctuations has contributions from a linear destabilization (the γ_0 term), the nonlinear saturation mechanism (the α_1 term), and the suppression by the electric field shear (the α_2 term). The second equation includes processes of the shear viscous damping (the first term) and the generation by turbulence (the α_3 term). In the equation of pressure gradient (22.12c), the α_4 term denotes the turbulent transport. This particular form was derived by dimensional arguments, so it is not necessarily unique. Its extension is given in [22.28, 22.29]. Similar, but not identical, dynamical models have been given by other groups [16.7, 19.7, 22.30–22.32]. This kind of dynamical model has two types of steady state, i.e.,

$$A = \frac{\gamma_0}{\alpha_1}, \qquad U = 0 \tag{22.13a}$$

and

$$A = \frac{\mu}{\alpha_3}, \qquad U = \left(\gamma_0 - \frac{\alpha_1 \mu}{\alpha_3}\right)\frac{1}{\alpha_2}. \tag{22.13b}$$

These two branches merge smoothly when $\gamma_0\alpha_3 = \mu\alpha_1$. Only a soft transition (pitchfork bifurcation) is predicted. These models can be used for the phenomena with soft transitions. A hard transition is presented in an alternative dynamical model [21.43].

It should be noted that the damping of the flow shear, which is symbolized by μ in this model, has an important role in determining the level of fluctuations in the state of equation (22.13b). In this state, the source of instability (denoted by γ_0) does not explicitly influence the fluctuation level A. The sources of fluctuations are transformed into the sheared poloidal flow via fluctuations. The poloidal flow shear (denoted by U) is enhanced by γ_0. The increment of U suppresses the growth of fluctuations, although the instability source γ_0 tends to enhance the fluctuation level.

The example here is taken from a numerical simulation. The evolution of the fluctuation energy in the mesoscale and micro-scale regimes is plotted in figure 22.7. Complex dynamics among fluctuations of different scale lengths is observed [22.33]. A survey of plasma turbulence from a complementary viewpoint is made in [22.34].

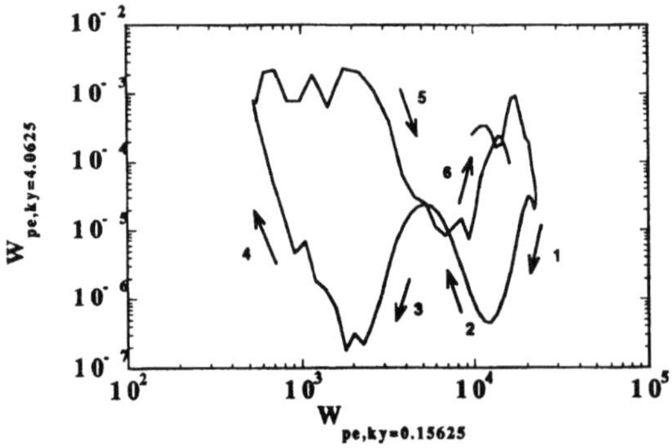

Figure 22.7. Energy in the long-wave length region (mesoscale) (horizontal axis) and that in micro-scale fluctuations (vertical axis) show complicated dynamics. Quoted from [22.33].

22.2 Self-Organized Dynamics

In inhomogeneous plasmas, nonlinearity drives self-organized dynamics. Systems with a few degrees of freedom, like the Lorenz model or the Volterra-Lotka model, allow a limit cycle oscillation under fixed external parameters. A structural transition can occur periodically as well. Self-organized dynamics related to edge localized modes (ELMs) [2.35, 2.38] are now surveyed.

22.2.1 Dithering ELMs

Hysteresis in the gradient–flux relation gives rise to the self-organized dynamics under a constant supply. An example is illustrated in figure 22.8(a). In a state of small loss rate the density increases (path A); once the critical condition is reached, the transport changes abruptly (jump B). On the branch of larger loss rate, the density decreases (path C); another transition takes place at the critical condition (jump D), closing a cycle. A solution of the set of equations like (22.6) and (22.7) shows that repetitive bursts of plasma flux occur and the turbulence level as well as the global plasma gradient change periodically [19.16, 22.35–22.38]. This self-organized dynamics is a typical model of the dithering ELMs that have been observed. Compound dithers have also been predicted theoretically [22.38].

Soft transition models also predict self-organized dynamics under the constant supply of the plasma [16.6, 16.7, 19.9, 22.39, 22.40]. Variety of limit cycle oscillation is possible in the plasma with hysteresis in the transport property.

22.2.2 Giant ELMs

Self-organized dynamics in the previous subsection occurs in a particular parameter space near the critical condition. In the higher pressure-gradient region, another type of repetitive structural transition takes place.

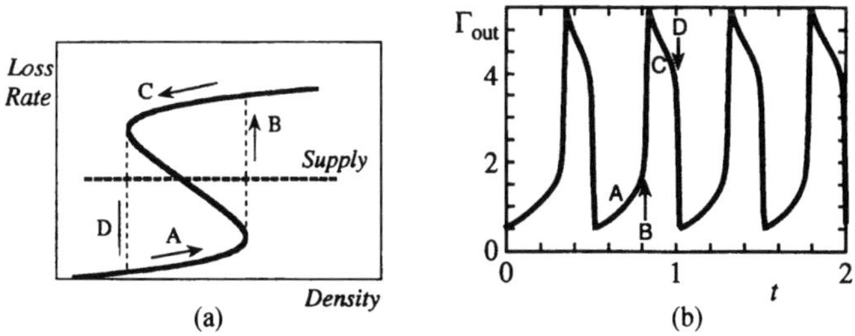

Figure 22.8. An example of self-organized dynamics in a structural transition. Hysteresis in the plasma parameter–loss rate space is shown in (a), and the temporal evolution of the plasma outflux across the surface is shown in (b).

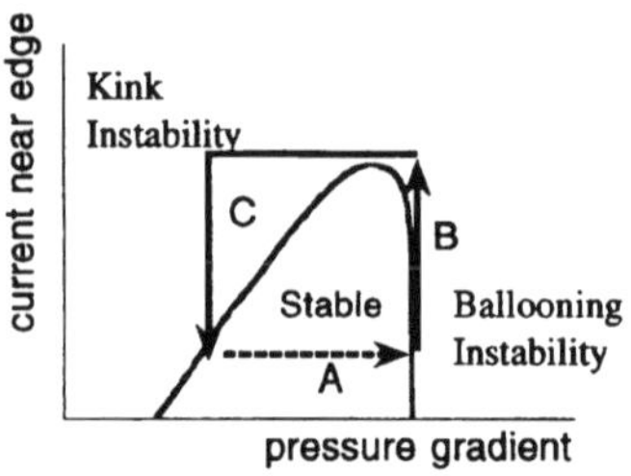

Figure 22.9. A pair of instabilities, under constant heating conditions, causes repetitive rise and decay in the pressure gradient. (After [22.42].)

A pair of instabilities which cause repetitive bursts have been proposed [22.41, 22.42] (figure 22.9). The ballooning instability is excited by the pressure gradient, while the surface kink instability [22.43] is driven by the current density near the plasma surface. The instability region is illustrated in figure 22.9. In the stable region, the pressure gradient increases due to the high heat flux (path A). Once the ballooning instability criterion is reached, the pressure gradient might stop increasing at the critical value even though the heating continues. The characteristic diffusion time of the magnetic flux and current is slower than that of energy, so that the current density near the surface increases (path B). When the threshold of the kink instability is reached, then the pressure gradient and the current density decay (path C) so as to close the loop.

Based on the transition resulting from the onset of magnetic stochasticity [21.41], a sequence for the structural transition has been predicted [22.44, 22.45]. This mechanism causes the periodic collapse of transport barriers and bursts of plasma loss. Figure 22.10 illustrates the transition of a turbulent transport coefficient as a function of the plasma pressure gradient $\alpha = -(q^2 R/a)(\mathrm{d}\beta/\mathrm{d}r)$. The plasma pressure increases on the H-branch by the strong heating power. The critical pressure gradient α_c is reached, at which point a transition to a strongly turbulent state takes place. (The critical

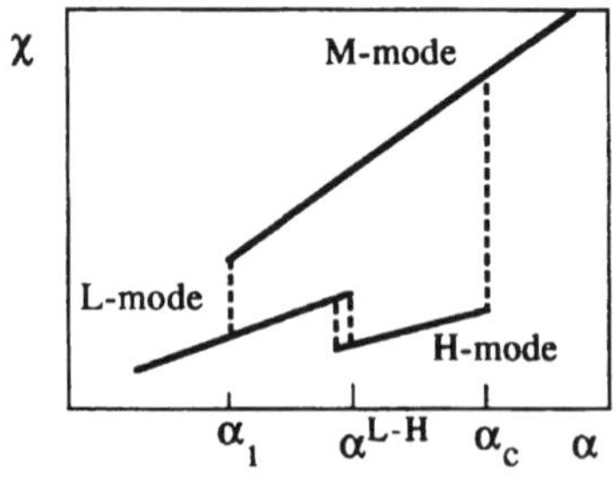

Figure 22.10. Transport coefficient as a function of the pressure gradient. Transition between various turbulent states is predicted.

pressure gradient α_c is illustrated in figure 21.7.) The enhanced transport coefficient causes the collapse of the pressure gradient. When the pressure gradient becomes low, the back transition to the lower branch occurs at $\alpha = \alpha_l$. The pressure gradient starts to increase again, closing the loop.

A very detailed survey of the theoretical work and comparisons with experimental observations are presented in [2.50].

References

[22.1] Itoh S-I and Itoh K 1990 *J. Phys. Soc. Jpn.* **59** 3815

[22.2] Fukuyama A, Itoh K, Itoh S-I, Yagi M and Azumi M 1995 *Plasma Phys. Contr. Fusion* **37** 611

[22.3] Zeiler A, Biskamp D, Drake J F and Rogers B N 1998 *Phys. Fluids* **5** 2654

[22.4] Rogister A L 1994 *Plasma Phys. Contr. Fusion* **36** A219

[22.5] Cohen R H and Xu X 1995 *Phys. Plasmas* **2** 3374

[22.6] Guzdar P and Hassam A B 1996 *Phys. Plasmas* **3** 3701

[22.7] Rogers B N and Drake J F 1997 *Phys. Rev. Lett.* **79** 229

[22.8] Scott B D 1998 *Plasma Phys. Contr. Fusion* **40** 823

[22.9] Chankin A V and Matthews G F 1998 *Contrib. Plasma Phys.* **38** 177

[22.10] Kerner W, Igitkhanov Yu, Janeschitz G and Pogutse O 1998 *Contrib. Plasma Phys.* **38** 118

[22.11] Hinton F L 1991 *Phys. Fluids B* **3** 696

[22.12] Hinton F L and Staebler G M 1993 *Phys. Fluids B* **5** 1281

[22.13] Taylor J B, Connor J W and Helander P 1998 *Phys. Plasmas* **5** 3065

[22.14] Hinton F L, Staebler G M and Kim Y-B 1994 *Plasma Phys. Contr. Fusion* **36** A237

[22.15] Fukuyama A, Itoh S-I, Yagi M and Itoh K 1995 *Nucl. Fusion* **35** 1669

[22.16] Kotschenreuther M, Dorland W, Beer M A and Hammett G W 1995 *Phys. Plasmas* **2** 2381

[22.17] Fukuyama A *et al* 1996 *Plasma Phys. Contr. Fusion* **38** 1319

[22.18] Staebler G M, Hinton F L and Wiley J C 1996 *Plasma Phys. Contr. Fusion* **38** 1461

[22.19] Connor J W *et al* 1996 *Proceedings of the International Conference on Fusion Energy* (Montreal, 1996) vol 2 p 935

[22.20] Kinsey J, Bateman G, Kritz A and Redd A 1996 *Phys. Plasmas* **3** 561

[22.21] Cherubini A, Erba M, Parail V, Springmann E and Taroni A 1996 *Plasma Phys. Contr. Fusion* **38** 1421

[22.22] Nordoman H, Strand P, Weiland J and Christiansen J P 1997 *Nucl. Fusion* **37** 413

[22.23] Zonca P, Frigione D, Marinucci M, Zanza V and Romaneli F 1997 in *Theory of Fusion Plasmas* ed J W Connor, E Sindoni and J Vaclavik (Bologna: Editrice Compositori) p 399

[22.24] Tendler M 1997 *Plasma Phys. Contr. Fusion* **39** B371

[22.25] Fukuyama A *et al* 1998 *Plasma Phys. Contr. Fusion* **40** 653

[22.26] Rosenbluth M N 1999 *Plasma Phys. Contr. Fusion* **41** A99

[22.27] ITER Physics Basis 1999 *Nucl. Fusion* **39**, chapter 2, section 8

[22.28] Carreras B A, Newmann D, Diamond P H and Liang Y-M 1994 *Phys. Plasmas* **1** 4014

[22.29] Carreras B A, Diamond P H and Vetoulix G 1996 *Phys. Plasmas* **3** 4016

[22.30] Dnestrovskij A Yu, Parail V V and Vojtsenkhhovich I A 1993 in *Plasma Physics and Controlled Nuclear Fusion Research 1992* (Vienna: IAEA) vol 2, p 371
[22.31] Horton W, Hu G and Laval G 1996 *Phys. Plasmas* **3** 2912
[22.32] Osipenko M V 1997 *Plasma Phys. Report* **23** 837
[22.33] Yagi M, Itoh S-I, Itoh K and Fukuyama A 1997 *Plasma Phys. Contr. Fusion* **39** 1887
[22.34] Spatschek K H 1999 *Plasma Phys. Contr. Fusion* **41** A115
[22.35] Itoh S-I, Itoh K and Fukuyama A 1993 *Nuclear Fusion* **33** 1445
[22.36] Zohm H 1994 *Phys. Rev. Lett.* **72** 222
[22.37] Vojtsekhovich L A, Dnestrovskij A Yu and Parail V V 1995 *Nucl. Fusion* **35** 631
[22.38] Toda S *et al* 1996 *Plasma Phys. Contr. Fusion* **38** 1337
[22.39] Lebedev V B, Diamond P H, Gruzinova I and Carreras B A 1995 *Phys. Plasmas* **2** 3345
[22.40] Takayama A and Wakatani M 1996 *Plasma Phys. Contr. Fusion* **38** 1411
[22.41] Hegna C, Connor J W, Hastie R J and Wilson H R 1996 *Phys. Plasmas* **3** 248
[22.42] Connor J W, Hastie R J, Wilson H R and Miller R L 1998 *Phys. Plasmas* **5** 2687
[22.43] Manickam J 1992 *Phys. Fluids B* **4** 1901; Huysmans G T A, de Blanck H J, Kerner W, Goedbread J P and Nave M F F 1992 *Proceedings of the 19th EPS Conference on Controlled Fusion and Plasma Physics* (Innsbruck, 1992) ed W Freysinger *et al* (Geneva: European Physical Society) part I, p 247
[22.44] Itoh S-I, Itoh K, Fukuyama A and Yagi M 1996 *Plasma Phys. Contr. Fusion* **38** 527
[22.45] Itoh S-I, Itoh K, Fukuyama A and Yagi M 1996 *Plasma Phys. Contr. Fusion* **38** 1367

Chapter 23

Statistical Picture of Bifurcation

Various examples of subcritical excitation of plasma fluctuations suggest that multiple kinds of fluctuations can be realized for a given set of global plasma parameters. The bifurcation between different self-sustained fluctuations can possibly occur. The statistical approach considering a noise is explained in this chapter.

23.1 Statistical Approaches for Bifurcation of Turbulence

Two approaches are often used to study the bifurcation of turbulence. One is based on the Langevin equation with a stochastic noise. Another is the Fokker–Planck equation. In the former approach, an explicit relation between the fluctuation characteristics (level, decorrelation rate, typical wavenumber, etc.) can be derived as a function of driving parameters (thermal excitation and instability drive). In the latter approach, a probability density function (PDF) is derived, and the statistical properties are examined. From both approaches, the extended fluctuation dissipation relation is obtained.

23.1.1 Fluctuation Dissipation Relation from Stochastic Equation

A closed set of equations is derived by use of a statistical theory in chapter 14. They consist of:

(i) the eddy-damping rate as a function of fluctuation spectrum (renormalization relation);
(ii) the decorrelation rate as a function of the global parameters and eddy-damping rate (nonlinear dispersion relation); and
(iii) the extended fluctuation dissipation (FD) relation (nonlinear balance).

A bifurcation between fluctuations can also be studied.

The projected variables of fluctuations in chapter 14 are used to study the fluctuation amplitude. A spectral function of the electrostatic potential is introduced as

$$I_k \equiv \langle f_{1,k}^{(1)*} f_{1,k}^{(1)} \rangle \tag{23.1}$$

where the superscript (1) denotes the least stable branch of oscillations, the suffix k stands for the wavevector, and f_1 denotes the electrostatic potential fluctuation. (See §14.4.4 for details of notation.)

23.1.1.1 Renormalization relation

The relation between the eddy-damping rate and the fluctuation amplitude is formulated as equations (14.48) and (14.49). With the Markovian approximation,

$$\tau_{j,kpq} \simeq \left[\bar{\gamma}_{j,k}\right]^{-1} = \left[\gamma_{j,k} + \gamma_{jc,k}\right]^{-1} \tag{23.2}$$

the renormalization relation is given as

$$\gamma_{i,k} = -\sum_{k'} M_{i,kk'q} M_{i,qkk'}^{*} \bar{\gamma}_{i,k'}^{*\,-1} \langle f_{1,k'}^{(1)^*} f_{1,k'}^{(1)} \rangle. \tag{23.3}$$

An order of the magnitude estimate of the summation

$$\sum_{k'} M_{i,kk'q} M_{i,qkk'}^{*} \langle f_{1,k'}^{(1)^*} f_{1,k'}^{(1)} \rangle \to k^4 I_k$$

yields an approximate evaluation of equation (23.3) as

$$\gamma_{v,k}(\gamma_{v,k}^{*} + \gamma_{vc,k}) \simeq k^6 I_k \tag{23.4}$$

where $\gamma_{v,k} = \gamma_{1,k}$ is the eddy-viscosity damping rate.

23.1.1.2 The extended FD relation

The extended fluctuation dissipation (FD) relation is given by equation (14.73). The noise source has two origins: the turbulent noise (equation (14.74)) and the thermodynamical noise which is given by

$$\langle \mathcal{S}_{\text{th}}^{*} \mathcal{S}_{\text{th}} \rangle = 2|A_{11}|^{-2} \mu_{vc} \hat{T} \tag{23.5}$$

where $\hat{T} = 2\mu_0 B_p^{-2} k_B T$ is the normalized temperature. (See the discussion in Appendix 23A.) Substitution of each noise into the total, $\langle \mathcal{S}^{(m)*} \mathcal{S}^{(m)} \rangle = \langle \mathcal{S}_k^{(1)*} \mathcal{S}_k^{(1)} \rangle + \langle \mathcal{S}_{\text{th}}^{*} \mathcal{S}_{\text{th}} \rangle$, gives an expression for the extended FD relation equation (14.73) as

$$I_k = \frac{C_0 \gamma_v}{2\,\Re(\lambda_1)} I_k + \frac{\mu_{vc}}{\Re(\lambda_1)} \hat{T} \tag{23.6}$$

where λ_1 is the eigenvalue for the least stable eigenmode of the renormalized operator (equation (14.50)) and $\Re(\lambda_1)$ corresponds to the decorrelation rate of fluctuations. A competition between the nonlinear and thermodynamical drive controls the bifurcation between the turbulent and thermodynamical fluctuations. Equation (23.6) is rewritten as

$$\left(\Re(\lambda_1) - \frac{C_0\gamma_v}{2}\right) I_k = \mu_{vc}\hat{T} \tag{23.7}$$

Equations (23.4) and (23.7) together with the solution of the nonlinear dispersion relation form a closed set of equations for the fluctuation level, the eddy-damping rate and the decorrelation rate [14.25].

23.1.2 Fokker–Planck Equation for Macrovariable (Coarse-Grained Quantity)

Statistical properties of coarse-grained variables (averaged, filtered and so on) should be examined to understand turbulent phenomena. The Langevin equations for each k component, equation (14.47), can be reduced to just one for single macrovariable as

$$\frac{\partial}{\partial t}\mathcal{E} + 2\Lambda\mathcal{E} = \tilde{g}w(t) \tag{23.8}$$

where the total fluctuating energy integrated over some finite-size volume (L^2 in the perpendicular cross-section),

$$\mathcal{E} \equiv \frac{1}{2}\sum_k k_\perp^2 |\phi_k^2|, \tag{23.9}$$

is introduced. In equation (23.8), Λ and $\tilde{g}w(t)$ indicates the damping rate and noise source term, respectively, and $w(t)$ represents the Gaussian white noise. The damping rate Λ is given as

$$\Lambda \equiv \mathcal{E}^{-1}\sum_k 2\lambda_{1,k} k_\perp^2 |\phi_k^2| \tag{23.10a}$$

($\lambda_{1,k}$ is the nonlinear decorrelation rate of the k-mode) and the magnitude of the statistical source term, $g^2 \equiv \langle \tilde{g}^2 \rangle$, can be written as

$$g^2 = 4\hat{T}\gamma_{\mathrm{m}}\mathcal{E} + \sum_k g_k^{(1)^2} k_\perp^4 |\phi_k^2| \tag{23.10b}$$

where the noise amplitude $g_k^{(1)^2}$ is given in equation (14.79). The first term in the right-hand side represents the thermal excitation, and the second term is due to the turbulent noise. The coefficient Λ in the turbulent state deviates from the value in the limit of thermodynamical equilibrium, which is given by the mean decorrelation rate at thermal equilibrium,

$$\gamma_{\mathrm{m}} \equiv \mathcal{E}^{-1}\sum_k \gamma_{vc}\mathcal{E}_k, \tag{23.11}$$

where $\gamma_{vc} = \mu_{vc}k_\perp^2$.

The Fokker–Planck equation for the PDF of the coarse-grained quantity, $P(\mathcal{E})$, is described by

$$\frac{\partial}{\partial t}P(\mathcal{E}) = \frac{\partial}{\partial \mathcal{E}}\left(2\Lambda\mathcal{E} + \frac{1}{2}g\frac{\partial}{\partial \mathcal{E}}g\right)P(\mathcal{E}). \tag{23.12}$$

(See also [14.27] for the basis of the reduction to the Fokker–Planck equation.)

23.1.3 Steady-State Probability Density Function

The steady-state equilibrium PDF, as the solution of equation (23.12), is governed by the effective potential [23.1]

$$S(\mathcal{E}) = \int^{\mathcal{E}} \frac{4\Lambda\mathcal{E}}{g^2}\,\mathrm{d}\mathcal{E}. \tag{23.13}$$

The steady-state PDF is given by

$$P_{\mathrm{eq}}(\mathcal{E}) = \bar{P}\,\frac{1}{g}\exp[-S(\mathcal{E})]. \tag{23.14}$$

Therefore, the minimum of the effective potential $S(\mathcal{E})$ predicts the probable state of the plasma turbulence, which is either given at

$$\mathcal{E} = 0 \quad \text{if } \Lambda(\mathcal{E} \to 0) > 0 \tag{23.15a}$$

or at

$$\Lambda(\mathcal{E}) = 0 \quad \text{if } \left.\frac{\partial\Lambda(\mathcal{E})}{\partial\mathcal{E}}\right|_{\Lambda=0} > 0. \tag{23.15b}$$

The former state corresponds to the thermodynamical fluctuations and the latter to the turbulent fluctuations. Both of them satisfy the nonlinear marginal stability conditions. A specification of the fluctuations provides $\Lambda(\mathcal{E})$.

23.2 Bifurcation Between Thermodynamical and Turbulent Fluctuations

A bifurcation phenomenon between the thermodynamical and turbulent fluctuations is a fundamental case. The subcritical excitation to microscopic CDIM turbulence is discussed here [23.2].

23.2.1 Example of CDIM and Extended FD Relation

First, an approach of the stochastic equation (Langevin equation) is applied to this problem. Equations (23.4) and (23.7) together with the solution of the nonlinear dispersion relation of the CDIM turbulence form a closed set of equations.

23.2.1.1 The decorrelation rate

The nonlinear dispersion relation, equation (14.50), has been obtained for the CDIM turbulence and gives the decorrelation rate analytically (for inhomogeneous plasmas in a sheared magnetic field) as

$$\lambda_1 = \gamma_v + \gamma_{v,c} - \hat{G}_0^{3/5}\gamma_{v,c}^{4/5}(\gamma_e + \gamma_{e,c})^{1/5}, \tag{23.16}$$

where $\hat{G}_0 = (\delta/as)^{2/3}k_\perp^{2/3}\gamma_{v,c}^{-4/3}G_0$ is a normalized gradient parameter. (See equation (12.24c) for the combined pressure gradient parameter.) The mechanism of subcritical excitation, for a proper range of G_0, is as follows: in the zero-amplitude limit of fluctuations, $\gamma_v \to 0$, the collisional damping (the second term in the right-hand side of equation (23.16)) determines the decorrelation rate, yielding a stability. For an intermediate amplitude of fluctuations, the third term in the right-hand side dominates so as to cause an instability. In the large amplitude limit, the turbulent eddy-damping rate (the first term in the right-hand side) again dominates the decorrelation rate.

Upon combining with equations (23.4), (23.7), and (23.16), one has a nonlinear equation for the fluctuation level, in terms of the eddy-damping rate, as

$$\left(\gamma_v + \gamma_{v,c} - \frac{C_0\gamma_v}{2} - \hat{G}_0^{3/5}\gamma_{v,c}^{4/5}(\gamma_v + \gamma_{e,c})^{1/5}\right)\gamma_v(\gamma_v + \gamma_{v,c}) = \gamma_{v,c}^3 T_n, \tag{23.17}$$

where

$$T_n \equiv \mu_{v,c}^{-2}\hat{T} = 2\mu_{v,c}^{-2}\mu_0 B_p^{-2}k_B T \tag{23.18}$$

is a normalized temperature.

Figure 23.1 shows the fluctuation level I_k, the eddy-damping rate γ_v and the decorrelation rate λ_1 as a function of the global inhomogeneity parameter

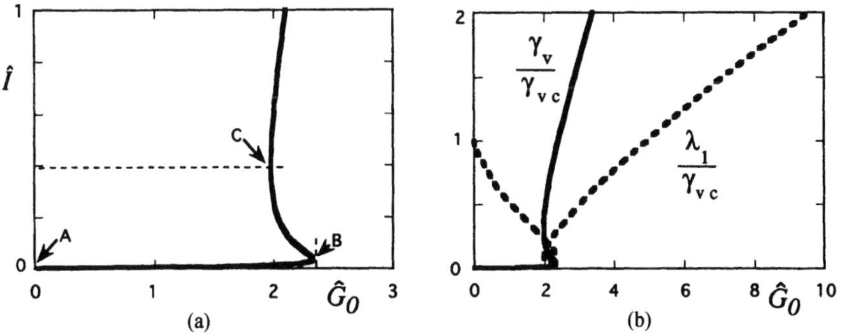

Figure 23.1. Fluctuation level, $\hat{I} = k_\perp^2\mu_{v,c}^{-2}I_k$, as a function of the gradient at low temperature (a), for the parameters of $T_n = 0.003$ and $C_0 = 0.6$. Eddy-damping rate γ_v (solid line) and the decorrelation rate λ_1 (dashed line) are shown in (b).

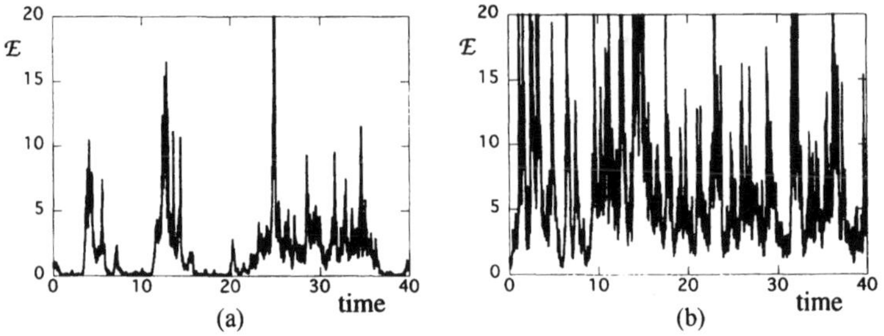

Figure 23.2. Temporal evolution of an energy of fluctuations (integrated over a certain volume). The cases in the hysteresis region and that in the turbulent state are shown in (a) and (b), respectively [23.3].

for a given temperature. In the region along A–B in figure 23.1(a), the thermodynamical fluctuation can exist. The critical value at the point B is less than that for the linear instability ($\hat{G}_0 \approx 3.5$ for this case). In the large pressure gradient region, a strongly turbulent state is self-sustained. Hysteresis is observed and a bifurcation can occur.

The decorrelation rate is shown in figure 23.1(b). The small gradient limit is given by the damping due to molecular viscosity. As the gradient increases, the correlation time becomes longer. Although a system is linearly stable, the presence of an instability source makes the amplitude of thermodynamical fluctuation slightly increased (figure 23.1(a)) and the decorrelation rate smaller (figure 23.1(b)). In contrast, the decorrelation rate of turbulent fluctuations increases as the pressure gradient becomes larger.

Transition between the thermodynamical and turbulent fluctuations is also shown by a direct solution of the Langevin equation. Figure 23.2 illustrates a time series of $\mathcal{E}(t)$ for the cases in the hysteresis region along C–B of figure 23.1(a) and in the region of large pressure gradient beyond B. Figure 23.2(b) show the turbulent fluctuations. In the region of multiple values of the hysteresis curve (along C–B of figure 23.1(a)), fluctuations are either in the turbulent state or thermodynamical equilibrium state, and continue jumping from one state to another as illustrated in figure 23.2(a). The average of the lower level corresponds to the thermodynamical branch, and that of the stronger fluctuations to the turbulent branch.

23.2.2 Phase Diagram for Thermodynamical and Turbulent Fluctuations

The width of the hysteresis is influenced by the temperature. The thermodynamical excitation becomes larger as the temperature increases, and the interval between points B and C in figure 23.1(a) becomes narrower. At a critical temperature, the hysteresis disappears and the fluctuation level becomes a monotonous function of the pressure gradient. The thermodynamical and turbulent fluctuations are connected via soft transition.

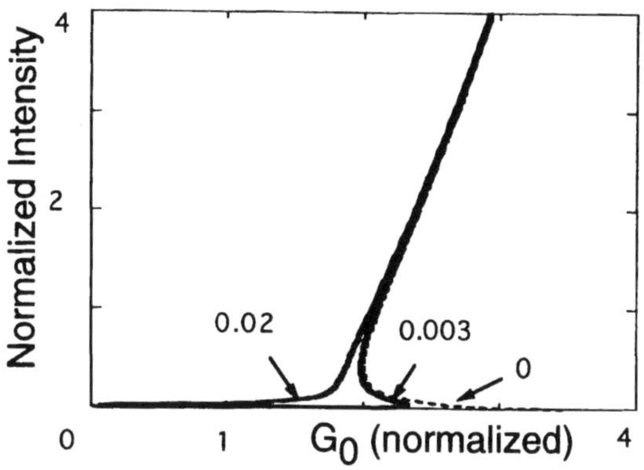

Figure 23.3. The fluctuation level as a function of the gradient, $\hat{I}[\hat{G}_0]$, is shown for various values of T_n. The dotted curve indicates the case of $T_n = 0$. As temperature increases, the hysteresis disappears.

Figure 23.3 illustrates the dependence of fluctuation level on pressure gradient for various values of the temperature.

Summarizing these dependences, a phase diagram of the fluctuations is drawn in figure 23.4. A cusp-type catastrophe is observed. In the region of the small pressure gradient the fluctuation is driven thermodynamically, and in the large pressure gradient limit the turbulence is self-sustained. In an intermediate region, multiple states are allowed and hysteresis exists.

23.2.3 Example of CDIM and Fokker–Planck Equation

The turbulent fluctuation and bifurcation are also approached by the Fokker–Planck equation. The steady-state PDF is governed by the effective

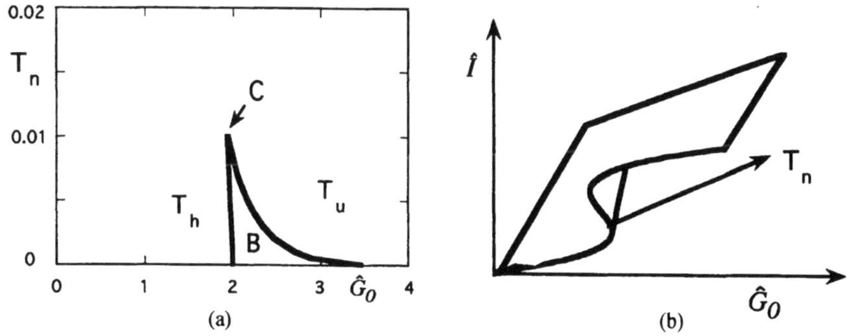

Figure 23.4. (a) Phase diagram on the gradient-temperature plane. Symbols T_h and T_u denote thermodynamical and turbulent fluctuations, respectively. In the region B, multiple states are allowed. C indicates the critical point. (b) Conceptual bird's-eye view of the fluctuation level as a function of the gradient and temperature. Cusp-type catastrophe is observed.

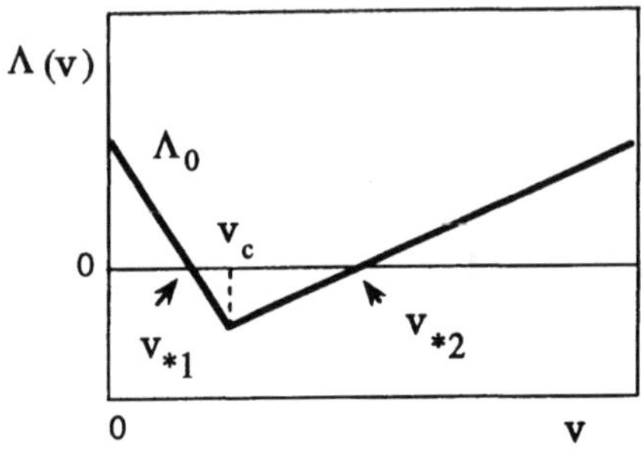

Figure 23.5. A model function of the decorrelation rate Λ as a function of the fluctuation velocity v.

potential defined by $S(\mathcal{E}) = \int^{\mathcal{E}} \mathrm{d}\mathcal{E}\, 4\Lambda\mathcal{E}g^{-2}$, whose minima and maxima are given by the zeros of $\Lambda(\mathcal{E}) = 0$.

In the parameter region where multiple states exist (i.e., the interval between B and C in figure 23.1(a)), zeros of $\Lambda(\mathcal{E})$ exist. A feature of the function $\Lambda(\mathcal{E})$ is illustrated in figure 23.5. A function of fluctuation velocity v

$$v = \sqrt{\mathcal{E}} \tag{23.19}$$

is introduced. In the low amplitude region, Λ is a decreasing function of v, representing a subcritical excitation owing to the nonlinear instability. In the large amplitude limit, Λ is an increasing function of v, and asymptotic dependence

$$\Lambda \propto v \tag{23.20}$$

holds. As a model function

$$\Lambda(v) = \Lambda_0 - \Lambda_0' v \quad (0 < v < v_c) \tag{23.21a}$$

$$\Lambda(v) = \Lambda_1' v - \bar{\Lambda}_0 \quad (v > v_c) \tag{23.21b}$$

is chosen here. At two values of v, v_{*1} and v_{*2}, Λ vanishes. In the figure, equation (23.15b) is satisfied at $v = v_{*2}$. The PDF has a peak at the value

$$\mathcal{E}_{\mathrm{eq}} = v_{*2}^2. \tag{23.22}$$

This is an estimate for the fluctuation level at the steady state. Figure 23.6 illustrates the shape of the effective potential and the PDF at the steady state for the case where a hysteresis exists.

The magnitude of the statistical noise source g^2 is given by equation (23.10b), and in a strong turbulent limit [14.24] the relation

$$\sum_k g_k^{(1)^2} k_\perp^4 |\phi_k^2| \propto |\phi|^5 \tag{23.23}$$

holds. Noting the relation $\mathcal{E} \propto |\phi|^2$, we may write an asymptotic form of the

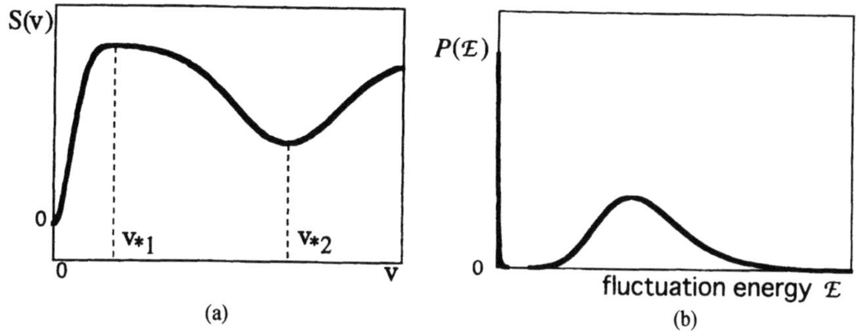

Figure 23.6. (a) An effective potential (i.e., an integral of renormalized dissipation rate) $S(v)$ as a function of the fluctuation amplitude. $S(v)$ has the local minima at $v = 0$ and at $v = v_{*2}$. (b) The schematic drawing of the steady-state PDF. P_{eq} has two peaks, i.e., one for the thermodynamical fluctuations and another for the self-sustained turbulence.

second term of equation (23.10b) as

$$\sum_k g_k^{(1)^2} k_\perp^4 \phi_k^2 = \bar{\bar{g}}_0^2 \left(\frac{\mathcal{E}}{\mathcal{E}_{eq}}\right)^{5/2}, \tag{23.24}$$

where an order of magnitude estimate of the coefficient $\bar{\bar{g}}_0^2$ is made at $\mathcal{E} = \mathcal{E}_{eq}$ as

$$\bar{\bar{g}}_0^2 \simeq \sum_k g_k^{(1)^2} k_\perp^4 \phi_k^2 \Big|_{\mathcal{E} = \mathcal{E}_{eq}}. \tag{23.25}$$

The amplitude of the noise source term is written then as

$$g^2 = 4\hat{T}\gamma_m \mathcal{E} + \bar{\bar{g}}_0^2 \left(\frac{\mathcal{E}}{\mathcal{E}_{eq}}\right)^{5/2}. \tag{23.26}$$

By using the forms of Λ and g^2, equations (23.21) and (23.26), the steady-state PDF is obtained as

$$P_{eq}(v) = \bar{P}_1 \frac{1}{g} \frac{(v+d)^{3b_1}}{(v^3+d^3)^{a_1}} \exp\left(-2\sqrt{3}b_1 \arctan\left(\frac{2v-d}{\sqrt{3}d}\right)\right) \quad (0 < v < v_c) \tag{23.27a}$$

$$P_{eq}(v) = \bar{P}_2 \frac{1}{g} \frac{(v^3+d^3)^{a_2}}{(v+d)^{3b_2}} \exp\left(-2\sqrt{3}b_2 \arctan\left(\frac{2v-d}{\sqrt{3}d}\right)\right) \quad (0 > v_c) \tag{23.27b}$$

where numerical coefficients are defined by

$$d^3 = \hat{T}\gamma_m \bar{\bar{g}}_0^{-2} \mathcal{E}_{eq}^{5/2}, \tag{23.28}$$

and

$$a_1 = (1 - 2d/v_{*1})b_1, \qquad b_1 = (\Lambda_0/3d)\bar{\bar{g}}_0^{-2}\mathcal{E}_{\mathrm{eq}}^{5/2},$$
$$a_2 = b_2(1 - 2d/v_{*2}), \qquad b_2 = (\bar{\Lambda}_0/3d)\bar{\bar{g}}_0^{-2}\mathcal{E}_{\mathrm{eq}}^{5/2},$$

and $\bar{P}_1$ and $\bar{P}_2$ are numerical constants.

Several features of the steady-state PDF are observed from equation (23.27). In the low amplitude region of $0 < v < v_c$, the distribution function deviates from the Gibbs distribution, i.e., it has a tail component. The exponential part has an argument $-2\sqrt{3}b_1 \arctan\{(2v - d)/\sqrt{3}d\}$; therefore, the characteristic value of v for the variation of $P_{\mathrm{eq}}(v)$ depends upon the values of $|d/b_1|$ and $b_1 - a_1$.

The second peak near $v = v_{*2}$ in the PDF represents the most probable turbulent state. The variation of the PDF in this region is controlled by the parameters $|d/b_2|$ and $b_2 - a_2$. Expanding the argument of the exponential part of equation (23.27b) in the vicinity of v_{*2}, the half-width of the peak Δv is approximately given as

$$\Delta v = \sqrt{\frac{\hat{T}\gamma_{\mathrm{m}} + \bar{\bar{g}}_0^2 \mathcal{E}_{\mathrm{eq}}^{-5/2} v_{*2}^3}{v_{*2}\Lambda_0'}}. \tag{23.29}$$

In a strong turbulence limit where the thermodynamical excitation is neglected, one has $\Delta v = \sqrt{\bar{\bar{g}}_0^2 \mathcal{E}_{\mathrm{eq}}^{-3/2} \Lambda_0'^{-1}}$. Finally, this probability density function equation (23.27b) has a tail in the large v limit as

$$P_{\mathrm{eq}}(v) \propto v^{3a_2 - 3b_2 - 5/2}. \tag{23.30}$$

This recovers the asymptotic form in a strong turbulence limit. For the case of CDIM turbulence, a tail component is given as,

$$P_{\mathrm{eq}}(\mathcal{E}) \propto \left(\frac{\mathcal{E}}{\mathcal{E}_{\mathrm{eq}}}\right)^{-\eta}, \tag{23.31}$$

with the power index given by

$$\eta = \frac{5}{4} + \frac{3}{2C_0}\frac{\bar{\bar{\Lambda}}}{\gamma_{k,\mathrm{eq}}} s \left(\frac{L}{\delta}\right)^2 G_0^{-1}, \tag{23.32}$$

where $\mathcal{E}_{\mathrm{eq}}$ is a statistical average of fluctuation energy, the asymptotic coefficient of nonlinear damping rate is $\bar{\bar{\Lambda}} = \lim_{\mathcal{E}\to\infty} \Lambda(\mathcal{E}/\mathcal{E}_{\mathrm{eq}})^{-1/2}$, and $\gamma_{k,\mathrm{eq}}$ is the nonlinear eddy-damping rate at $\mathcal{E} = \mathcal{E}_{\mathrm{eq}}$. For a fixed averaging-size L, the power index as

$$\left(\eta - \tfrac{5}{4}\right) \propto G_0^{-1} \tag{23.33}$$

depends on the inhomogeneity. The index of the power in the PDF is not constant. As the gradient increases, the tail component becomes larger.

23.3 Bifurcation Between Multiple Scale Length Turbulences

Many kinds of collective instabilities can grow in inhomogeneous plasmas. Fluctuations with the different scale lengths can coexist. As examples, microscopic fluctuations of the ITG mode, the CDIM and the ETG mode are explained in this monograph. They are characterized by the scale lengths of the ion gyroradius, ρ_i, the collisionless skin depth, $\delta = c/\omega_{\mathrm{p}}$, and the electron gyro radius, ρ_e, respectively. Interactions between the modes with different scale lengths are of importance. The mesoscale structure of a radial electric field causes various dynamics of the microscopic fluctuations; examples include the electric field domain interface [2.28], the zonal flow [19.25] and the streamer [21.25]. Nonlinear interplay between the different scale length fluctuations causes bifurcations. A statistical modelling is explained [23.4]. Hereafter we call the drift wave type fluctuations as 'semi-micro mode' and distinguish it from the 'micro mode' of the scale of the collisionless skin depth.

23.3.1 Scale Separation

The dynamical equations of fluctuations are symbolically written as

$$\frac{\partial}{\partial t} \boldsymbol{f} + \mathcal{L}^{(0)} \boldsymbol{f} = \mathcal{N}(\boldsymbol{f}, \boldsymbol{f}) + \tilde{\boldsymbol{S}}_{\mathrm{th}}, \tag{23.34}$$

where $\boldsymbol{f}$ denotes the fluctuating field. The linear operator $\mathcal{L}^{(0)}$ describes both the semi-micro and micro modes.

The Lagrangian nonlinearity term $\mathcal{N}(\boldsymbol{f}, \boldsymbol{f})$ gives three effects on a test mode $\boldsymbol{f}_k$ of the turbulent fluctuations. A part of Lagrangian nonlinearity is the coherent interactions to the test mode. This part causes the turbulent drag, which is written as $-\boldsymbol{\Gamma}_k \boldsymbol{f}_k$. The second effect causes a modification of the driving force. This is generated by the interactions between the modes of different scale lengths, and is written as $\mathcal{D}_k \boldsymbol{f}_k$ (the symbol '$\mathcal{D}$' stands for a 'drive'). The rest, the incoherent part, is considered to be a random self noise $\tilde{\boldsymbol{S}}_k$. Symbolically, we write

$$\mathcal{N}_k(\boldsymbol{f}, \boldsymbol{f}) = -\boldsymbol{\Gamma}_k \boldsymbol{f}_k + \mathcal{D}_k \boldsymbol{f}_k + \tilde{\boldsymbol{S}}_k. \tag{23.35}$$

A scale separation is assumed that there exists a mode number k_{sep} in Fourier representation to separate the semi-micro mode ($|k| < k_{\mathrm{sep}}$) and the micro mode ($|k| > k_{\mathrm{sep}}$). Figure 23.7 illustrates the schematic distribution of the spectrum in the wave-number space.

The renormalized nonlinear drag comes from fluctuations of the same-wavelength or of the shorter-wavelength modes. The nonlinear effects on the semi-micro and the micro modes are

$$\mathcal{N}^l(\boldsymbol{f}, \boldsymbol{f}) = -(\boldsymbol{\Gamma}^l_{(l)} + \boldsymbol{\Gamma}^l_{(h)}) \boldsymbol{f}^l + \tilde{\boldsymbol{S}}^l_{(l)} + \tilde{\boldsymbol{S}}^l_{(h)} \tag{23.36a}$$

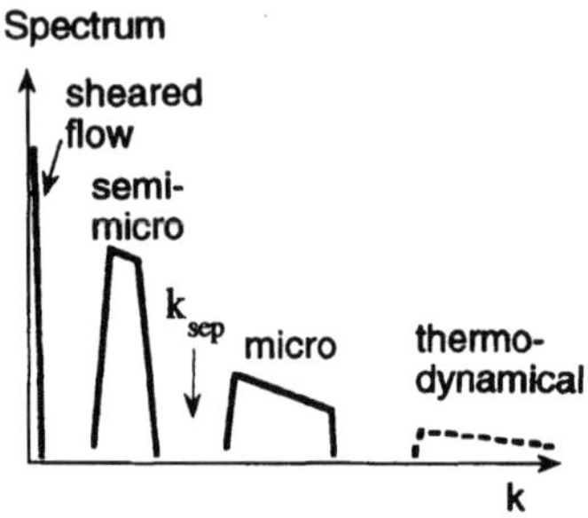

Figure 23.7. Schematic distribution of the symmetry-breaking structures in the wave-number space: the sheared global flow, the semi-micro mode fluctuations, the micro mode fluctuations and the thermodynamical fluctuations.

and

$$\mathcal{N}^h(\boldsymbol{f},\boldsymbol{f}) = -\boldsymbol{\Gamma}^h_{(h)}\boldsymbol{f}^h + \mathcal{D}^h_{(l)}\boldsymbol{f}^h + \tilde{\boldsymbol{S}}^h_{(h)}, \tag{23.36b}$$

respectively. The superscripts l and h denote the semi-micro and micro modes, respectively, and the subscripts (l) and (h) denote the contributions from semi-micro and micro modes, respectively. A set of the Langevin equations is formed as equation (14.47). For the semi-micro mode, one has

$$\mathcal{L} = \mathcal{L}^{(0)} + (\boldsymbol{\Gamma}^l_{(l)} + \boldsymbol{\Gamma}^l_{(h)}) \quad \text{and} \quad \tilde{\boldsymbol{S}} = \tilde{\boldsymbol{S}}^l_{(l)} + \tilde{\boldsymbol{S}}^l_{(h)} + \tilde{\boldsymbol{S}}_{\text{th}}. \tag{23.37a}$$

For the micro mode, one has

$$\mathcal{L} = \mathcal{L}^{(0)} + (\boldsymbol{\Gamma}^h_{(h)} - \mathcal{D}^h_{(l)}) \quad \text{and} \quad \tilde{\boldsymbol{S}} = \tilde{\boldsymbol{S}}^h_{(h)} + \tilde{\boldsymbol{S}}_{\text{th}}. \tag{23.37b}$$

23.3.2 Extended Fluctuation Dissipation Relation

An extended FD relation is derived according to §23.1.1; a closed set of equations consists of

(i) eddy-damping rate as a function of fluctuation spectrum (renormalization relation);
(ii) decorrelation rate as a function of the global parameters and eddy-damping rate (nonlinear dispersion relation); and
(iii) extended fluctuation dissipation (FD) relation (nonlinear balance).

This set dictates the fluctuation level and bifurcation. Using (i), (ii) and (iii) without the noise effect, the bifurcation between multiple scale-length turbulences is also predicted to occur. The total fluctuation amplitude of the jth field as coarse-grained variables for the micro mode and the semi-micro mode fluctuations are given, respectively, as

$$I_j^h = \sum_{k'} I_{j,k'}^h = \sum_{k'} \langle f_{j,k'}^{h*} f_{j,k'}^h \rangle, \tag{23.38a}$$

$$I_j^l = \sum_{k'} I_{j,k'}^l = \sum_{k'} \langle f_{j,k'}^{l*} f_{j,k'}^l \rangle. \tag{23.38b}$$

23.3.2.1 Eddy-damping rate

The damping of the micro mode is induced only by the micro mode, and the approximation (which yields equation (23.4)) gives

$$\gamma_i^h(\gamma_i^{h*} + \gamma_{ic}^h) \simeq (k_0^h)^4 I_i^h, \tag{23.39}$$

where γ_i^h is the characteristic eddy-damping rate and k_0^h is the characteristic wavenumber of the micro mode.

For the semi-micro mode, the self and the shorter scale-length modes, i.e., both the semi-micro and micro modes, contribute to the drag. A similar procedure gives

$$\gamma_i^l(\gamma_i^{l*} + \gamma_{ic}^l) \simeq (k_0^l)^4 I_i^l + (k_0^l)^2 (k_0^h)^2 \frac{\gamma_i^{l*} + \gamma_{ic}^l}{\gamma_i^{h*} + \gamma_{ic}^h} I_i^h, \tag{23.40}$$

where γ_i^l and k_0^l are the characteristic eddy-damping rate and the wavenumber of the semi-micro mode, respectively.

23.3.2.2 Decorrelation rate

A decorrelation rate λ is given by solving the nonlinear dispersion relation. An application is discussed in the next subsection.

23.3.2.3 Extended fluctuation dissipation relation

The turbulent self-noise is induced in the micro mode, so that the correlation function of the projected noise source $\langle S_{(h),k}^{h*} S_{(h),k}^h \rangle$ is evaluated in the same manner as in the preceding subsections. One has

$$I_1^h = \frac{C_0^h \gamma_v^h}{2\Re(\lambda^h)} I_1^h + \text{thermodynamical fluctuations}. \tag{23.41}$$

The turbulent noise to the semi-micro mode is induced from both the micro and semi-micro modes. One has

$$I_1^l = \frac{1}{2\Re(\lambda^l)} \left[C_0^l \gamma_v^l I_1^l + \hat{C}_0^h \gamma_v^h \left(\frac{k_0^l}{k_0^h} \right)^4 I_1^h \right] + \text{thermodynamical fluctuations} \tag{23.42}$$

(C_0 and $\hat{C}_0$ are numerical coefficients.)

23.3.3 Example of the ITG and the CDIM Turbulence

As an example, we choose a set of the ITG mode and the CDIM fluctuations. In the following, the level of fluctuating $E \times B$ motion $(I_1^{h,l})$ is taken as representative for fluctuation energy. $I^l = I_1^l$ and $I^h = I_1^h$ are used as an abbreviation.

The operator $\mathfrak{D}_{(l),k}^h$ includes the fluctuating field associated with the semi-micro perturbations. It fluctuates in time. In obtaining the solution of the eigenvalue equation with equation (23.37b), two limiting cases are considered. One is the case where the autocorrelation time of the semi-micro mode is much longer than that of the micro mode. In this case, the operator $\mathfrak{D}_{(l),k}^h$ is approximated as a constant in time in solving the dynamical equation of the microscopic fluctuations. We call this case a 'dc limit'. An analysis of chapter 20 is applied.

Another limit is that the autocorrelation time of the semi-micro mode is much shorter than that of the micro mode. The operator $\mathfrak{D}_{(l),k}^h$ includes a random oscillation effect. This case is called a 'random oscillator limit'. In this case, an analysis of §14.2.2 is applied. We take the 'dc limit' here, since the correlation time of the semi-micro mode is usually much longer than that of the micro mode, when analyzed separately. If one chooses the 'zonal flow' as a semi-micro component, its autocorrelation time could be equal to or shorter than that of the micro fluctuations [20.18]. Another limit equation (14.30) may be applied.

The semi-micro mode stabilizes the micro mode by the sheared flow. Another destabilizing effect of is the local steepening of the pressure gradient. Figure 23.8(a) illustrates the stretching of the flow pattern of the micro mode in the presence of semi-micro mode fluctuations. Figure 23.8(b) summarizes the nonlinear interactions in the turbulent fluctuations composed of the semi-micro and micro modes.

The dispersion relation of the ITG mode has been discussed in the literature. To model the stabilization due to the shear of a radial electric field, a

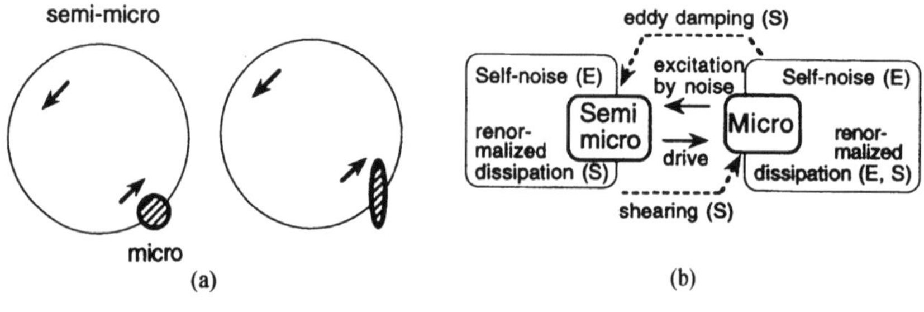

Figure 23.8. (a) Stretching of the micro mode by the semi-micro mode. (b) Mutual nonlinear interactions between the semi-micro mode and the micro mode. Symbols (E) and (S) denote the 'excitation' and the 'suppressing effect', respectively.

fitting function of the Lorentzian form is introduced. The nonlinear eigenvalue of the semi-micro mode is given as

$$\Re\lambda^l = -\frac{\gamma_0^l}{1 + (\omega_{E1}/\omega_{Ec}^l)^2} + \bar{\gamma}_v^l, \tag{23.43}$$

where

$$\gamma_0^l = \frac{r}{qR} s(1 + \eta_i)|\omega_{*i}| \tag{23.44}$$

is the characteristic growth rate and

$$\omega_{E1} = \frac{1}{B}\frac{\mathrm{d}}{\mathrm{d}r} E_\mathrm{r} \tag{23.45}$$

represents the shearing rate of the $E \times B$ flow by the global radial electric field. It is a limit of steep E_r gradient of equation (20.9). The critical value of the shearing rate for the stabilization is given by $\omega_{Ec}^l = 2\gamma_0^l/\sqrt{1 + \eta_i}$ for this case.

Taking into account a change of the local pressure gradient by the semi-micro mode, the characteristic growth rate of the CDIM fluctuations is given as

$$\gamma_{(l)\mathrm{driven}}^h \simeq \gamma_0^h \sqrt{1 + |k_0^{l2}/2\gamma_v^l|\sqrt{I^l}} \quad \text{and} \quad \gamma_0^h = \sqrt{G_0}\tau_{\mathrm{Ap}}^{-1}. \tag{23.46}$$

The shearing rate of $E \times B$ flow induced by the semi-micro mode is given by $|\omega_{E(l)}| = (k^l)^2\sqrt{I^l}$, and its influence on the driving nonlinearities $\mathfrak{D}_{(l)}^h$ is added. On the micro mode, one has

$$\Re\lambda^h = -\frac{\gamma_0^h \sqrt{1 + |k_0^{l2}/2\gamma_v^l|\sqrt{I^l}}}{1 + (\omega_{E1}/\omega_{Ec}^h)^2 + (k^l)^4(\omega_{Ec}^h)^{-2} I^l} + \bar{\gamma}_v^h, \tag{23.47}$$

where the critical shearing rate is given as $\omega_{Ec}^h \simeq s v_{\mathrm{th},i}/\sqrt{aR}$.

23.3.4 Bifurcation of Turbulence with Different Scale Lengths

Equations (23.40), (23.42), and (23.43) form a set of equations for the semi-micro mode, and equations (23.39), (23.41) and (23.47), for the micro mode.

From equations (23.40), (23.42), and (23.43), one has

$$\left(\frac{\sqrt{I^h} + \sqrt{I^h + 4I^l}}{2} - D^l\right) I^l = \varepsilon (I^h)^{3/2}, \tag{23.48}$$

where

$$D^l = \frac{2}{2 - C_0^l}\frac{1}{k_0^{l2}}\frac{\gamma_0^l}{1 + (\omega_{E1}/\omega_{Ec}^l)^2}, \tag{23.49}$$

and $\varepsilon = \hat{C}_0^h(2 - C_0^l)^{-1}(k_0^l/k_0^h)^2$ is a smallness parameter. D^l indicates the magnitude of the driving due to the global inhomogeneity. In the absence of this nonlinear interplay, D^l shows the characteristic value of the diffusivity by the semi-micro mode. The parameter ε represents the nonlinear noise of the micro mode, which couples to the semi-micro mode.

Equations (23.39), (23.41), and (23.47) provide a nonlinear equation for the level of micro mode fluctuations as

$$I^h = (D^h)^2 \frac{1 + |k_0^{l2}/2\gamma_v^l|\sqrt{I^l}}{(1 + I_{\mathrm{eff}}^{-1} I^l)^2}, \tag{23.50}$$

where

$$D^h = \frac{2}{2 - C_0^h} \frac{1}{1 + (\omega_{E1}/\omega_{Ec}^h)^2} \frac{\gamma_0^h}{k_0^{h2}}, \tag{23.51}$$

and

$$I_{\mathrm{eff}} \equiv [1 + (\omega_{E1}/\omega_{Ec}^h)^2](\omega_{Ec}^h)^2(k^l)^{-4}. \tag{23.52}$$

D^h is the characteristic diffusion rate in the absence of the coupling, representing the magnitude of the drive. At the level of the semi-micro mode I_{eff}, the suppression of the micro mode effectively takes place.

Equations (23.48) and (23.50) form a closed set of equations for the semi-micro and micro fluctuations including the nonlinear interplay between them. In the absence of the coupling, one has a solution for each fluctuation level as

$$I^h = (D^h)^2 \quad \text{and} \quad I^l = (D^l)^2. \tag{23.53}$$

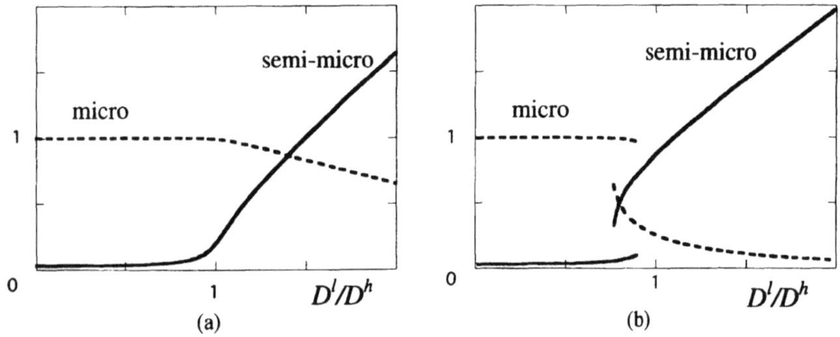

Figure 23.9. The fluctuation amplitude for the semi-micro mode $\sqrt{I^l}/D^h$ (full curve) and that for the micro mode $\sqrt{I^h}/D^h$ (dotted curve) as a function of the driving rate of the semi-micro mode D^l. The driving rate for the micro mode, D^h, is fixed. (a) The case of weak drive $D^h = \sqrt{0.3 I_{\mathrm{eff}}}$. (b) The case of strong drive, $D^h = 2\sqrt{I_{\mathrm{eff}}}$. For the intermediate value of D^l, multiple solutions are allowed and a hard transition takes place at a critical value of D^l.

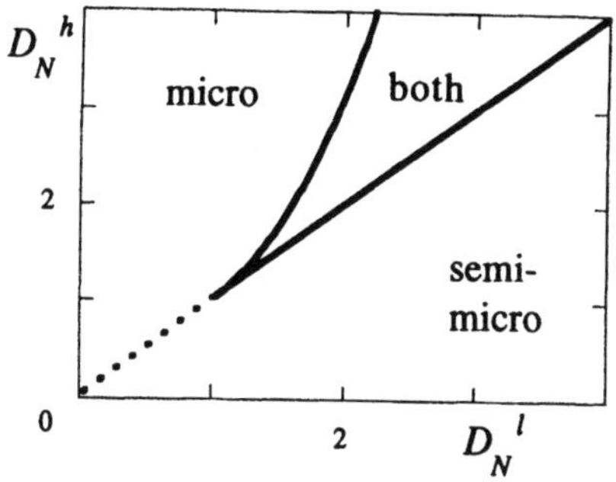

Figure 23.10. Phase diagram on the plane of global parameters represented by D^l and D^h. Coordinates are normalized to $\sqrt{I_{\text{eff}}}$ as $D_N^l \equiv D^l/\sqrt{I_{\text{eff}}}$ and $D_N^h \equiv D^h/\sqrt{I_{\text{eff}}}$. A cusp-type catastrophe is obtained. In the region of 'micro', the semi-micro mode is quenched. In the region of 'semi-micro', the micro mode coexists but is suppressed.

The nonlinear interaction between them causes the bifurcation. Equations (23.48) and (23.50) are solved for various driving parameters. The level of self-consistent solution is given in figure 23.9 as a function D^l. When D^l is small, the micro mode is excited, while the semi-micro mode is quenched, with amplitude I^l little influenced by the absolute value of the drive D^l. The threshold for the excitation of the semi-micro mode is observed at $D^l \simeq D^h$. As the drive D^l increases, I^l increases but I^h is more strongly suppressed. The coexistence of two branches exhibits a hysteresis in the relation between the gradient and fluctuation level. At the critical point, the transition from solution 'semi-micro' to 'micro' (or *vice versa*) takes place.

The presence of hard bifurcation in the fluctuation level forms a cusp-type catastrophe in the phase diagram represented by two parameters D^l and D^h. Figure 23.10 illustrates the phase diagram for the turbulence composed of the semi-micro and micro modes. The amplitude $\sqrt{I^h}$ or $\sqrt{I^l}$ is the measure of turbulent transport. Therefore figure 23.10 also represents the cusp-type catastrophe of the turbulent transport. Two solutions 'semi-micro' and 'micro' can appear simultaneously in the cusp region

$$D^l < D^h < \frac{1}{4D^l I_{\text{eff}}}(D^{l2} + I_{\text{eff}})^2. \tag{23.54}$$

The critical point of the cusp is given by

$$D^l = D^h = \sqrt{I_{\text{eff}}}. \tag{23.55}$$

Appendix 23A

Thermodynamical Equilibrium Limit

In the absence of the turbulence fluctuation, the result reduces to the fluctuation in a thermodynamical equilibrium. In the thermodynamical equilibrium, an equi-partition law for each $\boldsymbol{k}$ mode holds, and the spectrum is expressed in terms of the temperature. This limit is presented by focusing on the kinetic energy of fluctuations.

The fluctuation velocity of is given by the $\boldsymbol{E} \times \boldsymbol{B}$ velocity and the kinetic energy of the kth component per unit volume $\mathcal{E}_k$, is given in a dimensional form as

$$\mathcal{E}_k = \frac{\rho}{2} \frac{|\tilde{E}_{\perp,k}|^2}{B^2} = \frac{c^2}{v_A^2} \frac{\varepsilon_0}{2} |\tilde{E}_{\perp,k}|^2 \tag{23A.1}$$

or

$$\mathcal{E}_k = \frac{c^2}{v_A^2} \frac{\varepsilon_0}{2} k_\perp^2 \tilde{\phi}_k^2. \tag{23A.2}$$

where ρ is the mass density. In the thermodynamical equilibrium state, an equi-partition law holds, i.e.,

$$\langle \mathcal{E}_k \rangle = k_B T \tag{23A.3}$$

where the bracket $\langle \, \rangle$ means the statistical average, k_B is the Boltzmann constant, and T is the temperature. In the MHD normalization to study the plasma turbulence, i.e.,

$$\tilde{\phi}/v_{Ap} B_0 \to \phi \quad \text{and} \quad ka \to k,$$

the relations equations (23A.2) and (23A.3) are rewritten as

$$\langle k^2 \phi_k^2 \rangle = \hat{T} \tag{23A.4}$$

where the normalized temperature (with an additional dimension of a volume) is introduced as

$$\hat{T} = \frac{2\mu_0}{B_p^2} k_B T. \tag{23A.5}$$

Rewriting equation (23A.4), the thermodynamical excitation term $\langle I_{th}(k) \rangle = \langle \phi_k^2 \rangle$ in the extended fluctuation–dissipation theorem can be written by

$$\langle I_{th}(k) \rangle = \frac{\hat{T}}{k_\perp^2}. \tag{23A.6}$$

The noise source is also expressed in terms of the temperature. The relation between the random noise and the excited fluctuation is given from equation (14.73) as

$$I_{\text{th}}(k) = \frac{1}{2\Re(\lambda(\rightarrow 0))} |A_{11}|^2 \langle \mathcal{S}_{\text{th}}^* \mathcal{S}_{\text{th}} \rangle \tag{23A.7}$$

where $\lambda_1(\rightarrow 0)$ is the decorrelation rate in the absence of turbulence fluctuation which is given by the damping rate due to the molecular viscosity as

$$\lambda_1(\rightarrow 0) = \mu_{v,c} k_\perp^2 \tag{23A.8}$$

(in the case of $\mu_{v,c} = \chi_c$). Combining equations (23A.6)–(23A.8), one has an expression of the excitation rate of the thermodynamical noise as

$$|A_{11}|^2 \langle \mathcal{S}_{\text{th}}^* \mathcal{S}_{\text{th}} \rangle = 2\mu_{v,c} \hat{T}. \tag{23A.9}$$

This is the Einstein relation, i.e., it describes the relation between the thermodynamical excitation, the temperature and the molecular viscosity.

The magnitude of the fluctuation level of equation (23A.4) is compared with the thermal energy of the plasma, which is given as $nk_{\text{B}}T$ per unit volume where n is the number density of the plasma. The kinetic energy of the wave is given as

$$\text{density of kinetic energy of wave} \simeq N_{\text{th}}^{\text{wave}} k_{\text{B}} T \tag{23A.10}$$

where $N_{\text{th}}^{\text{wave}}$ is the number of eigenmodes in the unit volume. If one chooses the CDBM as an example, the relevant perpendicular wavenumber is of the order of $k_\perp \approx \delta^{-1}$ and the parallel wavenumber is of the order of $k_\parallel \approx a^{-1}$. An estimate for $N_{\text{th}}^{\text{wave}}$ is given by

$$N_{\text{th}}^{\text{wave}} \simeq \frac{\text{unit volume}}{\delta^2 a}. \tag{23A.11}$$

The ratio between the kinetic energy of the wave and the thermal energy is given as

$$\frac{\text{kinetic energy of wave}}{\text{kinetic energy of plasma}} \simeq \frac{1}{n\delta^2 a} \quad \text{(CDBM)}. \tag{23A.12}$$

For other low-frequency fluctuations of the ITG mode and the ETG mode, one has $k_\perp \approx \rho_i^{-1}$ and $k_\perp \approx \rho_e^{-1}$, respectively. The parallel wavenumber is of the order of $k_\parallel \approx a^{-1}$ for both the modes. One has

$$\frac{\text{kinetic energy of wave}}{\text{kinetic energy of plasma}} \simeq \frac{1}{n\rho_i^2 a} \quad \text{(ITG)}, \tag{23A.13}$$

$$\frac{\text{kinetic energy of wave}}{\text{kinetic energy of plasma}} \simeq \frac{1}{n\rho_e^2 a} \quad \text{(ETG)}. \tag{23A.14}$$

These levels of fluctuation energy, equations (23A.12)–(23A.14), are much lower than that of fluctuations in the range of Debye length, that causes the Coulombic collisions. For the latter fluctuations, the relation

$$\frac{\text{kinetic energy of wave}}{\text{kinetic energy of plasma}} \simeq \frac{1}{n\lambda_{\mathrm{D}}^3} \tag{23A.15}$$

holds where λ_{D} is the Debye length, $v_{\mathrm{th},e}\omega_{\mathrm{p}}^{-1}$. In the thermodynamical equilibrium, fluctuations are strongly excited in the branch of plasma oscillation than in the branches of low-frequency fluctuations.

References

[23.1] Itoh S-I and Itoh K 2000 *J. Phys. Soc. Jpn.* **69** 408

[23.2] Itoh S-I and Itoh K 2000 *J. Phys. Soc. Jpn.* **69** 427

[23.3] Kawasaki M, Furuya A, Yagi M, Itoh K and Itoh S-I 2002 *Plasma Phys. Contr. Fusion* **44** A473; Kawasaki M, Itoh S-I, Yagi M and Itoh K 2002 *J. Phys. Soc. Jpn.* **71** 1268

[23.4] Itoh S-I and Itoh K 2001 *Plasma Phys. Contr. Fusion* **43** 1055

Chapter 24

Transition Probability

Multiple states with different turbulence are allowed to exist in plasmas, and a bifurcation can occur. The model equations in chapter 18 are formulated as deterministic equations. However, the random source is also induced by the turbulence interactions. If one counts a statistical noise source in the turbulence, the transition from one turbulent state to another takes place as a statistical process. The transition occurs with finite probability even when a controlling parameter does not reach a critical value obtained by the deterministic analyses [23.2].

In a system near the thermodynamical equilibrium, a deviation from the equilibrium is originated from the thermal fluctuations, and the transition probability follows the Arrhenius law [24.1]. In turbulent plasmas, the probability density function (PDF) deviates from the Gibbs distribution; therefore the transition probability does not obey the Arrhenius law.

24.1 Transition by Noise

The transition between two different states has been studied by use of dynamical equations like equation (23.12). However, as is illustrated in figure 23.2, the observable quantity, say an average fluctuation energy $\mathcal{E}$, fluctuates around the most probable value. The statistical fluctuations induce the transition between the different states [23.2].

The steady-state PDF has been discussed in §23.2 for the system with hysteresis characteristics between the fluctuation level $\mathcal{E}$ and the control parameter G_0. The schematic relation between the thermodynamical branch and turbulent branch is drawn in figure 24.1. In the region $G_* < G_0 < G_c$, the system has multiple solutions of the thermodynamical and the turbulent branches. The steady-state PDF has been solved for a model function of $\Lambda(\mathcal{E})$.

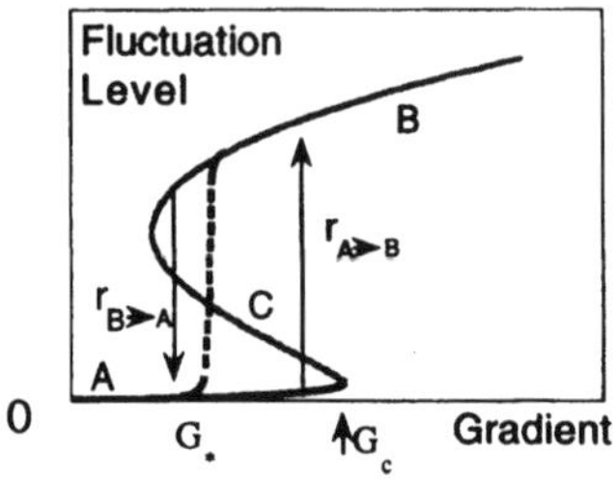

Figure 24.1. The transition probability between two states is calculated.

24.1.1 Rate Equation and Transition Probability

The effective potential $S(\mathcal{E}) = \int^{\mathcal{E}} \mathrm{d}\mathcal{E}\, 4\Lambda\mathcal{E}g^{-2}$, equation (23.13), is the integral of the renormalized dissipation rate $4\Lambda\mathcal{E}g^{-2}$, and is related with the steady-state PDF $P_{\mathrm{eq}}(\mathcal{E})$ in the absence of the flux of probability.

The number density of state A in the ensemble, N_{A}, of the finite energy width $\Delta\mathcal{E}_{\mathrm{A}}$ is expressed by the PDF, $P(\mathcal{E})$, by

$$N_{\mathrm{A}} = \int_0^{\Delta\mathcal{E}_{\mathrm{A}}} P(\mathcal{E})\,\mathrm{d}\mathcal{E} \tag{24.1}$$

and is evaluated as

$$N_{\mathrm{A}} = P(\mathcal{E}_{\mathrm{A}})\Delta\mathcal{E}_{\mathrm{A}}. \tag{24.2}$$

In the same way, the number density of state B in the ensemble is

$$N_{\mathrm{B}} = \int_{\mathcal{E}_{\mathrm{B}} - \Delta\mathcal{E}_{\mathrm{B}}/2}^{\mathcal{E}_{\mathrm{B}} + \Delta\mathcal{E}_{\mathrm{B}}/2} P(\mathcal{E})\,\mathrm{d}\mathcal{E}, \tag{24.3}$$

where $\Delta\mathcal{E}_{\mathrm{B}}$ is the width of energy of the state B. N_{B} is evaluated as

$$N_{\mathrm{B}} = P(\mathcal{E}_{\mathrm{B}})\Delta\mathcal{E}_{\mathrm{B}}. \tag{24.4}$$

The rate equations of N_{A} and N_{B} are written as

$$\frac{\partial}{\partial t} N_{\mathrm{A}} = -r_{\mathrm{A}\to\mathrm{B}} N_{\mathrm{A}} + r_{\mathrm{B}\to\mathrm{A}} N_{\mathrm{B}} + h_{\mathrm{A}} \tag{24.5a}$$

and

$$\frac{\partial}{\partial t} N_{\mathrm{B}} = r_{\mathrm{A}\to\mathrm{B}} N_{\mathrm{A}} - r_{\mathrm{B}\to\mathrm{A}} N_{\mathrm{B}} + h_{\mathrm{B}}, \tag{24.5b}$$

where $r_{\mathrm{A}\to\mathrm{B}}$ and $r_{\mathrm{B}\to\mathrm{A}}$ are the transition probability from the state A to B and that from the B to A, respectively, and h_{A} and h_{B} are the sums of source and sink for the states A and B, respectively. In the following, we call $r_{\mathrm{A}\to\mathrm{B}}$ 'forward-transition probability' and $r_{\mathrm{B}\to\mathrm{A}}$ 'back-transition probability'.

The transition probabilities $r_{\mathrm{A}\to\mathrm{B}}$ and $r_{\mathrm{B}\to\mathrm{A}}$ are calculated from the Fokker–Planck equation. Consider a stationary state with the source and the sink

$$\partial N_A/\partial t = \partial N_B/\partial t = 0. \tag{24.6}$$

If the sink of state B is large so that the density of state B vanishes,

$$N_B = 0, \tag{24.7}$$

then the transition condition satisfies the relation

$$r_{A\to B} = \frac{h_A}{N_A}. \tag{24.8}$$

The transition probability can be calculated by evaluating the necessary source term from the Fokker–Planck equation with the condition equation (24.7).

24.1.2 Flux of Probability and Probability of Transition

The source (sink) rate is connected to the flux of probability w in $\mathcal{E}$ space, which is governed by the Fokker–Planck equation (23.12). The Fokker–Planck equation, in the presence of an external source h, is rewritten as

$$\frac{\partial}{\partial t}P(\mathcal{E}) + \frac{\partial}{\partial \mathcal{E}}w = h \tag{24.9}$$

with the flux of probability

$$w = -\left(2\Lambda\mathcal{E} + \frac{1}{2}g\frac{\partial}{\partial \mathcal{E}}g\right)P(\mathcal{E}). \tag{24.10}$$

If one integrates equation (24.9) in the region $0 < \mathcal{E} < \Delta\mathcal{E}_A$, one has

$$\frac{\partial}{\partial t}N_A + w(\mathcal{E}_A) = h_A \equiv \int_0^{\Delta\mathcal{E}_A} h\,\mathrm{d}\mathcal{E}. \tag{24.11}$$

In a steady state, $\partial N_A/\partial t = 0$, one has a conservation law of probability from equation (24.11) as

$$w(\mathcal{E}_A) = h_A. \tag{24.12}$$

The flux of probability w is independent of $\mathcal{E}$ in the region where the source is absent. We consider the case where no source exists in the region $\mathcal{E} > \Delta\mathcal{E}_A$ for the condition of equation (24.7). By use of equation (24.12), equation (24.8) is written as

$$r_{A\to B} = \frac{w}{N_A}. \tag{24.13}$$

The flux of probability w is obtained from the probability by that amount the barrier at C in the effective potential is overcome.

By use of equation (23.13), Λ is eliminated from equation (24.10) and w of equation (24.10) can be rewritten in terms of $S(\mathcal{E})$, $P(\mathcal{E})$, and g as

$$w = -\tfrac{1}{2} g \exp[-S(\mathcal{E})] \frac{\partial}{\partial \mathcal{E}} \{gP(\mathcal{E}) \exp[S(\mathcal{E})]\}, \tag{24.14a}$$

or

$$w^{-1} \frac{\partial}{\partial \mathcal{E}} \{gP(\mathcal{E}) \exp[S(\mathcal{E})]\} = -2g^{-1} \exp[S(\mathcal{E})]. \tag{24.14b}$$

Since w is constant in a stationary state, we obtain the flux of probability from the integration of equation (24.14b) from state A to state B as

$$w = \frac{\{gP(\mathcal{E}) \exp[S(\mathcal{E})]\}|_{\mathrm{B}}^{\mathrm{A}}}{2 \int_{\mathrm{A}}^{\mathrm{B}} \frac{1}{g} \exp[S(\mathcal{E})] \, \mathrm{d}\mathcal{E}}. \tag{24.15}$$

Recalling the boundary condition at state B, $N_{\mathrm{B}} = 0$, i.e., $P(\mathcal{E}_{\mathrm{B}}) = 0$, one has an expression of the flux of probability density as

$$w = \frac{g(\mathcal{E}_{\mathrm{A}}) P(\mathcal{E}_{\mathrm{A}}) \exp[S(\mathcal{E}_{\mathrm{A}})]}{2 \int_{\mathrm{A}}^{\mathrm{B}} g^{-1} \exp[S(\mathcal{E})] \, \mathrm{d}\mathcal{E}}. \tag{24.16}$$

An analytic estimate of the denominator of equation (24.16) is possible by use of a method of the steepest descent in the vicinity of state C [24.2]. An expansion

$$\begin{aligned} \frac{1}{g} \exp[S(\mathcal{E})] &= \exp[S(\mathcal{E}) - \ln(g)] \\ &= \exp\left\{ S(\mathcal{E}_{\mathrm{C}}) - \ln[g(\mathcal{E}_{\mathrm{C}})] - \frac{1}{2\Delta_{\mathrm{c}}^2} (\mathcal{E} - \mathcal{E}_{\mathrm{C}})^2 + \cdots \right\}, \end{aligned} \tag{24.17}$$

where Δ_{c}^{-1} denotes the curvature of $S(\mathcal{E})$ at $\mathcal{E} \simeq \mathcal{E}_{\mathrm{C}}$, provides the evaluation

$$2 \int_{\mathrm{A}}^{\mathrm{B}} \frac{1}{g} \exp[S(\mathcal{E})] \, \mathrm{d}\mathcal{E} = \frac{2\sqrt{2\pi}\, \Delta_{\mathrm{c}}}{g(\mathcal{E}_{\mathrm{C}})} \exp[S(\mathcal{E}_{\mathrm{C}})]. \tag{24.18}$$

Upon substituting equation (24.18) into equation (24.16), the flux of probability associated with the A $\rightarrow$ B transition is given as

$$w_{\mathrm{A} \rightarrow \mathrm{B}} = \frac{1}{2\sqrt{2\pi}\, \Delta_{\mathrm{c}}} g(\mathcal{E}_{\mathrm{C}}) g(\mathcal{E}_{\mathrm{A}}) P(\mathcal{E}_{\mathrm{A}}) \exp[S(\mathcal{E}_{\mathrm{A}}) - S(\mathcal{E}_{\mathrm{C}})]. \tag{24.19}$$

The probability flux associated with the B $\rightarrow$ A transition (back transition) is obtained by the same procedure with the boundary condition

$$P(\mathcal{E}_{\mathrm{A}}) = 0. \tag{24.20}$$

The result is obtained by replacing A and B in equation (24.19) as

$$w_{B\to A} = \frac{1}{2\sqrt{2\pi}\,\Delta_c} g(\mathcal{E}_C)g(\mathcal{E}_B)P(\mathcal{E}_B)\exp[S(\mathcal{E}_B) - S(\mathcal{E}_C)]. \tag{24.21}$$

24.1.3 Transition Probability

The flux of probability from A to B in figure 24.1 is obtained. The saddle point is denoted by the suffix C. The function $\Lambda(\mathcal{E})$ is Taylor expanded in the vicinity of $\mathcal{E}_C$ as

$$\Lambda(v) = \Lambda(v_{*1}) - \Lambda_0'(v - v_{*1}) + \cdots \tag{24.22}$$

($v = \sqrt{\mathcal{E}}, v_{*1} = \sqrt{\mathcal{E}_C}$), and Δ_c in equation (24.19) is given as $\Delta_c = (\Lambda_0' v_{*1})^{-1/2}$. For the model like equation (23.21a), one has $\Delta_c = \Lambda_0^{-1/2}$. The transition probability is given as [23.2]

$$r_{A\to B} = \frac{\sqrt{\Lambda_0}}{2\sqrt{\pi}} \frac{1}{\Delta\mathcal{E}_A} g(\mathcal{E}_A)\exp[S(\mathcal{E}_A) - S(\mathcal{E}_C)]. \tag{24.23}$$

The transition probability from B to A is calculated in a similar manner by replacing A and B in equation (24.23).

24.2 Transition Between Thermodynamical and Turbulent Fluctuations

The time rate of the transition from thermal to turbulent fluctuations is calculated. The states of thermodynamical and turbulent fluctuations are labeled A and B, respectively.

24.2.1 Transition from Thermodynamical Fluctuations

In state A, the PDF of the fluctuation energy is approximated by the Gibbs distribution and the width is estimated by

$$\Delta\mathcal{E}_A = \hat{T}, \tag{24.24}$$

where the normalization of temperature is given in equation (23.5). The magnitude of the thermodynamical noise term is calculated from equation (23A.9) as $g(\mathcal{E}_A)^2 = 4\gamma_m \hat{T}^2$, i.e., $g(\mathcal{E}_A) = 2\sqrt{\gamma_m}\hat{T}$. Upon substituting this relation into equation (24.23) with the condition $S(\mathcal{E}_A) = 0$, one obtains

$$r_{A\to B} = \frac{\sqrt{\Lambda_0 \gamma_m}}{\sqrt{\pi}} \exp[-S(\mathcal{E}_C)]. \tag{24.25}$$

24.2.2 Thermodynamical Equilibrium Limit

This result is an extension of the theory based on the thermodynamical excitations. The Arrhenius law is a typical example for the transitions of the systems in thermodynamical equilibrium.

In the limit of thermodynamical equilibrium of equation (24.25), the noise term is expressed in terms of heat bath temperature

$$g^2 = 4\gamma_m \hat{T}^2. \tag{24.26}$$

Note that γ_m is the average of the eddy-damping rate due to the molecular viscosity γ_{vc} (see equation (23.11)). The decorrelation rate is also given by equation (23.11) as

$$\Lambda_0 = \gamma_m \tag{24.27}$$

and the effective potential $S(\mathcal{E})$ leads to

$$S(\mathcal{E}) = \frac{\mathcal{E}}{\hat{T}}. \tag{24.28}$$

The coefficient $\exp[-S(\mathcal{E}_C)]$ is expressed as $\exp(-\mathcal{E}_C/\hat{T})$. Upon substituting this relation into equation (24.25), the transition probability is given by

$$r_{A \to B} = \frac{1}{\sqrt{\pi}} \gamma_m \exp\left(-\frac{\mathcal{E}_C}{\hat{T}}\right). \tag{24.29}$$

Noting the strong exponential dependence $\exp(-\mathcal{E}_C/\hat{T})$, one finds the relation

$$\ln r_{A \to B} \propto -\frac{1}{\hat{T}}. \tag{24.30}$$

The Arrhenius law is recovered.

24.2.3 Transition to Turbulent Fluctuations

In the presence of turbulent self-noise, both the thermodynamical excitation and the self-noise are important in determining the transition probability. The cooperative effects are studied.

Based on the scaling property of the random self-noise term, g^2 is written as equation (23.26). The integral $S(\mathcal{E})$ is given as

$$S(\mathcal{E}) = \int_0^{\mathcal{E}} \frac{4\Lambda\mathcal{E}}{4\hat{T}\gamma_m\mathcal{E} + \bar{\bar{g}}_0^2 \mathcal{E}_{eq}^{-5/2} \mathcal{E}^{5/2}} \, d\mathcal{E}. \tag{24.31}$$

It is convenient to use the form

$$S(v) = \int_0^{v} \frac{2\lambda(v)v}{\hat{T}\gamma_m + \bar{\bar{g}}_0^2 \mathcal{E}_{eq}^{-5/2} v^3} \, dv \tag{24.32}$$

where v is the fluctuation velocity, $\mathcal{E} = v^2$.

The effective potential is calculated by specifying the formula of subcritical excitation. Equation (23.21a) is used; $\Lambda(v) = \Lambda_0 - \Lambda_0' v$ for $0 < v < v_c$. By substituting equation (23.21a) into equation (24.32), the transition probability is given as

$$r_{A \to B} = \frac{\sqrt{\Lambda_0 \gamma_m}}{\sqrt{\pi}} \frac{(v_{*1} d^{-1} + 1)^{3b_1}}{(v_{*1}^3 d^{-3} + 1)^{a_1}} \exp\left\{ -2\sqrt{3} b_1 \left[\arctan\left(\frac{2v_{*1} - d}{\sqrt{3} d} \right) + \frac{\pi}{6} \right] \right\} \tag{24.33}$$

where $\mathcal{E}_C = v_{*1}^2$, $a_1 = (1 - 2d/v_{*1}) b_1$, $b_1 = (\Lambda_0 / 3d) \bar{\bar{g}}_0^{-2} \mathcal{E}_{eq}^{5/2}$ and d^3 is given by equation (23.28), $d^3 = \hat{T} \gamma_m \bar{\bar{g}}_0^{-2} \mathcal{E}_{eq}^{5/2}$.

Equation (24.33) shows that the transition probability is controlled by the parameter v_{*1}/d. The parameter v_{*1} represents the threshold for the onset of nonlinear instability, and d is a combined parameter of thermodynamical and turbulent noises. The competition between the threshold and noise levels dictates the transition probability. When the threshold amplitude is low, $v_{*1}/d \ll 1$, a Taylor expansion of equation (24.33) with respect to v_{*1}/d to provides an expression

$$r_{A \to B} = \frac{\gamma_m}{\sqrt{\pi}} \left(1 - 3b_1 \frac{v_{*1}^2}{d^2} + \cdots \right) \quad (v_{*1}/d \ll 1). \tag{24.34a}$$

If the threshold amplitude is high, $v_{*1}/d \gg 1$, the limiting formula is deduced as

$$r_{A \to B} = \frac{\gamma_m}{\sqrt{\pi}} \exp\left(-\frac{4\pi}{\sqrt{3}} b_1 \right) \quad (v_{*1}/d \gg 1). \tag{24.34b}$$

In an intermediate regime, $1 < \mathcal{E}_C / d^2 < 100$, equation (24.33) gives a power-law dependence as

$$r \sim \frac{\gamma_m}{\sqrt{\pi}} \left(\frac{v_{*1}}{d} \right)^{-2b_1} = \frac{\gamma_m}{\sqrt{\pi}} \left(\frac{\mathcal{E}_C}{d^2} \right)^{-b_1}. \tag{24.34c}$$

The characteristic value of transition rate is proportional to the eddy-damping rate by molecular viscosity, γ_m, and the transition probability remains finite even in the limit of $v_{*1}/d \to \infty$. A small but finite tail in the PDF contributes to the occurrence of transition for the case of $v_{*1}/d \gg 1$.

24.2.4 Back Transition Probability

Back transition probability is given similarly to equation (24.25) as

$$r_{B \to A} = \frac{\sqrt{\Lambda_0 \gamma_B}}{4\sqrt{\pi}} \exp[S(\mathcal{E}_B) - S(\mathcal{E}_C)]. \tag{24.35}$$

For the model decorrelation rate of equation (23.21), the back transition probability is also calculated. One has an exponential factor $\exp[S(\mathcal{E}_B) - S(\mathcal{E}_c)]$ as

$$\begin{aligned}
&\exp[S(\mathcal{E}_B) - S(\mathcal{E}_c)] \\
&= \frac{(v_{*1}d^{-1}+1)^{3b_1}}{(v_{*1}^3 d^{-3}+1)^{a_1}} \frac{(v_{*2}d^{-1}+1)^{3b_2}}{(v_{*2}^3 d^{-3}+1)^{a_2}} \frac{(v_c^3 d^{-3}+1)^{a_1+a_2}}{(v_c d^{-1}+1)^{3b_1+3b_2}} \\
&\quad \times \exp\left\{-2\sqrt{3}b_1\left[\arctan\left(\frac{2v_{*1}-d}{\sqrt{3}d}\right) - \arctan\left(\frac{2v_c-d}{\sqrt{3}d}\right)\right]\right\} \\
&\quad \times \exp\left\{-2\sqrt{3}b_2\left[\arctan\left(\frac{2v_{*2}-d}{\sqrt{3}d}\right) - \arctan\left(\frac{2v_c-d}{\sqrt{3}d}\right)\right]\right\},
\end{aligned} \tag{24.36}$$

where $a_2 = b_2(1 - 2d/v_{*2})$, $b_2 = (\bar{\Lambda}_0/3d)\bar{\bar{g}}_0^{-2}\mathcal{E}_{eq}^{5/2}$ and d is given by equation (23.28), $d^3 = \hat{T}\gamma_m \bar{\bar{g}}_0^{-2}\mathcal{E}_{eq}^{5/2}$.

In the limit of strong turbulence,

$$v_{*2} \gg d, \qquad v_{*1} \gg d, \tag{24.37}$$

one has the form as

$$\exp[S(\mathcal{E}_B) - S(\mathcal{E}_c)] = \frac{v_{*1}^{3b_1-3a_1}}{v_c^{3b_1-3a_1}} \frac{v_{*2}^{3b_2-3a_2}}{v_c^{3b_2-3a_2}}. \tag{24.38}$$

Upon substituting equation (24.38) into equation (24.35), the probability for the back transition (in the strong turbulence limit) is obtained as

$$r_{B\to A} = \frac{\sqrt{\Lambda_0\gamma_B}}{4\sqrt{\pi}}\left(\frac{v_{*1}}{v_c}\right)^{\eta_1}\left(\frac{v_{*2}}{v_c}\right)^{\eta_2}, \tag{24.39}$$

with

$$\eta_1 = 2\Lambda_0'\bar{\bar{g}}_0^{-2}\mathcal{E}_{eq}^{5/2} \quad \text{and} \quad \eta_2 = 2\Lambda_1'\bar{\bar{g}}_0^{-2}\mathcal{E}_{eq}^{5/2}. \tag{24.40}$$

The back-transition probability has also a form of the power law. It is noticed that the time rate is given by

$$\sqrt{\Lambda_0\gamma_B},$$

which is accelerated in comparison with γ_m. The power indices η_1 and η_2 in the expression $r_{B\to A}$ do not explicitly include the temperature T. This is because the random noise is dominated by the turbulent self-noise in state B: the deviation from the mean value of fluctuations is not influenced by the thermodynamical excitations.

24.2.5 Example of Strong Turbulence

Statistical quantities in the turbulent state (state B) are calculated for the CDIM turbulence. The steady-state level of turbulent fluctuations and noise source are estimated as

$$\mathcal{E}_B = \frac{1}{4} G_0^2 \left(\frac{\delta}{sa} \right)^2 \left(\frac{L}{a} \right)^2 \tag{24.41}$$

and

$$\bar{\bar{g}}_0^2 \simeq \frac{8C_0}{3} \gamma_0 \mathcal{E}_{eq}^2 \left(k_0 \frac{L}{a} \right)^{-2}, \tag{24.42}$$

where k_0 is a typical wavenumber

$$k_0 \simeq G_0^{-1/2} \frac{sa}{\delta} \tag{24.43}$$

and $\gamma_0 = \gamma_B$ is the characteristic eddy-damping rate at state B

$$\gamma_0 \simeq G_0^{1/2} \tag{24.44}$$

(in MHD unit).

The threshold amplitude of the nonlinear instability corresponds to the energy at saddle point $\mathcal{E}_C$. For the interchange turbulence its value is given as

$$\mathcal{E}_C \simeq \frac{\mu_{ec}^2}{4} k_0^2 \left(\frac{L}{a} \right)^2 \left(1 - \frac{G_0}{G_c} \right) \tag{24.45}$$

in the vicinity of the linear stability criterion

$$G_0 \simeq G_c. \tag{24.46}$$

Away from thermodynamical equilibrium, the transition probability obeys a power law. In the vicinity of the critical linear stability boundary G_c in figure 24.1, the transition probability is given as a function of $G_c - G_0$ as

$$r_{A \to B} \approx \frac{\gamma_m}{\sqrt{\pi}} \left(\frac{\mu_{ec}}{2} \right)^{-2b_1} k_0^{-4b_1/3} \left(\hat{T} \gamma_m \frac{3}{16C_0} \right)^{2b_1/3} \left(1 - \frac{G_0}{G_c} \right)^{-b_1}, \tag{24.47}$$

where

$$b_1 \simeq \frac{1}{3} \left(\frac{3}{16C_0} k_0 \gamma_m \right)^{2/3} \hat{T}^{-1/3} \left(\frac{L}{a} \right)^2, \qquad d^3 = \hat{T} \gamma_m \frac{3}{16C_0} k_0 \left(\frac{L}{a} \right)^3, \tag{24.48}$$

and L is the size of the small region over which the fluctuation energy $\mathcal{E}$ is integrated.

Comparing equation (24.27) with Arrhenius law equation (24.49), the characteristic features of a turbulent transition are seen. First, the transition probability is explicitly expressed in terms of the plasma inhomogeneity parameter. Second, it is greatly enhanced, owing to a large statistical variance. Third, it has a power-law dependence on the plasma inhomogeneity parameter. Namely, the transition probability increases as

$$r_{\mathrm{A}\to\mathrm{B}} \propto (G_{\mathrm{c}} - G_0)^{-b_1} \tag{24.49}$$

when G_0 approaches G_{c}. These results, combined with the dynamical and spatial structural formations, are the distinct features of the plasma turbulence transition.

24.3 Phase Boundary in Statistical Theory

The multiple states can be a marginal-stability solution of the self-consistent equations. A transition can occur owing to the statistical noise source. A new time scale in the turbulence, a lifetime of the state, is an inverse of the transition probability. The lifetime of state B is given by $r_{\mathrm{B}\to\mathrm{A}}^{-1}$. This lifetime is much longer than the dynamical time scale of fluctuations, such as the decorrelation time. The relative magnitude between the lifetime and the time for the evolution of global parameters is a key issue. If the control parameters (e.g., gradient parameter, temperature, etc.) evolve much slower than the lifetime, the statistical equilibration between states A and B occurs in transitions, and the phase boundary is drawn in the region of hysteresis.

24.3.1 Phase Boundary

The 'phase boundary' between states A and B is defined by the condition

$$N_{\mathrm{A}} = N_{\mathrm{B}},$$

and is obtained by use of the transition probability. From equation (24.5), the ratio of number densities is given in a steady state as

$$\frac{N_{\mathrm{A}}}{N_{\mathrm{B}}} = \frac{r_{\mathrm{B}\to\mathrm{A}}}{r_{\mathrm{A}\to\mathrm{B}}} \tag{24.50}$$

if the source and sink are small. An equilibration takes place following the ratio between the forward (from A to B) and the backward (from B to A) transition probabilities given by

$$\frac{r_{\mathrm{A}\to\mathrm{B}}}{r_{\mathrm{B}\to\mathrm{A}}} = \frac{\Delta\mathcal{E}_{\mathrm{B}}g(\mathcal{E}_{\mathrm{A}})}{\Delta\mathcal{E}_{\mathrm{A}}g(\mathcal{E}_{\mathrm{B}})} \exp[S(\mathcal{E}_{\mathrm{A}}) - S(\mathcal{E}_{\mathrm{B}})]. \tag{24.51}$$

If this ratio exceeds unity, i.e., $r_{\mathrm{A}\to\mathrm{B}} > r_{\mathrm{B}\to\mathrm{A}}$, the transition from A to B dominantly occurs. In contrast, if $r_{\mathrm{A}\to\mathrm{B}} < r_{\mathrm{B}\to\mathrm{A}}$, the back transition is dominant. The equiprobability condition $N_{\mathrm{A}} = N_{\mathrm{B}}$ is given by

$$r_{\mathrm{A}\to\mathrm{B}} = r_{\mathrm{B}\to\mathrm{A}}, \tag{24.52}$$

i.e.,

$$\exp[S(\mathcal{E}_B) - S(\mathcal{E}_A)] = \frac{\Delta\mathcal{E}_B g(\mathcal{E}_A)}{\Delta\mathcal{E}_A g(\mathcal{E}_B)}, \tag{24.53a}$$

or

$$S(\mathcal{E}_B) - S(\mathcal{E}_A) = \ln[\Delta\mathcal{E}_B g(\mathcal{E}_A)] - \ln[\Delta\mathcal{E}_A g(\mathcal{E}_B)]. \tag{24.53b}$$

In equation (24.53), if an exponential dependence on $S(\mathcal{E}_A) - S(\mathcal{E}_C)$ dominates, the balance equation is approximated by

$$S(\mathcal{E}_A) = S(\mathcal{E}_B) \tag{24.54}$$

with a logarithmic correction. This is a generalization of 'Maxwell's construction rule' for the phase boundary in thermodynamics.

24.3.2 Averaging Time and Observation of Hysteresis

It is worth noting a statistically-expected value of the fluctuation level. The statistical average is taken over the time τ_{ave}, which is longer than the correlation time τ_{cor},

$$\tau_{ave} \gg \tau_{cor}. \tag{24.55}$$

Both the states A and B can change into B and A, respectively, with the characteristic lifetime of

$$\tau_{life} \approx r_{A\leftrightarrow B}^{-1}. \tag{24.56}$$

The temporal evolution of fluctuation amplitude is schematically shown in figure 24.2. Transition and back transition between states A and B are illustrated.

Depending on the choice of an average time τ_{ave}, the time average of a characteristic variable changes. If the average time τ_{ave} is not too long

$$\tau_{ave} \ll \tau_{life} \tag{24.57}$$

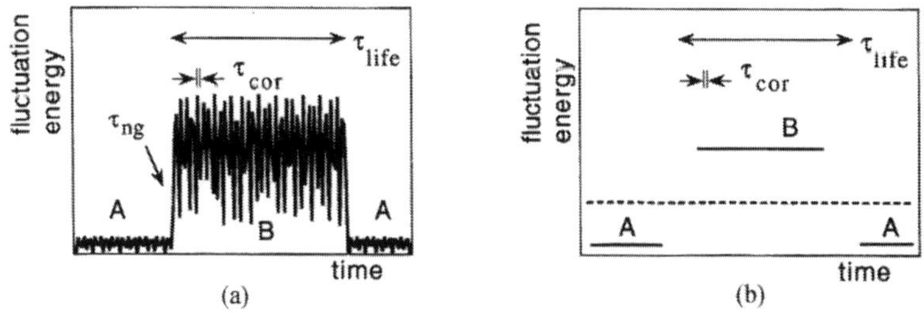

Figure 24.2. (a) Schematic drawing of a temporal evolution of fluctuation amplitude. (b) Average over a time that satisfies equation (24.55) (a) and equation (24.57) (b). Global parameters are fixed constant in time. The ensemble average for the case of equation (24.58) is shown by the dotted line in (b).

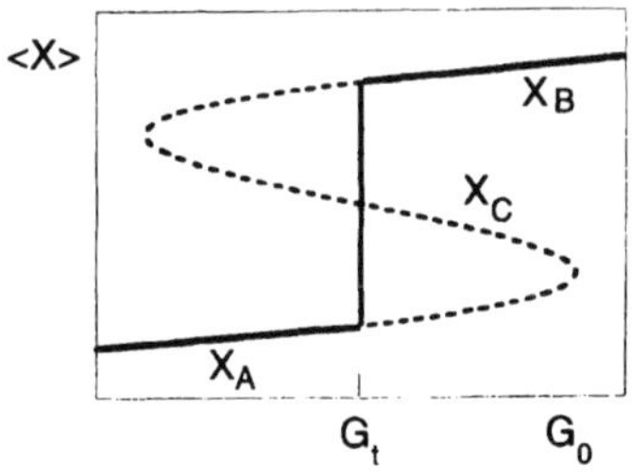

Figure 24.3. A schematic drawing of the statistical average in the system with the mechanism of hysteresis. X represents a statistical variable and a parameter G_0 is chosen as a control parameter. If $\tau_{\mathrm{ave}} \ll \tau_{\mathrm{life}}$ holds, the average $\langle X \rangle$ follows the dotted curve, showing the hysteresis. If $\tau_{\mathrm{ave}} \gg \tau_{\mathrm{life}}$, equilibration between two states is completed and the full curve is recovered.

but longer than τ_{cor}, the statistical average is observed as either the value of state A or that of state B. The average of fluctuation energy for the condition of equation (24.57) is schematically shown by solid lines in figure 24.2(b). In the limit of the long-time average

$$\tau_{\mathrm{ave}} \gg \tau_{\mathrm{life}}, \tag{24.58}$$

an equilibration of states A and B is satisfied, and they appear according to the probability of equation (24.50). This long-time average is shown by the dotted line between lines of A and B in figure 24.2(b), and agrees with the ensemble average.

The appearance of two averaged values with the condition of $\tau_{\mathrm{ave}} \ll \tau_{\mathrm{life}}$ shows a hysteresis as illustrated by the dotted curve of figure 24.3. In contrast, if one chooses the condition $\tau_{\mathrm{ave}} \gg \tau_{\mathrm{life}}$, the long-time average of fluctuation level behaves as the full curve in figure 24.3. The critical parameter G_t is given by the condition (24.54). Note that the PDF can have two peaks on both the solid and dashed lines of figure 24.3, as has been illustrated in figure 23.6.

As a result, whether a hysteresis is observed or not is determined by the competition between the lifetime τ_{life} and the evolution time of global parameters τ_{global}. If the temporal change of the control parameter (gradient, etc.) becomes rapid and comparable with the lifetime of the state, the equilibration between two states A and B does not occur, and the hysteresis is preserved. Comparing the two time scales τ_{global} and τ_{life}, the evolution is summarized as follows.

In a long lifetime limit (rapid change limit)

$$\tau_{\mathrm{life}} \gg \tau_{\mathrm{global}} \tag{24.59}$$

the system evolves according to a deterministic view as is discussed in §22.2. The selection of the state is determined by the memory effect.

In the other limit of a short lifetime (stationary limit)

$$\tau_{\text{life}} \ll \tau_{\text{global}} \tag{24.60}$$

the equilibration takes place as in §24.3. The long-time average of fluctuation level is seen as a full curve in figure 24.3 and the hysteresis is not observed in it. In an intermediate range,

$$\tau_{\text{life}} \approx \tau_{\text{global}}, \tag{24.61}$$

the probabilistic excitation of transition is important.

24.4 Probabilistic Transition

In the system with hysteresis, a self-generated oscillation is possible to occur under the condition of constant supply of energy and/or particle. The discussion in §22.2 is based on a deterministic picture, in which transitions take place when a certain critical condition is reached. As reviewed in §24.1, the turbulent state is associated with the random excitations which are much stronger than the thermodynamical fluctuations. The transition occurs according to the probabilistic law. The role of stochastic excitation of transition is important, and the deterministic view for the transition is not valid. The probabilistic view of transition and self-organized dynamics is necessary. The probabilistic excitation of transitions introduces a concept of expected lifetime during which the plasma is in a certain state of turbulence [24.3].

This process has been analyzed by including the noise term in model equations that describe the transitions. For instance, model equations take the form for the dithering ELMs as

$$\frac{\partial}{\partial t}\alpha = S - \gamma\alpha, \tag{24.62a}$$

$$\zeta\frac{\partial}{\partial t}\gamma = \alpha - 1 + a(\gamma - 1) - b(\gamma - 1)^3, \tag{24.62b}$$

where the variables α and γ represent the global plasma parameter (density, temperature, or their gradients) and the level of turbulence (turbulent transport), respectively. This set of equations is a simplified model of equations (22.1) and (22.4). The notation is as follows: S is the energy influx into the layer, ζ denotes the possible difference of dynamical time between α and γ; the cubic equation $a(\gamma - 1) - b(\gamma - 1)^3$ describes the shape of the hysteresis in the gradient–flux relation. This Ginzburg–Landau model describes the dithering ELMs under a given source magnitude S. Based on the fact that the turbulent state itself fluctuates in time, the coefficient a is modelled to include the fluctuating part as

$$a = a_0 + \varepsilon_a. \tag{24.63}$$

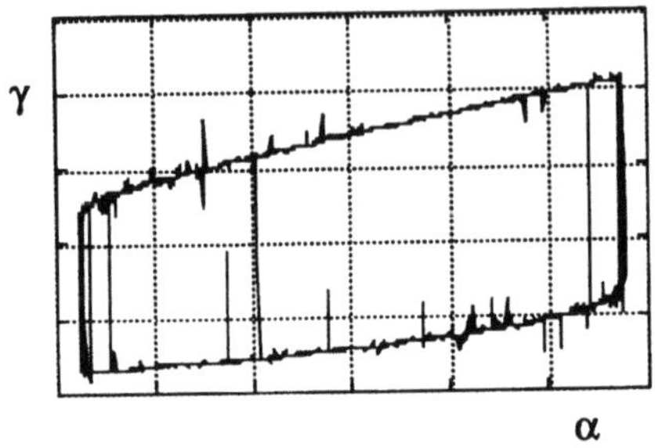

Figure 24.4. An example of the limit cycle evolution, in the presence of statistical noise, of the pressure gradient α and turbulent-loss rate γ. (Quoted from [24.4].)

A power-law noise is assumed, and the probability density function of ε_a is taken as

$$P(\varepsilon_a) \propto \varepsilon_a^{-k}. \tag{24.64}$$

Equations (24.62), (24.63), and (24.64) are solved, and the self-regulated oscillation and the transition probability are calculated. An example of the trajectory $[\alpha(t), \gamma(t)]$ is given in figure 24.4. The probability for a transition to occur at α_c^{obs} is found to follow

$$P(\alpha_c^{obs}) \propto (\alpha_{c0} - \alpha_c^{obs})^{-k} \tag{24.65}$$

where α_{c0} is the critical value of the deterministic picture, which is given in the absence of the statistical noise [24.4].

References

[24.1] Fowler R H 1936 *Statistical Mechanics* 2nd edition (Cambridge: Cambridge University Press) chapter 18
[24.2] Kitahara K 1994 *Science of Nonequilibrium Systems II* (Tokyo: Kodansha Scientific) [in Japanese] §3.3
[24.3] Itoh S-I, Toda S, Yagi M, Itoh K and Fukuyama A 1998 *J. Phys. Soc. Jpn.* **67** 4080
[24.4] Toda S, Itoh S-I, Yagi M, Itoh K and Fukuyama A 1999 *J. Phys. Soc. Jpn.* **68** 3520

Chapter 25

Transient Response and Transport

It is often assumed that a temporal and spatial scale separation holds between global transport and microscopic fluctuations. This leads to the picture that the change of global profile propagates in space with the diffusion time scale. Scale separation is not always satisfied in plasma turbulence, and the transient responses of inhomogeneous plasma could be different from those described by the diffusion process. See the reviews of [2.30, 2.31, 25.1] for the phenomena of transient responses. Theoretical modelling has been developed recently, and some efforts are illustrated in this chapter.

25.1 Long Scale Length of Fluctuations and Transient Response

If the correlation length of fluctuations includes a mesoscale or is comparable with the global gradient scale length (figure 25.1), the change in the transport properties propagates rapidly [25.2, 25.3]. In this case, the gradient–flux relation usually becomes an integrodifferential equation. (In some cases, however, Fick's law (i.e., the local equation) is valid for particle flux in a long mean-free-path regime [25.4].)

Amplitudes of fluctuations with long correlation length are not determined by the local plasma parameter. The long-wavelength modes can be excited by microfluctuations. Based on this process, the electron heat flux q_e may be modelled as a sum of the local diffusion flux and nonlocal turbulent flux as [25.5]

$$q_e(r,t) = -\int_0^a n_e(r',t)\chi_e(r',t)K_l(r,r')[c_{\rm nl}\nabla T_e(r,t) + (1-c_{\rm nl})\nabla' T_e(r',t)]\,{\rm d}r', \tag{25.1a}$$

$$K_l(r,r') \equiv \frac{r}{r'}\left\{C_{\rm local}\delta(r-r') + C_{\rm global}\frac{1}{\sqrt{\pi}l}\exp\left[-\left(\frac{r-r'}{l}\right)^2\right]\right\}, \tag{25.1b}$$

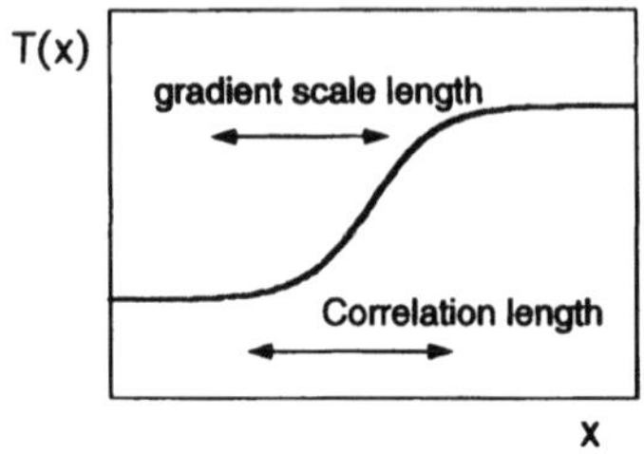

Figure 25.1. A schematic illustration of a correlation length compared with a scale length of inhomogeneity.

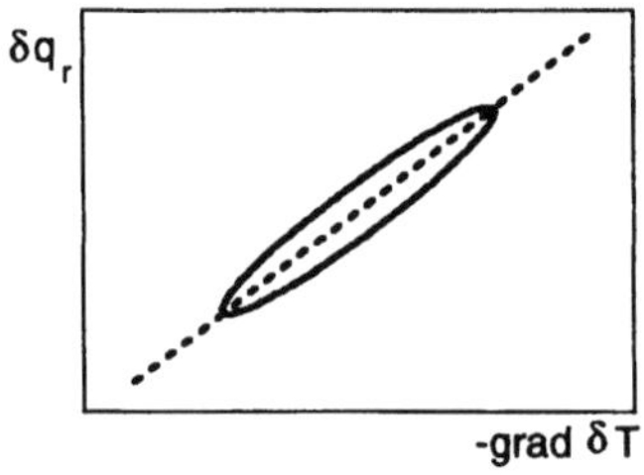

Figure 25.2. The perturbed heat flux and gradient on the same magnetic surface (near the half-radius, $r \approx a/2$) against a periodic modulation of the heat source at the center. The dotted curve indicates the response in a stationary state. (See [2.29].)

where l is the half-width of non-local interactions and represents the correlation length of the long-wavelength fluctuations that bear an additional nonlocal transport. $\delta(r - r')$ is the delta function, and $c_{\mathrm{nl}}(0 \le c_{\mathrm{nl}} \le 1)$; C_{local} and C_{global} $(C_{\mathrm{local}} + C_{\mathrm{global}} = 1)$ are numerical constants. The parameter c_{nl} represents a part due to the local temperature gradient in the non-local transport process.

According to this process, the heat flux at one location is influenced by the plasma inhomogeneity at another location. The gradient–flux relation is no longer a unique line. It sometimes constitutes a hysteresis loop, depending on the rapidity of the temporal change. A schematic diagram is given in figure 25.2 The transient response is different from the static response [25.6]. The change of flux *precedes* the change of gradient.

Other models, in which the global term is switched on/off, have also been applied to the study of transient response [25.7, 25.8].

25.2 Fluctuations with Long Correlation Length

Fluctuations with long correlation lengths are excited in turbulent plasmas. Two cases are discussed. One is the nonlinear excitation by the noise from microscopic fluctuations. The other is from a geometrical coupling. In toroidal plasmas, microscopic fluctuations may have long one-time correlation lengths.

25.2.1 Statistical Noise Excitation

Fluctuations of the long-wavelength mode of the order of the minor radius in the presence of background micro fluctuations are considered. The case is investigated where the long-wavelength modes are stable, and are nonlinearly excited by the noise from microscopic fluctuations.

According to the extended fluctuation dissipation (FD) relation, its fluctuation level and associated thermal flux are given as

$$\langle \tilde{\phi}_l^* \tilde{\phi}_l \rangle = \frac{1}{2\lambda_l} |A_{l,11}|^2 \langle \tilde{\mathrm{S}}^* \tilde{\mathrm{S}} \rangle \tag{25.2}$$

and

$$q_l = -\overline{\sum_l \frac{k_{l,y}^2}{\bar{\gamma}_{l,p} - \lambda_l} \frac{1}{2\lambda_l} |A_{l,11}|^2 \langle \tilde{\mathrm{S}}^* \tilde{\mathrm{S}} \rangle p'_0} \tag{25.3}$$

where the suffix l indicates the long-wavelength mode, λ_l is the nonlinear eigenvalue (the decorrelation rate), A_l is the projection operator, and $\tilde{\mathrm{S}}$ is the nonlinear noise for the long-wavelength mode, respectively [23.4]. The over-bar denotes integrals over the correlation length of the l mode. This level of equation (25.2) is the statistical average of the local amplitude, and the radial shape is given by the eigenmode in toroidal plasmas.

The noise which is nonlinearly driven by the background microscopic turbulence is discussed in §23.3. It is given as

$$\langle \mathrm{S}_l^* \mathrm{S}_l \rangle \simeq C_0 \gamma_k p^2 k^{-4} \chi_{\mathrm{turb}}^2, \tag{25.4}$$

where p and k are the wavenumbers of the long-wavelength and background microscopic fluctuations, respectively, $\gamma_v^h \simeq \chi_{\mathrm{turb}} k^2$, and χ_{turb} is the transport coefficient induced by the microscopic fluctuations. In deriving equation (25.4), the strong turbulence limit is used as $I^h \simeq \chi_{\mathrm{turb}}^2$.

Combining the noise source amplitude, equation (25.4), with the heat flux formula equation (25.3), one has the heat flux which is induced by the long-wavelength fluctuations as

$$q_l \simeq -\overline{\sum_l \frac{\hat{C}_0^h}{k^2} \chi_{\mathrm{turb}} p'_0} \approx -\overline{p^2 k^{-2} \chi_{\mathrm{turb}} p'_0}. \tag{25.5}$$

where an estimate of $\lambda_l \simeq \chi_{\mathrm{turb}} p^2$ is used. ($\hat{C}_0^h$ is a numerical constant in equation (23.42).)

The result, equation (25.5), gives an insight into the transient transport problems as follows.

This contribution is small in a stationary state compared with the local heat flux driven by the microscopic turbulence. The latter is given as $-\chi_{\mathrm{turb}} p'_0$. This nonlocal heat flux is about $p^2 k^{-2}$ times smaller.

However, the nonlocal component is influential in a transient response. Equation (25.5) shows that the heat flux change occurs over the distance of correlation length of the order of p^{-1}. When the noise to the p mode suddenly changes at one location, the change of the global mode amplitude appears across the plasma radius. The amplitude reaches the average within an auto-correlation time λ_l^{-1} as a response of a radial eigenmode. The long-wavelength mode induces a statistical change of heat flux of the magnitude $(p/k)^2(-\chi_{turb}p_0')$ at a distance of p^{-1} in a time interval of λ_l^{-1} after a local impulse. This change of heat flux does not occur as a diffusive process. Nevertheless, if one interprets it as the diffusion process, then this change of heat flux may be attributed to an effective diffusivity $\chi_{eff,l}$ as

$$\chi_{eff,l} \simeq |\ln(p/k)^2|^{-1}\chi_{turb}. \tag{25.6}$$

This value is modified by the factor $|\ln(p/k)^2|^{-1}$ from χ_{turb}. This is much larger than the heat diffusivity in a stationary state, since $|\ln(p/k)^2|^{-1} \gg p^2k^{-2}$ holds. (See also Appendix 25A.)

These analyses show that the contribution of the long-wavelength mode, which is excited by the statistical process of micro-turbulence, has a strong influence on the response of energy transport after the transient perturbation. This gives an insight into the problem of transient transport phenomena in toroidal plasmas, and provides a basis for making the nonlocal models for the quantitative comparison analysis with experimental observations.

25.2.2 Geometrical Effect and Long Correlation Length

Micro instabilities in tokamak plasmas have been analyzed, and the mode amplitude is known to be localized in a region of bad magnetic curvature, e.g., [17.26, 17.27]. The eigenmode with the toroidal modenumber n and central poloidal modenumber M, $\phi_{M,n}(r,\theta,\zeta)$ (where r,θ,ζ are the minor radius, the poloidal angle, and the toroidal angle, respectively) and its radial extent are given by the plasma parameters (such as the density gradient, the diamagnetic drift frequency ω_* and the gradient of ω_*) at the radius $r = r_{M,n}$ where $M = nq(r_{M,n})$ holds and $q^{-1} = RB_\theta/rB_\zeta$ is the safety factor. In the limit of short wavelength, the eigenmode is extended near the central mode rational surface as $r_{M,n} - l_E < r < r_{M,n} + l_E$. The radial correlation length of the mode l_E is given in the quasi-linear limit as

$$l_E \simeq \sqrt{s^{-1}a\rho_i}. \tag{25.7}$$

This is the hybrid scale length between the microscopic and global ones. Within this radial extent, the eigenmode evolves coherently. A one-time correlation length (the Eulerian view of correlation length) is evaluated by equation (25.7). Figure 25.3 illustrates the eigenfunction schematically.

The particle flux at a radius r by the kinetic ballooning mode has been derived as $\Gamma_r(r) = \sum_{M,n} F_{M,n}(r)$, where $F_{M,n}(r)$ is the transport quantity by

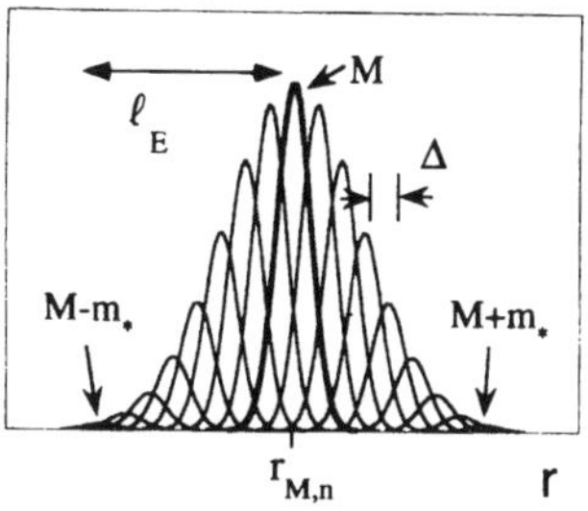

Figure 25.3. A schematic drawing of an eigenmode in a toroidal plasma. Each curve corresponds to one (m, n) Fourier component.

the (M, n) mode. (An explicit form of $F_{M,n}(r)$ is given in [25.9].) $F_{M,n}(r)$ is extended in the region $r_{M,n} - l_E < r < r_{M,n} + l_E$. The total transport at the radius r is given by the summation of $F_{M,n}(r)$ over (M, n) modes. The location of the mode has radial dependence through $r_{M,n}$. The M summation is approximately replaced by the integral over $r_{M,n}$ as

$$\Gamma_r(r) = \sum_n \int_{r-l_E}^{r+l_E} \mathrm{d}r_{M,n}\, \Delta^{-1} \hat{F}_{M,n}(r - r_{M,n}) \tag{25.8}$$

where we rewrite as $F_{M,n}(r) = \hat{F}_{M,n}(r - r_{M,n})$. A similar integral form is also derived for the energy flux. A nonlocal form of the transport flux is explicitly derived when the fluctuations have ballooning forms.

25.2.3 Kubo Number and Effective Transient Transport Coefficient

The Kubo number for the fluctuations in the range of drift-wave frequencies is of the order of unity in slab plasmas. The turbulence level of $e\tilde{\phi}/T_e \simeq 1/k_\perp L_n$ with $k_\perp \rho_i \simeq 1$ and the decorrelation rate of $\gamma_{\mathrm{dec}} \simeq c_s/L_n$ have been obtained. The thermal diffusivity has been given to be of the order of $\rho_i T/L_n eB$, i.e., the so-called gyro-Bohm diffusion. However, the Kubo number is much smaller than unity in toroidal plasmas with the eigenfunction shown in figure 25.3. The $E \times B$ velocity is estimated as $\tilde{V}_{E\times B} \simeq \rho_i c_s L_n^{-1}$, and the eddy-turnover time is given as

$$\tau_{\mathrm{et}} \simeq \frac{l_E L_n}{\rho_i c_s}.$$

The Kubo number $\mathcal{K} = \gamma_{\mathrm{dec}}^{-1} \tau_{\mathrm{et}}^{-1}$ is evaluated as

$$\mathcal{K} \simeq \rho_i / l_E \simeq \sqrt{s \rho_i / L_n}, \tag{25.9}$$

and is much smaller than unity. A one-time correlation length of fluctuation fields (the Eulerian view) is much longer than the correlation length of fluctuating plasma motion.

In the case with a small Kubo number, an upper bound for the transient transport coefficient is obtained as

$$\frac{\chi_{\text{eff}}}{\chi} \leq \mathcal{K}^{-2} \tag{25.10}$$

where χ is the one in the static limit.

The analysis on the transient transport problem illustrates an important role of influence of fluctuations at distance. The action at distance has been studied numerically in conjunction with the Bohm scaling [25.10]. Such a nonlinear simulation can be extended to the study of transient transport problems.

Another important mechanism in the nonlocal transport comes from the strongly inhomogeneous radial electric field. The radial correlation length of equation (25.7) is derived in the absence of an inhomogeneous radial electric field, which reduces the radial correlation length of fluctuations. An impact on transient transport must be simultaneously discussed with the suppression of turbulence in plasmas.

25.3 Memory Effects

The non-Markovian effect also affects the transient response. According to the arguments of §14.6, the memory effect makes the transport equation as an integrodifferential equation. Given a long-time assumption, the following transport equation is obtained [14.30]:

$$\frac{\partial}{\partial t} n(x,t) = \int_0^t \mathrm{d}t' C_L(t') \nabla^2 n(x, t - t'). \tag{25.11}$$

The contribution of the tail to the correlation function is separated out, and an approximate form is given as

$$\frac{\partial}{\partial t} n(x,t) = D_0 \nabla^2 n(x,t) - H_0 \int_{t_{\min}}^t (t')^{\alpha_{\text{tail}} - 2} \nabla^2 n(x, t - t') \,\mathrm{d}t' \tag{25.12}$$

where $\alpha_{\text{tail}} = 0.58$, D_0 is the conventional diffusion component, and the coefficient of tail, H_0, is discussed in [14.30, 14.31]. This transport equation predicts subdiffusive transport. The dispersion evolves in time as

$$\langle [x(t) - x(0)]^2 \rangle \propto t^{\alpha_{\text{tail}}}. \tag{25.13}$$

The relation $\alpha_{\text{tail}} < 1$ holds, and the response is different from a Gaussian packet. The central peak decays much more slowly than a Gaussian packet (i.e., subdiffusion), but the tail of packet propagates *much more quickly*. The long-time tail in Lagrangian correlation modifies the transient response from a simple diffusive process.

25.4 Fast Propagation of Bump

In many cases, the introduction of multiple scale lengths of fluctuation is necessary to describe a global plasma profile. For instance, bumps could be imposed on a smooth inhomogeneous profile. The temporal evolution of bumps can be different from the energy flow associated with the smooth gradient.

Fast (ballistic) propagation of a bump is possible if the plasma state is close to the critical condition for the onset of instability. The change of heat flux, δq_r, associated with a bump of pressure, δp, is modelled as follows [25.11]

$$\delta q_r = \frac{A_1}{2}\,\delta p^2 + A_2\left(\frac{\mathrm{d}}{\mathrm{d}r}\,\delta p\right)^2 + \cdots, \tag{25.14}$$

where A_1 and A_2 are coefficients. As a simple model, the case of

$$\delta q_r = \frac{A_1}{2}\,\delta p^2$$

was discussed. The temporal evolution of the perturbed profile is described by the energy balance equation

$$\frac{\partial}{\partial t}\,\delta p + \frac{\partial}{\partial x}\left(\frac{A_1}{2}\,\delta p^2\right) - D_0\frac{\partial^2}{\partial x^2}\,\delta p = \tilde{S}, \tag{25.15}$$

where D_0 is the diffusivity in the unperturbed state, and $\tilde{S}$ represents the source.

Equation (25.15) in a slab geometry is expressed as a form of Burger's equation, and represents a ballistic propagation of the bump at the velocity of

$$V_r = A_1\delta p_0, \tag{25.16}$$

where δp_0 is the height of the bump. The higher the bump is, the higher the propagation velocity is. It is shown that the transient perturbation may propagate much faster than diffusive propagation.

25.5 Plume, Avalanche and Self-Organized Criticality

Considerations in this chapter lead to a view that a mesoscale structure appears, from time to time, in inhomogeneous plasmas. A mesoscale structure may give rise to the rapid propagation of the energy through long correlation length or the propagation of large bumps. The localized structure of a radial electric field, on the other hand, can suppress turbulence (e.g., suppression by sheared flow). The dynamic evolution of the mesoscale structure is an essential element. The damping rate of the mesoscale structure, a simple form of which is given in equation (18.33), has an important role in the level of

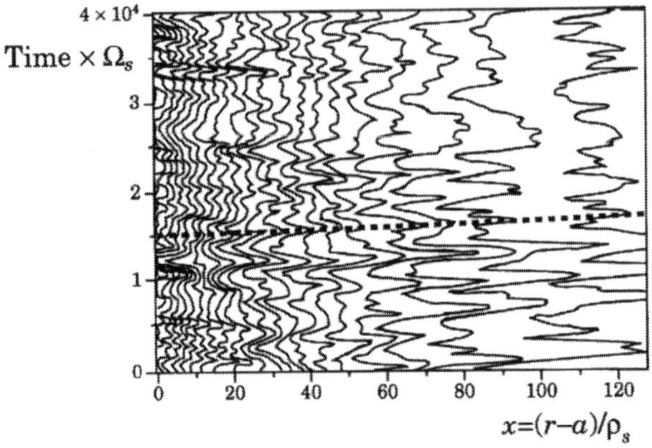

Figure 25.4. Nonlinear simulation of an avalanche with fixed flux from source. Isodensity contours, $\overline{n(x,t)}_y$ averaged over the y direction, are shown. A strong global modification of the profile appears in a short time, and is observed as an 'avalanche'. Reproduced from [25.18].

turbulence that drives turbulent transport [25.12, 25.13]. Such an influence of the damping rate of the mesoscale structure has been confirmed in direct nonlinear simulation.

Recent progress in nonlinear simulations also demonstrates the importance of the intermittent appearance of a strong flow. A strong global modification of the profile appears in a short time (figure 25.4). In other words, an 'avalanche' is observed [25.14–25.18]. An alternative type of nonlinear simulation has been performed by taking into account the mesoscale structure in tokamak plasmas [25.19]. It has been shown that symmetry-breaking structures with intermediate mode numbers have substantial effects on the evolution of plasma profiles. A hierarchical approach including the mesoscale structure and microturbulence has been proposed [25.20, 25.21].

An avalanche has also been predicted as a consequence of successive M-mode transitions, the onset of which is explained in §21.2.4. Consider a case where a transition from L-mode to M-mode (figure 22.10) occurs at a certain magnetic surface. A large amount of energy, transported from one transition region to the neighboring regions, causes pressure steepening in the neighboring regions, so as to reach the critical condition for the M-mode transition there. A successive M-mode transition occurs, and the location of the transition propagates in the radial direction. It has been shown that successive transitions propagate in space as an avalanche, not as diffusion [25.22, 25.23].

Theoretical study of plasma bifurcation presents the view that a plasma profile is rapidly modified when the parameter approaches the bifurcation boundary. The modification propagates rapidly in space. Analyses [25.24, 25.25] based on the sand pile model which gives rise to the self-organized criticality (SOC) [25.26] have been performed. This model introduces discrete automata which have number-conserving topping rules with a nonlinear threshold and exhibit avalanches at all scales.

Long-time power-law correlations of perturbations are deduced from this model. The experimental data were examined, and a long-time tail was observed. The importance of the SOC model of plasma transport has also been stressed [25.27]. The model might have relevance in much wider circumstances. However, some care is necessary when widening the applicability using the observed PDF [25.28]. Nonlinear theories can predict power-law distribution, even in the framework of the local model. For instance, an algebraic evolution is predicted for nonlinear drift-wave turbulence as equation (21.15), and strong plasma turbulence exhibits the power law in PDF and associated transition probability function as in equations (23.31), (24.39), and (24.49). Further analysis is required.

It is stressed that the transport properties can be different depending on whether the global gradient or the source is fixed constant. These differences are seen as the hysteresis, the transitions, the large statistical deviation from average, the long-range interactions, the avalanche, and others. This is a characteristic feature of far-nonequilibrium turbulent plasmas. In contrast, the materials show the unique transport property in thermal equilibrium, which is independent of the constraints of either fixed gradient or fixed flux.

Appendix 25A

Nonlocal Transport and Transient Response

The transient response of transport is analyzed based on a nonlocal expression of fluxes. An essential result in equation (25.8) is that a flux at r is influenced by a gradient at $r = r_{M,n}$, where $|r - r_{M,n}| < l_E$ is satisfied. A simple model which includes this interference is given as

$$q = -\int \mathrm{d}r' \chi K(r \quad r') T'(r'), \tag{25A.1}$$

where χ is a thermal transport coefficient, and

$$K(r - r') = \pi^{-1/2} l_E^{-1} \exp[-(r - r')^2 l_E^{-2}]. \tag{25A.2}$$

In general, χ can be a nonlinear function of the temperature gradient that introduces a difference between χ_{hp} (the thermal diffusivity deduced from a heat pulse response) and χ_{pb} (the thermal diffusivity deduced from a stationary power balance). A simple model of constant χ is used here, because we analyze a mechanism to cause the large deviation of χ_{hp} from χ_{pb}, which cannot be explained by the nonlinear dependence of χ on $\mathrm{d}T/\mathrm{d}r$.

A transient response is studied for the transport equation:

$$\frac{\partial}{\partial t}T = -\frac{\partial}{\partial x}q. \qquad (25A.3)$$

The study in a slab geometry suffices, because $l_E \ll a$ is satisfied. A temporal perturbation of the form $T(x,t) \propto \exp(-i\Omega t)$ is given at a boundary, and the asymptotic response in the downstream of the form

$$T(x,t) \propto \exp[-i\Omega t + ik(\Omega)x] \qquad (25A.4)$$

is searched for. Upon substituting equation (25A.1) into equation (25A.3), one has the equation for $k(\Omega)$ as

$$i\Omega = \chi k^2 \exp(-k^2 l_E^2/4). \qquad (25A.5)$$

If one regards the transport as a diffusive process, the phase of perturbation is given by use of the effective diffusivity $\chi_{\rm eff}$ as $T(x) \propto \exp(i\sqrt{\Omega/2\chi_{\rm eff}}x)$. By use of $\Re k(\Omega)$, $\chi_{\rm eff}$ is formally given as

$$\chi_{\rm eff} = \frac{\Omega}{2(\Re k)^2}. \qquad (25A.6)$$

A solution of the equation (25A.5) with $\chi_{\rm eff}$ is illustrated in figure 25A.1. In the limit of $\Omega \to 0$, we have $\Re k(\Omega) \to \sqrt{\Omega/2x}$ and $\chi_{\rm eff} \to \chi$. In the case of larger $\Omega, \Omega > \chi l_E^{-2}$, the numerical solution of equation (25A.5) provides a relation

$$\frac{\chi_{\rm eff}}{\chi} \simeq \frac{3\Omega l_E^2}{4\chi}. \qquad (25A.7)$$

The faster the temporal change is, the larger the effective diffusion coefficient becomes.

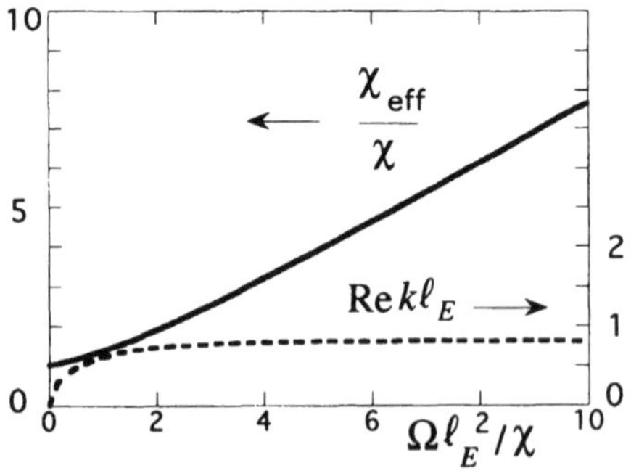

Figure 25A.1. Wave number of temperature perturbation Re $k(\omega)$ as a function of the modulation frequency Ω (dotted curve). Effective thermal diffusivity $\chi_{\rm eff}$ is illustrated by the full curve.

The upper bound of the effective transport coefficient is expressed in terms of the Kubo number. The form of the transport flux, e.g., equation (25.8), is derived by averaging over the period which is longer than the correlation time, τ_{cor}. That is, the oscillation frequency Ω must be lower than τ_{cor}^{-1}, $\Omega \le \tau_{cor}^{-1}$. Combining this bound with equation (25A.7), one has $\chi_{eff}/\chi < 3l_E^2/4\tau_{cor}\chi$. Noting that the thermal conductivity is given by an estimate of $\chi \simeq \tau_{cor}^{-1}k_\perp^{-2} \simeq \tau_{cor}^{-1}\rho_i^2$, this upper bound of the effective thermal conductivity is rewritten as

$$\frac{\chi_{eff}}{\chi} \le \mathcal{K}^{-2}, \tag{25A.8}$$

in the case of $\mathcal{K} \ll 1$. For the case of the ballooning mode, equation (25.9), one has

$$\frac{\chi_{eff}}{\chi} \le \frac{a}{s\rho_i}. \tag{25A.9}$$

References

[25.1] Callen J D and Kissick M W 1997 *Plasma Phys. Contr. Fusion* **39** B173

[25.2] Rosenbluth M N and Liu C S 1976 *Phys. Fluids* **19** 815

[25.3] Lazzara A and Putterman S 1986 *Phys. Rev. Lett.* **57** 2810

[25.4] Hazeltine R D 1999 *J. Plasma Fusion Res.* **2** 12

[25.5] Iwasaki T, Itoh S-I, Yagi M, Itoh K and Stroth U 1999 *J. Phys. Soc. Jpn.* **68** 478

[25.6] Iwasaki T, Toda S, Itoh S-I, Yagi M and Itoh K 1999 *Nucl. Fusion* **39** 2127

[25.7] Parail V V *et al* 1997 *Nucl. Fusion* **32** 1985

[25.8] Parail V V *et al* 1998 *Plasma Phys. Contr. Fusion* **40** 805

[25.9] Itoh S-I, Itoh K, Tuda T, Tokuda S and Nishikawa K 1981 *J. Phys. Soc. Jpn.* **50** 3503

[25.10] Garbet X *et al* 1999 *16th EPS Conference* (*Venice*) Part I 299

[25.11] Diamond P H and Hahm T S 1995 *Phys. Plasmas* **2** 3640

[25.12] Lin Z, Hahm T S, Lee W W, Tang W M and Diamond P H 1999 *Phys. Rev. Lett.* **83** 3645

[25.13] Hinton F L and Rosenbluth M N 1999 *Plasma Phys. Contr. Fusion* **41** A653

[25.14] Garbet X and Waltz R E 1996 *Phys. Plasmas* **3** 1898; 1998 *Phys. Plasmas* **5** 2836

[25.15] Carreras B A *et al* 1996 *Phys. Plasmas* **3** 2903

[25.16] Sarazin Y and Ghendrih Ph 1998 *Phys. Plasmas* **5** 4214

[25.17] Naulin V, Nycander J and Rasmussen J J 1998 *Phys. Rev. Lett.* **81** 4148

[25.18] Beyer P, Sarazin Y, Garbet X, Ghendrih P and Benkadda S 1999 *Plasma Phys. Contr. Fusion* **41** A757

[25.19] Thyagaraja A 2000 *Plasma Phys. Contr. Fusion* **42** B255; 1994 *Plasma Phys. Contr. Fusion* **36** 1037

[25.20] Yagi M 1999 *J. Plasma Fusion Res.* **75** 1097 (in Japanese)

[25.21] Yagi M, Itoh S-I, Itoh K and Fukuyama A 2000 *Plasma Phys. Contr. Fusion* **42** A133

[25.22] Kubota T, Itoh S-I, Yagi M and Itoh K 1998 *J. Phys. Soc. Jpn.* **67** 3100
[25.23] Matsukawa S, Itoh S-I and Yagi M 2000 *Contrib. Plasma Phys.* **40** 381
[25.24] Newmann D E, Carreras B A, Diamond P H and Hahm T S 1996 *Phys. Plasmas* **3** 1858
[25.25] Dendy R O and Helander P 1997 *Plasma Phys. Contr. Fusion* **39** 1947
[25.26] Bak P, Tang C and Wiesenfeld K 1987 *Phys. Rev. Lett.* **59** 381
[25.27] Carreras B A *et al* 1998 *Phys. Plasmas* **5** 3632; 1999 *Phys. Plasmas* **6** 1885
[25.28] Krommes J A 2000 *Phys. Plasmas* **7** 1752

Chapter 26

Thermodynamical Equilibrium Fluctuations and Far Nonequilibrium Turbulence

The statistical properties of turbulence in inhomogeneous plasmas have been discussed in preceding chapters. In this chapter, a brief survey of fluctuations in thermodynamical equilibrium is given. The statistical nature of strong turbulence is illustrated.

26.1 Thermodynamical Equilibrium

26.1.1 Neutral Fluid

In the invicid limit of fluid dynamics, the Hopf solution for the characteristic function $\Phi(u_k)$ is known as

$$\Phi(u_k) = \exp\left(-\frac{k_B T}{2} \int \mathrm{d}\boldsymbol{k}\, u_{-k,a}(\delta_{ab} - k_a k_b k^{-2}) u_{k,b} \right). \tag{26.1}$$

This solution describes a state satisfying the equi-partition law as

$$E(k) = 4\pi k_B T k^2. \tag{26.2}$$

Equation (26.2) shows the equi-partition of energy to all possible degrees of freedom.

This fluctuation spectrum, equation (26.2), is far from the Kolmogorov spectrum. In the thermodynamical equilibrium, all the degrees of freedom satisfy the local balance between the thermodynamical excitation and the drag originated from thermodynamical noises. In contrast, in the Kolmogorov, the excitation of the larger-scale mode balances with the viscous damping of the smaller-scale modes. In the inertial regime, the cascade dominantly controls its spectrum. The shape of the spectrum is determined explicitly by the energy input to the larger scale mode and the dissipation at the micro-scale. This is a characteristic feature of the state far from thermodynamical equilibrium. Interactions in the larger-scale-length modes are essential in understanding turbulence.

26.1.2 Description by the Hasegawa–Mima Equation

The equilibrium properties of the Hasegawa–Mima (HM) equation have been studied. A function which is expressed in terms of constants of motion is a particular solution of the Liouville equation. For the HM equation, the energy and enstrophy are conserved. In the Fourier representation, they are given as

$$\mathcal{E} = \frac{1}{2}\sum_k (1 + k_\perp^2)\phi_k^2 \quad \text{and} \quad \mathcal{U} = \frac{1}{2}\sum_k k_\perp^2(1 + k_\perp^2)\phi_k^2. \tag{26.3}$$

In thermodynamical equilibrium, the Gibbs distribution is sometimes assumed and the probability density function is given by [26.1, 26.2]

$$P[\{\phi_k\}] \propto \exp(-\alpha\mathcal{E} - \beta\mathcal{U}) = \prod_k \exp\left[-\left(\frac{\alpha}{2} + \frac{\beta}{2}k_\perp^2\right)(1 + k_\perp^2)\phi_k^2\right]. \tag{26.4}$$

The statistical averages of Fourier components of the fluctuation potential, energy and enstrophy are given as

$$\langle\phi_k^2\rangle = \frac{1}{(1 + k_\perp^2)(\alpha + \beta k_\perp^2)}, \tag{26.5a}$$

$$\langle\mathcal{E}_k\rangle = \frac{L^2}{2\pi}\frac{k}{(\alpha + \beta k_\perp^2)}, \tag{26.5b}$$

$$\langle\mathcal{U}_k\rangle = \frac{L^2}{2\pi}\frac{k(1 + k_\perp^2)}{(\alpha + \beta k_\perp^2)}, \tag{26.5c}$$

where L^2 is the plasma volume per unit length in the $\hat{z}$ direction (this coordinate can be neglected). The energy is larger in the longer-wavelength region, and the enstrophy is dominant in the short-wavelength region.

26.1.3 Description by the Hasegawa–Wakatani Equations

The cross-correlation between the electrostatic potential and pressure, for instance, dictates the energy flow across the magnetic field and can be essential in sustaining the turbulence as is shown by equation (11.27). A simple model to incorporate the phase relation and associate transport is the Hasegawa–Wakatani (HW) equations.

In strong collision limit, equation (11.19), two equations in HW equations decouple. The internal energy

$$\mathcal{E}^{(n)} = \frac{1}{2}\sum_k n_k^2 \tag{26.6a}$$

and the cross-correlation between density and vorticity

$$\mathcal{C} = \frac{1}{2}\sum_{k} k_{\perp}^2 n_k \phi_k \tag{26.6b}$$

are additional integrals of motions. The thermodynamical equilibrium density function is given by

$$P[\{\phi_k\}] \propto \exp(-\alpha\mathcal{E} - \beta\mathcal{U} - \alpha'\mathcal{E}^{(n)} - \beta'\mathcal{C}) \tag{26.7}$$

The functional and the asymptotic dependencies of $\langle\mathcal{E}_k\rangle$ and $\langle\mathcal{U}_k\rangle$ on $k_{\perp}^2$ are similar to those in equations (26.5b) and (26.5c). They are

$$\langle\mathcal{E}_k\rangle \propto k_{\perp}^{-1} \tag{26.8a}$$

and

$$\langle\mathcal{U}_k\rangle \propto k_{\perp} \tag{26.8b}$$

in the large $k_{\perp}^2$ limit. It is noted in [26.3] that the 'temperature' of quasiparticles could become negative if one employs the Gibbs canonical ensemble. Instead, the microcanonical ensemble is used in [11.17].

26.1.4 Representation by Use of Beltrami Functions

Equation (26.3) is expressed in terms of Fourier space. The choice of other functional space is possible. The Beltrami field

$$\nabla \times \boldsymbol{u} = \chi \boldsymbol{u}, \tag{26.9}$$

where $\{\boldsymbol{u}\}$ constitutes a complete set of orthogonal eigenfunctions, has been used to analyze the thermodynamical equilibrium [26.4]. In this study, the distribution function is expressed in terms of energy and helicity. A similar power-law distribution is obtained. The two-dimensional and three-dimensional cases are explained in [21.2].

26.1.5 Correlation Functions and Plasma Property

The particular solution equation (26.4) diverges in the short-wavelength limit; that is, it is not a physically observable state. (For instance, the integral of equation (26.5b) over k space shows divergence.)

A large amplitude in the long-wavelength region is often called the 'inverse-cascade of energy'. Clusters of charge in the long-wavelength region dominate the fluctuation energy. There, a positive vortex is surrounded mainly by positive vortices; a negative vortex is surrounded mainly by negative vortices. In other words, the system of equation has a property of anti-shielding. (This is in contrast to shielding phenomena in quiescent plasmas, like the Debye-shielding and the shielding of current perturbation in the scale length of collisionless skin-depth.) The correlation function of the vorticity is obtained by using the microcanonical ensemble as

$$\langle \rho(0)\rho(l) \rangle \propto K_0 \left[\frac{l}{l_s} \exp\left(-\frac{1}{2} \frac{\mathcal{E}}{Ne^2} \right) \right] \tag{26.10}$$

where K_0 is the zeroth-order modified Bessel function of the second kind, l_s is the screening length in the HM equation, N and e are the number and charge of quasi-particles (vortices), respectively, and $\mathcal{E}$ is the total Coulomb energy among interacting quasi-particles. The correlation function $\langle \rho(0)\rho(l) \rangle$ decays at distance

$$l \simeq l_s \exp(\mathcal{E}/2Ne^2). \tag{26.11}$$

This length is considered to be the size of a vortex (clump) in thermodynamical equilibrium. As the total interacting energy $\mathcal{E}$ becomes larger, the correlation length (clump size) becomes exponentially longer [11.17].

This influence of plasma property is also analyzed in the study of the PDF of fluctuations which are driven by a white noise source (thermodynamical excitation). In an HM equation model of magnetized inhomogeneous plasma, the modon (see §13.2.3) is a particular solution. Responses of modons to thermodynamical noise are calculated by use of the method of instanton, and a PDF of momentum flux is obtained [26.5]. A non-Gaussian exponential tail was obtained. Influences of the drift–wave vortex on transport in nonlinear simulation have been discussed in, e.g., [26.6].

26.2 Nonequilibrium Property and Intermittency

The establishment of the Kolmogorov spectrum for homogeneous turbulence is based on the equilibration of the nonlinear energy transfer. As is discussed in §5.2.3, the nonlinear energy transfer from the pumped (large-scale) mode to microscopic modes takes place within a finite time. When the change of the pump takes place within this finite time, the relaxation to the Kolmogorov spectrum is not completed. The deviation as an intermittency appears, as is shown in equations (5.180) and (5.181).

The finite time of nonlinear transfer can be an origin of intermittency. Upon considering the fluctuating deviation from the Kolmogorov spectrum, the nonequilibrium properties are studied. The statistical average of the deviation remains finite in a stationary turbulence.

Analysis of [3.17] is one example; it emphasizes that the transferred energy per unit volume, ε, fluctuates, even with a constant input in time, $\varepsilon_{\text{stationary}}$. If ε fluctuates, a statistical deviation of the spectrum can appear. ($\varepsilon_{\text{stationary}}$ is the dissipation rate in the stationary state.) The stationary turbulence is defined by

$$\langle \varepsilon \rangle = \varepsilon_{\text{stationary}}, \tag{26.12}$$

where $\langle\ \rangle$ denotes an ensemble average. The argument of Kolmogorov scaling law imposes

$$\varepsilon = \varepsilon_{\mathrm{stationary}}. \tag{26.13}$$

Condition (26.12) is relaxed from equation (26.13), and this allows an extension of the statistical deviation of spectrum. The scale invariance method in chapter 15 is extended. The transformation equation (15.4) is employed, with a notation change of (a_1, a_2) to (a_1, α) as $a_2 = a_1^{-\alpha/3+1}$,

$$\mathcal{T}_{\mathrm{K}}:\quad x \rightarrow a_1 x \quad \text{and} \quad t \rightarrow a_1^{-\alpha/3+1} t. \tag{26.14}$$

Under this transformation, the energy transfer ε and energy spectrum $\mathcal{E}(k)$ of fluctuations scale as

$$\varepsilon \rightarrow a_1^{\alpha-1} \varepsilon \tag{26.15a}$$

and

$$\mathcal{E}(k) \rightarrow a_1^{2\alpha/3+1} \mathcal{E}(k). \tag{26.15b}$$

In the presence of statistical noise, equation (26.12), ε is no longer scale-independent, Hence, the value α is not constrained by

$$\alpha = 1 \tag{26.16}$$

but deviates from unity. With this deviation $\alpha - 1$, $\mathcal{E}(k)$ deviates from the $k^{-5/3}$ dependence. In contrast, the exact stationarity, equation (26.13), requires that ε is independent of the scale transformation, and equation (26.16) holds. In this case, equation (26.15b) shows that $k^{5/3}\mathcal{E}(k)$ is independent under scale transformations. That is, $\mathcal{E}(k) \propto k^{-5/3}$ holds and the Kolmogorov law is deduced (see figure 26.1). To address this statistical property, a probability density function $P(\alpha)$ is introduced. Reference [3.17] introduced the picture that the distribution of singularities obeys the generalized entropy (extensive Renyi's entropy or non-extensive Tsallis' entropy), and discussed the probability density function. Fractal dimension $F(\alpha)$,

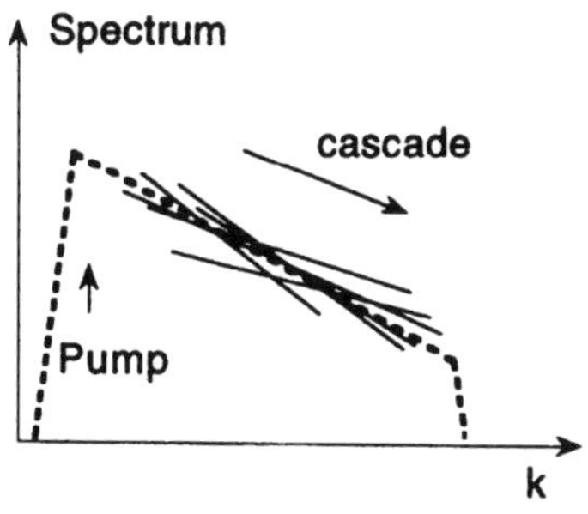

Figure 26.1. Schematic drawing of the spectrum with statistical deviations from the Kolmogorov spectrum. Solid lines represent samples with different values of α. The dashed curve denotes the Kolmogorov spectrum.

with which the singularities in velocity gradient fill the physical dimension of d, is given as

$$F(\alpha) = d + (1 - q)^{-1} \log_2[1 - (\alpha - \alpha_0)^2 (\Delta\alpha)^{-2}]. \tag{26.17}$$

Forms of q, α_0 and $\Delta\alpha$ are discussed in [3.17].

26.3 Comparison of Cases for Strong Turbulence and Thermodynamical Equilibrium

In the usual thermodynamical equilibrium statistics, the principle of 'minimum entropy production rate' is employed [26.7]. In fluids, the cascade process dominates in the inertial range and an ansatz of 'minimum enstrophy' proposed for the most probable state. If large-scale and strong MHD turbulence occurs in plasmas, a similar argument holds, i.e., minimum magnetic energy with the constraint of magnetic helicity [10.23]. For the interaction of charged clumps (vortex, quasi-particle) in plasmas, an inverse cascade of energy and a direct cascade of enstrophy occur. Another minimax principle is derived. The 'maximum energy/enstrophy' hypothesis is the reference for the limit of thermodynamical equilibrium.

Strong turbulence is not in thermodynamical equilibrium, and an equipartition is not established. The balance between the pumping and nonlinear damping expresses the extended FD relation. A power-law dependence

$$\langle \phi_k^2 \rangle \propto k_\perp^{-6}, \quad \text{i.e.,} \quad \langle \mathcal{E}_k \rangle \propto k_\perp^{-3}, \tag{26.18}$$

has been obtained in many cases (see, e.g., [2.28, 11.16]). This form is far from the state of equipartition of energy.

The following principle for strong turbulence was proposed. The effective potential $S(\mathcal{E})$, which is given as an integral of renormalized dissipation equation (23.13)

$$S(\mathcal{E}) = \int_0^{\mathcal{E}} \frac{4\Lambda\mathcal{E}}{g^2} \, \mathrm{d}\mathcal{E}, \tag{26.19}$$

must be a minimum, where Λ is the nonlinear damping rate of the dressed test mode (quasi-particle) and g^2 stands for the statistical noise. The source g^2 includes the self-noise of the turbulent fluctuations. The specification of the turbulence is expressed by the nonlinear damping rate Λ.

The *minimum* S principle reduces to the *maximum* entropy S_{ent} in the limit of thermodynamical equilibrium.

This result is also a natural extension of the 'principle of minimum dissipation rate' in the near equilibrium. Prigogine's 'minimum principle of the entropy production rate' [26.7] is recovered if the noise is governed by the equilibrium temperature. In the absence of turbulent noise, we have $g^2 = 4\hat{T}\gamma_m\mathcal{E}$ in equation (26.19), and γ_m is independent of the fluctuation energy. The limiting formula is given as

$$S(\mathcal{E}) = \int_0^{\mathcal{E}} \frac{4\Lambda\mathcal{E}}{g^2}\, d\mathcal{E} \to \frac{1}{\hat{T}\gamma_m}\int_0^{\mathcal{E}} \Lambda\, d\mathcal{E}. \tag{26.20}$$

The integral in equation (26.20), $\int_0^{\mathcal{E}} \Lambda\, d\mathcal{E}$, is the dissipation rate of the fluctuating energy. One can define the entropy production rate as

$$\left.\frac{\partial S_{\text{ent}}}{\partial t}\right|_{\text{irr}} = \frac{1}{T}\int_0^{\mathcal{E}} \Lambda\, d\mathcal{E}. \tag{26.21}$$

Table 26.1. The principles of statistical theory for strong plasma turbulence are compared with the principles of thermodynamical equilibrium. (Example of CDIM/CDBM turbulence [14.23–14.25, 23.1, 23.2].)

	Near thermodynamical equilibrium	Far-nonequilibrium
Basic assumption	Stosszahl ansatz; $1/\Omega$ expansion	Large number of degrees of freedom with positive Lyapunov exponent
Damping	Molecular viscosity $\gamma_c = \mu_c k_\perp^2$	Nonlinear (eddy) damping $\gamma_N \approx \tilde{\phi} k_\perp^2 / B$
Micro versus macro	$\mu_{\text{micro}} = \mu_{\text{macro}}$ Onsager's ansatz	Scale-dependent
Excitation (random)	Thermal excitation	Thermodynamical and nonlinear drives
(coherent)	None	Instability drive
Decorrelation rate	γ_c	Nonlinear decorrelation λ_1
Balance	FD theorem Einstein's relation	Extended FD relation $I \approx \dfrac{\text{nonlinear noise}}{\text{nonlinear decorrelation}}$
Partition	Equipartition $E_k \approx Tk$	Nonlinear balance $E_k \approx \lvert\nabla p_0\rvert k^{-3}$
Probability density function	Boltzmann $P(\mathcal{E}) \approx \exp(-\mathcal{E}/k_B T)$	Integral of renormalized dissipation $P(\mathcal{E}) \approx \exp[-S(\mathcal{E})]/g$ power-law tail
Min./max. principle	Maximum entropy/ minimum entropy production rate	$S(\mathcal{E})$ minimum
Phase boundary	Maxwell's construction	$S(\mathcal{E}_A) = S(\mathcal{E}_B)$
Transition probability	$\ln(K) \approx -\Delta Q/T$ Arrhenius law	$K \propto \exp[-S(\mathcal{E}_{\text{saddle}})]$ (power law)
Transport matrix	Onsager's symmetry	Not necessarily symmetric
Interference of fluxes	Curie's principle	Interferences between heat, particle and momentum
Transport coefficients	Independent of gradient	Dependent on gradient

If one uses this definition, one finds from equation (26.20) that the relation

$$S(\mathcal{E}) \propto \left.\frac{\partial S_{\mathrm{ent}}}{\partial t}\right|_{\mathrm{irr}} \qquad (26.22)$$

holds near the thermodynamical equilibrium.

An entropy and an entropy production rate are difficult to define in the turbulent state. The integral of renormalized dissipation rate $S(\mathcal{E})$ plays a role in controlling the probability density function. The minimum principle for S is a generalization of the thermodynamical principle to its turbulence counterpart.

The transition between different turbulent states, in general, is discussed in terms of the transition probability. The Arrhenius law in the thermodynamical equilibrium is a typical example. In strong plasma turbulence, the transition probability follows a power law.

Properties of transport differ in turbulent plasmas from those in thermodynamical equilibrium states. The symmetry relations like Onsager's relation or Curie's principle, established for the classical systems in the thermodynamical equilibrium, do not always hold. Kinetic coefficients, together with transport coefficients, depend on the system's size and the magnitude of inhomogeneity.

Comparison between the statistical principle for turbulence and for thermodynamical equilibrium is made in table 26.1. These are the characteristics features of media far from thermodynamical equilibrium.

References

[26.1] Fyfe D and Montgomery D 1979 *Phys. Fluids* **22** 246

[26.2] Gang F Y *et al* 1989 *Phys. Fluids B* **1** 1331

[26.3] Edwards S F and Taylor J B 1974 *Proc. R. Soc. A* **336** 257

[26.4] Ito N and Yoshida Z 1996 *Phys. Rev. E* **53** 5200

[26.5] Kim E-J and Diamond P H 2002 *Phys. Plasmas* **9** 71

[26.6] Koniges A E, Crotinger J A and Diamond P H 1992 *Phys. Fluids B* **4** 2785.

[26.7] Prigogine I 1961 *Introduction to Thermodynamics of Irreversible Processes* 2nd edition (New York: Interscience Publishers)

Summary

Theoretical methods for fluid turbulence and plasma turbulence have shown remarkable progress recently. Recognition of the turbulence has matured: understanding has advanced concerning, e.g., excitation of fluctuations by instabilities that are caused by inhomogeneities, energy partition via cascade and enhanced dissipation, formation of mesoscale structure by inverse cascade or parametric excitation, transport and destruction of inhomogeneous structure by turbulence, suppression of turbulence by inhomogeneity, and subcritical excitation owing to nonlinear instabilities. These processes are either cooperating or competing with each other, and constitute dynamics of turbulence and turbulent state. The subject of this monograph is to illustrate turbulence theories which have given analytic insights into these essential features of turbulent media.

Among the various kinds of nonlinearly interacting dynamics of fluid and plasma, the turbulence is characterized by the large degree of freedom. One of the main progresses is seen in the statistical description of turbulence, which is discussed in this monograph. Contrary to the media that is in or near the thermodynamical equilibrium, which is described by well-established statistical laws (Maxwell–Boltzmann statistics, Fermi–Dirac statistics or Bose–Einstein statistics), the turbulent state is in the far-nonequilibrium state. Statistical theories have progressed, and various useful closure theories have been investigated. On the basis of this, hierarchical modelling like the two-scale direct-interaction approximation and the mean field approach like the K-ε model are developed. The renormalized dielectric is derived in plasma turbulence theory, which describes the self-consistent equation for the turbulent spectrum. These methods allow investigation of the dynamics of generation and nonlinear decorrelation of turbulence, providing a key for the understanding of turbulence and turbulent transport. At the same time, statistical physics of the nonequilibrium state is generalized. Theoretical formulae are surveyed.

Another outstanding progress in the turbulence theories is the understanding of mutual interactions between the perturbations and inhomogeneities with various scale lengths. This mechanism introduces the various characteristics of turbulent media. Inhomogeneous structure (e.g., sheared flow, pressure gradient, localized electromagnetic fields, etc.) generates turbulence so that the ordered structure is destroyed through turbulent transport. Simultaneously, global or ordered structures are generated from the nonlinear interactions among turbulent fluctuations. Theories have shown these two competing processes (i.e., the destroying and constructing mechanisms) are equally important, providing an explanation of why nature is full of observable structures although the Reynolds number is extremely large so that fully developed turbulence might mask possible inhomogeneities. Structural transitions of turbulent media, of which the H-mode transition is the leading example in plasma physics, are described. Dynamo is another challenging example, theories for which are addressed.

In this monograph, we have surveyed methods in turbulence theories. A lot of common physics has flourished in fluid dynamics and plasma physics. Therefore, emphasis is made on plasma turbulence theories that are based on the fluid description of plasmas, and little is discussed about kinetic turbulence theories.

Another approach to investigating turbulence is a direct nonlinear simulation. This approach is indispensable, and is complementary to the methods which are explained in this book. Quantitative description requires direct nonlinear simulations, because exactly solvable systems are very limited in the field of turbulence problems. Owing to the selection of focus of this monograph, description of direct numerical simulation is limited here. The turbulence theories provide understanding of the physics mechanisms and could provide a law that governs the turbulent phenomena. Turbulence theories of fluid and plasma, cooperating with simulation study, will continue to challenge the frontier of physics, i.e., the problem of the millenaries, '*all things flow*'.

Index

Printed and bound by CPI Group (UK) Ltd, Croydon, CR0 4YY
23/10/2024
01778238-0014